THE WIZARD AND THE PROPHET

"The most persuasive writers on the environment punctuate their big-picture theses with telling details that bring the relevant issues to life. Like Elizabeth Kolbert and Tim Flannery, Charles C. Mann is one of the masters of this art. . . . A stimulating, thoughtful, balanced overview of matters vital to us all." —*The Boston Globe*

"Fascinating. . . . An inquisitive and gifted science writer." —*Science*

"Charles C. Mann specializes in deep, comprehensive looks at the past that better elucidate the present." —*San Francisco Chronicle*

"An elegantly written, devoted testimonial to the art of the possible." —*Sierra*

"Bestselling author and journalist Mann tackles the thorny problem of humankind's future through the lens of two twentieth-century visionaries. . . . A sweeping, provocative work of journalism, history, science, and philosophy." —*Library Journal* (starred review)

"Without taking sides, Mann delivers a fine examination of two possible paths to a livable future." —*Publishers Weekly* (starred review)

"An insightful, highly significant account that makes no predictions but lays out the critical environmental problems already upon us." —*Kirkus Reviews* (starred review)

"Charles Mann provides a deeply corrugated, richly nuanced, and highly entertaining narrative to make sense of the most consequential decisions facing civilization. Read, think, and enjoy." —Ruth DeFries, author of *The Big Ratchet: How Humanity Thrives in the Face of Natural Crisis*

Charles C. Mann

THE WIZARD AND THE PROPHET

Charles C. Mann, a correspondent for *The Atlantic*, *Science*, and *Wired*, has written for *Fortune*, *The New York Times*, *Smithsonian*, *Technology Review*, *Vanity Fair*, and *The Washington Post*, as well as for the TV network HBO and the series *Law & Order*. A three-time National Magazine Award finalist, he is the recipient of writing awards from the American Bar Association, the American Institute of Physics, the Alfred P. Sloan Foundation, and the Lannan Foundation. He is the author of *1491*, which won the National Academy of Sciences' Keck Award for best book of the year, and *1493*, a *New York Times* bestseller. He lives with his family in Amherst, Massachusetts.

www.charlesmann.org

THE WIZARD AND THE PROPHET

THE
WIZARD
AND THE
PROPHET

TWO REMARKABLE SCIENTISTS AND THEIR

DUELING VISIONS TO SHAPE TOMORROW'S WORLD

Charles C. Mann

VINTAGE BOOKS
A Division of Penguin Random House LLC
New York

FIRST VINTAGE BOOKS EDITION, APRIL 2019

Portions of this work first appeared, sometimes in significantly different form,
in *The Atlantic, National Geographic, The New York Times Magazine, Orion,
Pacific Standard, Science, Vanity Fair*, and *Wired*.

The Library of Congress has cataloged the Knopf edition as follows:
Names: Mann, Charles C., author.
Title: The wizard and the prophet : two remarkable scientists and their dueling visions
to shape tomorrow's world / by Charles C. Mann.
Description: First edition. | New York : Alfred A. Knopf, 2018.
Includes bibliographical references and index.
Identifiers: LCCN 2017024776
Subjects: LCSH: Vogt, William, 1902–1968. | Borlaug, Norman E. (Norman Ernest),
1914–2009. | Environmental sciences—History—20th century. | Food security. |
Water security. | Energy security. | Climatic changes. | Environmentalists—
United States—Biography.
Classification: LCC GE56.V64 M36 2018 | DDC 363.70092/273—dc23
LC record available at https://lccn.loc.gov/2017024776

Vintage Books Trade Paperback ISBN: 978-0-345-80284-2
eBook ISBN: 978-0-307-96170-9

www.vintagebooks.com

Printed in the United States of America
10   9   8   7   6   5   4   3   2

To Ray—

Just two words,

because a thousand would not be enough

No wonder they disagreed so endlessly;

they were talking about different things.

—ROBERT L. HEILBRONER

Contents

ONE FUTURE

THE WIZARD AND THE PROPHET

All parents must remember the moment when they first held their children—the tiny crumpled face emerging, an entire new person, from the hospital blanket. I extended my hands and took my daughter in my arms. I was so overwhelmed that I could hardly think.

After the birth, I wandered outside for a while so that mother and child could rest. It was three in the morning, late February in New England. There was ice on the sidewalk and a cold drizzle in the air. As I stepped from the curb, a thought popped into my head: when my daughter is my age, almost 10 billion people will be walking the earth.

I stopped in midstride. I thought: How is *that* going to work?

Like other parents, I want my children to be comfortable in their adult lives. But in the hospital parking lot this suddenly seemed unlikely. Ten billion mouths, I thought. How can they possibly be fed? Twenty billion feet—how will they be shod? Ten billion bodies—how will they be accommodated? Is the world big enough, rich enough, for all these people to flourish? Or have I brought my children into a time of general collapse?

When I began as a journalist, I envisioned myself, romantically, as an eyewitness to history. I wanted to chronicle the important events of my time. Only after I began work did the obvious question occur: What *are* those important events? My first article, essentially the caption for a photograph of a bad automobile accident, certainly didn't document one. But what was the standard? Hundreds of years from now, what will historians view as today's most significant developments?

For a long time I believed that the answer was "discoveries in science and technology." I wanted to learn about the curing of diseases, the rise of computer power, the unraveling of the mysteries of matter and energy. Later, though, it seemed to me that what was important was less the new knowledge than what it had enabled. In the 1970s, when I was in high school, about one out of every four people in the world was hungry—"undernourished," to use the term preferred by the United Nations. Today, the U.N. says, the figure is one out of ten.* In those four decades, the global average life span has risen by more than *eleven years*, with most of the increase occurring in poor places. Hundreds of millions of people in Asia, Latin America, and Africa have lifted themselves from destitution into something like the middle class. In the annals of humankind, nothing like this surge of well-being has occurred before. It is the signal accomplishment of this generation, and its predecessor.

This enrichment has not occurred evenly or equitably; millions upon millions are not prosperous, and millions more are falling behind. Nonetheless, on a global level—the level of 10 billion—the increase in affluence is undeniable. The factory worker in Pennsylvania and the farmer in Pakistan may both be struggling and angry, but they are also, by the standards of the past, wealthy people.

Today the world has about 7.3 billion inhabitants. Most demographers believe that around 2050 the world's population will reach 10 billion or a bit less. About this time, human numbers will probably begin to level off—as a species, we will be around "replacement level," each couple having, on average, just enough children to replace themselves. All the while, economists say, the world's development should continue, however unevenly, however slowly. The implication is that when my daughter is my age a sizable percentage of the world's 10 billion souls will be middle class. Jobs, homes, cars, fancy electronics, a few occasional treats—these are what the affluent multitudes will want. (Why shouldn't they?) And though the lesson of history is that the great majority of these men and women will make their way, it is hard not to

---

* In absolute terms, the decrease seems less impressive. Several hundred million still live in destitution. In recent years, moreover, the number of hungry has risen a bit. Researchers disagree on whether this reversal is a long-term problem or a temporary blip due to violence (Southwest Asia, parts of Africa) and falling commodity prices, which have lowered national incomes in some places. Nonetheless, a child born in the twenty-first century has less chance of emerging into a life of absolute want than at any other century in known history.

be awed by the magnitude of the task facing our children. A couple of billion jobs. A couple of billion homes. A couple of billion cars. Billions and billions of occasional treats.

Can we provide these things? That is only part of the question. The full question is: Can we provide these things without wrecking much else?

As my children were growing up, I took advantage of journalistic assignments to speak, from time to time, with experts in Europe, Asia, and the Americas. Over the years, as the conversations accumulated, it seemed to me that the responses to my questions fell into two broad categories, each associated (at least in my mind) with one of two people, Americans who lived in the twentieth century. Neither is well known to the public, yet one man has often been called the most important person born in that century and the other is the principal founder of the most significant cultural and intellectual movement of that time. Both recognized and tried to solve the fundamental question that will face my children's generation: how to survive the next century without a wrenching global catastrophe.

The two people were barely acquainted—they met only once, so far as I know—and had little regard for each other's work. But in their different ways, they were largely responsible for the creation of the basic intellectual blueprints that institutions around the world use today for understanding our environmental dilemmas. Unfortunately, their blueprints are mutually contradictory, for they had radically different answers to the question of survival.

The two people were William Vogt and Norman Borlaug.

Vogt, born in 1902, laid out the basic ideas for the modern environmental movement. In particular, he founded what the Hampshire College demographer Betsy Hartmann has called "apocalyptic environmentalism"—the belief that unless humankind drastically reduces consumption its growing numbers and appetite will overwhelm the planet's ecosystems. In best-selling books and powerful speeches, Vogt argued that affluence is not our greatest achievement but our biggest problem. Our prosperity is temporary, he said, because it is based on taking more from Earth than it can give. If we continue, the unavoid-

William Vogt, 1940

able result will be devastation on a global scale, perhaps including our extinction. *Cut back! Cut back!* was his mantra. *Otherwise everyone will lose!*

Borlaug, born twelve years later, has become the emblem of what has been termed "techno-optimism" or "cornucopianism"—the view that science and technology, properly applied, can help us produce our way out of our predicament. Exemplifying this idea, Borlaug was the primary figure in the research that in the 1960s created the "Green Revolution," the combination of high-yielding crop varieties and agronomic techniques that raised grain harvests around the world, helping to avert tens of millions of deaths from hunger. To Borlaug, affluence was not the problem but the solution. Only by getting richer, smarter, and more knowledgeable can humankind create the science that will resolve our environmental dilemmas. *Innovate! Innovate!* was Borlaug's cry. *Only in that way can everyone win!*

Both Borlaug and Vogt thought of themselves as environmentalists facing a planetary crisis. Both worked with others whose contributions, though vital, were overshadowed by theirs. But that is where the similarity ends. To Borlaug, human ingenuity was *the* solution to our problems. One example: by using the advanced methods of the Green Revolution to increase per-acre yields, he argued, farmers would not have to plant as many acres. (Researchers call this the Borlaug hypothesis.) Vogt's views were the opposite: the solution, he said, is to get smaller. Rather than grow more grain to produce more meat, humankind should, as his followers say, "Eat lower on the food chain." If people ate less beef and pork, valuable farmland would not have to be devoted to cattle and pig feed. The burden on Earth's ecosystems would be lighter.

I think of the adherents of these two perspectives as Wizards and Prophets—Wizards unveiling technological fixes, Prophets decrying the consequences of our heedlessness. Borlaug has become a model for the

Wizards. Vogt was in many ways the founder of the Prophets.

Borlaug and Vogt traveled in the same orbit for decades, but rarely acknowledged each other. Their first meeting, in the mid-1940s, ended in disagreement. So far as I know, they never spoke afterward. Not one letter passed between them. They each referred to the other's ideas in public addresses, but never attached a name. Instead, Vogt rebuked the anonymous "deluded" scientists who were actually aggravating our problems. Meanwhile, Borlaug derided his opponents as "Luddites."

Norman Borlaug, 1944

Both men are dead now, but their disciples have continued the hostilities. Indeed, the dispute between Wizards and Prophets has, if anything, become more vehement. Wizards view the Prophets' emphasis on cutting back as intellectually dishonest, indifferent to the poor, even racist (because most of the world's hungry are non-Caucasian). Following Vogt, they say, is a path toward regression, narrowness, and global poverty. Prophets sneer that the Wizards' faith in human resourcefulness is unthinking, scientifically ignorant, even driven by greed (because remaining within ecological limits will cut into corporate profits). Following Borlaug, they say, at best postpones an inevitable day of reckoning—it is a recipe for what activists have come to describe as "ecocide." As the name-calling has escalated, conversations about the environment have increasingly become dialogues of the deaf. Which might be all right, if we weren't discussing the fate of our children.

Wizards and Prophets are less two ideal categories than two ends of a continuum. In theory, they could meet in the middle. One could cut back here à la Vogt and expand over there, Borlaug-style. Some people believe in doing just that. But the test of a categorization like this one is less whether it is perfect—it is not—than whether it is useful. As a practical matter, the solutions (or putative solutions) to environmental problems have been dominated by one of these approaches or the other. If a government persuades its citizenry to spend huge sums revamping offices, stores, and homes with the high-tech insulation and low-water-use plumbing urged by Prophets, the same citizenry will resist ponying up for Wizards' new-design nuclear plants and monster desalination facilities. People who back Borlaug and embrace genetically modified, hyper-productive wheat and rice won't follow Vogt and dump their steaks and chops for low-impact veggie burgers.

Moreover, the ship is too large to turn quickly. If the Wizardly route is chosen, genetically modified crops cannot be bred and tested overnight. Similarly, carbon-sequestration techniques and nuclear plants cannot be deployed instantly. Prophet-style methods—planting huge numbers of trees to suck carbon dioxide from the air, for instance, or decoupling the world's food supply from industrial agriculture—would take equally long to pay off. Because backtracking is not easy, the decision to go one way or the other is hard to change.

Most of all, the clash between Vogtians and Borlaugians is heated because it is less about facts than about *values.* Although the two men rarely acknowledged it, their arguments were founded on implicit moral and spiritual visions: concepts of the world and humankind's place in it. Entwined with the discussion of economics and biology, that is, were whispers of "ought" and "should." As a rule, these views were articulated more explicitly by those who followed Vogt and Borlaug than by the two men themselves. But they were there from the beginning.

Prophets look at the world as finite, and people as constrained by their environment. Wizards see possibilities as inexhaustible, and humans as wily managers of the planet. One views growth and development as the lot and blessing of our species; others regard stability and preservation as our future and our goal. Wizards regard Earth as a toolbox, its contents freely available for use; Prophets think of the natural world as embodying an overarching order that should not casually be disturbed.

The conflict between these visions is not between good and evil, but between different ideas of the good life, between ethical orders that give priority to personal liberty and those that give priority to what might be called connection. To Borlaug, the landscape of late-twentieth-century capitalism, with its teeming global markets dominated by big corporations, was morally acceptable, though ever in need of repair. Its emphasis on personal autonomy, social and physical mobility, and the rights of the individual were resonant. Vogt thought differently. By the time he died, in 1968, he had come to believe that there was something fundamentally wrong with Western-style consumer societies. People needed to live in smaller, more stable communities, closer to the earth, controlling the exploitative frenzy of the global market. The freedom and flexibility touted by advocates of consumer society were an illusion; individuals' rights mean little if they live in atomized isolation, cut off from Nature and each other.

These arguments have their roots in long-ago fights. Voltaire and Rousseau disputing whether natural law truly is a guide for humankind. Jefferson and Hamilton jousting over the ideal character of citizens. Robert Malthus scoffing at the claims of the radical philosophers William Godwin and Nicolas de Condorcet that science could overcome limits set by the physical world. T. H. Huxley, the famed defender of Darwin, and Bishop Samuel Wilberforce of Oxford, contending whether biological laws truly apply to creatures with souls. John Muir, champion of pristine wilderness, squaring off against Gifford Pinchot, evangelist for managing forests with teams of experts. The ecologist Paul Ehrlich and the economist Julian Simon betting whether ingenuity can outwit scarcity. To the philosopher-critic Lewis Mumford, all of these battles were part of a centuries-long struggle between two types of technology, "one authoritarian, the other democratic, the first system-centered, immensely powerful, but inherently unstable, the other man-centered, relatively weak, but resourceful and durable." And all of them were about, at least in part, the relationship of our species to Nature—which is to say they were debates about the nature of our species.

Borlaug and Vogt, too, took sides in the dispute. Both believed that *Homo sapiens*, alone among Earth's creatures, can understand the world through science, and that this empirical knowledge can guide societies into the future. From this point, though, the two men diverged. One of them believed that ecological research has revealed our planet's inescapable limits, and how to live within them. The other believed that science could show us how to surpass what would be barriers for other species.

Who is right, Vogt or Borlaug? Better to have your feet on the ground or live, however chancily, in the air? Cut back or produce more?

Wizard or Prophet? No question is more important to our crowded world. Willy-nilly, our children will have to answer it.

. . .

What this book is not: a detailed survey of our environmental dilemmas. Many parts of the world I skip over completely; many issues I do not discuss. The subjects are too big and complicated to fit in a single book—at least, not a book that I can imagine anyone reading. Instead I am describing two ways of thinking, two views of possible futures.

Another thing this book is not: a blueprint for tomorrow. *The Wizard and the Prophet* presents no plan, argues for no specific course of action. Part of this aversion reflects the opinion of the author: in our Internet era, there are entirely too many pundits shouting out advice. I believe I stand on firmer ground when I try to describe what I see around me than when I try to tell people what to do.

In the first chapter I step back to consider what biology suggests about the trajectory of any species—that is, why one would imagine that *Homo sapiens* actually *has* a future. Biologists tell us that all species, if given the chance, overreach, overreproduce, overconsume. Inevitably, they encounter a wall, always to catastrophic effect, and usually sooner rather than later. From this vantage, Vogt and Borlaug were equally deluded. Here I ask here whether there is reason to believe the scientists are wrong.

Next I turn to Vogt and Borlaug themselves. I follow Vogt from his birth in pre-suburban Long Island to his near death from polio to his ecological conversion experience off the coast of Peru. I close the first part of his story with the publication of his tract *The Road to Survival* (1948), the first modern we're-all-going-to-hell book. *Road* was meant as a warning bell, based on objective science, but it was also an implicit vision of how we should live: a moral testament. Vogt was the first to put together, in modern form, the principal tenets of environmentalism, the twentieth century's only successful, long-lasting ideology.

Borlaug's tale begins with his birth into a poor Iowa farming community. Borlaug was released from what he saw as endless toil by the great good fortune of having Henry Ford invent a tractor that could be built and sold cheaply enough to replace his labor on the farm. Allowed to attend college, he labored through the Depression until a concatenation of accidents put him into the research program that led to the Green Revolution. In 2007, when Borlaug was ninety-three, *The Wall Street Journal* editorialized that he had "arguably saved more lives than anyone in history. Maybe one billion."

In the middle section of this book, I invite the reader to put on, as it were, Vogtian and Borlaugian spectacles and look at four great, oncoming challenges: food, water, energy, and climate change. Sometimes I think of them as Plato's four elements: earth, water, fire, and air. Earth represents agriculture, how we will feed the world. Water is drinking water, as vital as food. Fire is our energy supply. Air is climate change, a by-product, potentially catastrophic, of our hunger for energy.

Earth: If present trends continue, most agronomists believe, harvests will have to rise 50 percent or more by 2050. Different models with different assumptions make different projections, but all view the rise in demand as due both to the increase in human numbers and the increase in human affluence. With few exceptions, people who became wealthier have wanted to consume more meat. To grow more meat, farmers will need to grow more grain—much more. Wizards and Prophets have radically different ways of approaching these demands.

Water: Although most of Earth is covered by water, less than 1 percent of it is accessible freshwater. And the demand for that water is constantly increasing. The increase is a corollary of the rising demand for food—almost three-quarters of global water use goes to agriculture. Many water researchers believe that as many as 4.5 billion people could be short of water by as early as 2025. As with food, the disciples of Borlaug tend to react in one way to this worry; those of Vogt, in another.

Fire: Predicting how much energy tomorrow's world will need depends on assumptions about, for instance, how many of the roughly 1.2 billion people who do not have electricity will actually get it, and how that electricity will be provided (solar power, nuclear power, natural gas, wind, coal). Still, the main thrust of every attempt to estimate future requirements that I am aware of is that the human enterprise will require more energy—probably quite a lot more. What to do about it depends on whether you ask Borlaugians or Vogtians.

Air: In this list, climate change is odd man out. The other three elements (food, freshwater, energy supply) reflect human needs, whereas climate change is an unwanted consequence of satisfying those needs. The first three are about providing benefits to humankind: food on the table, water from the tap, heat and air-conditioning in the home. With climate change, the benefit is invisible: avoiding problems in the future. Societies put their members through wrenching changes and then, with a bit of luck, nothing especially noteworthy occurs. Temperatures don't rise much; sea levels stay roughly where they are. Little wonder that Wizards and Prophets disagree about what to do!

Climate change is different from the others in a second way, too. It is rare to encounter people who don't accept that the world's increasingly prosperous population will ratchet up demand for food, water, and energy. But a significant minority believe that climate change is not real, or is not attributable to human activity, or is so minimal as to be not worth bothering about. The disagreement is so passionate that it's easy for one side to say, "Well, if he gives any credence to this claim, then he belongs on the other team, and forget about anything else he reports!" Hoping to avoid this fate, I split the discussion of climate change in two. In the first section, I ask the skeptics to accept—just for the moment— that climate change is a real future problem, so I can look at how Borlaugians and Vogtians would address it. In an appendix, I address in what ways some of the skeptics could be correct.

The question this book asks is not "How will we resolve these four challenges?" but "How would a Vogt or Borlaug approach them?" I close with the last years of both men, melancholy in both cases. Tying up a loose philosophical end, an epilogue returns to the discussion of why one might believe that our species could succeed, and even thrive.

The Wizard and the Prophet is a book about the way knowledgeable people might think about the choices to come, rather than what will happen in this or that scenario. It is a book about the future that makes no predictions.

In college I read two Vogtian classics: *The Population Bomb* (1968), by the ecologist Paul Ehrlich, and *The Limits to Growth* (1972), by a team of computer modelers. Famously, *The Population Bomb* begins with a thunderous claim: "The battle to feed all of humanity is over." Matters go downhill from there. "Sometime in the next fifteen years, the end will come," Ehrlich told CBS News in 1970. "And by 'the end' I mean an utter breakdown of the capacity of the planet to support humanity." *The Limits to Growth* was a bit more hopeful. If humankind changed its habits completely, it said, civilizational collapse could be avoided. Otherwise, the researchers argued, "the limits to growth on this planet will be reached sometime within the next one hundred years."

The two books scared the pants off me. I became a Vogtian, convinced that the human enterprise would fall apart if our species didn't abruptly reverse course. Long afterward, it occurred to me that many of the Prophets' dire forecasts had not come true. Famines had occurred in the 1970s, as *The Population Bomb* had predicted. India, Bangladesh, Cambodia, West and East Africa—in that decade all were wracked, horribly, by hunger. But the death tally was nowhere near the "hundreds of millions" predicted by Ehrlich. According to a widely accepted count by the British development economist Stephen Devereux, starvation claimed about 5 million lives during that period—with most of the deaths due to warfare, rather than environmental exhaustion. Compared to the past, in fact, famine has not been increasing but has become rarer. Nor did anything like Ehrlich's planetary breakdown occur by 1985, though there have been awful losses that will not be easy to set right. Similarly, pesticides did not lead to lethal epidemics of heart disease, cirrhosis, and cancer, a prospect Ehrlich warned of in 1969. Farmers continued ✓ to spray their fields, but U.S. life expectancy did not fall to "42 years by 1980."

In the mid-1980s I began work as a science journalist. I met many Wizard technologists and grew to admire them. I became a Borlaugian, scoffing at the catastrophic scenarios I had previously embraced. Cleverness will get us through, I thought, as it had in the past. To think anything else, given recent history, seemed foolishly pessimistic.

Nowadays, though, worrying about my children, I am waffling. My daughter, in college as I write, is headed into a future that seems ever more jostling and contentious, ever closer to overstepping social, physical, and ecological margins.

Ten billion affluent people! The number is unprecedented, the difficulties like nothing before. Maybe my optimism is as ill-founded as my previous pessimism. Maybe Vogt was right after all.

Thus I oscillate between the two stances. On Monday, Wednesday, and Friday, I think Vogt was correct. On Tuesday, Thursday, and Saturday, I go for Borlaug. And on Sunday, I don't know.

I wrote this book to satisfy my own curiosity, and to see if I could learn something about the roads my children could take.

ONE LAW

ONE

State of the Species

Special People

Begin with an image, a man alone on a tract of land near a city. The man is thirty years old and just beginning to discover his own ambition. His name is Norman Borlaug, the Wizard of my title. His greatest advantage is a remarkable capacity for hard technical work. The land, on the periphery of Mexico City, is badly damaged; Borlaug has been assigned the task of coaxing something to grow on it. To most of the people whom Borlaug is likely to know, the task and place seem remote and inconsequential. Borlaug, the Wizard, will change that view.

It is April 1946, the perfervid months after the end of the Second World War. Most people in North America and Europe are wholly caught up in the shocking changes that follow the conflict—the onset of the atomic age, the beginning of the Cold War, the disintegration of colonial empires. Borlaug, the hardworking man, is not. Newspapers and radios are not readily available where he works. He spends his days staring at dying plants. Years later, some people will say that the work he began there was more important than any of the occurrences in the newspapers.

Now on this land appears a second man. This second man, the Prophet of my title, is twelve years older, light-haired and blue-eyed. He walks with a pronounced limp, the legacy of polio. His name is William Vogt. He, too, is discovering the extent of his ambitions—maybe it would be better to say that he is at last *admitting* them.

Borlaug's project is housed at a university in Chapingo, a settlement east of Mexico City. Built in a former hacienda, the university has been transformed from a private rural backwater into a crowded expression of the contemporary state, desperately sought after and grossly underfunded in the modern mode. Among its glories is a set of huge, brilliantly colored murals by the celebrated Mexican painter Diego Rivera. Vogt is on his honeymoon; he and his wife visit the murals. But Vogt is also traveling in his official capacity, as the head of the Conservation Division of the Pan American Union. He is deeply interested in agriculture and its effects on the landscape.

At this time, Borlaug and three Mexican assistants are the only people working on the land. Vogt spends a big chunk of his daylong visit there. Inquisitive and gregarious, he surely walks to the sweating people in their dusty khakis and boots and asks what they are doing with this 160 acres of starveling wheat and maize at the edge of campus.* Vogt has no idea that this plain-faced man, lean and taciturn, will become an enduring international symbol of technical prowess and a way of thinking that Vogt will come to regard as dangerous to human survival. Borlaug does not guess that the visitor, this limping man with his wife in tow, will spark a movement that Borlaug will come to regard as blinkered, when not duplicitous, effectively an enemy to human well-being. From the evidence left behind, Vogt doesn't say much during his visit. One imagines him watching and listening as Borlaug explains his ideas.

This is the beginning, these two men looking over the parcel of damaged land near the city. All the rest of their lives begin at this place, in what they see and how they choose to think about it. Things skirl out from Chapingo, they sprawl across the world and pass decades into the past and future, they involve millions upon millions of people who have never heard the names of Borlaug or Vogt. But always it opens here: two men, a parcel of bad soil, the nearby city.

Before the Spanish conquest, Chapingo, and Mexico City were on opposite sides of a lake that was more than thirty miles wide, rich with fish, and lined with prosperous villages. On the fringes of this great lake were hundreds of small artificial islands called *chinampas*. Made by piling up lake muck, *chinampas* were used as farms. Yielding multiple harvests in a year, they were among the most productive farms in the world.

---

* I use "maize," rather than "corn," because Mexican maize—often multicolored and mainly eaten after drying and grinding—is strikingly unlike the sweet, yellow kernels evoked in the United States by the name "corn."

All this is gone. Persistent mismanagement has over the generations drained the lake and wiped out the *chinampas* and turned the good soil into something that is cracked and lifeless.

Vogt and Borlaug have the same mission: to use the discoveries of modern science to spare Mexico from a future of poverty and environmental degradation. But prospects are unlikely, in Mexico in 1946, for this to happen; indeed, Vogt and Borlaug believe that the situation grows direr by the day.

Not much later both men will realize that the challenges they see before Mexico actually confront all of humankind. Vogt and Borlaug are among the few who have some glimpse then of the magnitude of the tests that face our species today, as we move ever closer to 2050, when the world will hold 10 billion souls. But their understanding of how to resolve them differs, as do their views on their causes.

Vogt sees the city reaching across the dry lake bed to engulf the last fields and streams and says: Hold it back! We cannot let our species overwhelm the natural systems on which we all depend! Borlaug sees the pitiful scrim of wheat and maize on the tract of land and says: How can we give people a better chance to thrive? Vogt wants to protect the land; Borlaug wants to equip its occupants.

Which is correct? To Vogt, the fields of maize and wheat throughout the dry hills of central Mexico are a plague that will lead in the end to destruction. He calls for more sustainable, land-sparing agriculture, to keep people from trying to use this fragile, depleted soil. One can imagine his reaction to learning that Borlaug hopes to develop new strains of maize and wheat that will better allow humans to exploit that land. It is, from Vogt's perspective, like trying to fight arson with gasoline.

Later critics will call Vogt and people like him names like "tree-hugger" and say they are apostles of a new religion, an irrational cult that fetishizes Nature. In Vogt's assessment he is simply speaking from the tradition of ecology (or what he understands to be ecology)—a holistic view that seeks to place humanity within a framework of over-arching natural law. It asks: How can we best fit into the world, and not overstep our bounds? Even to ask such a question calls for a reordering of society.

Borlaug, by contrast, speaks from the point of view of genetics—an effort that seeks to break organisms into their smallest components so that they can be harnessed for human benefit. Of Vogt's natural bounds, it asks, How can we leapfrog them altogether? Critics will call this "techno-optimism," an advocacy of "salvation by technical advance," and accuse its Wizard advocates of being apologists for economic systems that are fundamentally at odds with the ability to sustain life on Earth. Nature knows best! Anything else is hubris and folly.

One wants the two men to have a ringing debate, like Abraham Lincoln and Stephen Douglas. That doesn't happen. Instead, a few months after his trip to Mexico, Vogt tries to get Borlaug shut down.

Largely as a result of Vogt's advocacy, the Mexican government has adopted new soil and water conservation laws. But there is more to be done, he thinks, and his money is running out. Vogt works for the Pan American Union, but his work in Mexico is supported by several small, poorly funded conservation groups, including the New York Zoological Society, the International Committee for Bird Preservation, and the American Wildlife Institute. Saving the world, he thinks, will require deeper pockets.

Borlaug, by contrast, is supported by the Rockefeller Foundation, based in New York City, long the world's biggest private charity. The Rockefeller Foundation in 1946 is like the Bill & Melinda Gates Foundation today—an international symbol of largesse. Vogt seems to spend his life scrambling after pennies to fund his world-important tasks. How it must have haunted him to see Borlaug treading into his terrain, concerned about the right problem, backed by powerful money—and proceeding, in Vogt's opinion, in exactly the wrong way!

Even as Vogt and his wife spend a month in Guatemala and then travel to El Salvador and Venezuela, he drafts and redrafts a letter to the Rockefeller Foundation. It finally goes out on August 2, 1946. It bears the signature of L. S. Rowe, director general of the Pan American Union, but Vogt has written every word. The letter has a delicate task: to say, tactfully but clearly, that Rockefeller (1) is doing everything wrong and (2) should put Vogt in charge of doing it right. Graciously it salutes the foundation's history of fighting disease, then turns in a different direction. "Millions of dollars, from [the foundation], are being used to reduce mortality rates—in other words, to increase populations. Little thought is being given to the feeding of those populations." In Mexico, Rockefeller is backing efforts to grow more wheat and maize. But, the letter says, boosting agriculture and industry is not the answer, because the resources necessary for both "are being wiped out through destruction of watersheds, raw materials, and purchasing power." Simply giving people better tools, Vogt believes, will only help people hit limits faster. If just ten fish remain in a pond, the solution to running out of fish is not more efficient nets.

Instead, what is needed, above and beyond all else, is a change in our relationship with Nature. If people understood the value of the ecosystems in which they are embedded, society would be profoundly different. In the past, Mexico could exist with this incorrect understanding of the world, but soon there will be no more margin for error. The city is rushing in to cover the land. Matters must change in the next few decades. "It is doubtful that in the entire Western Hemisphere there exists a problem that is more important or more pressing," Vogt says.

His letter to the foundation begins a long argument that continues to the present day.

The World Is a Petri Dish

Nonsense! I hear Lynn Margulis say. Poppycock! Or, rather, I hear her say something more pungent.

A researcher who specialized in cells and microorganisms, Margulis was one of the most important biologists in the last half century—she literally helped to reorder the tree of life, convincing her colleagues that it did not consist of two kingdoms (plants and animals), but five or even six (plants, animals, fungi, protists, and two types of bacteria).* Until her death in 2011, she lived in my town, and I would bump into her on the street from time to time. She knew I was interested in environmental issues, and she liked to needle me. Hey, *Charles*, she would call out, are you still all worked up about protecting endangered *species*?

Margulis was no apologist for unthinking destruction. Still, she

---

* The definition of "kingdoms" remains contentious. The "two types of bacteria" I refer to are bacteria proper and archaea, which resemble bacteria physically but have different biochemical pathways. "Protist" is a catch-all for everything else, including amoebas, slime molds, and single-celled algae—anything that is not an animal, plant, fungus, bacterium, or archaeon. Viruses are not usually included in these lists, because they are so simple that most biologists don't view them as a form of life.

Lynn Margulis, 1990

couldn't help regarding conservationists' fixation on birds, mammals, and plants as evidence of their ignorance about the greatest source of evolutionary creativity: the microworld of bacteria, fungi, and protists. More than 90 percent of the living matter on Earth consists of micro-organisms, she liked to remind people. Heck, there are as many bacterial cells in our body as there are human cells!

Microorganisms can do things undreamed of by clumsy mammals like us; they can form giant super-colonies, reproduce asexually, swap genes with wildly different species, and accomplish all sorts of chemical feats that human beings can only do in big laboratories—the list is as endless as it is amazing. Microorganisms have changed the face of the earth, crumbling stone and even giving rise to the oxygen we breathe. Compared to this power and diversity, Margulis liked to tell me, pandas and polar bears were epiphenomena—interesting and fun, perhaps, but not actually *significant*.

I never told her of my image of the two men in Mexico, but I am quite sure what she would have said about it. *Homo sapiens*, she once told me, is an unusually successful species. And it is the fate of every successful species to wipe itself out—that is the way things work in biology. By "wipe itself out" Margulis didn't necessarily mean extinction—just that something comprehensively bad would happen, wrecking the human enterprise. Borlaug and Vogt might have wanted to stop us from destroying ourselves, she would have said, but they were kidding themselves. Neither conservation nor technology has anything to do with biological reality.

Margulis explained these ideas to me while talking about one of her scientific heroes, the Russian microbiologist Georgii Gause. Born in 1910, Gause was a prodigy: he published his first scientific article at the age of nineteen (it appeared in *Ecology*, the premier journal in the field). Like Vogt, Gause looked with envy at Rockefeller's funds, so much greater than anything available to him in the Soviet Union. Hoping to impress the foundation, he decided to perform some experiments and include the results in a grant application.

Gause knew just what to do. In 1920, two Johns Hopkins biologists, Raymond Pearl and Lowell Reed, had published a mathematical formula that described the rate at which the population of the United States grew over time. Their argument was almost completely theoretical. They imagined what the rate of growth should look like, given their knowledge of biology, and sought to match their hypothetical curve to the actual population of the United States as recorded in census data. The two matched well enough that Pearl and Reed believed they were on to something. Pearl was especially excited; he had been conducting parallel research with fruit flies, locking a male and female in a bottle full of food and observing how many flies would be produced in the next few generations. The results looked so similar to his U.S. Census data that he was convinced that he had found a universal law, applicable to fruit flies in bottles and humans in North America alike. "The growth of populations of the most diverse organisms," he said, "follows a regular and characteristic course."

An expert in self-publicity, Pearl proclaimed the new law in a dozen articles and three books. But the onslaught failed to prevent critics from attacking his ideas. Pearl had begun, the critics said, by assuming his hypothesis might be true, then looking for a match in his data; when he found one, he claimed that the match proved him correct. Pearl's detractors argued that this procedure missed an essential step: demonstrating that no other hypotheses also fit the data. Worse, the law didn't work very well—Pearl had to wave his hands at the numbers to make them come out right.

To win Pearl's support for a Rockefeller grant, Gause decided to try to nail down the case with a series of experiments on fruit flies. He soon discovered that the flies moved around so much that they were hard to count. To obtain better results, Gause decided to work with microorganisms. These could be spread across a microscope slide and counted.

By today's standards, his methodology was simplicity itself. Gause placed half a gram—that is, just a pinch—of oatmeal in one hundred milliliters (about three ounces) of water, boiled the results for ten minutes to create a broth, strained the liquid portion of the broth into a container, diluted the mixture by adding water, and then decanted the contents into small, flat-bottomed test tubes. Into each he dripped five *Paramecium caudatum* or *Stylonychia mytilus*, both single-celled protozoans, one species per tube. He stored the tubes for a week and observed the results. The conclusions appeared in a 163-page book, *The Struggle for Existence*, published in 1934.

Today *The Struggle for Existence* is viewed as a scientific landmark, one of the first successful marriages of experiment and theory in ecology. But it was not enough to get Gause a fellowship; Rockefeller turned down the twenty-four-year-old student as insufficiently eminent. Gause did not visit the United States for another twenty years, by which time he had indeed become eminent. But he had also left microbial ecology and become an antibiotics researcher.

What Gause saw in his test tubes—and what Pearl had theorized before him—is often depicted in a graph, time on the horizontal axis, the number of protozoa on the vertical. By squinting a bit, it is possible to imagine that the curve forms a kind of flattened *S*, which is why scientists often refer to Gause's curve as an "*S*-shaped curve." At the beginning (that is, the left side of the *S*-shaped curve), the number of protozoans grows slowly, and the graph line slowly ascends to the right. But then the line hits an inflection point, and suddenly rockets upward—a frenzy of growth. The mad rise continues until the organism begins to run out of food, at which time there is a second inflection

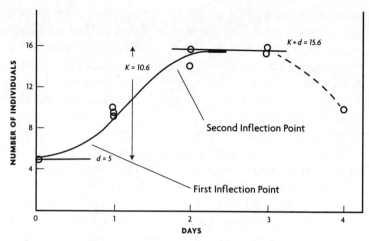

One of Gause's diagrams of his S-shaped curve, with labels modified by the author

point, and the growth curve levels off again as protozoa begin to die. Eventually the line descends, and the population falls toward zero.

Years ago I watched Margulis demonstrate Gause's conclusions to one of her classes with a time-lapse video of *Proteus vulgaris*, a bacterium that resides in the intestinal tract. To humans, she said, *P. vulgaris* is mainly notable as an occasional cause of hospital infections. Left alone, it divides about every fifteen minutes, producing two individuals where before had been one. Margulis switched on the projector. Onscreen was a tiny dot—*P. vulgaris*—in a shallow, circular glass container: a petri dish, its bottom covered with a layer of reddish nutrient goo. The students gasped. In the time-lapse video, the colony seemed to pulse, doubling in size every few seconds, rippling outward until the mass of bacteria filled the screen. In just thirty-six hours, she said, this single bacterium could cover the entire planet in a foot-deep layer of single-celled ooze. Twelve hours after that, the ball of living cells would be the size of Earth.

Such a calamity cannot happen, Margulis said, because rival organisms and lack of resources prevent the vast majority of *P. vulgaris* from reproducing. This is natural selection, Darwin's great insight. All living creatures have the same purpose: to make more of themselves, ensuring their biological future by the only means available. And all living creatures have a maximum reproductive rate: the greatest number of offspring they can generate in a lifetime. (For people, she told the class, the maximum reproductive rate is about twenty children per couple per generation. The potential maximum for dachshunds is around 330: eleven pups per litter, three litters a year, for roughly ten years.) Natural selection ensures that only a few members of each generation manage to reach this rate. Many individuals do not reproduce at all; blocked, they fall by the wayside. "Differential survival is really all there is to natural selection," Margulis said. In the human body, *P. vulgaris* is checked by the size of its habitat (portions of the human gut), the limits to its supply of nourishment (food proteins), and other, competing microbes. Thus constrained, its population remains roughly steady.

Things are different in the petri dish. From *P. vulgaris*'s position, the dish initially seems limitless, a boundless ocean of breakfast, no storm on the horizon, no competition for sustenance. The bacterium eats and divides, eats and divides. Racing across the nutrient goo, it passes the first inflection point and hurtles up the left side of the curve. But then its colonies slam into the second inflection point: the edge of the petri dish. When the food supply is exhausted, *P. vulgaris* experiences a vest-pocket apocalypse.

By luck or superior adaptation, a few species manage to escape their limits, at least for a while. Nature's success stories, they are like Gause's protozoans; the world is their petri dish. Their populations grow at a terrific rate; they take over large areas, engulfing their environment as if no force opposed them. Then they hit a barrier. They drown in their own wastes. They starve from lack of food. Something figures out how to eat them.

When I lived in New York City, zebra mussels invaded the lower Hudson River, the western boundary of Manhattan island. An inch or two long, their shells patterned with wriggly bands of brown and white, zebra mussels are capable of spitting out a million eggs a year apiece. The species originated in the Azov, Black, and Caspian seas on Europe's Russian- and Turkic-speaking periphery. Globalization has been good to it. Escaping their native waters, zebra mussels hitchhiked around the world in ship bilges and ballast water. They have been recorded in Europe since the eighteenth century. The Hudson first saw them in 1991. Within a year zebra mussels constituted half the mass of living creatures in the river. In some places tens of thousands carpeted every square foot. They covered boat bottoms, blocked intake tubes, literally smothered other species of shellfish with a blanket of striped shell. Zebra mussels were shooting up the S-shaped curve.

Bust followed boom; the population collapsed. In 2011, two decades after the mussel was first sighted in the Hudson, its survival rates were "1% or less of those in the early years of the invasion" (the quote is from one long-range study). Unlike Gause's protozoa, the mussels had not run into a physical wall—the physical world is always more complex than a test tube. They did exhaust their food supply, but they also were attacked by a local predator, the blue crab, which had learned to eat the new-comers. Their S-shaped curve wiggled more than those in Gause's book, but the result was the same. Fifteen years ago, when I went to a park at the edge of the Hudson, I couldn't step into the river—the sharp edges of open mussel shells were too thick underfoot. Nowadays at the park the creatures are mostly gone. Children splash happily in the shallows. Crumbled shells lie in the sediment, testament to the mussel's collapse.

Humans are no different, Margulis believed. The implication of evolutionary theory is that *Homo sapiens* is just one creature among many, no different at base than *P. vulgaris*. We and they are controlled by the same forces, produced by the same processes, subject to the same fate. When Borlaug and Vogt stood on the tract of bad land, looking at the city, they were on the edge of the petri dish. Wizard or Prophet, it didn't matter. *Homo sapiens*, in Margulis's eyes, was just another briefly successful species.

Of Lice and Men

Why and how did humankind become "successful"? And what, to an evolutionary biologist, does "success" mean, if self-destruction is part of the definition? Does that self-destruction include the rest of the biosphere? What are human beings, anyway? With more than 7 billion of us crowding the planet, it's hard to imagine more vital questions.

One way to begin answering them came to Mark Stoneking in 1999, when he received a notice from his son's school warning of a lice outbreak in the classroom. Stoneking was a researcher at the Max Planck Institute for Evolutionary Biology, in Leipzig, Germany. He didn't know much about lice. As a biologist, it was natural for him to noodle around to find information about them. The most common louse to afflict humans, he learned, is *Pediculus humanus*, an insect that lives on human bodies, as its name suggests. *P. humanus* has two distinct subspecies: *P. humanus capitis*, head lice, which feed and live on the scalp; and *P. humanus corporis*, body lice, which feed on skin but live in clothing. In fact, body lice are so dependent on the protection of clothing that they cannot survive more than a few hours away from it.

It occurred to Stoneking that the difference between the two subspecies could be used as an evolutionary probe. *P. humanus capitis*, the head louse, could be an ancient annoyance, because human beings have always had hair for it to infest. But *P. humanus corporis*, the body louse, must not be especially old, because its dependence on clothing meant that it could not have existed when humans went naked. Humanity's great cover-up had created a new ecological niche, and some head lice had rushed to fill it. Natural selection thereupon did its magic; a new subspecies arose. While Stoneking couldn't be sure that this scenario had taken place, it seemed likely. If his idea was correct, then discovering when the body louse diverged from the head louse would provide a rough date for when people first wore clothing.

With two colleagues, Stoneking measured the difference between snippets of genes in the two louse subspecies. Because genetic material picks up small, random mutations at a roughly constant rate, scientists use the number of differences between two populations to tell how long ago they diverged from a common ancestor—the more differences, the longer the separation. In this case, the body louse seemed to have separated from the head louse about 107,000 years ago. This meant, Stoneking hypothesized, that clothing also dated from about 107,000 years ago.

The subject was anything but frivolous: donning a garment is a complicated act. Clothing has practical uses—warming the body in cold places, shielding it from the sun in hot places—but it also transforms the appearance of the wearer, something of inescapable interest to a visually oriented species like *Homo sapiens*. Clothing is ornament and symbol; it separates human beings from their earlier, unself-conscious state. (Animals run, swim, and fly without clothing, but only people can be *naked*.) The arrival of clothing was a sign that a mental shift had occurred. The human world was becoming a realm of complex, symbolic artifacts.

It was not only clothing. As scientists have painstakingly established, a host of innovations were occurring around that time. Human beings were engraving pieces of ochre and ostrich shells in southern Africa. They were carving elegant harpoons from bone in central Africa. They were making ornamental beads in northwestern Africa. They were burying the dead with care in the Levant, just across from northeast Africa. They were, in sum, becoming human.

In these discussions "human" has many meanings. One is scientific: relating to or characteristic of our species, *Homo sapiens*, a bipedal primate. A second, somewhat different meaning is also scientific: relating to or characteristic of our genus, *Homo*. (A genus is a group of closely related species.) Today there is little distinction between the two meanings, because the genus *Homo* contains only one species, *H. sapiens*. But 300,000 or so years ago, when *Homo sapiens* emerged, the meanings were different. Several species of *Homo*—the exact number is as uncertain as the next archaeological find, as the next anthropological quarrel over taxonomy—were scattered around the world. *Homo sapiens* (us), *Homo neanderthalensis* (Neanderthals), *Homo denisova* (Denisovans), *Homo naledi*, *Homo heidelbergensis*, *Homo floresiensis* (nicknamed "hobbits," because of their small stature). All were human. Nobody knows how all these humans behaved when they met, whether they were amicable, antagonistic, or aloof. At least some of these ancient types of humans bred with our ancestors—*Homo sapiens* with *Homo neanderthalensis*, for example—leaving scattered traces of their coupling in our genes. But whatever the sequence of interactions, we do know the outcome. For better or worse, only one species of human now walks the planet.

But there's a third sense of "human," one captured in the phrase of "being human." Human-ness is the quality—a mix of creativity, drive, and moral awareness—that transforms humans into persons. It is a special spark or spirit, unique among living creatures, a flame possessed in abundance by our heroes, possessed in small amounts by all. It is what makes *Homo sapiens* want to believe they are special, to believe they are unlike the other members of the genus *Homo*.

Humans weren't always human in this third sense, as far as we can tell. In the beginning, *Homo sapiens* seems not to have created art, played music, invented new tools, worked out the motions of the planets, or worshiped gods in the celestial sphere. These capacities accumulated slowly, over tens of thousands of years. Sometimes a new trait—a new kind of art, a new kind of construction—arose, only to fade out. But over the long run, as the other human species disappeared, these attributes built up in us, until perhaps fifty thousand years ago something resembling modern humankind—"behaviorally modern" humans, in the jargon—was loose in the world. Only as humans were reduced to humankind did humans gain their humanity. And only then did we truly leave Africa, a conquering horde, carrying our lice into every corner of the world.

This army, the human army, was an army of similars, its soldiers remarkable in their genetic uniformity. DNA, the material of which genes are made, consists of long, skinny, string-shaped molecules. Each molecule is composed of two chains that are twisted around themselves to create the famous double helix. Individual links in the chains are called "bases" or "nucleotides." Arrays of links—segments of the DNA chain, so to speak—form individual genes. The totality of the genetic information in an individual or species is that individual or species' "genome." The lineup of bases and genes in one person—that person's genome—is barely distinguishable from the lineup in the next person. This similarity is striking to geneticists but hard to describe exactly. Roughly speaking, two peoples' genomes differ in only about one out of every thousand bases. This is like having two pages in two different books differ by a single letter. The equivalent figure for two *Escherichia coli*, the most common bacterium in the human gut, might be one out of fifty. By this measure, the bacteria in people's intestines are twenty times more diverse than their hosts.

These comparisons are incomplete. In addition to differences in single bases, organisms also vary in terms of duplicated or deleted segments of DNA. These disparities are larger than single bases and usually more important. They are also hard to quantify: if one member of a species has ten copies of a particular gene variant and another has twenty copies of that variant, are they alike because they have the same gene variant or different because they have different numbers of it? Still, the overall point remains: compared to bacteria, humans are genetically similar to the point of tedium. Bacteria may not be the best comparison. They are so diverse genetically that researchers into the microworld often object to classifying them as a "species," because that implies they share single, identifiable pools of DNA. It may be better to look at mammals, which are closer to us. In general, apes are far on the low end of mammal diversity, and humans are less diverse than almost all other apes. The genetic differences between one chimpanzee and its neighbor on a single hillside in central Africa can be greater than those of two humans in central Asia and Central America. When scientists list mammals in order of their genetic diversity, humans are at the bottom, along with endangered species like wolverines and lynxes.

Genetic uniformity is usually a legacy of small population size—the descendants of a small group have only the genes bequeathed to them by their few founders. Reasoning backward from humankind's sparsely filled genetic larder, some researchers have argued that at some point our numbers must have fallen dramatically, perhaps to a breeding population of as few as ten thousand people—the size of a midsize university. (The actual population would have been bigger; this estimate is of the number of people who successfully produced children.)

When a species shrinks in number, chance can alter its genetic makeup with astonishing rapidity. New mutations can arise and spread; a snippet of scrambled DNA in a single gene in a single member of the small group that populated Ice Age Europe apparently led to the blue eyes that predominate in most of Scandinavia. Rare genetic variants that are already present can suddenly become more common, effectively transforming the species within a few generations as once-unusual traits proliferate. Or common genetic variants can, by chance, fall to the wayside. For these and other reasons, researchers have often speculated that in the short span of some tens of thousands of years—a lightning flash in the history of life—something happened in our DNA, something unprecedented, something special, something that made us human. Changes accelerated. And about seventy thousand years ago, perhaps a bit less, our species took a fateful step.

One way to illustrate the impact of this change is to consider *Solenopsis invicta*, the red imported fire ant. Geneticists believe that *S. invicta* originated in southern Brazil, an area with many rivers and frequent floods. The floods wipe out ant nests. Over the eons, these small, furiously active creatures have evolved the ability to respond to rising water by knitting their bodies together into floating swarm-balls—workers on the outside, queen in the center—that can ride on the flood for days.

Once the waters recede, colonies swarm back onto previously sub-merged land so rapidly that *S. invicta* can use the devastation to increase its range. Like criminal gangs, fire ants thrive on chaos.

In the 1930s *Solenopsis invicta* was transported to the United States, probably in ship ballast, which often consists of haphazardly loaded soil and gravel. An adolescent bug enthusiast named Edward O. Wilson, later a famous biologist, spotted the first colonies in the port of Mobile, Alabama. From the ant's vantage, it had been dumped onto an empty, recently flooded expanse. *S. invicta* took off, never looking back.

More than likely, the initial incursion seen by Wilson was just a few thousand individuals—a number small enough to hint that random, bottleneck-style genetic change played a role in what happened next. (The evidence is not yet conclusive.) In its homeland, fire ant colonies constantly fight each other, reducing their numbers and creating space for other types of ant. In North America, by contrast, the species forms cooperative super-colonies, linked clusters of nests that can spread for hundreds of miles, wiping out competitors along the way. Remade by chance and opportunity, new-model *S. invictus* needed just a few decades to conquer much of the southern United States.

A primary obstacle to its expansion is another imported South American ant, *Linepithema humile*, the Argentine ant. After escaping its natal territory more than a century ago, *L. humile* formed its own super-colonies in the United States, Australia, New Zealand, Japan, and Europe (the European colony stretches from Portugal to Italy). In recent years researchers have come to believe that these huge, geographically sepa-rate ant societies in fact may be part of a single intercontinental unit, a globe-spanning entity that exploded across the planet with extraordi-nary speed and rapacity, and is now the most populous society on Earth.

Homo sapiens did something similar as it became human. Our spe-cies first clearly appears in the archaeological record about 300,000 years ago (though we may well have emerged before then). Until about 75,000 years ago—that is, for the majority of our existence on Earth—humankind was restricted to Africa, though we sent out occasional for-ays into the rest of the world, almost all unsuccessful, all limited in scope. Around 70,000 years ago, everything changed. People raced across the continents like so many imported fire ants. Human footprints appeared in Australia within as few as ten thousand years, perhaps within four or five thousand. Stay-at-home *Homo sapiens* 1.0, a wallflower that would never have interested Lynn Margulis, had been replaced by aggressively expansive *Homo sapiens* 2.0. Something happened, for better and worse, and we were born.

No more than a few hundred people initially left Africa, if geneticists are correct. And for a long time their geographic expansion was not matched by an increase in population. As recently as ten thousand years ago we numbered perhaps 5 million, about one human being for every five square miles of Earth's habitable surface. *Homo sapiens* was a scarcely noticeable dusting on the surface of a planet dominated by microbes.

At about this time—10,000 years ago, give or take a millennium—our species swung around the first inflection point, with the invention of agriculture. The wild ancestors of cereal crops like wheat, barley, rice, and sorghum have been part of the human diet for almost as long as there have been humans to eat them. (The earliest evidence comes from Mozambique, where researchers found tiny bits of 105,000-year-old sorghum on ancient scrapers and grinders.) In some cases people may have watched over patches of wild grain, returning to them year after year. Yet despite the effort and care, the plants were not domesticated. As botanists put it, wild cereals "shatter"—individual grain kernels fall off as they ripen, making it impossible to harvest the plants systematically. Only when an unknown genius discovered naturally mutated grain plants that did not shatter—and purposefully selected, protected, and cultivated them—did true agriculture begin. Planting great expanses of these altered crops, first in southern Turkey, later in almost a dozen other places, early farmers created landscapes that, so to speak, waited for hands to harvest them.

Farming transformed our relationship to nature. Foragers manipulated their environment with fire, burning areas to kill insects and encourage the growth of useful species—plants we liked to eat, plants that attracted the other creatures we liked to eat. Nonetheless, their diets were largely restricted to what the world happened to provide in any given time and season. Agriculture gave humanity the whip hand. Instead of natural ecosystems with their haphazard mix of species, farms are taut, disciplined communities dedicated to the maintenance of a single species: us. Before agriculture, the Middle West, Ukraine, and the lower Yangzi Valley had been sparsely populated domains of insects and grass; they became breadbaskets, as people scythed away suites of species that used soil and water we wanted to control and replaced them with maize, wheat, and rice. To Margulis's bacteria, a petri dish is a uniform expanse of nutrients, all ready for the taking. For *Homo sapiens*, agriculture transformed the planet into something like a petri dish.

As in a time-lapse movie, we divided and multiplied across the newly opened land. It had taken *Homo sapiens* 2.0, aggressively modern humans, barely fifty thousand years to reach the farthest corners of the globe. *Homo sapiens* 2.0A—A for agriculture—took a tenth of that time to subdue the planet.

Since the beginning of agriculture, farmers have plowed manure and compost into their soil to promote plant growth. They didn't know it, but the chief reason manure and compost helped their crops was that they replenished a key plant nutrient, the nitrogen in the soil. But this method of recharging the soil had drawbacks. In most places, the supply of manure and compost was limited; importing them from elsewhere was impossibly expensive.

In the early twentieth century, two German chemists, Fritz Haber and Carl Bosch, discovered the key steps to making synthetic fertilizer. Suddenly farmers could go to a store and buy all the fertilizer they wanted—factory-made, cheap, and plentiful. Haber and Bosch are not nearly as well known as they should be; their discoveries, linked into what is called the Haber-Bosch process, have literally changed the chemical composition of the earth. Farmers have injected so much synthetic fertilizer into their fields that soil and groundwater nitrogen levels have risen worldwide. Today, almost half of all the crops consumed by humankind depend on nitrogen derived from synthetic fertilizer. Another way of putting this is to say that Haber and Bosch enabled our species to extract an additional 3 billion people's worth of food from the same land.

Synthetic fertilizer is not alone in its impact. The improved wheat, rice, and (to a lesser extent) maize varieties developed by Borlaug and other plant breeders in the 1950s and 1960s drove up yields greatly. Antibiotics, vaccines, disinfectants, and water-treatment plants pushed back humankind's bacterial, viral, fungal, and protozoan enemies. All allowed humankind ever more unhindered access to the planet.

Rocketing up the growth curve, humankind every year takes ever more of the earth's richness. An often quoted estimate by a team of

Stanford biologists is that humans grab "about 40% of the present net primary production in terrestrial ecosystems"—40 percent of the entire world's output of land plants and animals. This assessment dates from 1986. Ten years later, a second Stanford team calculated that the figure had risen to "39 to 50%." (Others have suggested a figure closer to 25 percent, still very high for a single species.) In 2000, the chemist Paul Crutzen and the biologist Eugene Stoermer awarded a name to our time: the Anthropocene, the era in which *Homo sapiens* became a force operating on a planetary scale.

Lynn Margulis, it seems safe to say, would have rolled her eyes at these statements, which in every case that I am aware of do not take into account the enormous impact of the microworld. But she would not have disputed the central idea: *Homo sapiens* has become a successful species.

As any biologist would predict, this success led to an increase in human numbers—slow at first, then rapid, tracing Gause's S-shaped curve. We began rising up the steepest part of the slope in the sixteenth or seventeenth century. If we follow Gause's pattern, growth will continue at delirious speed until the second inflection point, when we have exhausted the global petri dish. After that, human life will be, briefly, a Hobbesian nightmare, the living overwhelmed by the dead. When the king falls, so do his minions; it is possible that in our desperation we might consume most of the world's mammals and many of its plants. Sooner or later, in this scenario, Earth will again be a choir of microorganisms, as it has been through most of its history.

It would be foolish to expect anything else, Margulis thought. More than that, it would be *strange*. To avoid destroying itself, the human race would have to do something deeply unnatural, something no other species has ever done or could ever do: constrain its own growth (at least in some ways). Brown tree snakes in Guam, water hyacinth in African rivers, rabbits in Australia, Burmese pythons in Florida—all these successful species have overrun their environments, heedlessly wiping out other creatures. Not one has voluntarily turned back. When the zebra mussels in the Hudson River began to run out of food, they did not stop reproducing. When fire ants relentlessly expand their range, no inner voices warn them to consider the future. Why should we expect *Homo sapiens* to fence itself in?

What a peculiar thing to ask! Economists talk about the "discount rate," which is their term for the way that humans almost always value the local, concrete, and immediate over the faraway, abstract, and distant in time. We care more about the broken stoplight up the street now than social unrest next year in Chechnya, Cambodia, or the Congo. Rightly so, evolutionists say: Americans are far more likely to be killed at that stoplight today than in the Congo next year. Yet here we are asking governments to focus on potential planetary boundaries that may not be reached for decades or even centuries. Given the discount rate, nothing could be more understandable than a government's failure to grapple with, say, climate change. From this perspective, is there any reason to imagine that *Homo sapiens*, unlike mussels, snakes, and moths, can exempt itself from the fate of all successful species? This is what Borlaug and Vogt were asking people to do in their very different ways.

To a biologist like Margulis, who spent her career arguing that humans are simply part of evolution's handiwork, the answer should be clear. All life is similar at base, she and others say. All species seek to make more of themselves—that is their goal. By multiplying until we reach our maximum possible numbers, we are following the laws of biology, even as we take out much of the planet. Eventually, in accordance with those same laws, the human enterprise will wipe itself out. Shouting from the edge of the petri dish, Borlaug and Vogt might as well be trying to hold back the tide.

From this standpoint, the answer to the question "Are we doomed to destroy ourselves?" is "Yes." That we could be some sort of magical exception—it seems unscientific. Why *should* we be different? Is there *any* evidence that we are special?

TWO MEN

TWO

The Prophet

Heaps of Nitrogen

In the southern Pacific is a great circular moil of trade winds and currents, the South Pacific Gyre, which rotates counterclockwise between New Zealand and the western coast of South America. The piece of the gyre that runs along the South American coast is known as the Humboldt Current. Its near-constant winds, dragging against the shore, push away the warm surface water, drawing colder water up from below. The upwelling water is thick with nutrients: organic matter that has sloughed off the coast and sunk to the ocean floor. The soup feeds huge blooms of plankton, which in turn feed great schools of fish, especially mackerel, sardines, and anchovetas (one of the many species of anchovy). By some measures the Humboldt Current is the most productive marine ecosystem on Earth.

The richness of the ocean is not matched on land. To the east, the Andes Mountains shield the shore from the warm, wet winds that blow in from Brazil; to the west, the Humboldt Current is so cold that the air above it cannot hold much moisture. Sandwiched between these two barriers, the Peruvian coast is dry to the point of desolation. In many places it receives less than an inch of rain per year. Equally barren are Peru's thirty-nine offshore islands. Hot, small, and almost waterless, they are unsuitable for human habitation. But the abundant anchovetas and sardines in the Humboldt Current make the islands attractive to seabirds, which have roosted there for millennia. Like all living things, the seabirds excrete wastes. The parched islands rarely receive enough rain to dissolve them. Over time the bird feces—guano—has accumulated into deposits as much as 150 feet high.

The guano banks have a powerful smell reminiscent of bus-stop bathrooms. This is because as much as a sixth of the guano consists of uric acid, a main ingredient in human urine. Farmers have known for thousands of years that adding urine and feces, animal or human, to the soil helps crops to grow. In the past, Europeans often used *poudrette*, a cocktail of human excrement, charcoal, and gypsum. Other soil additives included ashes, compost, blood from slaughterhouses, and (in China) cakes of soybean residue. Only in the mid-nineteenth century did scientists learn that these substances benefit crops because they put nitrogen into the soil. Soon after, a chemist in Peru informed the government that guano had very high nitrogen levels. The nation's barren, excrement-laden islands were, so to speak, a guano gold mine.

A few bags of Peruvian guano went to Europe. Farmers sprinkled the contents in their fields, saw harvests rise, and demanded more. It was the world's first high-intensity commercial fertilizer. European ships flocked to the barren Peruvian littoral and filled their holds with ancient excrement. To satisfy the demand, Lima gave guano-mining concessions to European companies. They stripped the islands as fast as they could, importing bondsmen from China to do the actual mining. Birds from South America were supercharging plant growth in Europe via slave labor from Asia. Guano dust is laden with toxic ammonia and potassium chloride; slaves wrapped their faces in cloths but still died in droves. Meanwhile the Peruvian government cashed checks. Despite the islands' tiny size, they were responsible for as much as three-quarters of government revenue. To capture the guano trade, Spain seized the most important islands from Peru in 1864. Fearful of losing their guano supply, Britain and the United States threatened to retaliate. At the last moment a global war over fertilizer was avoided.

The entire crazy system depended on the birds that actually produced the guano, of which the most important was the Guanay cormorant. Long-winged and long-necked, black on the back and white on the breast, adorned with an orange-red bandit mask around the eyes, Guanays are noisy creatures, gregarious beyond imagining; their colonies, clustered on the sea, form dark, raft-like masses hundreds of yards on a side. Robert Cushman Murphy, an ornithologist who traveled through Peru in the 1920s, saw Guanays returning to their home islands: "a solid river of birds, which streams in a sharply-marked unbroken column, close above the waves, until an amazed observer is actually wearied as a single formation takes four or five hours to pass a given point." A drizzle of excrement rained below. "The Guanay is in effect a machine for converting fish into guano," Murphy wrote. Every year, a typical cormorant produces about thirty-five pounds of it.

At the end of the nineteenth century the guano-bird population declined. By 1906 their numbers were so low—and the guano industry in such a panic—that Peru sought advice from a U.S. fisheries scientist. He recommended that Peru turn the islands and the surrounding waters into sanctuaries, protecting both anchovetas and guano birds. Following this advice, the islands were nationalized in 1909; control was awarded to the newly formed Compañía Administradora del Guano. Entry to the area was forbidden for months on end, with armed guards stationed on the major islands. The measures were one of the first programs of sustainable management in the world. And they worked—until they didn't. In the 1930s the number of birds again fell rapidly. Apprehensive guano administrators again sought help from the United States.

Robert Cushman Murphy, who had seen the "solid river" of Guanay cormorants, was the world's leading authority on South American seabirds. The Compañía Administradora del Guano naturally approached him. Would he be interested in discovering what was happening to the birds? Murphy declined—he was happy in his position as curator of birds at the American Museum of Natural History in New York City. Instead he recommended a friend who had recently lost his job. The Compañía took Murphy's advice. The friend took a ship from New York and arrived on the guano islands on January 31, 1939. His name was William Vogt.

From today's perspective, Vogt was an improbable candidate for the task of teasing out seabird population dynamics on remote islands. Thirty-six years old, he was a handsome, bushy-haired man with a resonant, actorish voice, piercing blue eyes, and a manner that was confident to the point of imperiousness. But he had no academic training or credentials as a biologist—indeed, he had sedulously avoided mathematics and barely passed his required science courses in college. He did not speak Spanish. He had never been to a foreign country. He had never seen a Guanay cormorant. He didn't even have a sun hat. He was a French literature major who had fallen into birdwatching and

Vogt in a promotional image for the guano firm taken soon after his arrival

befriended a number of professional ornithologists, Murphy among them. Fired from his previous two jobs, he had accepted the guano position in part because he had no alternative.

All of the above is true, but unfair. Vogt was more than an unemployed bird fancier. He was a man who was feeling the first stirrings of a mission—a spark at the nape of the neck that would become a flame. In the previous few years he had become convinced that something was wrong with the way humankind viewed the natural world, or at least the way Americans viewed the eastern half of the United States. To make people listen to his message, Vogt was trying to transform himself into a professional ecologist by performing research so stunning that it would win him a doctoral degree even though he had not attended any classes. He had three years to do all this before his contract with the Compañía Administradora del Guano expired.

Improbably, he succeeded. Although he would never win his doctorate or even formally publish his research, this amateur scientist's time on these faraway islands nonetheless led to what the historian Gregory Cushman has called "an astounding application of advances in ecological theory to practical problems . . . one of the pillars of modern environmental thought." What Vogt saw in Peru would crystallize his picture of the world and the human place in it—a vision of *limitation*. It would bring him to the Prophet's essential belief: humans have no special dispensation to escape biological constraints.

Remake the world! That became his goal, which he pursued with remarkable single-mindedness. To change how people think. To impart a message. He sounded the alarm for years, but died in the conviction that nobody had listened and that he had failed—an astonishing notion, in retrospect. His work marks the beginning of modern environmentalism, the most abiding intellectual and political ideology of Vogt's century and ours.

"The Pleasures of Solitude"

Everything began with the birds, Vogt would admit, but he also drew inspiration from his childhood home in central Long Island. He was born, he liked to say, on a different Long Island, a Long Island of fields and pastures, a Long Island before "the automobiles, the airports, the mosquito control commissions, the shopping centers, the billboards, and the hot dog joints." (This quotation, and others like it, come from unpublished autobiographical jottings in his papers.) Manhattan was just twenty miles away, but young Bill saw it only once a year, during a ritual visit to Santa Claus. The crowds in the stores frightened him—an early memory was of being mashed into the back of an elevator—and he returned with relief to his home.

Vogt spent his first years in a cluster of three small, interconnected villages—Mineola, Garden City, and Hempstead. He lived in a row house in the center of Garden City, two blocks from the train station. A planned community built by a wealthy dry-goods merchant, Garden City had the station, four stores, a tall Episcopal cathedral and its associated school, and the opulent Garden City Hotel, designed by the famed architect Stanford White. To the north was the county fairgrounds, with its horse track and touts. To the south a scatter of homes quickly gave way to farms and ranches. To the west was the cathedral, its acres of greensward tended by the ladies' auxiliary. To the east was a spur rail line that went north to Mineola, and then to Long Island's North Shore. On the other side of the tracks was the green-gold expanse of the Hempstead Plains, mile upon mile of butterflies, birds, and tasseled grass—a vastness, he said later, that gave him "something of the sense of limitless space felt by the newcomers to the plains of Nebraska." Shy and solitary, the boy was ever going into the fields, walking for hours, his only companion his grandmother's St. Bernard.

It is possible that the fields were a refuge from the tensions of his history. Vogt's father, also named William Walter Vogt, was the son of a warehouse clerk in Louisville, Kentucky. The family was respectable: upwardly mobile German immigrants. Vogt senior took a different path. He was charming and fun and untroubled by scruple. He joined the naval hospital corps during the Spanish-American War and lolled about occupied Cuba after his discharge. In August 1900 he traveled with two navy pals to New York. Within hours of their arrival they were arrested, drunk, in the Tenderloin, Manhattan's red-light district. Shrugging off the arrest, Vogt quickly acquired a job and a fiancée. The job was working behind the counter at a drugstore in Mineola. The fiancée was a high school junior from Garden City named Frances Bell Doughty. Fannie, as she was known, had just turned eighteen. Liking the drug business—in those days, selling patent medicines of dubious efficacy—Vogt borrowed money from friends and set up his own drugstore in the first weeks of 1901. He married Fannie, suddenly and quietly, at her mother's home the following Halloween. The sole guest was the bride's mother. Fannie dropped out of school. Six and a half months after the wedding, on May 15, 1902, Vogt's son came into the world. Twelve days later the proud father disappeared.

The impetus was a visit from one of Vogt's friends and financial backers. Eleven days after the birth, the friend showed up at Vogt's store to collect a twenty-dollar payment. Explaining that he needed to pick up the money in New York City, Vogt asked his friend to mind the counter while he got the cash. Early the next morning, without telling Fannie, Vogt took the ferry to Manhattan. He failed to return that night, or the next.

That same morning a woman named Mary Schenck drove a horse and buggy to her brother's house on Long Island's North Shore. Then, she, too, left for Manhattan. She, too, neglected to inform her family— her husband, a prosperous meatpacker, and their three children—of her plans. And she, too, failed to return that night, or the next.

Naturally concerned, her husband asked his friends if they had seen her. Some suggested, as the *Brooklyn Daily Eagle* put it, "that the simultaneous leavetaking of young Voght [*sic*] and Mrs. Schenck is at least a coincidence that Mr. Schenck should investigate."

The friends were correct. Vogt had long been interested in this woman—wealthy, bored, ready for adventure. He had already obtained from her what the *Eagle* called "a substantial loan." A week after their disappearance they were spotted in New York City. Then they were seen in Washington, D.C. To pay off Vogt's debts, the sheriff seized his drugstore and auctioned its contents.

Schenck returned in October, begging her husband to reconcile. Vogt had taken her from New York to Havana, Schenck paying all the bills. In Cuba her money had run out, and with it, from Vogt's point of view, her appeal. He left her without regret, and she came back to Long Island. Soon after, Vogt wrote to his mother-in-law, Clara Doughty, asking for a ticket home.

Clara Doughty was proudly descended from Francis Doughty, a seventeenth-century preacher to whom the king had granted much of what is now Flushing, in Queens—the site, Vogt junior later bragged, of the 1964 World's Fair. The Doughty farm, east of Garden City, had been in the family for five generations. Clara herself was a hardworking woman who had not taken a vacation from her work as a postmistress for twenty-two years. When Fannie became pregnant, she had rammed through the wedding. Now, though, Clara had had enough. She did not reply to her son-in-law's letter. In a huff, Vogt wrote Fannie that he was never leaving Cuba. It was the last time he contacted his family. Even decades later, his unforgiving son told others that his father had died when he was a baby.

Fannie was stranded. She was still legally married, which under the laws of the day meant that she did not have clear title to her income or custody of her son. Yet obtaining a divorce was difficult, because in New York the sole legal basis for ending a marriage was adultery. Vogt had vanished, so Fannie couldn't prove adultery unless Mary Schenck was willing to testify about it. One can imagine the pressure that Fannie and her mother exerted to force Schenck to appear at *Vogt v. Vogt* in March 1908. On the stand Schenck's memory fled; she couldn't recall how long she had known Vogt, where they went, or what they did. But she tersely admitted adultery. Two months later Schenck's husband sued her for divorce.

The scandal had erupted when Vogt's son, William, was twelve days old. Its last ripples did not subside until he was seven. For that place and time—the early twentieth century, a semi-rural village—the whole business must have been deliciously tawdry, fodder for dinner-table conversations in the village, something to shame the family, to mock the boy.

The temptation in tracing anyone's line of thought is to look for explanations in their early life. Always there is the risk of overinterpretation. But it is easy to imagine that a child marked by scandal might end up spending time alone and come to view the natural world as a source of solace and meaning. In any case, these were the circumstances that began the life of one of the century's great crusaders against careless human breeding.

Soon after his father left, Bill and his mother moved into his grandmother's home in Garden City. Money was tight, but the family was not actually poor. In addition to Clara's salary as a postmistress, Fannie worked as a part-time clerk; later she taught at a private kindergarten. Despite the scandal, despite his lack of friends, Vogt always described these years as happy—a childhood idyll in a household of women, surrounded by "what seemed an almost unbounded sea of grass." Walking alone in the Hempstead Plains, he later wrote,

> I learned the pleasures of solitude, the unbroken freedom to see, smell, and listen. These hours alone, though never many at a time, nonetheless sensitized me to the open countryside and prepared me for the enjoyment of winds and skies, plains, mountains, forests and the sea, for the rest of my life.

The idyll didn't last. In 1911 Vogt's mother married Lewis Brown, a carpet-lining salesman. The family's new home was in an industrial district in southeast Brooklyn that was becoming residential. Sandwiched between two major streets, the area was everything Vogt hated: noisy, crowded, enveloped by pavement. He was soon "held up at knife point" in a park and relieved of his "total wealth of 17 or 27 cents."

"As is invariably noted at the beginning of positively all literary biographies," Vladimir Nabokov tells us, "the little boy was a glutton for books." Vogt was no exception; he learned to read early and in Brooklyn found solace by imagining himself somewhere else. Especially resonant were the animal tales of Ernest Thompson Seton. Seton's writing, sentimental and overwrought, has not aged well. Even at the time, his portrayal of animals' abilities—foxes deliberately poisoning their captive offspring rather than allowing them to live in chains, crows that counted to thirty and marched according to a leader's instructions—dismayed scientists. Still, Vogt later recalled, the stories "fired my imagination, as they did that of practically every budding naturalist of my generation."

Seton's bestiary alive in his mind, Vogt hunted for bits of nature in the city: beaches on Staten Island, oak and hickory copses in the Palisades, the dairy farm in Westchester County, just north of the city, where his stepfather had grown up; the chicken farm run by his cousins on Long Island. Mainly, though, Vogt plunged into Seton's own organization, the Boy Scouts of America. (Seton was one of its founders and the first U.S. Chief Scout.) "I was an instructor before being old enough to get the freshman badge, almost immediately assistant patrol leader, and long before old enough was running the Scout troop," he wrote. "I have been running something ever since." Spending as much time in the woods as he could, Vogt was a healthy, rather bossy fourteen-year-old in August 1916, when he came down with polio at Scout camp.

Contagious and incurable, the polio virus attacked much of the United States that summer. New York City alone had almost nine thousand cases; about one out of four was fatal. A disproportionate number of the victims were children; the disease was then known as "infantile paralysis." To prevent infection, the city had shut down schools, colleges, theaters, and playgrounds. Many of the boys at Vogt's camp—Camp Leeming, in the Hudson River Valley—had been sent there expressly to avoid polio. Now in the person of adolescent Bill Vogt the disease had appeared. The local health officer, Vogt wrote later,

> was about as anxious as anyone could be to get rid of me. My family agreed to have me sent to New York and he promised to move me in an ambulance. He failed to keep his word, and the Camp

doctor carried me on his lap for 50 miles in a 1916 car over 1916 roads, and when I got into the Willard Parker Hospital, which was the New York pest house in those days, after having my head bob up and down on the back seat of the car for most of the distance, I was in such bad shape that a wire was sent to my mother saying I might not live until morning.

Vogt survived but was confined to bed for a year: standard treatment for polio at the time. A young librarian in a nearby branch library heard of his housebound situation and sent him *White Fang*, by Jack London. The other staple of literary origin stories, Nabokov might have observed, is the child turning to the world of books when illness forces inactivity (viz., L. Frank Baum, Elizabeth Bishop, Julio Cortázar, Yukio Mishima, Virginia Woolf). Following the classic pattern, *White Fang* led Vogt to the rest of Jack London, which in turn introduced him to other writers. The boy spent hours upon hours immersed in everything from mysteries to George Bernard Shaw, from Rousseau to Turgenev. Constant reading, he later claimed, ruined his eyesight. Thick eyeglasses were "well worth it." The librarian, he said, was "the most influential woman in my life."

Gradually he recovered enough to return to school, though he was "still flopping about rather badly." Hauling himself on a cane, he continued hiking in the hills. When he climbed Whiteface Mountain in the Adirondacks, he didn't care that he had been forced to crawl up the steep trail. Favoring his frail left leg would over the years twist his spine. His lungs were so weak he sometimes had trouble breathing. He never liked being told he was courageous.

The first member of his family to finish high school, Vogt was also the first to attend college: St. Stephen's College, eighty miles north of New York City. (In 1934 the school changed its name to Bard College, after its founder, John Bard.) A Brooklyn minister who admired his Boy Scout leadership found him a scholarship. Avoiding courses in mathematics and science, Vogt majored in French literature and focused on the theater and writing. He had been president of his high school literary club; now he co-edited the St. Stephen's literary magazine, contributing poems and stories. He won the college poetry prize.

Because poetry would not pay the rent, he went to work after gradu-

Vogt (circled) was president of the Brooklyn Manual High School literary magazine,
the Scribe.

ation as an insurance investigator, with a sideline in freelance drama criticism. The insurance job was short-lived; his boss fired him because of his disability. "The next thing I knew," Vogt wrote, "I was editing the publication of the New York Drama League, and a few months later was its Executive Secretary." His title was impressive; his salary was not. On the make but not getting anywhere, trading on flashing blue eyes, an engaging baritone, and a sympathy-inducing limp, Vogt talked his way into multiple jobs on the fringes of publishing and theater. By 1928 he was the editor of a new monthly targeted at boys. Before it could appear, the magazine morphed into *The Funnies*, a comic-strip-filled tabloid that was an early precursor of today's comic books. Vogt lost the job. By then he was married.

Juana Mary Allraum was born in southern California on April 27, 1903. Six months later, the baby's parents tied the knot. Soon afterward her father decamped, leaving no forwarding address. Although her mother had little money, Juana was able to attend the Southern Branch of the Los Angeles State Normal School, precursor to today's University of California at Los Angeles. Her interest in drama grew after she trans-

Juana Allraum, 1922

ferred to the University of California at Berkeley. Petite and vivacious, clever and adaptable, she won the lead in the Berkeley Parthenaia within weeks of her arrival. The biggest social event of the year, the Parthenaia was an outdoor pageant set in ancient Greece. Several hundred female students gamboled about a college garden dressed as nymphs, dryads, and other mythological entities. Juana played Marpessa, a symbol of maidenhood wooed by a randy sun god. In her slightly risqué costume, she was twice featured on the front page of the *San Francisco Chronicle*. After graduation, she moved to New York City, her mother in tow, intending to become an actress. One can picture how the aspiring actress and the freelance drama critic might have met. They wed on July 7, 1928, and moved into a tiny bungalow on the Hudson River, about twenty miles north of the Broadway theaters that fascinated them both.

"The Mosquito Racket"

Ornithology is an outlier in the history of science. At a time when physics and chemistry were transforming themselves from amateur endeavors into professions that were inaccessible to the lay public, bird scientists were crowd-sourcers. Ornithologists could not keep track of millions of birds by themselves, so they sought to harness the energy of amateur birders. It was a good match. In those days, Vogt recalled, "the really active bird-watcher was still considered something of an odd-ball." The American Ornithologists' Union allowed hobbyists to become full members and asked them to contribute field observations. In return, the experts provided encouragement and acceptance—a balm, perhaps, to wounded egos.

Vogt was one of these amateurs. With his limited mobility, Vogt's passion for long hikes had shifted at St. Stephen's into a less arduous devotion to birdwatching. After graduation his passion grew into something close to obsession—he was always out on the Hudson or in the parks, binoculars clapped to his face. It was as if his childhood love of landscape were concentrated in the bodies of the birds. Zeal gained Vogt entrée at the American Museum of Natural History, where he befriended researchers like Frank Chapman, the museum's curator of birds; Ludlow Griscom, arguably the progenitor of organized birdwatching in the United States; Ernst Mayr, an ornithologist who became a luminary in evolutionary biology; and, most important to Vogt, Robert Cushman Murphy, the seabird expert who had visited the guano islands. Going from college birding to ornithological fieldwork with Mayr and Murphy was like going from a small-town orchestra to playing with the Vienna Philharmonic. Vogt was thrilled to be taken into the inner sanctum of Science—who wouldn't be? In return, the researchers acquired an assistant who had no scientific training but was boundlessly energetic and willing to work for no pay. The offers flowed in. Secretary of the Linnaean Society (an amateur natural-history group); editor for the New York Academy of Sciences; proofreader for the Audubon Society—Vogt was always ready to take on new duties.

One of Vogt's closest friends was another amateur, an art student named Roger Tory Peterson. The child of poor immigrants, Peterson had been fascinated by birds since boyhood, not least because the hours spent in their company were a refuge from the drunken, violent rants of his father. At seventeen, Peterson dropped out of high school and moved to New York City. He met Vogt at a birdwatching club in the Bronx. On one of their walks Peterson amazed Vogt by identifying pine siskins from barely audible snippets of their song. How do you *do* that? Vogt asked. It turned out that Peterson had figured out a set of rules to distinguish one bird from another quickly and easily. Vogt told Peterson he should write and illustrate a book about his techniques, grandly promising that he could ensure its publication.

Peterson produced hundreds of simple drawings depicting the features he looked for in birds. A novice writer, he relied on Vogt as his muse, cheerleader, and editor. When the guide was finished, Vogt charged out to sell it.

I again demonstrated the incompetence in salesmanship that has been a characteristic all my life. I took the manuscript and drawings to virtually every well-known publisher in New York. They were all sure it would not sell.

Eventually the book was picked up by Houghton Mifflin in Boston for a small advance and a low royalty rate (the numerous pictures increased production costs). Dedicated to Vogt, Peterson's *Field Guide to the Birds* (1934) introduced a generation of children to the environment and was long the most popular book Houghton Mifflin ever published.

As Vogt rose in the bird world, his wife floundered in the theater world. Juana Vogt had climbed the traditional ladder from regional stages to tiny off-Broadway productions to the Great White Way. Her first Broadway show, a George S. Kaufman–Alexander Woollcott vehicle called *The Channel Road*, opened on October 17, 1929. Twelve days after her debut the stock market crashed, lighting the fuse of the Great Depression. The U.S. job market collapsed with frightening speed, as did the job markets elsewhere in the developed world. With families no longer able to buy goods and services, businesses went bust across the nation. The theater industry was hit as badly as everything else. Juana's career hopes turned to ash.

Vogt's drama-criticism income fell for the same reason. But he was rescued by his birdwatching friends, who had him appointed as director of the Jones Beach State Bird Sanctuary, an extension of a newly established park on the South Shore of Long Island. Created by Robert Moses, the "master builder" of twentieth-century New York City, the park was part of an ambitious plan to transform Long Island into a playground for the city's middle-class families. The four-hundred-acre sanctuary was intended to introduce urbanites to the wonders of Nature.

Living with Juana in a former hunting shack on the grounds, Vogt took Mayr's advice—even amateurs, the scientist said, should "have a problem"—and used his new job for research. He counted eggs and hatchlings. He recorded mating rituals (he won a prize for a study of courting behavior in the eastern willet). He made a census of the 270 bird species or subspecies that lived at the sanctuary. Often he was visited by Robert Cushman Murphy. A fourth-generation Long Islander, Murphy had a passion for understanding and conserving the creatures of his childhood. In the winter of 1932 he joined Vogt to observe an amazing, heartbreaking sight: hundreds of dovekies—eight-inch, black-and-white seabirds, round and puffy as plush toys, usually found only in the Arctic—being driven in from the sea by winter gales. Exhausted birds dropped out of the sky onto homes and yards all over Long Island—a phenomenon Murphy, for all his expertise, had never seen. Long Island was not the only venue: the birds rained down on the streets of New York City and washed ashore as far south as Florida. The two men collected hundreds of amateur reports about dovekie deaths for an article in *The Auk,* the journal of the American Ornithologists' Union. Published in July 1933, it was Vogt's first appearance in a peer-reviewed scientific journal.

As a rule, dovekies live in huge colonies atop protective cliffs on Arctic islands. Why, Vogt and Murphy wondered, were these creatures showing up thousands of miles from their icy homelands? Could the dovekies have overwhelmed their native habitats, forcing them into a blind search for new breeding grounds? Was the dovekie onslaught, the two men asked, "a peculiar herd psychosis," a "rush emigration" driven by overpopulation?

The theory was gloomy and overwrought. But Vogt himself was gloomy and overwrought. The primary focus of the Jones Beach sanctuary was mainly the scope of duck species native to Long Island, especially the black duck, prized by local hunters. Many duck species were more common than ever at Jones Beach. But the increase, Vogt suspected, was less because conditions at the sanctuary were improving under his stewardship than because conditions on the rest of their range were deteriorating. The crowd of birds at his sanctuary was like the crowd of patients at a hospital during an epidemic: nothing to celebrate. Overall, Vogt told *The New York Times*, black duck numbers were "dangerously low."

To Vogt, the cause of the decline was obvious. Long Island was one of the first places in North America to undergo extensive suburbanization. The landscapes of Vogt's childhood—landscapes that had been, he thought, little changed for centuries—were being flattened by real-estate developers. The sweeping meadows outside his door in Garden City had been sliced into lots for commuter homes. The cathedral gardens had become a golf course. Highways were grinding across the land.

The ducks were collateral damage. And all of this, Vogt realized, was fostered by his own boss, whom he came to call "the bulldozer-subdivider, Robert Moses."

Worse still, Vogt had lost his job, despite his reputation for competence. Moses had leased the sanctuary land for a dollar a year from the adjacent township of Oyster Bay. To raise funds to maintain Jones Beach, he put a toll on the roads leading to it. Suddenly people in Oyster Bay had to pay to visit a beach that had previously been free. In a fit of pique, the Oyster Bay town board revoked the sanctuary lease on May 20, 1935, shutting it down. Vogt's job went with it. Again he was rescued by a birdwatching contact.

In 1934 John H. Baker, an investment banker and amateur ornithologist, became director of the National Association of Audubon Societies, the umbrella organization for the many state and local birdwatching groups using the Audubon name. A brusque, forceful organizer and fund-raiser, he bought the journal *Bird-Lore*, made it the official Audubon publication, and asked Vogt to become editor. Unemployed at the height of the Depression, Vogt happily accepted, returning to Manhattan. Juana decided to reinvent herself and taught speech at Queens College while studying for a master's degree at Columbia Teachers College.

Vogt quickly shook up *Bird-Lore*, adding new artists and writers like his friends Peterson and Murphy, as well as Aldo Leopold, founder of the U.S. wilderness movement. Vogt was a charismatic figure at the office: five foot ten and 175 pounds, theatrically exuberant, wielding his cane like an actor's prop. Constantly busy, he gave speeches to local Audubon groups, published a new version of John James Audubon's *Birds of America* (a commercial hit), and helped to organize the society's annual breeding-bird census, a major effort in which thousands of amateur birdwatchers write down their observations of nests and fledglings along randomly selected roadside routes throughout North America. He set up the Committee on Bird Protection, an early effort to protect the bald eagle, drawing in Leopold. As he hopscotched from place to place, Vogt learned that bird populations were falling not only on Long Island, but all along the East Coast. The decline had many causes, but he came to believe that one predominated: mosquito control.

Mosquito control was another way of saying malaria control. Malaria today is confined to poor, hot places, but in the 1930s it afflicted huge numbers of people on every continent but Antarctica—5 million in

North America alone. Malaria is caused by a single-celled parasite that is spread by mosquitos. Because no treatment existed for the parasite, researchers believed that it could best be fought by eliminating the wetlands that were the breeding grounds for its mosquito hosts. During the First World War, according to the historian Gordon Patterson, workers in eastern states dug ditches to drain ponds and marshes, then sprayed heavy oil or insecticide into the water to poison remaining mosquito larvae. In the Depression, Patterson writes in *The Mosquito Crusades* (2009), Washington took over the campaign; ditch-digging was instant work for the jobless. Thousands of newly hired mosquito-fighters cut and poisoned tens of thousands of miles of drainage ditches. So many ditches were dug so fast that in some places local governments lost track of them and begged Washington to conduct surveys to identify them.

Long Island was a focus of this haphazard crusade. Nassau County, where Vogt was born, had slashed more than a thousand miles of ditches through its meadows and marshes since mosquito control began. Suffolk County—the poorer, more rural eastern portion of the island—hadn't begun ditch-digging until 1934, when Washington opened the cash spigot. But its mosquito-fighters made up for lost time. Draining and poisoning marshes ruined so much bird habitat that the county tried to bring back the missing birds with a new program of excavating artificial ponds in the ditch-crossed marshes.

Vogt was appalled. The meadows of his childhood were already being buried beneath parkways and suburban development. Now the mosquito brigades were slicing up the remains. He came to view his Audubon job as a chance to do something about it. In 1937, he wrote "Thirst on the Land," a pamphlet intended to rally Audubon Society members against mosquito control. Its tone was lacerating: mosquito-control programs, "perilously close to destructive government-sponsored rackets," were spreading "like some form of terrestrial erysipelas" (an infection that causes skin sores). Beneath the bluster, though, "Thirst on the Land" was prescient: Vogt anticipated the arguments that Rachel Carson—who later became a friend—made famous in *Silent Spring*.

Strikingly, "Thirst on the Land" reintroduced an idea that mostly had lain dormant since its introduction in the 1860s by the innovative geographer George Perkins Marsh: landscapes and the species that live on them perform useful functions—purifying water, decomposing wastes, nourishing crops, housing wildlife, regulating air temperature—

which are both free to the beneficiaries and costly to replace. (Today these functions are called "ecosystem services.") If the economic benefits of mosquito control were weighed against the ecological costs, as Vogt believed they should be, mosquito control failed. Drained marshes were no longer able to store and filter stormwater; to replace them, towns would have to build protective dikes, impoundment reservoirs, and water-treatment facilities. If the costs of the dikes, reservoirs, and water plants were taken into account, Vogt argued, the untouched marsh often was worth more than the newly drained land. The government should leave marshes alone rather than destroy them to kill mosquitos.

Stung by the attacks, federal mosquito-control officials in March 1938 faced off against Vogt and a new ally, U.S. Biological Survey researcher Clarence Cottam, at the Third North American Wildlife Conference, a gathering of field biologists and land managers. Speaking before an audience full of mosquito-control officials, Cottam lambasted their projects as "ill-advised and woefully misdirected." Vogt was harsher still: "Since every drainage ditch robs the land of its life blood, often wastes large sums of your money and mine, and wipes out wildlife, it seems to me there is considerable justification for contending that *something is damnably wrong with mosquito control*" (italics in transcript). The mosquito officials spluttered protests, but nobody in the room supported them.

Despite winning (in his view) the debate, Vogt wasn't satisfied: ditching and spraying went on just as before. What was needed, he told friends like Leopold, was to mobilize Audubon members into a mass anti-drainage movement. Advised by scientists, thousands of militant birders could act as an environmental warning system: a union of scientifically informed amateurs and politically informed scientists that would rise in defense of Nature. Vogt was groping toward the idea of changing Audubon from a circle for upper-middle-class hobbyists into what would now be called a large-scale, broad-based environmental organization—a pioneering step.*

---

* No previous environmental group was quite like what Vogt envisaged. The oldest, now called the Société Nationale de Protection de la Nature et d'Acclimatation de France, was founded in Paris in 1854. It tried to protect rare plants and animals, especially birds, as did similar societies in Sweden (established in 1869), Germany (1875), and Britain (1889). Other

The tool to create the movement, Vogt thought, was *Bird-Lore*. Month after month, issue after issue, Vogt railed against destroying wetlands, polluting rivers, and overusing insecticides ("insist that poison be kept off the dinner table"). Later, all would become a focus of the green movement. But the magazine's swerve from tales about spotting the elusive Eskimo curlew to jeremiads against real-estate tycoons dismayed subscribers. Readers, Vogt's friend Roger Tory Peterson wrote later, "wanted to be diverted, instructed, and entertained—not preached to." In a single month, he recalled, *Bird-Lore* featured an essay "about the suicidal tendencies of muskrats under population pressures; another described the difficulties of winter survival of quail; a third dealt with the ecological effects of poisons—the entire issue reeked of death and destruction."

Unsurprisingly, Audubon president Baker told Vogt to change the tone. Vogt believed he was afraid of upsetting the society's wealthy donors. But Baker may simply have been worried that Vogt was alienating members or annoyed that a subordinate was changing *Bird-Lore*'s focus without consulting him. Stressed from fighting mosquito control, running *Bird-Lore*, and clashing with Baker, Vogt was hospitalized for "nervous exhaustion." While he recovered, Vogt—"stormy petrel that he was," as Peterson said—came up with a scheme. Profiting from staffers' dislike of Baker's rigid style, Vogt planned to lead them in a strike, which he believed would induce the board of directors to force Baker to resign. Baker caught word of the plot and took the conflict to the board in late 1938. Vogt was fired at once. Four months later he was in Peru.

Don Guano

As a new employee of the Compañía Administradora del Guano, Vogt based his operations on the Chincha Islands, three granitic outposts thirteen miles off the southwest coast of Peru. Named, unexcitingly, North, South, and Central Chincha, they were each less than a mile across, ringed by hundred-foot cliffs, and completely covered in

---

groups focused on individual landscapes, like the British National Trust (set up in 1895 to safeguard the Lake District) and the French Société pour la Protection des Paysages de France (1901). The biggest U.S. conservation organization in Vogt's time, the Sierra Club, concentrated on outdoor recreation for upper-middle-class businesspeople. ∧∨

heaps of bird excrement—treeless, gray-white barrens of guano. Atop the guano, shrieking and flapping, were millions of Guanay cormorants, packed together three nests to the square yard, sharp beaks guarding eggs that sat in small guano craters lined by molted feathers. The birds' wings rustled and thrummed; multiplied by the million, the sound was a vibration in the skull. Fleas, ticks, and biting flies were everywhere. So was the stench of guano. By noon the light was so bright that Vogt's photographic light meter "often could not measure it." Vogt's head and neck were constantly sunburned; later his ears developed precancerous growths.

Vogt worked, ate, and slept in the bird guardians' barracks on North Chincha, remaining offshore for weeks on end (he was also given an apartment in the nearby shore town of Pisco). His quarters on the island were almost without furniture, covered with guano dust, alive with flies and roaches. Birds mated, fought, and raised their offspring on the roof overhead, leaving so much guano that the building had to be shoveled off periodically to avoid collapse. Vogt's "laboratory," a bare room with a battered table, had no electricity or running water. There was no scientific equipment other than a thermometer, binoculars, and a camera, all of which he had brought from New York. (Later he lost even these, and spent weeks waiting for replacements to arrive from the United States.) "I'm doubtless one of the few men who ever spent three years on a manure pile in the interest of science," he said.

Vogt loved it. He was delighted by the staff, who cooked, cleaned, arranged transport, assisted in research, and gave free Spanish lessons to the man they called "Doctor Pájaro"—Dr. Bird. (His American friends had another nickname: Don Guano.) Twice a week the barracks manager roasted coffee beans in a cast-iron pan. As the beans' smell filled the room, he gently coated them with sugar and clarified butter, Vietnamese-style. To Vogt's joy, on North Chincha he became the proprietor "of a luxury I am not likely to have again—a private scallop bed." Night after night, the cook prepared ceviche, *parihuela* (the Peruvian answer to bouillabaisse), sea-turtle stew, avocados stuffed with prawns, and other Peruvian marine delights, transforming Vogt into that most forgivable of annoyances, a food snob.

More important, the stark coastal environment enraptured him. He was thrilled by the brilliantly clear night skies, the endlessly variable ocean, the muted browns and yellows of the arid, foggy shore, and, above all, the profusion of living creatures in this apparently inhospitable zone. "It is worth traveling thousands of miles to see but it is the sort of place only a naturalist would entirely appreciate," he wrote.

> As I watched the flocks, day after day, against gray skies, against blue skies and blue sea or, more often, the dark green sea rich in plant food, or against the varied, muted colors of the desert and coastal range that edge the Humboldt Current, I could feel myself part of that cosmos. The stuff of my bones was the stuff of their bones. Through their metabolic system coursed primeval molecules, perhaps used over and over again; they were transported to the ancient, irrigated field of the coast, and through plants and flesh back to our table on the island.

It was a place where he could create a new life for himself. He didn't mind the smell.

The guano company had hired Vogt to solve the riddle of the birds' shrinking numbers. Its goal, as an ornithologist friend put it, was "to augment the increment of excrement." Vogt was not interested in whether the company's profits rose or fell, an attitude that would soon create friction with his supervisors. But to understand the population decline he would need to investigate a host of scientific questions that *did* interest him: What are the maximum and minimum ages at which Guanays can reproduce? Are Guanays monogamous? What factors limit their reproduction? Do the islands have a maximum capacity? And, of course, he wanted to use the answers to safeguard the birds.

To study the cormorants without disturbing them, Vogt and the guano guardians built a burlap blind on North Chincha. From its shelter he spied on the "love life" of "11,000,000 guano birds"—courting, fighting, mating, nesting, and feeding offspring. To avoid having to kill and autopsy birds to inventory their insect parasites, he settled for counting the bugs that tried to feast on him in the blind. (He dressed in white so the insects would be visible.) With the help of island guards and local fishing families, he banded tens of thousands of birds—thirty-nine thousand in 1940 alone. He measured air and water temperatures. He counted eggs. He weighed baby birds, live and dead. He sampled

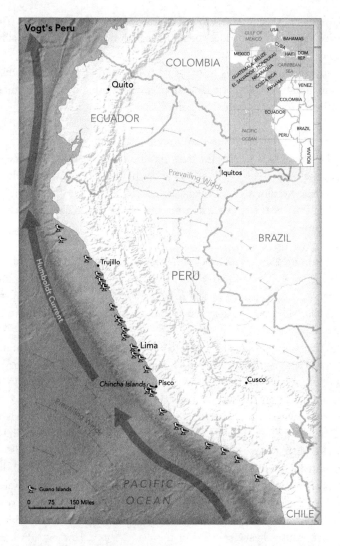

Masked and goggled against the stench of millions of cormorants, Juana Vogt (opposite, top) stands on the dock at the entrance to North Chincha; the "craters" by her feet are old nests. On the other side of the island was Vogt's bird blind (opposite, middle), in which he spent countless hours in the baking sun. Because access to the guano islands was restricted, Vogt had to carry a special permit (opposite, bottom) identifying him as an employee of the guano company. The firm had near-absolute control of Peru's thirty-nine guano islands (map, above).

plankton and anchovetas. When he returned from the boat to the barracks, Vogt with his bad legs sometimes had to be hauled up the cliffs in a basket.

Vogt tore himself away from the work in June 1939 to attend Juana's graduation from Columbia. Traveling with her husband to South America the next month, she was initially dismayed. Peru, she thought,

> is the dirtiest place I have ever seen. The adobe is a dirty color, the ground is the same, the trees and the plants are covered in the adobe colored dust, and the legs and arms and faces of the people are encrusted with dirt. The tablecloth [at a restaurant] was filthy. The shirt of the construction foreman was filthier. They all pick their teeth.

Soon, though, she came to love the "pure good luck" of being able to live where sea-lion families "sleep and breed and carry on riotous family feuds." Like Vogt, she ended up liking the harsh plenitude of North Chincha. "You seem to get close to the secret places of the universe in such a spot," she wrote.

By chance, the Vogts had come to North Chincha near the beginning of what Vogt called an "ecological depression on the Peruvian coast." Andean peoples had long known that every few years the coastal climate shifted dramatically, with warm downpours inundating the cold, dry coast. Because the rains usually began around Christmas, Peruvians referred to them as El Niño, a Spanish nickname for the Christ Child. In 1891 three Peruvians—an engineer, a geographer, and a naturalist—separately figured out how El Niños worked. During these times, the Humboldt Current abruptly weakens, allowing warm equatorial water to surge close to the coast; the warm water heats up the normally cold coastal air, which allows it to hold more moisture than usual, which, in turn, causes heavy rainfall on the desert shore. Few outside Peru learned of these findings until, by chance, Robert Cushman Murphy visited the country during the severe El Niño of 1925. Collecting his own and others' observations, Murphy realized he was in the middle of a climatic system that extended across much of the Pacific and influenced the weather as far north as Canada. But the worst effects occurred in coastal Peru, where floods washed away railroads, wiped out farms, and destroyed power stations, blacking out cities. Thousands of "dead guano birds" were incidental damage. El Niño, Murphy said, "brings sickness and death to the population of the Humboldt Current."

The El Niño Vogt experienced had begun quietly, probably a month or two before his arrival. Slowly the water temperature, which typically hovered around 60°F, rose to 77°F. Temperatures on the islands themselves reached as high as 122°F. On June 2, 1939, the day before he left to attend Juana's graduation, Vogt estimated that the eighty-six acres of nesting ground on North Chincha held an astonishing 5,250,000 birds—an entire Chicago's worth of animals, packed into an area the size of a small-town fair. But when he returned with Juana the next month, he wrote, the adult cormorants "had *all* gone." And their chicks—the offspring they devoted so much effort to feeding—had been left behind to die. The sight tore at Vogt's heart; even decades later, he couldn't forget walking among "the horde of downy babies" as they starved.

> They would flap their unfledged wings, while they gave their hunger call, at the feet of this strange, uncormorant-like creature. . . . There was not a thing one could do for them. Day by day there were fewer begging, more staggering about and listlessly drooping. And then more—hundreds of thousands more—of the pitiful, collapsed, downy clumps that were the dead. . . . Somehow, ever since, it has been possible to understand more fully the famines of China and India.

Where were the adult birds? For weeks Vogt combed the coast, hunting for cormorants by plane, boat, and car. Taking advantage of the ornithological tradition of employing amateurs, he mobilized a network of birdwatchers to aid in the hunt. But nobody saw the Guanays.

On October 7 the cormorants abruptly came back, hundreds of thousands of them, only to disappear after a week. On the 20th the birds returned, then vanished on the 24th. By November 7 they were back—only to bolt a few days later.

In 1940 the warm waters came again. And in 1941. And they showed up earlier, at the beginning of nesting, so the birds then fled their nesting grounds and didn't reproduce. Entire generations were not being born. Vogt was looking at a demographic collapse.

But why were the Guanays fleeing? The temperature was not enough to hurt them directly; if they got hot, they could always take a swim. Nor did the birds' returns correlate with colder weather. They suffered from no obvious disease. What was going on?

The key to the puzzle, Vogt thought, was the condition of the few adults that *didn't* leave the Chinchas: hungry. The remaining Guanays left every morning to hunt for fish. But they returned ever later in the day, and their crops were often empty, which meant they couldn't feed their offspring. The lack of food, he concluded, was due to El Niño. Warmer water on the surface acted as a cap that blocked cold water from rising from the depths of the Humboldt Current, which set off a cascade of horribles: no upwelling meant no nutrients for plankton, which meant no plankton for anchovetas, which meant no anchovetas for Guanays.

Vogt was unable to test this hypothesis until late in 1940, when he persuaded the guano company to measure plankton abundance by dragging the sea at multiple locations with a fine silk net. He examined the samples with the sole tool available, a magnifying glass he had managed to acquire on a trip to Lima. Despite the crude equipment he was able to gather enough data to see what was happening. The "general tendency," he wrote, was for "falling temperature to be accompanied by increasing plankton, and vice versa"—an inverse relationship. Abrupt water-temperature rises "resulted in wholesale destruction" of plankton. Desperately hungry, the Guanays had scattered in every direction to search for food.

What did this mean for the Peruvian government, which wanted to maximize the guano-bird population? Vogt spelled out his answers in a 130-page report in October 1941. Written while racing between island bird blind, offshore guano boat, and coastal observation post, his report today seems unexceptional. But at the time his ideas were at the forefront of a wave of theories about the human relationship with nature largely associated with Aldo Leopold, whom Vogt had known since he wrote for *Bird-Lore* at Audubon, and who would for Vogt become an important friend, inspiration, and intellectual sounding board.

Fifteen years older than Vogt, Leopold had been raised in Burlington, Iowa, in a big house on a bluff above the Mississippi River. A shy boy who became an avid hunter, he reveled in solitary tramps through field and forest with his gun. Leopold went to Yale, graduating in 1909

Aldo Leopold, Vogt's mentor, friend, and sounding board, in the 1940s

from its forestry school, the first in the nation. Like much of his class, he went into the U.S. Forest Service, which sent him to New Mexico. After a bad infection, Leopold was forced to recuperate for a year and a half, a period in which he rethought his views. At Yale he had been taught that the goal of land management was to wring the maximum volume of some resource—timber or deer or fish—from a given piece of property. Now Leopold became skeptical of humankind's ability to understand the complexities of nature well enough to guide them. He came to think that ecosystems needed more to be *protected from* humans than *managed by* them—a stance that complicated his move, in 1933, to the University of Wisconsin at Madison, where he directed the first U.S. academic program in wildlife management.

Leopold's career coincided with the rise of a new scientific discipline: ecology. In 1905, Leopold's first year at Yale, the first ecology textbook was published. The author was Frederic Clements, whose ideas heavily influenced Leopold—and then, through Leopold, William Vogt and the global environmental movement. Clements's masterwork, *Plant Succession* (1916), contended that natural ecosystems developed in a predictable pattern over time. Much as a person begins as a baby, then passes through childhood to become a mature adult, Clements said that ecosystems also go through distinct growth stages, finishing with a mature "climax" state. Building on these ideas, many ecologists maintained that the climax represents an ultimate "balance of nature," a community of species that endures with little change until it is disturbed, sometimes by natural events like floods or fires, often, destructively, by humans.

Each species in this community, Clementsian ecologists believed, had gradually adapted to fill a specific ecological niche—a role played by it, and it alone. The relations among these niches were governed by the available resources—the *biotic potential*, as Vogt later called it—and the constraints imposed by the physical setting—the *environmental resistance*, as he put it. Biotic potential and environmental resistance were in a constant tension, one lifting up, one pushing down. The climax community was like a network of forces that canceled each other out, allowing the whole complex, diverse structure to maintain itself in rough equilibrium.

In some ways this vision goes back to the ancient Greeks, who saw nature as a balance maintained by the gods. Putting these ideas in modern terms, Clements claimed that natural communities function as a kind of "superorganism," with their different species standing in relation to them like the different organs of a single animal. When people killed off a species or destroyed its habitat, they were, in effect, attacking the vital organs of this superorganism. They were tipping the balance of nature, which could bring down the whole community.

Many of Clements's colleagues attacked his ideas, but to little effect. "'The balance of nature' does not exist, and perhaps has never existed," snarled the English ecologist Charles Elton. Because each rise or fall in the population of one species affects its fellows, Elton said, and because those species are also constantly changing in number, the result is that ecosystems do not form a stable climax community but exist in continuous turbulence—"the confusion is remarkable." Elton pioneered the study of how energy flows through these chaotic assemblages. But despite Elton's insistence that talk of "superorganisms" was mystical hogwash, Clements quickly incorporated his energy ideas into his theory. Energy moves through stable ecological communities, Clements said, the way blood circulates through an animal—its flow sustains the superorganism. Despite repeated, powerful critiques from Elton and his disciples, Clements's vision of natural systems as self-contained and dynamically stable continued to govern the field.

In Wisconsin, Aldo Leopold in effect tried to reconcile Clements and Elton. Leopold initially believed, like Clements, that ecological communities are an organism-like "collective total" in which all the constituent species have "some utility." Like Elton, though, he came to disagree that ecosystems are "normally static." Instead, Leopold said, they constantly change over time. But—a fundamental idea for Leopold—the changes are usually limited in rate and scope, which allows ecosystems to maintain their fundamental qualities. Much as a human community can maintain its essential identity even as residents move in and out, ecological communities can preserve their basic character despite fluctuations in its species populations—provided that the changes aren't too fast or radical. As a rule, Leopold said, humans act too quickly and clumsily, inadvertently destroying ecosystems' basic identities and functions.

Not quite Clementsian, not quite Eltonian, focused on practical landscape management rather than abstract science, a believer in hard data who also thought that morality and the spirit were essential to ecology, Leopold often felt that his colleagues could not understand him. In Vogt he was grateful to find a fellow thinker. The contrast between the two men was striking: Leopold soft-spoken and ruminative, unfailingly courteous, formal, and sometimes aloof, a physically adept man who hated to travel away from his five children (all of whom became biologists and conservationists); Vogt brash and theatrical, often acerbic, frequently disheveled but a bit of a dandy, ever struggling with his bad legs, a wanderer who gave no thought to having offspring. But on a deeper level they were colleagues, running toward the same goal. Of necessity, they had an epistolary friendship; letters flowed between Peru and Wisconsin. When Vogt returned from Peru, Leopold was both unsurprised and pleased to find that he and Vogt still "had identical views, even tho[ugh] we had hardly seen each other for years." Vogt, he joked to a friend, "is the leading exponent of my thought."

Although Vogt wrote and signed the report to the guano firm, it reads like a collaboration between the two men. If the company wanted to maximize guano production, the report said, it should be very careful about interfering with natural processes—reducing the anchoveta population by fishing, for instance. Almost any drastic change would upset the existing network of ecological relations, lowering biotic potential. As

Despite great effort by Peru, the guano trade collapsed in the 1950s, undone by overharvesting and artificial fertilizers. Today all of Peru's guano islands together house about 4 million birds, far fewer than the 60 million in the nineteenth century. As shown in these 2014 photographs by the artist Dinh Q. Lê (Central Chincha, top; North Chincha, bottom), the islands are mostly deserted today; the Peruvian government is leaving them alone in the hope that they will someday recover.

a result, Vogt believed, the goal for Peru's guano managers should be to keep the islands in the optimal, climax state. In practice, this meant that they should be returned as close as possible to the pristine wilderness that Vogt believed had existed before the arrival of Europeans.

For example, Vogt recommended that the Compañía Administradora del Guano eliminate the non-native rats, cats, and chickens on the islands; that it should stop killing native saltojo lizards, because they ate insects that attacked birds; and that it should ban low-flying aircraft, because their alien noise panicked the cormorants, which fled "so frantically that the rush is likely to expel eggs and chicks from their nests." Because the islands' surface had been flatter before mining, Vogt suggested that the company use explosives to re-level them, increasing potential nesting sites. And he thought that some artificial breeding zones could be created on the shore that would replace islands ruined by human activity.

But about El Niños the company should do . . . nothing. During El Niños, cormorants left their nesting grounds and didn't have offspring, reducing their population and thus their guano output. But these losses were not actually a problem, Vogt thought. They were natural changes, contained in scale and scope. They were a safety valve, not a risk—a feature, not a bug.

> In this world, death is as important as life. In the periods without [El Niños], the population of guano birds approximately doubles every year. If there were no way of limiting the population, there would in little time be no space or food on the west coast, and not only for the birds. . . . To me it seems that this disorder is biologically necessary, even for the welfare of the birds themselves.

Indeed, Vogt argued, trying to lift the cormorant population artificially beyond its climax level would lead only to higher death rates in the next El Niño. Guano output might increase for a while, but the long-term results would be worse than if humans had never inflated bird populations. In the end, Vogt told the Compañía Administradora del Guano, Peru could not "augment the increment of excrement." It could only "help conserve the balance between species continually sought by Nature." It would have to live within ecological limits.

Living within ecological limits! Vogt glumly told Leopold he was certain he would have to "jam" this idea "down the throat" of his superiors. As it turned out, he would be doing this to many other people, too.

Mosaic of Ruin

Final report in hand, fortieth birthday approaching, Vogt had to make decisions about his life. As his research had progressed, he had hatched a scheme: he would use his still-incomplete guano-bird data to lever himself into a Ph.D., probably at the University of Wisconsin with Leopold, without having to attend classes (or taking as few as possible). The credential would gain him standing in his goal to mobilize forces for conservation, though writing the dissertation would take several years. Should he stay in Peru to complete his research or go directly to Wisconsin and start his Ph.D.? Vogt was gearing up to make some decisions when, in December 1941, Japanese planes bombed Pearl Harbor.

A patriot, Vogt wanted to be of service in the coming war, but worried that he was too old and too lame. To his surprise, the U.S. State Department asked him to leave the guano company and use his scientific contacts and now-expert Spanish to travel through Chile, Colombia, and Ecuador, reporting on the level of sympathy there for Germany and Japan. Vogt immediately agreed. Juana was asked to snoop out Hitler's fans on the embassy cocktail-party circuit. In the hunt for Nazis, Vogt gave talks at Latin American Rotary clubs, hobnobbed with Peruvian agricultural and trade officials, spoke to Ecuadorian academics, tramped through Chilean national parks, and attended countless soirées with Juana. Along the way, he decided he would study with Leopold in Wisconsin. Jubilant, Leopold obtained a fellowship for Vogt. Ready to resume studies, the Vogts returned to the United States on May 2, 1942.

They spent a week briefing Washington on the "strategically placed Nazis" they had identified in Latin America. Impressed, Army Intelligence and the State Department asked the couple to do more intelligence work there. Vogt could not refuse. Two weeks after arriving in the United States, he and Juana returned to South America for a three-month tour. Vogt sadly wrote to Leopold that it would be impossible for him to go to Wisconsin.

After another round of information gathering, a grateful State Department ensured Vogt's appointment, in August 1943, as head of the newly created Conservation Section of the Pan American Union. Created in 1890 to promote cooperation in the hemisphere, the union was based in Washington, D.C. (After the war it was reconstituted as the Organization of American States, its current name.) Spurred by the Audubon Society, most of its members in 1940 had signed the Washington Convention, an agreement to protect endangered species. The treaty was a pioneering recognition of the value of biodiversity. It was also toothless—it required no particular actions, and there was no provision for enforcement of its terms. The Pan American Union established an office—the Conservation Section—to monitor whatever treaty-related events might occur. The U.S. State Department, which was bankrolling the Conservation Section, wanted a U.S. citizen to lead it, and Vogt was among the few U.S. ecologists with experience in Latin America and fluent Spanish. As a Nazi-hunter, Vogt had good anti-fascist credentials, a must-have in Washington in 1943. He was given the vague task of examining the relationship of climate, resources, and population to economic development.

After studying seabirds on Peruvian islands, Vogt was being asked to move his purview to human beings across an entire hemisphere. But he didn't see it as a huge shift. Ecology, he believed, provided a basic intellectual framework for understanding both birds on small islands and humans on big continents. It told him that both species were part of ecosystems ruled by biological law and shaped by their environment. Understand the rules and measure the environment, and you could comprehend the future. That is what he planned to do.

During the next five years, Vogt visited all twenty-two independent nations in the hemisphere. Again and again, he flew to cities, ate in nice restaurants, stayed in lovely colonial hotels, then went outside the cities—to discover wreckage. Eroded foothills in Mexico. Poisoned rivers in Argentina. Devastated fisheries in Venezuela. Drained aquifers in El Salvador and Honduras. Perhaps worst was the deforestation. Forests were vanishing across the Americas, promoting erosion, which led to floods that ruined fields, which in turn pushed farmers to clear new land.

In Vogt's discussions with Leopold, he had absorbed the other man's teaching that civilization depended on the health of its sustaining eco-systems, which in turn depended on the soil. A thin but immeasurably rich skin on Earth, the soil was quite literally the foundation of the human enterprise. As Vogt toured the Americas he saw this foundation eroding away everywhere, sliding down slopes in nation after nation, impoverishing ecosystems from the Mississippi headwaters to Patago-nia. Soon, he believed, the destruction would be unfixable.

Vogt spent ten months in Mexico, much of it with Juana, at the behest of the Mexican agricultural ministry, writing a guide to conservation for Mexican schoolchildren, and struggling with his cane and braces through twenty-six national parks. Although statistics showed that the country was Latin America's wealthiest, its landscapes were enshrouded by suffering: impoverished subsistence farmers, scratching at depleted soils, taking down the last stands of timber for their cookstoves. "Unless land-use patterns are radically altered," Vogt said, "most of Mexico will be virtually desert within a hundred years." Vogt laid out his case in a "confidential memorandum" to the Mexican government in November 1944. It began bluntly:

> Mexico is a sick country. It is rapidly losing the soil on which its very existence depends.... Its forests ... are being destroyed far faster than they are being replaced.... Its ecology—the inter-relationship of all the environmental factors—has been so thrown out of balance that many important land values are being wasted. As a result, living standards are constantly being depressed, and disaster lies ahead.

The wreckage was not confined to Mexico. Vogt prepared reports on each country he visited. Guatemala, he testified, was just a smaller ver-sion of Mexico—"Here again, there is little ground for optimism." Chile was a carnival of deforestation; "the point of view of many Chileans is that ... the forest is an enemy to be got rid of by any possible means." In Venezuela the situation was "daily growing worse." Colombia was "in an equally bad situation." From the air, Ecuador "looks as though it suf-fered from some horrible skin disease."

El Salvador was for Vogt the sharpest example of the problem, a har-binger of what lay in store for much of the world. The poorest, most densely inhabited country in the Americas, it was, Vogt believed, "face to face with a crisis" that other places were just approaching. In El Salvador, he insisted, a growing population was colliding with "the progressively rapid destruction of its natural resources, especially its cultivable land." The country's people and its resources were like trains racing toward each other on the same track. "El Salvador should act—and act at once." If it did not, he said, a future of poverty, political violence, and environmental ruin was inevitable.

What was driving the destruction in the hemisphere, Vogt thought, was *consumption*. Ceaselessly striving to satisfy their needs, people were stripping nature bare. The consumption had two causes. One was population growth—new mouths meant new demands on the land. Mexican couples, for instance, were having more than six children apiece, and the numbers were rising. The second, equally pernicious cause was the attempt to maximize economic growth.

Although economists since Adam Smith had championed growth, governments had typically focused instead on promoting national security or economic stability. Indeed, some thinkers feared that untrammeled growth would lead to despotism by concentrating wealth; others argued that continued economic expansion was impossible in developed, mature economies like those in Europe and North America. During the Depression, many U.S. New Deal programs actually were anti-productivity; in the hope of driving up farm incomes by artificially creating shortages, farmers were paid to plow under millions of acres of cotton fields and slaughter huge numbers of pigs. Partly influenced by the British economist John Maynard Keynes, some officials fought back against "scarcity economics" (to use the term of the historian Robert M. Collins). The Second World War strengthened their hand, as the United States pursued all-out production in the name of victory. The Employment Act of 1946 formalized the shift, declaring that Washington was committed to "promot[ing] maximum employment, production, and purchasing power." Galvanized by the example of the United States, other Western nations also embraced growth as an overriding social goal. "Government and business must work together constantly to achieve more and more jobs and more and more production," proclaimed President Harry S. Truman.

Vogt heard these pronouncements with horror. Ecological law was clear, he argued. The "books must balance." Truman's call for "more and more production" was intentionally *un*balancing the books. "Growthmaniacs" were, Vogt said, warring against the natural systems that nourished them—a war they had no idea they were fighting.

In May 1945 Vogt laid out his views publicly for the first time, writing an article, "Hunger at the Peace Table," for the *Saturday Evening Post,* a popular weekly magazine. Published at the height of the Pacific campaign, "Hunger" was an unsparing look at a future that its author believed would be as dark as the present. It laid out the Prophet's central tenet: humankind, though "apt to forget it, is a creature of the earth. 'Dust thou art' and 'All flesh is grass' were not said by scientists, but they are sound biology." When lower creatures exhaust their resources, Vogt argued, bad things happen. Exactly the same is true for *Homo sapiens*. The article tallied example after example of overreaching, most drawn from Vogt's travels in Latin America. But then, provocatively, he switched to the United States' current enemy, Japan: "Many explanations have been offered for Japanese aggression," he argued. But, he asked, "can anyone deny that population pressures set off the explosion?" Unless humankind controlled its appetites for procreation and consumption, Vogt said, "there can be no peace." Disturbed by his message, a dozen U.S. senators asked to meet with Vogt. Interest in his ideas grew further after nuclear weapons were used on Japan in August 1945. The obliteration of whole cities made Vogt's warning that humankind might destroy itself seem terrifyingly plausible.

Few felt, then or now, the overriding importance of sustaining the land in the way Leopold and his disciples did. The knowledge made them feel both superior and isolated—the only people with eyes in a world of the willfully blind. All around them others were marching about their business as if the ground were not literally disappearing under their feet. So much folly! So much waste! They were as oblivious to the consequences of their actions as so many protozoa.

One of the penalties of an ecological education [Leopold later wrote] is that one lives alone in a world of wounds. Much of the damage inflicted on land is quite invisible to laymen. An ecologist must either harden his shell or make believe that the consequences of science are none of his business, or he must be the doctor who sees the marks of death in a community that believes itself well and does not want to be told otherwise.

How to ring the bell? That was the ever-urgent question. Vogt, Leopold, and their friends saw themselves as benign conspirators, working to awaken a world that was obliviously promoting its own destruction by its quest for growth. But their efforts had little impact. Their voices were small and tinny, drowned out in the roar of war and consumption. People were heedlessly flooding the world, a moronic tide rushing out to the edge of the petri dish.

Malthusian Interlude

Vogt's views have been called "Malthusian," a term that has become a weapon of abuse. In its most neutral sense the word refers to ideas espoused by the Reverend Thomas Robert Malthus (1766–1834). Because Vogt shared some of Malthus's beliefs, the claim that Vogt was a Malthusian is accurate. But it is not especially *useful*; Malthus's perspective shifted over the years, and "Malthusian" only sometimes refers to anything he wrote.

Robert—never Thomas—Malthus was a Cambridge-educated cleric, a dashing figure despite a pronounced speech impediment (he had a cleft palate, untreatable at the time). Born in 1766, he married late, had few children, and was never overburdened with money. He held the first university position in economics—that is, he was the first professional economist—in Britain, and probably the world. As a young man, he was hungry for distinction. He found little opportunity to acquire it at his home in rural Surrey until he took up an argument with his father. Daniel Malthus, something of a freethinker, believed humankind could create a Utopia. His son disagreed, at length. *An Essay on the Principle of Population, as It Affects the Future Improvement of Society*, published anonymously when its author was thirty-two, created a furor. Malthus followed with a second edition that was five times longer than the first. This edition was signed; Malthus had grown more confident.

At bottom Malthus's argument was simple:

Human populations will reproduce beyond their means of subsistence unless they are held back by practices like celibacy, late marriage, or birth control. But the reproductive urge is so strong that people at some point will stop restricting births and have children willy-nilly. When this happens, populations inevitably grow too large to feed. Then disease, famine, or war step in and brutally reduce human numbers until they are again in balance with their means of subsistence—at which stage they will increase again, beginning the unhappy cycle anew.

Here is Malthus's argument in more detail:

Today, Malthus said, British farmers produce a given amount of grain every year—X million tons, let us say. If the nation transformed its remaining forests into farmland and used fertilizer more effectively, Malthus posited, it might double its harvests in twenty-five years—that is, Britain would grow $2X$ million tons of grain. But even with heroic efforts, Malthus wrote, "it is impossible to suppose" that the harvest then could be doubled *again*, to $4X$ million, in another twenty-five years. "The very utmost we can conceive, is, that the increase in the second twenty-five years might equal the present produce"—that is, production could go from $2X$ to $3X$ million tons. If Britain kept steadily increasing its harvests by X million tons every 25 years, it would produce $4X$ million in 75 years, $5X$ million in 100 years, $6X$ million in 125 years, and so on. In mathematics, this kind of regular increase, from 1 to 2 to 3 to 4 to 5 and beyond, is called *arithmetic* or *linear*.

Next Malthus examined the rate at which the human species could increase. The most rapidly growing society he knew of was the United States. The best demographic thinker there was Benjamin Franklin, a Pennsylvania polymath. Back in 1755, Franklin had studied the population of what was then England's North American colonies. As Malthus put it, Franklin argued that the population there "has been found to double itself, for above a century and a half successively, in less than twenty-five years." In other words, if the U.S. population today were Y million, it would be about $2Y$ million in 25 years, $4Y$ million in 50 years, $8Y$ million in 75, and so on. This kind of increase, from 1 to 2 to 4 to 8 to 16 and beyond, is called *geometric*.

Simple mathematics dictates that geometric increases (1–2–4–8–16) always outpace arithmetic increases (1–2–3–4–5). Human reproduction is geometric, Malthus said; all else is arithmetic. Populations, if allowed to grow freely, always overwhelm their food supply.

People can avoid reproducing at the maximum rate, Malthus admitted, by using what he called "preventive checks"—practices that lower birth rates. Among these are celibacy, birth control, delayed marriage, reduced wages (so would-be parents can't afford children), and increased education (which Malthus thought would make couples aware of the risks of reproduction). But because preventive checks are difficult, costly, and unpopular, people inevitably stop using them. Population booms.

When that happens, "positive checks" kick in. Positive checks are the opposite of preventive checks; rather than lowering birth rates, they raise death rates. Positive checks begin with social violence and go on to "epidemics, pestilence, and plague." Should these fail to reduce numbers enough, "gigantic inevitable famine stalks in the rear, and with one mighty blow, levels the population with the food of the world."

Humanity, Malthus thought, will always be one breath from calamity. Permanent victory over deprivation is impossible; prosperity, fleeting, is doomed to vanish. "Misery and the fear of misery," he said, are "the necessary and inevitable results of the laws of nature." No matter how good-willed, charity cannot help; aiding the poor leads only to more babies and more hunger. The rules of biology cannot be defeated by ingenuity or virtue. "We cannot lower the waters of misery by pressing them down in different places, which must necessarily make them rise somewhere else."

The *Essay* was a jolt. Unflinchingly logical, elegantly mordant, it seemed to show that hopes for a better future were delusional. Previous thinkers had anticipated these ideas. But Malthus had the luck to publish at an opportune moment. England had been beset in the 1790s by bad harvests, which led to food riots. Rebellion was flaring in Ireland and India, the army was losing a war in Haiti, and neighboring France was undergoing a reign of terror and the rise of a megalomaniac dictator. Malthus's gloomy book suited its gloomy time.

As the British economy recovered and the empire expanded, Malthus stood in the way of celebrating the good times, prophesying that they couldn't last. Progressives, conservatives, nationalists—none liked his message. Inevitably, the attacks moved from Malthus's ideas to Malthus himself. The poet Samuel Taylor Coleridge dismissed him as "so contemptible a wretch." Robert Southey, a future poet laureate, called him a

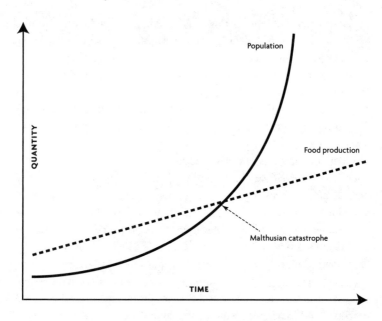

Malthus's argument is often summarized in a graph, with food production rising linearly and population rising geometrically. Eventually the two lines cross, and the horsemen of the apocalypse pay a call.

"fool" and a "booby"; Malthus's supporters, he jeered, were "voiders of menstrual pollution." Oddly, Marx called Malthus a "master of plagiarism." Equally oddly, Percy Bysshe Shelley described him as "a eunuch and a tyrant." "From the inmost Soul I abhor them," Coleridge said of Malthus's beliefs. "[W]ith all the energies of my Heart, Mind, and Spirit, I *defy* them!"

Nonetheless, they endured, most notably in biology, where they inspired both Charles Darwin and Alfred Russel Wallace, co-creators of the theory of evolution.* If populations always threaten to exceed resources, Darwin and Wallace separately realized, their members must be locked into "perpetual struggle, species against species, individual against individual." In this "struggle for existence," not every organism

---

* Darwin and Wallace independently came up with the idea of evolution by natural selection, but Darwin today gets most of the credit, partly because he, unlike Wallace, laid out a detailed case for evolution in a long book.

has an equal chance; the less fit are more likely to fail. The victors of this natural war will be, in general, the best adapted. Over time, species will change to become ever more suited to their conditions; they will evolve.

Overstating the impact of Darwin, Wallace, and Malthus is not easy. Their ideas passed rapidly beyond economics and biology to become models for society. Some thinkers viewed evolution as a process of competition leading to progress, and viewed it as justifying an unfettered market. Others saw the races of humanity fighting for resources, and sought victory for their group; foreign peoples, spilling over their borders, had to be choked back.

In the late nineteenth century, Europeans and North Americans watched their societies acquire colonies all over the world. Conscripting Darwin and Malthus, many people in these places concluded that these triumphs reflected the white race's innate superiority. All human groups had distinct, heritable physical and mental characters, they said. Some were better than others, and groups with better characters had won the struggle for existence. Racial superiority, these people claimed, was why Europe had colonized Asia and Africa, and not the other way around.

But this lofty reasoning did not bring ease of mind. Perched atop their empires, Europeans and Americans feared losing their thrones to swarming mobs of inferiors. Voices shouted that rich nations, torpid with prosperity, were allowing lesser races to overwhelm them by unfettered breeding. So low were European and U.S. rates of reproduction that the West was said to be committing "race suicide."

Among the most influential of these voices was Lothrop Stoddard, author of *The Rising Tide of Color Against White World-Supremacy*, a big hit in 1920. Stoddard's father was a famous photographer who introduced millions of middle-class Americans to images of people from faraway places. Stoddard himself became a fierce anti-immigration advocate, demanding that the United States close its borders to people from faraway places. "If the present drift be not changed, we whites are all ultimately doomed," he wrote in *Rising Tide*. The disappearance of Caucasians "would mean that the race obviously endowed with the greatest creative ability, the race which had achieved most in the past and which gave the richer promise for the future, had passed away, carrying with it to the grave those potencies upon which the realization of man's highest hopes depends." Civilization (by which Stoddard meant

Lothrop Stoddard, 1921

white civilization) must "either fully adapt or *finally perish.*"

Rising Tide was only one alarm. After the First World War, warnings about overpopulation poured off the presses, selling briskly across Europe and North America: *The Passing of the Great Race* (1916), by Madison Grant; *Birth-Rate and Empire* (1917), by the Reverend James Marchant; *Uncontrolled Breeding, or Fecundity Versus Civilization* (1917), by Adelyne More; *Mankind at the Crossroads* (1923), by Edward Murray East; *Standing Room Only?* (1927), by Edward Alsworth Ross; *Danger Spots in World Population* (1929), by Warren S. Thompson; *Asia's Teeming Millions and Its Problems for the West* (1931), by Étienne Dennery.

Strikingly, this jubilee of dismay was led by prominent intellectuals. East was a Harvard plant geneticist; Thompson, director of the world's first demographic research center; Dennery, an economist at the respected Paris Institute of Political Studies. Grant, a well-known New York lawyer, was close to President Theodore Roosevelt. Adelyne More ("add-a-line more") was the flippant pen name of the English linguist-philosopher Charles Kay Ogden. Marchant, head of the British National Council of Public Morals, was famed for his campaign against impure thoughts. Ross was a sociologist at the University of Wisconsin (he had been fired from Stanford for urging the U.S. Navy to blow up "every vessel bringing Japanese to our shores rather than to permit them to land"). Even Stoddard, a mere freelance journalist, had a Harvard Ph.D.

More striking still, many of the racial alarmists were also leaders in the nation's new conservation movement. The blue-blooded toffs who feared that the noble and superior white race was menaced by unwashed rabble also saw wild landscapes as noble and superior wildernesses menaced by the same rabble. Prizing the expert governance of resources, they found little difference between protecting forests and cleaning up the human gene pool.

Madison Grant was an example. Born into a patrician New England family and raised in a turreted mansion, Grant co-founded the Bronx Zoo, organized the preservation of the California redwoods, helped create the national park system, played a central role in saving bison from extinction, and wrote ecological texts that anticipated Aldo Leopold. He also spent decades trying to protect his privileged cohort from the rising lower classes—indeed, he wrote *The Passing of the Great Race* to decry "the transfer of power from the higher to the lower races." Among Grant's most zealous fans was Adolf Hitler, who (Grant boasted) sent him a fan letter; *The Passing of the Great Race* was the first foreign book published by the Nazis after they took power.*

Madison Grant, ca. 1925

So ineradicable was the elitist mark on conservation that for decades afterward many on the left scoffed at ecological issues as right-wing distractions. As late as 1970, the radical Students for a Democratic Society protested the first Earth Day as Wall Street flimflam meant to divert public attention from class warfare and the Vietnam War; the left-wing journalist I. F. Stone called the nationwide marches a "snow job."

Vogt, Leopold, Murphy, and their associates were not truly in this elitist company; in fact, they helped begin the transformation in which environmental issues switched from being a cause of the right to one of the left. Nonetheless, they shared much of the racial alarmists' intel-

---

* In what the Yale historian Timothy Snyder describes as "an extreme articulation of the nineteenth-century commonplace that human activities could be understood as biology," Hitler viewed our species as a group of genetically distinct races warring for survival. Echoing Malthus, Hitler insisted that "regardless of how [any race] raises the productivity of the land . . . , the disproportionate population in relation to the land . . . remains." As a result, the duty of Hitler and every other racial leader was to "reestablish an acceptable ratio between population and land area"—that is, to feed their ever-growing races by seizing an ever-greater share of the planet's finite resources.

lectual framework and often dismissed nonwhites in terms that read uncomfortably today. Vogt, for instance, was loudly scornful of the "unchecked spawning" and "untrammeled copulation" of "backward populations"—people in India, he sneered, breed with "the irresponsibility of codfish." But this second wave of conservationists rarely claimed that one race or culture was intrinsically superior to another. Vogt, again, is an example. No apologist for his own stock, he reserved special ire for "American vandals abroad," the "despoilers" and "parasites" who ruin foreign landscapes and exploited foreign people in the name of "that sacred cow Free Enterprise." In his view, "we be of one blood."

Most important, Vogt and his friends—like Stoddard, like East, like Hitler—viewed people as biological units, ruled by the same laws as bacteria and fruit flies. ("Nature is inexorable," wrote Stoddard. "No living being stands above her law; and protozoan or demigod, if they transgress, alike must die.") Like cormorants and anchovetas, like dovekies, willets, and mosquitoes, humans are evolutionary actors, packets of genetic drives, their courses in the world fixed by environmental limits. And these ideas—different from those of the race theorists, but equally based on a view of people as biologically fated—lurked in Vogt's mental background, formed part of his intellectual toolkit, as he shifted focus from birds to humanity.

Road to Survival

In the summer of 1945 Vogt's life changed dramatically. He was presented with a new chance to be heard, and he acquired a new traveling companion. The chance to be heard was the opportunity to write a book. The companion was his second wife, Marjorie Elizabeth Wallace. Vogt's book would set down, for the first time, the intellectual framework of today's green movement—ideas now so pervasive that many people don't even recognize them as ideas, or that people didn't always think them. None of this, he said, would have been possible without his new wife.

Born in 1916 to a British father and an American mother, Marjorie Wallace was raised in San Mateo, California. Like Juana, she attended Berkeley, where she caught the eye of George Devereux, a young, brilliant, mercurial European sociologist. The couple married after Marjorie's graduation in 1938, then moved to a mental hospital in Worcester, Massachusetts, where Devereux researched insanity, wrote studies of the Mojave Indians, and had affairs; Marjorie produced a master's thesis at Boston University.

In 1943 Devereux got a job in Wyoming, closer to his research subjects. Marjorie instead went home to California, where she apparently met Vogt, fourteen years her senior, who was futilely trying to convince Walt Disney to make an animated movie about soil. It seems evident that they began a relationship. Juana had spent much of the previous two years alone in Latin America, trolling the embassy circuit for Nazi gossip. In June 1945 the couple rendezvoused in California. The marriage collapsed. Two months later Juana went to Reno, Nevada, to obtain one of the city's famous quick divorces. Early in 1946 Marjorie also went to Reno, and for the same reason. Marjorie filed for divorce from Devereux, appeared before the court, received her decree, and married Bill on the same day: April 4, 1946.

The newlyweds flew to Mexico City, where, among other things, Vogt visited a new agricultural project backed by the Rockefeller Foundation—Borlaug's project. Few records survive of the trip. What we do know is that Vogt was aghast. He agreed with Rockefeller that it was important to "use our enormous capacity to produce food as a means of getting the world back on its feet." But it was "imperative," as he later put it, "that we recognize this as an emergency measure and do not try to put a million pairs of feet under our dinner table." In funding Borlaug's project, Vogt believed, Rockefeller was going exactly the wrong way.

The day after their visit to the Rockefeller project, Marjorie and Bill flew to Guatemala for a month-long honeymoon. Then, as if to make up for the idleness, the couple pressed on to present Vogt's environmental reports to skeptical Central and South American officials. When Vogt wasn't meeting bureaucrats, he was taking long expeditions up rutted roads into the country, a constant strain for his bad legs and weak chest. In the evening, fatigued, Vogt fell into bed, dictating notes and letters to Marjorie far into the night. Among them was a letter to the Rockefeller Foundation, asking them to change course. In El Salvador his hotel mail contained a book contract. A New York publisher had asked him to write an environmental book for a "whopping big advance." Despite his exhaustion, Vogt was jubilant. He had been thinking about writing a book since the mid-1930s—indeed, he had been offered a contract by a publisher who had attended one of his talks—but had always been overwhelmed by work and travel. Now the Pan American Union looked like it might give him enough time. He was finally going to be able to make a loud, public noise.

Vogt's publisher was William M. Sloane, a science fiction writer who had been an editor at the Henry Holt publishing company until March 1946, when he and four other Holt staffers left to launch a new firm, William Sloane Associates. Housed in a single room on the third floor of a walkup building, it had limited funds; Sloane himself was so deep in debt that he despaired of paying his grocery bills. Vogt had never written a book before and was unknown to the public. He was constantly traveling, was overburdened with his day job, and had the sort of prickly, imperious personality that resists editing. Nonetheless, Sloane bet heavily on him. Vogt's book would be among the first his company released.

To Vogt's surprise, he found himself in friendly book-writing competition with one of his associates: Fairfield Osborn. Unlike Vogt, Osborn was one of the old-school, upper-crust conservationists. His father, a wealthy paleontologist, had been president of both the American Museum of Natural History and the New York Zoological Society; his uncle was president of the Metropolitan Museum of Art. The two brothers were happy bigots; Osborn's father wrote a glowing preface to *The Passing of the Great Race*. After sixteen years on Wall Street, Osborn retired and switched his focus to conservation. Like his father, he was president of the New York Zoological Society. His main task was operating the Bronx Zoo, which he did with theatrical flair; often he brought a sparrow hawk to speaking engagements, released a cageful of moths, and set the bird to picking off the insects above the audience.

Both the First and Second World Wars, Osborn believed, were set off by environmental degradation—they were, at bottom, resource wars. In those days, he liked to say, humankind "was involved in *two* major conflicts—not only the one that was in every headline. The other war . . . contains potentialities of disaster greater even than would follow the misuse of atomic power. This other war is man's conflict with nature." Hoping to forestall further destruction, Osborn worked with Leopold and Vogt in 1948 to create the first global ecological organization: the Conservation Foundation. At the same time, he began writing a book "to show that mankind is a part of the earth's biological system and is not a form of genii that can successfully provide substitutes for the processes of nature."

If Vogt was dismayed by the thought of competing with the wealthier, more socially prominent Osborn, he never showed it. Letters volleyed back and forth between them, Osborn writing from the plush Fifth Avenue headquarters of the Zoological Society, Vogt from hotels in Latin America or his government-issue office at the Pan American Union. Osborn credited Vogt and Leopold for giving him the correct "philosophical approach to the problem." Vogt told Osborn that his manuscript had kept him up "until I finished it about 2:00 a.m." And he paid the author a writerly compliment: "More than once, as I went through your text, I said to myself, 'I wish I had thought of that.'" The two books came out within months of each other in 1948, Osborn's *Our Plundered Planet* on March 25, Vogt's *Road to Survival* on August 5.

Both were enormously successful. *Our Plundered Planet* was reprinted eight times and translated into thirteen languages. *Road to Survival* was a main selection of the Book-of-the-Month Club (a nationwide subscription service that automatically sent books deemed worthy to 800,000 American subscribers) and *Reader's Digest* (the biggest-circulation magazine in the world, which published a condensed version of the book for its 15 million worldwide subscribers). *Road* was translated into nine languages. Vogt was given awards by the Cranbrook Institute of Science and the Izaak Walton League, and the book was adopted within weeks by twenty-six colleges and universities as a textbook (many more would do so later).

The early reviews for both books were overwhelmingly positive. *Our Plundered Planet*, reported the *San Francisco Chronicle*, was "the most important word of warning delivered to the human race in the present century." Vogt's book was "controversial, exciting, dismaying and yet hopeful," said *The Boston Globe*. To *Saturday Review*, *Road* was "the most eloquent, provocative and informative book that has been written thus far in the United States on conservation—or the lack of it." *The Washington Post* was particularly enthusiastic:

Within the lifetime of many living men, there will not be enough to eat anywhere on the planet. . . . We are living on borrowed time, or more accurately on rapidly dwindling capital. . . . This is easily the most important book of 1948. It is also one of the best written.

Probably most important to Vogt were the personal congratulations. Roger Tory Peterson, a bird man through and through, had been skeptical about Vogt's decision to paint on a broader canvas. *Road*, Peterson now said, was the culmination of everything Vogt had been working toward. Speaking for her husband, Robert Cushman Murphy's wife, Grace, wrote, "Your book is the new Bible." Early on, Aldo Leopold had praised Vogt's proposed book outline as "excellent." Now he told Vogt that *Road* was "the most lucid analysis of human ecology and land use that I have yet encountered."

As Vogt's message sunk in, though, he was denounced. "Real scientists take a dim view of *Road to Survival*," scoffed *Time,* the world's best-selling newsmagazine. Every aspect of Vogt's "creed," wrote its anonymous reviewer, "is either false or distorted or unprovable." *Road* was condemned by Roman Catholics, because it advocated birth control; by conservatives, because it supported state regulation; and by business interests, because it attacked capitalism (the *Reader's Digest* condensation excised Vogt's critiques of the free market). But the greatest ire came from the left. Denouncing the book as a "totally unaware," "incoherent," "screaming" "bill of goods," *The Nation* magazine called Vogt's ecology proof of "science's bankruptcy in the face of pressing modern problems." At a world Communist summit in Paris, a Soviet novelist drew cheers when he damned *Road* as "merely a gross means of corrupting the American people." *Our Plundered Planet* was equally bad. Reading Vogt and Osborn, the novelist insisted, had caused "a serious increase in their crime wave."

Unsurprisingly, the two books were in some ways quite similar—and not just in that both railed against a new chemical called DDT. They were jointly inventing a new literary genre: *the concerned report on the global condition.* They were the first to portray our ecological worries as a single Earth-sized problem for which the human species is to blame. And by stating that the problem is one interconnected, worldwide issue, rather than something local or national, they implicitly argued that ecological issues could only be solved by a unified global effort, administered by global experts—by people, that is, like Vogt and Osborn.

Vogt and Osborn were also the first to bring to a wide public a belief that would become a foundation of environmental thought: consumption driven by capitalism and rising human numbers is the ultimate cause of most of the world's ecological problems, and only dramatic reductions in human fertility and economic activity will prevent a worldwide calamity.

Of the two, *Road* had greater impact. In a way its author could not have imagined, it became the blueprint for today's environmental movement. The book inspired both Rachel Carson, who later wrote *Silent Spring*, and Paul Ehrlich, author of *The Population Bomb*—the two most important environmental books of the 1950s and 1960s. "Every argument, every concept, every recommendation made in *Road* *to Survival* would become integral to the conventional wisdom of the post-Hiroshima generation of educated Americans," the historian Allan Chase has written. Vogt's ideas "would for decades to come be repeated, and restated, and incorporated again and again into streams of books, articles, television commentaries, speeches, propaganda tracts, posters, and even lapel buttons."

Part of the impact was due to Vogt's arrestingly harsh tone. Both Osborn and he lamented "the American way of doing business," which had ruined land (they said) since the time of the Pilgrims. But only Vogt described all of U.S. history as little more than a "march of destruction," in which colonists "chopped, burned, drained, plowed, and shot" their way from the Atlantic to the Pacific. "Our forefathers," he thundered, were "one of the most destructive groups of human beings that have ever raped the earth. They moved into one of the richest treasure houses ever opened to man, and in a few decades turned millions of acres of it into a shambles." Unlike Osborn, a product of Wall Street, Vogt scorned capitalism. He imagined his critics wailing,

> "Free enterprise has made our country what it is!" To this an ecologist might sardonically assent, "Exactly." For free enterprise must bear a large share of the responsibility for devastated forests, vanishing wildlife, crippled ranges, a gullied continent, and roaring

flood crests. Free enterprise—divorced from biophysical understanding and social responsibility.

Ecological collapse tomorrow would lead the day after to nuclear war. If humankind kept ignoring ecological realities, he warned,

> there is little probability that mankind can long escape the searing downpour of war's death from the skies. And when this comes, in the judgement of some of the best informed authorities, it is probable that at least three-quarters of the human race will be wiped out.

Road laid out the basic tenets for a now-common way of thought: *environmentalism*. Environmentalism is more than the simple recognition that polluting a neighbor's well or destroying a bald eagle's nest is a bad idea. In most cases that recognition can be viewed as a function of property rights. By poisoning a well, a polluter is, in effect, seizing the water without its owner's permission. (More precisely, it is seizing use of the water.) The eagles, too, are being taken from their owner, the public. Environmentalism, by contrast, is a political and moral movement based on a set of beliefs about nature and the human place within it.

Environmentalists want to stop polluting wells and protect bald eagle nests. But they see the well water not so much as property but as part of a natural cycle with its own value that needs to be maintained. The eagle, for its part, is a constituent of an ecosystem that has an essential integrity that should be protected. Any set of beliefs about the workings of the world is perforce a statement about what is good and important in that world. Environmentalism is an argument that respecting the rules of nature is indispensable to having a good society and living a good life. Leopold was among the first to set this idea down. "A thing is right when it tends to preserve the integrity, stability, and beauty of the biotic community," he wrote in "The Land Ethic," one of his most celebrated essays. "It is wrong when it tends otherwise."

Road to Survival had two main innovations. The first, as the environmental historians Paul Warde and Sverker Sörlin note, was introducing "the idea of *the* environment." The old idea of "environment" dates back at least to the ancient Greeks. It meant the external natural factors—climate, soil, altitude, and so on—that affected both individual people's lives and (it was thought) characters. Thus, for example, Hippocrates believed that fertile, well-watered terrain created people who were "fleshy, ill-articulated, moist, lazy, and generally cowardly." Hippocrates, raised on the Mediterranean coast, claimed that its environment produced tall, beautiful, intelligent people—people, presumably, like himself and his readers. Variants of this idea continued well into the twentieth century.

In this context, "environment" referred to a single type of place—forests, shorelines, marshes, and so on—that acted on people. As Warde and Sörlin emphasize, Vogt turned the word around. In *Road to Survival,* "environment" meant not the external natural factors that affected humans but the external natural factors that *were affected by* humans. Instead of Nature molding people, Vogt envisioned people molding Nature, usually negatively. And by "environment" he meant not a particular place, but a global totality. A statement about the effects of local conditions on people in the past and present had become a vision of humankind's impact on the entire Earth, with a focus on the future.

Defining a word in a new sense seems academic and abstract, but its consequences are not. Until something has a name, it can't be discussed or acted upon with intent. "People, by naming the world, transform it," wrote the Brazilian educator Paulo Freire. Without "*the* environment," there would be no environmental movement.

In the second of *Road*'s main innovations, Vogt summed up the relationship between humanity and this global environment with a single concept: *carrying capacity.* It is hard to overstate the importance of this. There are two ideas at the base of today's globe-spanning environmental movement. One is that *Homo sapiens,* like every other species, is bound by biological laws. The second is that one of these laws is that no species can long exceed the environment's carrying capacity.

Coined in the early nineteenth century, the term "carrying capacity" initially referred to the amount of cargo that a ship could hold. Over time it came to refer to the weight or volume of material that could be transported by some type of vehicle—the supplies that a mule train could take into the mountains, for example. In the 1880s the definition expanded to the number of grazing animals that could live in a given range. Working in the Forest Service's Office of Grazing during the First

World War, Aldo Leopold encountered the concept; by transferring it from cattle on pastures to game animals in forests, he made carrying capacity a basic ecological tool. In his textbook, *Game Management* (1933), Leopold argued that land managers' task was "enhancing productivity," which meant manipulating the landscape until it reached its maximum carrying capacity.

In *Road*, Vogt defined carrying capacity by means of a formula: $B - E = C$. In this equation, B was the biotic potential, the theoretical ability of that piece of land "to produce plants for shelter, for clothing, and especially for food." E was the environmental resistance, the practical limitations on the theoretical biotic potential. The actual carrying capacity, C, was never as high as the theoretical biotic potential, because there was always some environmental resistance. Hence $B - E = C$. Vogt argued that people were degrading the environment (*the* environment!) so much that E was rising worldwide. In consequence, C—the planet's capacity to support life—was shrinking.

With carrying capacity, Vogt rewrote Malthus. As the Harvard historian Joyce Chaplin has observed, Malthus offered no evidence in his *Essay* to prove that farm harvests could only increase arithmetically. Indeed, Malthus's theory could be restated as the claim that one species (humans) reproduces at a geometrically increasing rate, but other species (farm crops) cannot. No obvious reason exists for this to be true—for humans to be special in this way. As one recalls, Malthus's evidence for the rapidity of human reproduction came from an article by Benjamin Franklin. But, Chaplin notes, Franklin made clear in the same article that he believed that plants and people reproduced at comparable rates—the opposite of Malthus's contention.

Instead of concocting some reason why farm crops must be less fecund than people, Vogt used carrying capacity to reset the argument. Carrying capacity was a threshold that could not be surpassed by *any* species. Yes, Vogt conceded, scientists might be able to use technology to boost harvests enough to outstrip population growth. But the short-term triumph of raising farm output would lead to a long-term calamity. Our species would surpass Earth's carrying capacity, which would destroy the ecosystems that support us. Carrying capacity could not be avoided. Either people would reduce their numbers and consumption to stay below the world's carrying capacity—or the ecological devastation wrought by overpopulation would do it for them. In the end, as the ecologist-activist Paul Ehrlich later put it, "Nature bats last."

Vogt's argument was intuitively powerful but intellectually shaky. As the Berkeley geographer Nathan Sayre has highlighted, carrying capacity began as a concrete quantity that one could measure. If an individual ship had a carrying capacity of X tons, X was a number that could not change unless the ship was rebuilt. But as the notion of carrying capacity expanded to other forms of transportation, then environments like pastures and forests, and then the entire planet, it stopped being something that one could enumerate easily. It was unclear whether the carrying capacity of an ecosystem was actually a static, measurable entity or if it had a meaningful upper bound. An idea that could be useful on a small scale could become untenable if it was stretched like taffy to wrap over the entire world.

Was the carrying capacity of an individual ecosystem a rule of thumb—the way things happened to work out a lot of the time—or a biological law, something that reflected an underlying physical reality? Was an environment's biotic potential (and thus its maximum theoretical carrying capacity) a fixed, absolute limit, a value set by Nature, or was it a quantity that could change over time, and thus be influenced by people?

Vogt didn't answer these questions. But at about the same time that he was writing *Road*, the Georgia ecologist Eugene P. Odum did answer them, in *Fundamentals of Ecology* (1953), the first widely used ecology textbook. Yes, Odum said, carrying capacity is a concrete number determined by physical law and measurable in the field. Turning to the *S*-shaped growth curve found by Gause and others, Odum argued in *Fundamentals of Ecology* that the carrying capacity is simply the highest bit of the graph—"the upper level beyond which no major increase can occur." Environments have limits that cannot be ignored or overcome. The walls of the petri dish are real and cannot be surpassed.

Odum's definition of carrying capacity, novel at the time, today "is virtually universal in textbooks," wrote the Harvard ecologist James Mallet in 2012, "and has been taught to generations of undergraduates." I am one of them. In college I learned about carrying capacity from the third edition of Odum's book, published in 1971; its fifth edition, published in 2009 after his death, was in my daughter's high school biology classroom.

Today the concept of global carrying capacity has evolved into the idea of *planetary boundaries*. The boundaries set the environmental terrain "within which we expect that humanity can operate safely," a team of twenty-nine European and American scientists argued in an influential report from 2009. (It was updated in 2015.) To prevent "non-linear, abrupt environmental change," they said, humankind must not transgress nine global limits. That is, people must not

1. use too much fresh water;
2. put too much nitrogen and phosphorus from fertilizer into the land;
3. overly deplete the protective ozone in the stratosphere;
4. change the acidity of the oceans too much;
5. use too much land for agriculture;
6. wipe out species too fast;
7. dump too many chemicals into ecosystems;
8. send too much soot into the air; and
9. put too much carbon dioxide into the atmosphere.

The researchers provided specific figures for these boundaries; as an example, the ozone in the upper atmosphere (Limit No. 3) should not fall to less than 95 percent of its pre-industrial level. I have omitted the numbers to highlight that the basic argument is as simple as it was in Vogt's day. Stay within the limits, and people can develop freely. Go beyond the boundaries—exceed carrying capacity—and trouble will ensue.

Vogt wanted "intelligent reasonably literate people with little or no knowledge of ecology or conservation" to read his book. He got more than he had dreamed. His ideas, taken up and reframed by later generations, set the tone for environmental crusades for decades to come. Consume less! Eliminate toxins! Turn down the thermostat! Eat lower on the food chain! Reduce and recycle waste! Protect biodiversity! Live close to the land! Protect local communities! Small is beautiful! All have their root in Vogt's call to live lightly on the soil and work with nature, instead of overwhelming it.

Like many of his followers, Vogt believed that the kind of humble, locally focused, community-oriented life that he invoked was a logical consequence of the recognition of limits, both environmental (the need to respect global carrying capacity) and human (lack of knowledge about ecological interactions). But these injunctions are also inextricably bound to a conception of the good life—a particular way of living that critics mock with epithets like "tree-hugging" and advocates invoke with terms like "sustainability."

Vogt's ideal society—a network of self-sufficient citizens guided by ecological precepts—harkens back to Thomas Jefferson, who saw virtue arising from agrarian villages, as opposed to the market and the city. It is not a criticism to say that these Jeffersonian views celebrate the rural over the urban, husbandry over industry, intense local connection over mobile liberty, thrifty independence over opulence and commerce. Nor is it a criticism to note that Vogt's vision of the good life was much like the childhood idyll he lost when he moved to Brooklyn. But it is important to note that others believe that the goal of living well on the planet can be met in different, even opposing ways. Much as Alexander Hamilton opposed Jefferson's beliefs as unsound, these people see the best way to live with nature as gathering into big cities (which are said to use less resources than spread-out local communities), increasing productivity (because fewer people directly work the land, maximizing output per person), and growing more prosperous (because affluence makes societies better able to clean up environmental mishaps).

These arguments—the fundamental disputes between Wizard and Prophet—did not take form quickly, partly because Vogt's life was as unstable as ever. While he wrote *Road to Survival*, Aldo Leopold made plans to hire him as an ecological economist—a newly created position at the University of Wisconsin, and possibly the first such position in the world. But in April 1948, a few months before *Road* was published, Leopold suffered a fatal heart attack while helping a neighbor fight a brush fire. A small part of the loss was Vogt's job.

Oxford University Press had just agreed to publish what would become Leopold's masterpiece, *A Sand County Almanac*. Still mourning, Vogt went to New York to assure the firm that Leopold's son Luna could and would edit the manuscript (Vogt also went through it). As intellectually indebted to Vogt as Vogt's book had been to Leopold, *A Sand County Almanac* appeared with an enthusiastic blurb from Vogt on its cover. Despite Leopold's elegant prose, the book attracted little initial attention, but in the 1960s it became an environmental scripture, selling hundreds of thousands of copies and introducing a new generation to Vogt's message of carrying capacity and global limits.

Two months after Leopold's death, as part of the publicity push for *Road to Survival*, Vogt published an excerpt in *Harper's* magazine. The article set off alarms in the Manhattan headquarters of the Rockefeller Foundation. In its pages was a veiled but unmistakable attack on the foundation's Mexico project, work that would come to be associated with a man named Norman Borlaug.

THREE

The Wizard

More

Many years later, after he won the Nobel Prize, Norman Borlaug would look back on his first days in Mexico with incredulity. He was supposed to breed disease-resistant wheat in Mexico's central highlands. Only after he arrived, in September 1944, did he grasp how unsuited he was for the task—almost as unqualified in his own way as Vogt had been when he set sail for Peru. He had never published an article in a peer-reviewed, professional journal. He had never worked with wheat or, for that matter, bred plants of any sort. In recent years he had not even been doing botanical research—since winning his Ph.D., he had spent his time testing chemicals and materials for industry. He had never been outside the United States and couldn't speak Spanish.

The work facilities were equally unprepossessing. Borlaug's "laboratory" was a windowless tarpaper shack on 160 acres of dry, scrubby land on the campus of the Autonomous University of Chapingo. ("Autonomous" refers to the university's legal authority to set its curriculum without government interference; Chapingo was the name of the village outside Mexico City where it was located.) And although Borlaug was sponsored by the wealthy Rockefeller Foundation, it could not provide him with scientific tools or machinery; during the Second World War, such equipment was reserved for the military.

Borlaug had spent his childhood working in the fields of an impoverished family farm. He had regarded it as drudgery and wanted to get away. In Mexico he was back to hand tools and draft animals. During the day, the heat was unrelenting; at night, cold damp winds came from the hills. No hotels were near, so Borlaug slept on the shack's dirt floor. Dinner was a can of stew heated over a fire made from corncobs. Flies were a constant irritant; mice ran over his sleeping bag in the dark. Water came from a bucket; he boiled it before drinking but was often sick.

Worst of all was the consuming worry that he would prove unfit for the task—that he had, as he said later, "made a dreadful mistake" in coming to Mexico. He wanted through his work to help feed the hungry—a vision that been gathering slowly, almost without his knowing it, since adolescence. But he feared he would fail even to take the first step. Nothing he tried had worked. The plants were dying and he was at odds with his superiors. He had not felt so badly at sea since the first bewildering days after he left his family farm.

Despite his forebodings, he succeeded. The work that sprang from this neglected patch of Mexico would reach across the world and change lives from Bolivia to Bangladesh. He would be celebrated and denounced, but even his enemies would credit him with fundamentally transforming the human prospect. His supporters would say that he saved a billion souls from starvation, though he always demurred at the total.

The parallels between Borlaug and Vogt are inexact. Borlaug never wrote a manifesto and mostly declined the roles of theorist and exponent. Instead he became, by the example of his life, the emblem of a way of thought—the Wizard's way. His success would show, at least to Wizards, that science and technology, properly applied, could allow humankind to produce its way into a prosperous future. To the question of how to survive, his work said: be smart, make more, share with everyone else. It said: we can build a world of gleaming richness for all. And the concomitants of this world—the giant installations, the whirring machinery in the garden, the glare of artificial light in the night sky—are to be embraced, not feared.

Like any symbol, the image of Borlaug as an apostle of Science simplifies events and flattens ambiguities. But it still captures something about the man—his tenacity, his conviction that logic, knowledge, and hard work would pay off in the end. And it has been a rallying cry. I have met many people who have been inspired by him. When I asked these people about the future, they gave me diverse answers. But often they amounted to "What would Borlaug do?"

Norm Boy

In his long life Norman Borlaug would live in foreign lands for decades, but Iowa would always be home. Armies of prairie grass there had been conquered by his ancestors' plows. Here and there stood great isolated rocks, castoffs from ancient glaciers. Non-Indians came into the area in large numbers only in the mid-nineteenth century. Ole Olson Dybevig and his wife, Solveig Thomasdatter Rinde, emigrants from Norway, were among them. Ole, born in 1821, had grown up on a bend of Sognefjord, the second-longest fjord in the world, in a little cluster of farms too small to be called a village. The land there consisted of beautiful, rich fields above the placid water, but the Dybevigs didn't have much of it and what they did have was wracked by potato blight. Solveig, four years younger, was from an equally poor family on a neighboring farm. In 1854, less than a fortnight after their wedding, the couple set off for the United States. Along the way, they changed their surname to Borlaug, which they hoped Americans would have less trouble pronouncing. It was the name of the home settlement they would never see again.

After a short stay in Wisconsin, the Borlaugs moved west, to the banks of the Missouri River. The territory was contested by Indians—Dakota Sioux who had been cheated for years by both the U.S. Congress and the territorial government. In 1862 the Dakota lashed back. Enraged beyond measure, they killed hundreds of immigrants and defeated territorial militias in a series of battles before falling to the U.S. Army. Ole and Solveig Borlaug fled the slaughter, driving a covered wagon to Saude, in northeast Iowa.

A cluster of perhaps forty farming families, Saude had a general store, a feed mill, a part-time blacksmith, a cooperative creamery, and two churches. There were more trees when the Borlaugs arrived than now, stands of oak and poplar, but the low slopes of the landscape still stretched to the horizon as they do today. Visiting the area when the light of the winter sunset was the color of ale, I could picture it as the Borlaugs must have, a landscape at once chillingly vacant and full of promise.

The new arrivals built cabins from logs chinked with mud; grew clover, wheat, maize, and oats; pastured a few milk cows; let their dogs run free. Half of the area's inhabitants were Norwegian; most of the rest were Czech—Bohemian, as people said then. Relations between the enclaves were friendly but distant. In Saude I talked to three older men who had grown up in its Norwegian half. All had been told by their parents not to date Bohemian girls.

Each community gathered on Sundays in its church—Lutheran for Norwegians, Roman Catholic for Bohemians. In the Norwegian church men sat on one side, women on the other. Ministers wore white ruffs and black satin stoles. Services were in Norwegian until the early 1920s. At Christmas the congregation placed a tree in the church entrance, lighted candles tied to the branches. After the service everyone unwrapped presents together.

The Borlaugs picked Saude for its close-knit Norwegian community rather than for the soil, which was shallow and poorly drained. The wet conditions fostered crop diseases; stem rust attacked wheat so often that most local farmers, the Borlaugs among them, gave up planting it. Poor soil translated into poverty for all and early death for many. The Norwegian church held thirty funerals in 1877 alone—9 percent of its membership. By the beginning of the twentieth century Saude was slowly losing population to Cresco, the biggest nearby town. The grandchildren of the original settlers were abandoning the land their forebears had labored over. Saude was too poor, too lonely, too far from markets.

None of the Borlaugs ever left for long. For most of their lives all five of Ole and Solveig's offspring lived within walking distance of their parents' home. By the first years of the twentieth century, the middle child, Nels, had built up the biggest holding: 165 acres. In August 1913 his second son, Henry Oliver, became the first of his children to marry. Henry's wife, Clara Vaala, had been raised a dozen farms away.

Henry and Clara's first child, Norman Ernest Borlaug, came into the world seven months later, on March 25, 1914. His parents were living in Nels's home, sharing quarters with Henry's two youngest siblings. At harvest time the whole family—all of Ole and Solveig's children, all their offspring and their families—would work together, thirty or so Borlaugs singing Norwegian songs and wrestling with borrowed equipment in the fields. After sunset they gathered for dinner, tired and hungry, warming the house with their bodies. Potatoes from Nels's patch, beef from their cattle, home-baked bread, hard-boiled eggs, pies with apples from Ole's orchard. Everyone called the baby "Norm boy," never Norman. It wasn't until the boy was eight that his family—by then it included two younger sisters—moved into their own home, half a mile away. Purchased from a Sears, Roebuck catalogue, Modern Home No. 209 (price: $981.00) was built in a boxy, self-contained style—the "foursquare," to architectural historians—intended to signify solid American values. There was no insulation or plumbing, but it kept out the wind.

Saude was insular in a way that is hard to imagine now. The Norwegian families clung together so tightly that Norm boy's parents spoke English with thick Norwegian accents although they had never set foot in Norway. There were no telephones, radio, or television, no mass media of any kind but Nels's *Cresco Plain Dealer*, an eight-page weekly concerned almost exclusively with local doings. Sometimes on still winter nights Norm boy and his sisters would sit outside, wrapped in blankets, waiting to hear the Milwaukee train as it pulled into Cresco. "It was the one time we sensed we were part of a wider sphere," he said. "Those sound waves were our sole connection to the world." (This quote, like others to follow, are from oral history interviews with Borlaug.)

A three-mile walk down Saude's dirt roads was the community school, a one-room structure painted white. Built in 1865, it was lighted by oil lamps and heated by a potbellied stove. Facing the rows of worn desks was a slate blackboard; near it was a bookshelf with a dictionary, an encyclopedia, and a few old children's books. Outside were two privies, one for boys and one for girls. Norm boy began his education in the fall of 1919; he quickly found his father's initials scratched into a desk. A single teacher taught the entire school—eight grades, ten to twenty children in all, packed into one 28-by-24-foot space. All students were Caucasian; of the county's almost fourteen thousand inhabitants, only four were African-American. Mornings began with a ritual bellow of the Iowa Corn Song:

> We're from I-O-way, I-O-way
> State of all the land
> Joy on ev'ry hand
> We're from I-O-way, I-O-way
> *That's* where the tall corn grows!

When snowstorms came students rushed home, trying to beat the gale. Once during Borlaug's first winter a blizzard rose so fast that by the time the children put on their coats the snow and wind were like an assault. Arms shielding their faces, the oldest boys broke the path; five-year-old Norm boy followed.

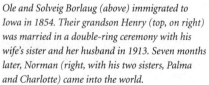

Ole and Solveig Borlaug (above) immigrated to Iowa in 1854. Their grandson Henry (top, on right) was married in a double-ring ceremony with his wife's sister and her husband in 1913. Seven months later, Norman (right, with his two sisters, Palma and Charlotte) came into the world.

Borlaug's childhood world was focused on his home (center, bedroom on left), church (top, Borlaug family graves in foreground), and school (right). The nearest town, Cresco, was thirteen miles away—far enough that it could be visited only once or twice a year.

We trudged through the swirling whiteness, leaning on the wind, blinded by the sleet and struggling against the clinging waist-deep snow. I was miserable. Icy drafts slipped through my clothes. . . . Snow clung to my face, mittens, jacket. The melt inside my boots numbed my feet. I began stumbling. Soon it became too much to bear. . . . There was just one thing to do: I lay down to cry myself to sleep in the soft white shroud nature had so conveniently provided. Then a hand yanked my scarf away, grabbed my hair, and jerked my head up. Above me was a face tight-lipped with anger and fright. It was my [twelve-year-old] cousin Sina. "Get up!" she screamed. "Get up!" She began slapping me over the ears. "Get up! Get up!"

Sobbing, the boy allowed Sina to lead him home. When Borlaug staggered through the door his grandmother had just pulled bread from the oven. Shamed by his display of weakness, he sat down. A slice of bread appeared in front of him, hot enough to melt butter. "No food was ever as sweet as those loaves Grandmother baked the day when I was five years old and nearly died."

When not in school the Borlaug children did chores, rising before dawn and working until after sunset. Boys hoed weeds, dug potatoes, milked cows, stacked hay, hauled wood and water, fed chicken, cattle, and horses. Girls tended the vegetable garden, worked the washboard, cleaned house, mended clothing, cooked meals. The toil never ended but complaint was rare. The Borlaugs were subsistence farmers, and if they wanted to eat there was no alternative.

Norm boy worked dutifully but without enjoyment. He particularly detested harvesting maize. Every ear had to be sliced from the plant, husked on the spot, and flung into a wagon. The sharp leaves cut through gloves and clothes; the boy was scraped and bleeding by the end of the day. According to Noel Vietmeyer, a longtime co-worker who wrote a biography of Borlaug, an Iowa family with a forty-acre maize plot, typical for the time, hand-picked half a million ears every fall. It was, Borlaug told Vietmeyer, a "two-month horror."*

---

* Vietmeyer's book began as a ghostwritten autobiography, then morphed into a three-volume, self-published biography with long quotations supposedly from Borlaug but actually drawn from the initial, ghostwritten manuscript. Although it is unclear whether Borlaug ever

Norm boy saw himself as a worker, not a scholar. But his grandfather Nels passionately believed in education. He had been able to attend school for only three years, but he had pushed his son Henry as far as the sixth grade. Now Nels insisted that his grandson get more. *You must have an education!* he told the boy. *Your knowledge is the only protection you have in this world! Fill your head now to fill your belly later!* Norm boy knew he would become a farmer like Henry and that education was unlikely to change this. Nonetheless, he did his homework.

In seventh grade, he acquired a new teacher: nineteen-year-old Sina Borlaug, the cousin who had saved his life in the blizzard. A few weeks before Norm boy's graduation from eighth grade, Sina took it upon herself to tell his parents that their son should go to the high school in Cresco. For Henry and Clara, the decision was difficult. Cresco, thirteen miles from Saude, was too far to commute. If Norm went to high school, his family would not only lose his labor but have to pay for his room and board. Nels's words about education echoing in their heads, the Borlaugs sent their son to Cresco.

Borlaug's worries about depriving the family of his labor were relieved when Nels and Henry bought a tractor. Manufactured by the Ford Motor Company, the Fordson Model F was the Model T of tractors: a simply designed, solidly built machine that induced huge numbers of people to acquire a tool they had previously regarded with suspicion. As Borlaug noted, the tractor's 20-horsepower, four-cylinder engine "worked without a feed of oats morning, noon, and night." Its steel body did not need salves, rubdowns, or currycombs. On small farms like Henry Borlaug's, as much as half of the land was devoted to providing feed for the animals used to cultivate the rest. No longer needing draft animals, writes biographer Vietmeyer, Henry Borlaug sold most of his cattle and horses, planted what had been their pasture, and changed the oats to maize. The extra production meant extra money, which allowed him to buy more fertilizer and better seed, further increasing production. Ultimately, Henry's harvest quadrupled—on the same land. The extra money let him send his children to school without regret.

"Man minus the machine is a slave," proclaimed Henry Ford, touting

---

spoke the words attributed to him, he exerted so much editorial control that the book is as close to a Borlaug autobiography as we will ever see. I take occasional Borlaug "quotes" from it, believing that readers will understand their uncertain provenance.

Cresco, Iowa, 1908

his new tractor. "Man plus the machine is a free man." Decades afterward, looking back on the Model F, Borlaug agreed entirely. "Relief from endless drudgery," he said, "equated to emancipation from servitude."

Second Baseman

Compared to Saude, Cresco was enormous: more than three thousand inhabitants in the city proper, several hundred more commuting in from the countryside to work there. Borlaug reveled in Cresco's size, its frenetic activity, its sense of a wider world. Its elm-lined, stone-paved streets took him past tall churches, smoke-belching factories, and crowded stockyards. Walking about in a daze, the boy encountered example after example of amazing urban exotica. A hospital. A courthouse. An opera house. Several banks. A huge house owned by an actual millionaire, the president of one of the banks. And the high school! Built during the frenzied years when Cresco was competing for county seat, it was a three-story, vaguely Romanesque monster with a heavy stone foundation and a tall hipped roof. Its ninth-grade class, eighty-eight strong, was five times bigger than the entire student body at Saude.

Borlaug continued to be a dutiful student, but spent every spare moment on athletics. Despite his thin frame—five foot ten, 140 pounds—he played football every fall, and as a senior became team captain. In the winter, he wrestled, though for his first two years skin boils

prevented him from entering most competitions. But his greatest love was baseball. In a development almost as momentous as the purchase of the tractor, Grandfather Nels had acquired a radio, powered by a small windmill he installed on the roof. A Chicago station, WGN, broadcast Chicago Cubs games. Listening during summer vacation, Borlaug conceived an ambition. "Second baseman for the Chicago Cubs," he said. "That was my objective." Alas, Cresco had no high school baseball team—bats, balls, and gloves were too expensive.

Borlaug had a penchant for hastily deciding on some goal, heedless of its plausibility, then working relentlessly to achieve it. On impulse, he decided to organize his own baseball league, pitting kids from one local community against the next, Norwegians from Saude (Borlaug, at second base, was captain) against Bohemians from Spillville. Although the teams played in cow pastures and used burlap sacks for bases, the games quickly attracted spectators. By the time Borlaug was a senior, the Saude-Spillville baseball rivalry had become part of Spillville's annual Fourth of July celebration.

In Borlaug's junior year, a new principal came to Cresco. Husky and intense, David C. Bartelma had been an alternate on the 1924 U.S. Olympic wrestling team. He took over the Cresco wrestling program, as one would expect. A coach of the passionate, high-volume variety, he constantly exhorted his team to "give the best that God gave you. If you won't do that, don't bother to compete." Inspired, Cresco went 8–0 in Borlaug's senior year; Borlaug won third place in the state meet. Despite the success, Borlaug was coming to realize that he would not be able to play for the Cubs. Instead he decided to become a teacher-coach like Bartelma. Bartelma had attended Iowa State Teachers College, in Cedar Falls. After graduating in May 1932, Norm boy decided to go there, too. He worked odd jobs for a year to save up the money.

About a week before Borlaug was to leave for college, a strange automobile pulled into Henry and Clara's driveway. George Champlin Jr. was at the wheel. A recent Cresco graduate, Champlin had been a halfback on the best football team in the school's history, captain of the basketball team, and editor of the school newspaper. Now he was a star running back at the University of Minnesota. In Saude terms, it was as if the pope had suddenly dropped by.

Champlin had been charged by his football coach to look for players for the freshman squad. Having heard about Borlaug from Bartelma, Champlin suggested that Borlaug think about the University of Minnesota. "Ride up with me tomorrow morning," he proposed, in Borlaug's recollection. "It won't cost you anything."

"What's the purpose? I'm going to Iowa State Teachers a week from Friday."

"I can get you a job for your board and room." If Borlaug didn't like it, Champlin said, he could "hitch-hike back and go to Iowa State Teachers."

Impetuous again, Borlaug said yes. A dormant hope had rekindled in his heart. The University of Minnesota, unlike Iowa State Teachers, had a strong baseball team. If he could win a spot on the team, he might go on to the major leagues. He threw a few clothes in a bag and left with Champlin the next morning.

In Minneapolis, he stayed in a tiny boardinghouse room with Champlin and two other Cresconians. Champlin helped him win a job as a waiter at a diner. Borlaug's salary: one free meal per hour worked. Champlin also found Borlaug a second job parking cars. The work was unpaid but he was allowed to keep tips. Combined with Borlaug's savings, the food from one job and the money from the other would provide just enough for him to eat, rent a room, and pay tuition for the first year. Excited, Borlaug wrote to Iowa State Teachers College to say he would not be going to Cedar Falls.

Borlaug couldn't attend class, though, until he passed an admissions test at the end of the month. To fill the hours, he spent his days walking around the city. Minneapolis was to Cresco as Cresco was to Saude. Three-quarters of a million people! The sheer scale of the city—two cities, really, the conjoined metropolis of Minneapolis–St. Paul—was shocking. More shocking still was seeing the effects of the Great Depression. Saude had been insulated from the disaster because most of its people were subsistence farmers with little connection to the cash economy. But Minneapolis in the fall of 1933 was deep into it. The streets were lined with abandoned, ransacked buildings. On the sidewalks were homeless people wrapped in blankets. Some who could not afford blankets wrapped newspapers around their bodies. Many of these people, Borlaug learned, were dairy farmers who had lost their land and animals.

The crisis in cattle country had been building for a long time. During the First World War Washington had asked farmers to produce as much milk as possible for the troops and paid high prices for it. Duly incentivized, many farmers had increased their herds and invested in new tractors and milking machines; new safety regulations meanwhile forced them to buy pasteurization systems. Production rose, as did debt. After the war, milk prices fell, but the debt remained. Then came the Depression, and prices dropped again. Milk producers were selling every gallon at a loss. Foreclosures were pushing farmers off their land from Ohio to Nebraska; a swarm of dispossessed families had ended up in Minneapolis.

Eastern Wisconsin was consumed by a milk strike in May 1933. Strikers overturned milk trucks, beating "scabs" who tried to sell milk. Police and National Guard troops rode on milk convoys, using clubs, rifles, and tear-gas grenades to beat through farmer blockades. The federal government began controlling farm prices nationally but left the dairy and meat industries out of its plans. Instead smaller, regional organizations—cartels in effect if not name—were supposed to set price floors to protect dairy and meat producers. The effort failed; prices continued to drop, unrest to rise.

Fighting over milk erupted in Chicago in mid-September 1933. Scattered violence occurred as far away as Minneapolis, where it was witnessed by nineteen-year-old Norman Borlaug. Walking through a zone of shuttered factories, he saw a throng of gaunt, ragged people encircling a line of milk trucks, blocking their progress. The trucks were guarded by men toting baseball bats. Protesters were berating them. Not all of the shouting men were farmers, Borlaug realized. Some of them were just hungry—famished men, women, and children, almost maddened by want. "Suddenly, a cameraman tried to get up [on a car] to get a better picture with his tripod and his foot went through the canvas top on the car and then all hell broke out," he remembered. The guards "beat him up and busted his camera, and that triggered it."

As if the violence were a signal, the guards rushed the protesters, bringing their clubs down in a coordinated attack. Cries of pain rose as bloodied men collapsed. Others grabbed at the milk canisters in the trucks, pulling them to the ground, splashing milk on the cobblestones. Borlaug was terrified. Abruptly the milk trucks lurched forward, into the mêlée; people fell back shouting, a panicky shuffle that pinned Borlaug against a factory wall. He couldn't see but he could hear truck engines heaving as they drove through the gathering. The wailing was like nothing he had ever heard. When the crush eased, he ran shaking through the fight to his boardinghouse. The wounded were lying untended on the ground.

Something must be done, he thought. Those famished people were ready to tear apart the world, and who could blame them? Here began, or so he said afterward, the work that would make him the original Wizard. Everything commenced with the terrible fathomless hunger he saw explode in the street.

"I Just Liked the Outdoors"

After the riot, his life went awry, one bit after another. Borlaug showed up at the football tryouts and instantly realized that he was too small to make the team. He tried out for wrestling—and learned that the University of Minnesota had only an inexpert part-time coach who held practice for an hour or two each week. And then Borlaug flunked the university entrance exam.

Crestfallen, Borlaug was preparing to hitchhike home when Champlin pulled him into the university admissions office and asked officials there if they had some program for his friend. Lucky for Borlaug, the university had just launched a junior college for deprived and underprepared youths, and it was desperate for students. Borlaug glumly signed up for what he viewed as a "place for misfits." Determined to lever himself into the main university, he worked hard at his courses and did well enough to be allowed to transfer. When accepted, he was told to pick a major. Borlaug chose forestry, because, he later admitted, he "just liked the outdoors."

The University of Minnesota forestry school, founded in 1904, was one of the nation's oldest. Minnesota's big forest-products industry wanted the school to supply it with trained employees. In consequence, the curriculum focused almost exclusively on timber management. Students learned how to use a surveyor's transit; construct logging roads; plant and thin tree plantations; and grade wood products. They were taught to view forests not as wild ecosystems but as slow-growing farms: organic factories for wood. Trees were a crop, one species per tract, grown for harvest. A few hours away in Wisconsin, Aldo Leopold was delivering lectures on the conservation ethic and teaching land managers about ecosystems. Meanwhile, the Minnesota forestry department did not offer a single class in conservation or ecology. Unthinkable to a Leopold disciple, there was no course on *soil*. Holistic perspectives were not in evidence.

Nor was Borlaug likely to acquire one on his own; he was too busy scrambling. To earn money for tuition and rent, he worked forty or more hours a week, hopscotching between janitorial duties in university laboratories, serving meals at a sorority house (the diner had folded), and working for tips at the parking lot. So much time was devoted to scraping up money that Borlaug was forced to quit the university baseball team in his freshman year. Turning in his uniform "was one of the most difficult decisions I've ever made," he told Vietmeyer, his biographer. He remained on the college wrestling team, which demanded less time. Happily for Borlaug, he was able to persuade the university to hire a new coach—Dave Bartelma, his coach at Cresco. With Bartelma, the team improved radically; Borlaug made it to the Big Ten semifinals (he was elected to the National Wrestling Hall of Fame in 2002).

Despite all the effort, however, Borlaug ran out of funds. But the demand for foresters meant that he "could drop out of school and make enough money so I could go back and live a little better." Overall, he spent a year and a half on forest projects, often spending weeks alone in the woods. After he spent the summer of 1937 fighting forest fires in Idaho, the U.S. Forest Service offered Borlaug a job there following his graduation in December. He accepted immediately. Forestry school had

Borlaug as a college wrestler

paid off; as of January 15, 1938, he would have a steady job. He was not going to be forced to return to Saude and farming. Brimming with happiness, he returned to Minneapolis in a borrowed car. At last he would be financially stable. At last he could get married.

Borlaug had met his fiancée within weeks of leaving Saude. Margaret Grace Gibson was another waiter at the diner. Dark-haired and direct, she had the pale, lightly freckled complexion of her Scottish ancestors. Her father, Thomas Randall Gibson, was born in Glasgow in 1865, immigrated as a toddler to upstate New York, and married a girl from another Scottish immigrant family, thirty-one-year-old Isabella Skene, in 1903. At the time, homesteaders were pouring into the new state of Oklahoma, drawn by its nascent oil industry and unoccupied real estate (much of the state was created by breaking up tribal reservations). In 1910 the Gibsons moved to Medford, Oklahoma, a fast-growing town on former Cherokee land near the Kansas border. A year after their arrival most of Medford burned to the ground. Margaret, the sixth and youngest child, was born six weeks after the fire, on August 30, 1911.

From their first conversation, Margaret was intrigued by the quiet, wiry Borlaug and amused by his shyness around women. Margaret was his first girlfriend. Like him, she was from a poor family and had trouble paying for school. Borlaug smuggled food to her from the sorority when he could, but she often went to bed hungry. For her part, she was frightened when Borlaug's coat was stolen and he had to wear a thin jacket through the Minnesota winter. Six months shy of graduating, Margaret quit school; her older brother Bill, who edited the Minnesota alumni magazine, found her a job as a proofreader. Someday, she hoped, she would complete her degree.

The couple decided to put off marriage until they were financially stable. With the Forest Service job promising that stability, Borlaug proposed within hours of returning from Idaho. They set the date for the following Friday: September 24, 1937. Bill Gibson lent his sitting room for the ceremony; Borlaug's sisters came up by train. There was no money for a honeymoon; Norm and Margaret told themselves that they would soon be living among the stunning vistas of the Rocky Mountains, a honeymoon in itself. On the wedding night Norm moved into his wife's apartment, a one-room flat with a daybed and a shared bathroom in the hall.

Borlaug's graduation three months later should have been a capstone. Instead it coincided with the receipt of a letter from the Forest Service. Budget cuts had eliminated his job. If Borlaug wanted to reapply, he might get hired in the summer. It was precisely the situation Borlaug had tried to avoid: married, with no obvious means of support. Margaret told him that she could support them both on her salary. While waiting for the Forest Service, she suggested, Borlaug could take a semester of graduate school classes.

Maybe, she said, he could work with that man Stakman.

"Execute This Criminal Bush"

In one of Borlaug's senior-year seminars the teacher had passed out samples of fungus-infected wood and asked the class to examine them. As the students were leaning over their microscopes, a middle-aged man barged into the room, sucking on a pipe. Without a word of introduction or explanation, as Borlaug later remembered it, the man began quizzing students,

> not [about] the wood fungus we were looking at, but the damn structure of the wood and what species of wood was it and what's this and what's the other thing and why and started giving us a preliminary Ph.D. exam. . . . He first started working on the guy next to me, and he had him all confused, and then he came over and started in on me. I never did know who the man was. I suspected, though. After he had left and thrown the whole place into confusion by his barbing and prodding, I suspected that this was Dr. Stakman.

Borlaug's suspicion was correct: the man with the pipe was Elvin Charles Stakman, who would become a friend and lasting influence. Like Borlaug, Stakman (pronounced STAKE-man) had been raised poor in the hinterlands of the Middle West and got his degree at the University of Minnesota. He had become one of the first professors in the university's newly established department of plant pathology (the study of plant diseases). By the time Borlaug encountered Stakman, he was a campus legend. Charismatic, ambitious, rarely modest, Stakman did not view science as a disinterested quest for knowledge. It was a tool—maybe *the* tool—for human betterment. Not all sciences were equally valuable, as he liked to explain. "Botany," he said, "is the most important of all sciences, and plant pathology is one of its most essential branches."

After being grilled by Stakman in the lab, Borlaug attended one of his lectures. The subject was Stakman's special passion, the black stem-rust fungus, a parasite that attacks wheat. Stem rust is little known today outside agriculture, but it was long one of humankind's worst afflictions, responsible for millennia of famine. Borlaug knew it well; a stem-rust outbreak had driven his grandparents out of the wheat business in 1878. Epidemics in 1904 and 1916 had led to misery throughout the Middle West and northern Europe. Stakman had been fighting the fungus for more than two decades. Years later, Borlaug would join his anti-rust crusade—and find himself in the blasted field outside Mexico City.

Stem rust was long so pervasive and unstoppable that the Romans viewed it as a malign deity, and sacrificed rust-colored dogs to appease it. Only centuries later did scientists learn that stem rust is actually a fungus, not a supernatural force. The term "fungus" makes rust sound simple. In fact, stem rust is a wildly complex creature, a triumph of evolutionary guile. "All *five* types of spore reproduction!" Lynn Margulis once told me, her eyes agleam. "What's not to like?" In her opinion, the fungus was vastly more interesting than the wheat it infects.

Rust-infected wheat plants are stippled like smallpox victims with countless small, rust-colored pustules, each packed with thousands of spores. The spores, a thousandth of an inch long, cannot be seen by the naked eye. The slightest wind carries them in a thin mist high into the atmosphere. They saturate the air in unbelievable profusion and can spread for miles. In banner years, write the environmental historians Garnet Carefoot and Edgar Sprott, "more rust spores are produced in the world than there are blades of grass or grains of sand in all the world's beaches or stars in all the galaxies of the universe." Carefoot and Sprott were exaggerating, but less than one might expect; a single acre of "moderately rusted wheat," Stakman once estimated, can produce 50 *trillion* spores.

Stem rust has a complicated path through life that involves no fewer than four developmental stages. The most consequential for humankind is the stage in which it attacks wheat. But the most important for the fungus is the stage in which it feeds on an entirely different plant: European barberry. A spiny, shoulder-high shrub, barberry has small red berries that make a tart jam. European migrants carried the plant across the Atlantic, and spread it through most of the United States and Canada. On wheat, stem rust produces spores asexually; the spores grow into fungi that are genetically identical to their parent. On barberry, stem rust produces two different types of spores—"male" and "female," so to speak—that combine sexually to produce spores that are not genetically identical to either parent. Some of Stakman's earliest research showed how this sexual-style reproduction, which rapidly creates new combinations of gene variants, allows stem rust to exist in scores of different strains, each with its own abilities.

In colder parts of North America and Europe, the fungus cannot survive winter in the wheat fields. It has two means of returning. The first occurs in places like Mexico or North Africa, where the climate is never cold enough to kill the fungus. Wind perpetually blows spores from these warm reservoirs of infection into colder regions in the north. In late spring, when temperatures rise, the spores can survive this journey and afflict young wheat plants. *Puccinia graminis* is the scientific name for black stem rust. Its south-to-north pathways are known as "Puccinia highways." The second mode of attack occurs within colder areas themselves. In early winter, before the weather turns deeply cold, the fungus creates a type of spore that afflicts barberry. These can survive throughout the winter. In early spring the barberry-type spores germinate, spread through barberry leaves and stems, and erupt through the surface to produce yet another type of spore, this one capable of infecting wheat. Every year, both types of spore—those blown up the Puccinia highway, and those spread from barberry—assault northern farms in a double attack.

In 1916 stem rust ravaged North America, wiping out almost a third of the harvest. Bread prices shot up in response. A year later, the United States entered the First World War. Fearing food shortages, Washington quickly moved to bolster farm production. Plow more land! Harvest more wheat! it exhorted U.S. farmers. Posters appeared: FOOD WILL WIN THE WAR. Stakman became head of the War Emergency Board for Plant Pathology. He used his new pulpit to broadcast a message: *Puccinia graminis* was the greatest threat to U.S. grain, and the best way to fight it would be to wipe out barberry.

As U.S. troops went overseas Stakman persuaded Washington to let him run a nationwide campaign of barberry eradication—a pioneering act of large-scale environmental management. The goal was to exterminate the plant from Colorado to Virginia, from Missouri to North Dakota. Hundreds of thousands of handouts, posters, and leaflets described barberry as an "outlaw," a "menace," a "dangerous enemy alien." Brochures exhorted farmers to "execute this criminal bush wherever it is." Barberry, government press releases claimed, is "pro-German." Boy Scouts, church groups, Future Farmers of America, federal agents, elementary school classes—all were instructed to search for, dig out, and poison barberry bushes. "Rustbuster" clubs gave medals to children who informed on their bush-owning neighbors. Stem-rust vigilantes tore out barberry with tractors, then poured salt or kerosene into the hole to kill remaining roots. The Great Barberry Zap destroyed more than 18 million bushes in seventeen states in twelve years.

Removing barberry eliminated the fungus's sexual reproduction, which slowed the pace at which it could evolve new strains. In previous decades plant breeders had developed rust-resistant varieties of wheat, only to discover that *P. graminis* had adapted to them within a year or two and was as potent as ever. Expunging barberry threw sand in the fungus's evolutionary gears. As the barberry campaign crested, Stakman led a team of researchers that developed a new type of rust-resistant wheat. Thatcher, as it was called, made its debut in 1934. Without barberry, *P. graminis* wasn't able to overcome it for almost thirty years.

To Stakman, the anti-barberry campaign showed the power of science to improve human lives. It was a lesson for the future—one that paid off in 1943, when he was asked to develop scientific agriculture in Mexico. The request originated with U.S. vice president Henry Wallace. The Iowa-raised son of a secretary of agriculture, Wallace began breed-

Stakman examining wheat seedlings in his lab in about 1918

ing experiments as a child, discovering for himself the phenomenon of "hybrid vigor"—that some hybrid organisms can outperform their parents by mixing their genetic inheritances. A polymath and eccentric, Wallace also edited the family newspaper, studied Christian mysticism, made contributions to statistics and economics, and tested fad diets on himself. In college he spent weeks eating only a puree of soybeans, rutabaga, maize, and butter, making himself so sick that he had to leave school to recuperate; at other times he consumed nothing but oranges, or milk, or an experimental cattle feed.

Government scientists had bred the first successful hybrid maize in 1918. Wallace studied their results, developed them further, and in 1926 founded the company that would become Pioneer Hi-Bred, a major vendor of hybrid maize. Despite Wallace's idiosyncrasies, President Franklin Delano Roosevelt appointed him in 1933 to be secretary of agriculture, the post once held by his father. Seven years later Roosevelt selected Wallace to be his vice president. After Roosevelt's re-election in November 1940 Wallace went to Mexico, driving his own car to set a humble tone. The visit was dangerous; the country was locked in conflict between left and right. Fascists rioted in front of the U.S. embassy and attacked the inauguration motorcade; left-wing death threats accompanied Wallace's every move. Still, he hoped that the visit could soothe the strained relations between the United States and Mexico.

After the ceremony, Wallace and incoming Mexican agriculture secretary Marte Gómez spent three weeks inspecting rural villages, Wallace insisting on speaking to farmers in his fascinatingly imperfect Spanish. He was appalled to see them planting maize with pointed sticks, weeding by hand, and carrying the harvest to market on their backs. To Wallace, who deeply believed in the Christian message of compassion, it was obvious that something should be done. Conversations with Gómez made clear that direct assistance from Washington to Mexico would be seen as Yankee meddling and was therefore politically unpalatable. Indirect assistance was another matter. A month after his return to Washington, the vice president summoned the head of the Rockefeller Foundation to a meeting.

Created in 1913 by Standard Oil owner John D. Rockefeller Sr. and his son, John D. Rockefeller Jr., the foundation had an initial endowment of $100 million, an unheard-of sum at a time when the annual federal budget was less than a billion dollars. An early Rockefeller initiative had been to establish the General Education Board (GEB), which disseminated better farming techniques in the U.S. South—methods to prevent the spread of boll weevils and other cotton pests, for example. So successful were GEB programs that in 1914 Congress used them as a model for creating a national network of extension agents: technicians who transmitted the latest agricultural research to local farmers. (The network still exists and is a critical component of the U.S. agricultural system.) In the 1930s the longtime U.S. ambassador to Mexico had begged Rockefeller to replicate the GEB in Mexico. Foundation leaders resisted, fearing Mexico's long history of antipathy toward U.S. interference. Now Wallace pressed it to reconsider. If maize harvests could be raised, Wallace said, "it would have a greater effect on the national life of Mexico than anything [else] that could be done." Rockefeller, a private entity, could work quietly, avoiding politics.

Anxious foundation officials sought advice from experts, among them Carl O. Sauer, a Berkeley geographer who had studied Latin America for decades. Sauer told the foundation that the possibilities were "enormous"—but so were the risks.

Five to ten thousand years ago, indigenous geniuses in south-central Mexico developed the first maize from a much smaller wild plant, a grass called teosinte. Since that time Indian farmers had bred thousands of varieties of maize, each chosen for its taste, texture, color, and suitability for a particular climate and soil type. Red, blue, yellow, orange, black, pink, purple, creamy-white, and multicolored—the jumble of colors of Mexican maize reflects the nation's jumble of cultures and environmental zones. The small, varied plots in Mexico were like the anti-matter version of the huge, uniform maize fields in the U.S. Midwest.

Maize is open-pollinated—it scatters pollen far and wide. (Wheat and rice plants, by contrast, typically pollinate themselves.) Because wind often blows pollen from one small Mexican maize field onto another, varieties are constantly mixing. Over time, uncontrolled open pollination would create a few, relatively homogeneous populations of maize. But the pollination is not uncontrolled, because Mexican farmers carefully select the seed to sow in the next season, and generally do not choose obvious hybrids. Thus there is both a steady flow of genes among maize varieties and a force counteracting that flow. This roughly balanced genetic sea, maintained by farmers' individual choices, is a resource not only for Mexico but the entire world; it is the genetic endowment of one of Earth's most important foodstuffs.

Helping the farmers who maintain this resource, Sauer agreed, would be a good thing—but not if it destroyed their way of life and reduced the diversity of maize. "A good aggressive bunch of American agronomists and plant breeders could ruin the native resources for good and all by pushing their American commercial stock," he growled. He added, "Unless the Americans understand that, they'd better keep out of this country entirely."

Sauer's warnings thundering in the air, the foundation dispatched three scientists to Mexico: Paul C. Mangelsdorf, a Harvard plant geneticist who specialized in Latin American maize; Richard Bradfield, a Cornell soil expert; and Elvin C. Stakman, who had a long-standing interest in Mexico, the continental reservoir of stem rust. The three men spent six weeks in the summer of 1941 inspecting maize fields from the Rio Grande to the Guatemalan border. What they saw was a human catastrophe: the abridgment of hope on a massive scale. "The great majority of Mexican people are poorly fed, poorly clad, and poorly housed," they wrote afterward. "The general standard of living of the Mexican people is pitifully low." And things were getting worse. In 1940 the country harvested a third less maize than it had in 1920, even though it planted almost a million more acres of the crop. Meanwhile, the population had risen by more than 5 million.

The Rockefeller Foundation was in a position to provide assistance, Bradfield, Mangelsdorf, and Stakman said. Helping was the good—the *decent*—thing to do. It could station a small group of researchers in the country to offer "a little judicious advice" to Mexico's newly established corps of agricultural researchers. These could then pass on the results to farmers, mirroring the U.S. extension system.

The scientists had taken a route through Mexico that was much like the route taken by Bill and Marjorie Vogt two years later. Both groups wrote reports documenting the same terrible poverty and eroded land, but their ideas about the remedy were starkly different. To Vogt, the basic problem was land degradation, and the primary cure was to ease the burden on the land. By contrast, the scientists believed that Mexico's issues were caused, at bottom, by lack of knowledge and tools. The difference between these two approaches is profound, and at the heart of the split between Wizards and Prophets.

At the same time, the two reports had a striking similarity: neither attempted to understand how Mexican farmers had got into these straits. When Mexico won its independence in 1821, most of the citizens of the new country were landless peasants who worked on giant estates in conditions little different from slavery. Over time the situation worsened: under the dictator Porfirio Díaz, who controlled Mexico from 1876 to 1910, wealth and land were concentrated among a few hundred aristocratic families, foreign companies, and the Catholic Church. A bloody civil war led in 1917 to a new constitution that promised to redistribute land. Early efforts to fulfill this promise set off such violent resistance by the rich and the Church that the government pulled back. In 1934 a new president, Lázaro Cárdenas, tried again. The Cárdenas administration seized almost 50 million acres from estates and awarded them to *ejidos*, peasant-run collectives. (About 4 million acres of this land was owned by U.S. companies, which led to diplomatic squabbles between Mexico City and Washington, D.C.) As before, landholders fought back, some plotting coups and assassination attempts. Others ensured that the *ejidos* were forced to accept bad land—plots that were too dry or steep to cultivate. By 1940 the eleven thousand new *ejidos* were working almost 2.5 million acres of land that had been left alone ten years before. Unsurprisingly, the consequences were often destructive; erosion and soil depletion soared. Much of the devastation that Vogt saw as the unavoidable consequence of high birth rates was tied to political events that were anything but inevitable; much of the poverty that Stakman, Mangelsdorf, and Bradfield saw as lack of access to knowledge was the result of efforts by wealthy elites to maintain their position.

A few months after the scientists gave their report to Rockefeller, Japan bombed Pearl Harbor. As the United States mobilized, the advantages of having a calm and prosperous southern neighbor suddenly seemed large. The prospect of helping the war effort nudged the foundation to agree to step into waters it had once avoided. Stakman was asked if he would head the effort. He demurred; he had been swept up by the War Emergency Board of the American Phytopathological Society, an ad hoc group that the military had asked to prevent fungi, molds, and mildew from destroying equipment in the Pacific war theater. Instead Stakman recommended one of his students, J. George Harrar.

Slight in build, fluent in Spanish, both combative and charming, "Dutch" Harrar was an odd choice: he was a city kid who had never worked on a farm, and had never studied maize (like Stakman, he was a wheat-disease specialist). Nonetheless, the choice was successful; Harrar's affable relentlessness proved to be the right temperament for bilingual negotiation among Mexican bureaucrats, U.S. researchers, and indigenous farmers. Ultimately, Harrar's Mexican term was so successful that he was promoted to foundation president in 1961. But in the beginning he spent almost a year extracting the necessary permits from the two governments.

Expectations for the project were multiple and overlapping. Mexican officials wanted the program to help modernize the nation, but worried about its political impact—they could not be seen as allowing the United States to control a vital economic sector. To contain these anxieties, Harrar agreed that the project would work only in the Bajío, the highlands northwest of Mexico City. The Bajío had been the nation's agricultural heartland in colonial times but was now wretchedly poor and unproductive. In the United States, Rockefeller Foundation leaders, project scientists, and some politicians—Wallace, especially—hoped that the agriculture project would alleviate hunger. But most U.S. officials viewed it as a mechanism to help stabilize Mexico politically, suppressing the threat of unrest across the border. And scientists, foundation officials, and politicians alike hoped it would rebut Communist claims that poor countries' poverty and hunger were caused by Western capitalism. "What now are the great enemies of mankind?" Stakman, Bradfield, and Mangelsdorf asked as the program was under way.

> Hunger, the incapacity of the hungry, the resulting general want, the pressures of expanding and demanding population, and the reckless instability of people who have nothing to lose and perhaps something to gain by embracing new political ideologies designed not to create individual freedom but to destroy it. . . . Whether additional millions in Asia and elsewhere will become Communists will depend partly on whether the Communist world or the free world fulfills its promises. Hungry people are lured by promises, but may be won by deeds. Communism makes attractive promises to underfed peoples; democracy must not only promise as much, but must deliver more.

The project focused on improving the production of maize. But Stakman had wangled permission to devote a small part of its resources to a subsidiary goal: attacking the stem-rust reservoir south of the U.S. border. If Stakman could breed resistant wheat varieties in Mexico, the fungus would lose its base of support. No longer would it travel the Puccinia highway to the United States. Harrar at its head, the unimaginatively named Mexican Agricultural Program launched in February 1943.

Everything went wrong. Mindful of Sauer's strictures, the small Rockefeller team—a handful of researchers and technicians—decided to breed improved versions of local maize varieties that Bajío farmers could plant as they always had. Once these were developed, they would coordinate with Mexico's own agricultural researchers to disseminate them. But Mexican scientists had a different goal: modernization. They wanted their own hyper-productive hybrids, their own mechanized, high-producing farms. The U.S. Midwest was the model: uniform fields of optimized hybrid maize. Mexican scientists were determined to do away with the chaotic jumble of "contaminated" maize in the hinterlands.

To these researchers, the Rockefeller program's plans seemed absurd, even insulting—the *norteamericanos* were trying to saddle Mexico with the sort of second-rate methods that the United States had already abandoned. The small-scale Mexican farmers that Sauer exalted could better be employed in factories that made goods for middle-class consumers. Forget scratching at the dirt with sticks! We want modern science! For their part, the U.S. scientists thought that the Mexicans were chasing a chimera. Even if the nation's poor farmers somehow managed to buy and plant hybrid maize, Mexico didn't have markets organized well enough for them to sell it. "In practice," the historian Karin Matchett has written, "the attempted collaboration irritated all parties involved."

With little cooperation from Mexican officials, the maize researchers had no way to distribute their advances to rural farmers, a critical component of their plan. Under pressure to show results, the scientists eventually gave up on the rural poor and targeted the big commercial farms that Sauer had warned against. Ten years after the program began, it was clear that it had failed to accomplish its original goals. But few at Rockefeller were upset. To general astonishment, Stakman's stem-rust program, almost an afterthought, had become a world-altering success. Stakman was as surprised as anyone else. Having commitments in Minnesota, he had delegated responsibility for stem rust to Harrar, who in turn had given it to local staff. That staff consisted of a single person: Norman Borlaug.

In the Bajío

Borlaug's initial encounter with Stakman had not gone well. Following Margaret's suggestion, he asked Stakman if he could study forest pathology for a few months while waiting for the Idaho job to materialize. "Fill in a month or two—between jobs?" Stakman replied, in Borlaug's recollection. Graduate school, he told Borlaug, is "not like a novel you can pick up and put down. You'll have to be a bit more serious about it than that, my boy." And he refused to countenance forest pathology as a subject; Borlaug shouldn't lock himself into a single discipline. If Borlaug took general crop science, Stakman could offer an assistantship to defray costs. The position: counting stem-rust spores on glass slides. Borlaug accepted his terms. He ended up spending so many hours squinting at spores that he permanently damaged the vision in his right eye.

The Forest Service job in Idaho never came through. Borlaug researched a fungal disease in box elder and obtained a master's degree in 1941. Instead of looking for a job after completing his master's degree, Borlaug's original plan, he found himself working toward a Ph.D. For his dissertation, he investigated a soil fungus that attacks flax. The subject had a threefold appeal: first, Stakman had grant money to support the research; second, the fungus was a cousin of the box-elder fungus he had already worked with successfully; third, the subject had nothing to do with wheat or stem rust. So many of Stakman's students were studying both that Borlaug was certain that he would get lost in the crowd. Stakman rolled his eyes but accepted Borlaug's decision; beneath the brusque, cigar-chomping exterior, Borlaug was discovering, was an exceptionally kind man. Stakman pushed his students hard—but picked them up if they fell. Think big, he said. And he made sure that Borlaug, despite his misgivings, learned something about wheat.

In the fields next to the departmental building Stakman maintained 40 acres of diseased wheat plants, and made us pit our wits against smut, scab, rust, blast, blight, bunt, wilt, mildew and the rest. Every Saturday afternoon the students and faculty trailed through those acres of sick and dying plants. Stopping at each one, Stakman would stimulate arguments over what we were seeing. The sessions sometimes went for hours; and sometimes got very heated. That was his style. His laboratories and classrooms were open intellectual forums full of fire and light.

There was a limit to the openness, though. As an undergraduate, Borlaug had taken the usual smattering of general-education courses: English literature, practical psychology, even a course called "How to Study." As a graduate student, that changed. Except for auditing a semester of beginning French, Borlaug did not take a single course outside of botany and plant pathology. He learned no ecology, no agronomy, no soil science, no hydrology, no geography, no agricultural economics or history. Strikingly, Borlaug took no class in plant breeding, even though Herbert K. Hayes, perhaps the nation's most eminent plant breeder, taught at Minnesota. Borlaug, like Vogt, became a science guy. But the ecology Vogt focused on was vastly different from Borlaug's plant pathology.

As practiced by Leopold and his followers, ecology had a mission: to protect the integrity of the ecosystem by the holistic study of the network of interactions among species. Plant pathology had a completely different mission: removing pests and diseases that impeded human needs. Vogt's ecology was an exercise in humility and limits; Borlaug's plant pathology was a methodology of *extension*. Isolate the subject of study, perform the experiment over and over, then push the result as far as possible—this, Stakman told him, was the path to knowledge that could benefit people.

Following these principles, Borlaug accumulated more than a thousand samples of fungus-infected flax. From these he obtained pure cultures, which he injected into four-inch pots full of sterile soil. In a dozen experiments reported in his Ph.D. thesis, he planted as many as twenty varieties of flax in the inoculated soil and looked for those that resisted the fungus. Alas, not one flax variety was immune. He was forced to stop while the work was still incomplete—he had taken a job.

In October 1941, Stakman asked him to come by. Waiting in Stakman's office was one of Borlaug's undergraduate professors. The man had left Minnesota for several years to run a laboratory at E. I. du Pont de Nemours, a huge chemical company in Wilmington, Delaware. Now returning to Minnesota, he had a question for Borlaug: What would he think about replacing him at DuPont? The pay was $2,800 a year, much more than Margaret's proofreading salary. Stakman told Borlaug that he could finish his research in a month, move to Delaware, then write his thesis by night.

Borlaug was uncertain. DuPont, founded in 1802, had begun by making explosives for the military; recently it had developed synthetic fabrics like nylon, rayon, and orlon. What could an agricultural scientist contribute? In addition, Borlaug still hoped to live in the forests of the Rocky Mountain West. But when he told Margaret about the offer, her reaction was blunt: What alternative do we have? Other firms were not clamoring to hire him. On December 1, 1941, Norm and Margaret left Minneapolis in their only major possession, a 1935 Pontiac sedan they had bought from her parents.

Margaret and Norman Borlaug at about the time of their wedding

In those days the drive to Wilmington, much of it on rutted country roads, lasted almost a week. Passing through Philadelphia, Norm and Margaret saw agitated crowds on the street. Borlaug asked a passerby what was happening. "Pearl Harbor's been bombed!" the man said. Borlaug had never heard of Pearl Harbor. Mystified, he drove on to Delaware. The next day, December 8, was his first day at DuPont. Only then did he learn that the nation was at war.

Like Vogt, Borlaug wanted to serve his country. But his attempt to enlist in the army was rejected—the military wasn't interested in married twenty-seven-year-olds with bad vision. In any case, he was soon classified by the War Manpower Commission as "essential to the war effort." He had to ensure that DuPont's bacterial cultures were protected from Nazi saboteurs. Borlaug had become, in effect, an employee of the U.S. military.

After Borlaug safeguarded the company's petri dishes, he was presented with a long list of tasks. He tested the durability of camouflage paint. He examined the effect of water-purification chemicals on pathogens. He figured out what to spray on cardboard ration cartons to help them survive being dumped in the ocean. He built a "jungle room" of high heat and humidity to assess the rate at which mildew attacked military uniforms. He investigated protective packaging for electronic equipment. He invented a new method for sealing condom wrappers against mold.*

Borlaug was reticent about his feelings, except with family; he left no diary and few intimate letters. As with Vogt, one must infer his thoughts from scraps of evidence. He seems to have been discovering that testing paint, however useful to the war effort, had no ambition or grandeur. When he met Stakman and Harrar at a botany conference late in 1942, they told him about work in Mexico, a real long shot but interesting, something that might change the lives of millions. Was Borlaug interested in it? I can't leave my job, he told them. I'm Essential to the War Effort. Harrar thought there was regret in his voice: Borlaug was bored at DuPont.

Stakman kept talking to Borlaug about the Mexican Agricultural Program. The project was having trouble finding staff. Stakman and Harrar had interviewed scientist after scientist, and all were too old, or too hard to work with, or too likely to annoy the Mexicans. Borlaug, by contrast, was young, ready for adventure, and altogether genial. Except for his lack of expertise in the subject and nonexistent professional reputation, he was perfect. As the other candidates dropped out, Borlaug's virtues shone brighter. In June 1943 Harrar asked Borlaug again if he would take charge of stem rust. After consulting his wife, Borlaug agreed. If the Rockefeller Foundation hadn't asked him to join the Mexican Agri-

---

* Borlaug was an early tester of the insecticide DDT. According to Vietmeyer, Borlaug said that samples of DDT came to DuPont in 1942 from ICI, the English chemical giant, which had obtained them from the Soviets, who in turn had taken them from captured German soldiers. The Russians noticed that these POWs were not crawling with lice and found that they were carrying a bug-killing powder. The powder was DDT, developed in the 1930s by the Swiss dye firm Geigy and sold to the Nazis. Borlaug tested the powder on garden pests. For years, Borlaug told Vietmeyer, "I had the powder all over my hands and clothes and was as thoroughly exposed as anyone on earth. Yet I had no adverse health consequences then or since. Nor did I ever see evidence of environmental damage. . . . This is why I've always been skeptical of the claims of calamity that today surround DDT." Borlaug's story about the captured Germans may be wrong; Edmund P. Russell, a Boston University historian who has studied pesticides, told me that he had never heard of it. The standard history, supported by archival evidence, is that Geigy itself sent DDT samples to the U.S. government. At DuPont, Borlaug probably worked with the samples but may have misremembered their origin. If Vietmeyer's quotation is accurate, Borlaug's later skepticism about DDT's impact stemmed from this work. He was a smart man, but claiming that his personal experience demonstrated that DDT poses few risks is like claiming that Cousin Tillie smoked for fifty years and never got sick so therefore smoking does not cause lung cancer.

cultural Program, he told Harrar, he would have applied for a commission in the navy. Extracting himself from DuPont, arranging to hire a successor, signing a contract with Mexico, obtaining the relevant visas and wartime permissions, and establishing an office in Mexico City took Borlaug, Harrar, and the foundation more than a year—a test of Borlaug's patience. He left for Mexico on September 11, 1944. Margaret, heavily pregnant with their second child—their first child, Norma Jean (Jeanie), had been born a year before—remained in Wilmington, intending to join him after the birth.

Borlaug's first sight of the project was a shock. Two years' worth of planning and negotiation by the world's biggest charitable enterprise had established a small main office in a rundown northern suburb of Mexico City; a couple of rooms downtown, borrowed from the Rockefeller Foundation's far larger anti-malaria program; and 160 acres of scrubby, infertile land an hour east of town, at the Autonomous University of Chapingo. The main office held the Mexican Agricultural Program's four full-time U.S. employees: Borlaug; Harrar; Edwin Wellhausen, a highly regarded maize breeder; and William Colwell, a newly graduated soil scientist from Cornell. The downtown office was staffed by a single receptionist who controlled the project's most valuable physical asset: a telephone line capable of reaching the United States. The Chapingo testing ground not only didn't have a greenhouse or laboratory, it didn't have *fields*. One of Borlaug's first tasks was to lay out boundaries on the site for fields, roads, and (potential) irrigation lines.

In part, the foundation had posited that a small research group could have a big impact because it would need only to introduce Mexicans to superior U.S. methods. In the spring of 1944 Harrar had planted some of the most advanced U.S. hybrid maize, wheat, and beans on a plot at Chapingo. Borlaug saw the results in October, after his arrival. All three crops had been nearly wiped out by disease, insects, and unseasonable frost. A few wheat plants had survived, but they had produced almost no grain—for some reason, the northern varieties couldn't bear in southern conditions. "This was our first inkling that raising crops in Mexico might differ from anything we expected," Borlaug told Vietmeyer. "We'd assumed that our seeds would perform as they did back home. Suddenly it seemed we shouldn't be so sure of ourselves. This place was smarter than we thought." Harrar told Borlaug to drive further outside the city, find a farm with better soil, obtain permission to work there from the landowner, and try again with more wheat.

Borlaug did what Harrar asked, but his lack of preparation was coming home to him. "I was frightened," he said later. "I got sick, and I stayed sick for about three weeks or a month with the usual tourist thing, except that I seemed to get them twice as hard as anybody else. And I'm sure, many times during that first month, if I could have gotten my job back at DuPont, I would have left and returned to DuPont."

Adding to his unhappiness was news from Wilmington. On November 9, 1944, Margaret had given birth to a boy with spina bifida, a birth defect in which the spinal cord fails to close properly, remaining open to the air in an exposed bulge on the baby's lower back. In severe cases the flow of cerebrospinal fluid is blocked, to fatal effect. The baby had a severe case. (Today spina bifida can usually be treated.) Margaret had never been allowed to touch him; visits consisted of staring through glass at the unresponsive, intubated child. When Borlaug arrived in Wilmington, he told his wife he would quit the project and return to DuPont. She would have none of it. "My husband has a future," Margaret said, according to Vietmeyer. "My baby has none. You go back; I'll come when I can." Two days after Christmas, grieving and guilty, Borlaug returned to Mexico.

In February doctors urged Margaret to leave for the sake of her daughter. Clinging to Jeanie, she took the train south. Norm had found an apartment in the center of the city. Margaret found some relief in cleaning and arranging it. Jeanie had a bedroom of her own for the first time. And the whole family loved the sun that washed through the windows and the busy street life and the markets with their smells of chiles and churros and spicy Mexican chocolate. Still, the thought of their second child was a terrible, constant weight.

Work was a solace. In March 1945 Harrar told Rockefeller that he couldn't simultaneously direct the entire program and work with Borlaug to create rust-resistant wheat. Nor could Stakman pick up the slack. Despite his inexperience and poor Spanish, Borlaug would have to take charge of the wheat program—Harrar would help when possible.

In most cases, farmers in the Bajío planted their wheat in October or November, a few weeks before winter frosts. The seeds then sprouted and developed into four- or five-inch seedlings that remained dormant in cold weather. After winter, when temperatures rose, the plants resumed growth, flowered, and produced grain; they were ready for harvest by late spring or early summer. Wheat produced in this manner is called *winter* wheat. In other parts of the world, farmers also grow *spring* wheat: sown in the spring, harvested in the fall. Winter wheat varieties can't flower until they are exposed to a period of cold weather—a process called "vernalization"—whereas spring wheat varieties flower as soon as possible, without needing to experience cold weather. Winter wheat is typically more productive and nutritious than spring wheat. But spring wheat can thrive in places where winter is too cold or dry for winter wheat; it also grows more quickly, giving farmers time after harvest to plant a second crop (maize, say, or potatoes) in the same field.

Conditions in the high mountain valleys of the Bajío were less than ideal for winter wheat (cold, dry winters) but favorable for spring wheat (warm summer days, plentiful rainfall). Nonetheless, farmers rarely grew spring wheat there, because *P. graminis* rose up in the summer rains. Winter wheat could be harvested in the spring, before the annual onslaught of stem rust. Even with this precaution, the fungus wiped out as much as a fifth of the harvest every year. Wheat farming in Mexico, Borlaug had come to realize, was basically an exercise in rust management.

Unsure of how to proceed, Borlaug decided to tour the high plains southwest of Mexico City, looking for local wheat strains and farming methods that seemed able to fend off stem rust. He went in March 1945, soon after being entrusted with the wheat program. Dismayingly, the local wheat he saw proved to be almost as disease-stricken as the wheat planted by the Rockefeller scientists. Farmers planted a mix of varieties—tall and medium, red and white grain, early and late ripening, often ten to fifteen types in a field—hoping that a few would escape the fungus. And they sowed the crop sparsely, the plants widely separated, hoping to slow down stem rust's spread. Incredibly, Borlaug learned, some farmers refused to water their fields, "to minimize the losses caused by rust." Sowing thinly, planting a random jumble of varieties, and deliberately inducing drought conditions were terrible practices— it was like trying to prevent a heart attack by starving yourself to death. Yet he could see why Mexican farmers did it. Some farmers' plots were irrigated and covered by thick, uniform plantings of high-yielding varieties. And *P. graminis* had engulfed them all.

Wandering through villages, Borlaug and two Mexican research assistants snipped off about eight thousand heads of wheat that looked different from each other—that might be separate varieties. Back in Chapingo, they hand-harvested the grain. A typical head of wheat consists of twelve to fourteen "spikelets," each with two to four seeds. Borlaug thus had a bank of about 100,000 seeds. Helpful officials in the U.S. government had sent more than 600 foreign wheat varieties to the project: another 10,000 seeds. The three men put each batch from each variety in labeled envelopes. The plan was to plant all 8,600 varieties—in spring, the season farmers avoided—and watch rust attack them. The hope was that some might be resistant, and that Borlaug could use these as a base to breed better varieties.*

With little equipment, Borlaug and his Mexican co-workers had to borrow a cultivator and do the plowing by hand. Tying straps around their waists, the three men took turns pulling the plow, one man walking behind to steer. Borlaug's two main assistants were Pepe Rodríguez and José Guevara, both trained as agronomists at Chapingo. In that time their advanced degree was a point of pride, something that set them above the peasant farmer. They came to work in suits and ties and neatly shined shoes. They resisted getting their hands dirty. Borlaug took a grim amusement in sharing the plow duties with them, watching them struggling in their dark suits through Mexico City's intense sun, the plow sending up dust in the wind. Eventually Rodríguez and Guevara, like Borlaug, put on work clothes. Khaki pants, lace-up boots, and sweat-stained baseball caps became the project uniform. The three men set to work in April 1945. They hacked out two short rows for each of the 8,600 varieties—more than five miles of rows, 110,000 plants, all put in by hand. In addition, they planted a second, smaller field with another ninety-nine wheat samples sent by Stakman and Edgar McFadden, a U.S. Department of Agriculture plant breeder in Texas whom Borlaug had never met.

Hardly had Borlaug finished sticking the wheat in the ground when Wellhausen asked him to help build a breeding nursery for maize in the Bajío. The two men arrived in May, working on a hillside outside the city of Celaya, in Guanajuato state. The rainy season had just finished

---

* The wheat was of several types: bread wheat (the most common); durum wheat (used for pasta); and emmer wheat (an older form of wheat also used for bread). Although these are distinct species, I lump them together here for simplicity.

and the hills were prostrate beneath an unseasonable heat wave. At the same time the Celaya city power plant broke down. Unable to eat in hot, dark restaurants with no fans, Borlaug and Wellhausen cooked meals over fires fed by maize cobs. The same fire boiled their water. Despite the precautions, Borlaug recalled, he got sick. He would spend the day planting maize in the heat-dazed hills, then stagger to his hotel "writhing with pain and faint with nausea."

Much worse was the poverty. Borlaug had been poor all his life but always well fed and decently clad. In the Bajío he first encountered destitution on a geographic scale. Women walked for miles to carry water from contaminated wells. Men scratched at the earth with wooden hoes and slashed at weeds with sickles as ancient as time. Plumbing was a distant dream. Children died from diseases that were treatable nuisances in richer places. Again and again, he encountered people who had been so badly abused by authority that they clung to beliefs Borlaug found irrational. If Borlaug offered to procure steel plows and hoes, they told him that metal siphoned "heat" from the soil. If he asked about fertilizer, they told him it was a government plot against the farmer. Each conversation was like being thrust into a wildland of confusion and despair.

For most of his life, Borlaug had had little focus beyond getting off the farm. Now, in the Bajío, something larger was stirring in his heart. He wrote to Margaret:

> These places I've seen have clubbed my mind—they are so poor and depressing. The earth is so lacking in life force; the plants just cling to existence. They don't really grow; they just fight to stay alive. The levels of nourishment in the soil are so low that wheat plants produce only a few grains. . . . Can you imagine a poor Mexican struggling to feed his family? I don't know what we can do to help but we've got to do something.

We've got to do something. Much like Vogt, Borlaug was acquiring a sense of mission—a shiver in the spine that would drive him for the rest of his life. Vogt's nascent concerns, first awakened on Long Island, had been crystallized in the Mexican countryside in 1943 and 1944. The same landscape at about the same time sharpened and focused Borlaug's view. His concerns, initially provoked by food riots in the Depression, coalesced in the Bajío.

But the two men drew different conclusions from the same picture; they disagreed about which elements were figure and which were ground. Vogt saw the land behind and beneath as the protagonist of the story—the origin of both problem and solution. With his ecologist's eye, he viewed the fundamental issue as one of carrying capacity. People, biological agents like any other, had to fit in.

By contrast, Borlaug saw the farmers as the central characters. Their suffering was caused not by overshooting the capacity of the land but by their lack of tools and knowledge. With industrial fertilizer, advanced irrigation techniques, and the finest new seed stock, they could transform the landscape, making it more productive and themselves wealthy. Fitting in with their world would be a human catastrophe. Instead they needed to reconstruct that world on more useful principles.

To Borlaug, the people he saw struggling in the Bajío were like the farmers who had rioted in Minnesota, driven by need and helplessness almost to madness. The solution was straightforward: bigger harvests. More food would mean more money would mean less hunger and poverty. Neither Vogt nor Borlaug gave any thought to how the consequences of their ideas would ripple across the world.

Shuttle Breeding

Borlaug spent every day in the late spring and summer of 1945 with his thousands of wheat varieties, knee-walking through the rows of young plants in a search for the powdery blisters that signaled stem rust. If he saw them on leaves or stem, he yanked out the plant and discarded it. When he finished inspecting all 110,000 plants, he began anew.

The work seemed endless, even though he now had two more assistants in Chapingo (the new assistants, at Harrar's insistence, were women, a break with tradition). If each person spent ten seconds examining each plant, the five sets of eyes and hands would take two weeks to finish a single round of inspection. That wasn't fast enough to catch stem rust; they worked longer hours.

To avoid the hour-and-a-half commute from his apartment to Chapingo, Borlaug laid a sleeping bag on the dirt floor of the research shack. After a while he stopped being bothered by the *rodeodores*—"tiny flies that imbibed so much of our blood they could no longer fly; we'd watch them roll off our arms and flop to the ground." He stopped being bothered by the heat and dust and the taste of water boiled in beef-stew cans. He stopped being bothered by only being able to change his clothes and shower once a week, when he returned to Margaret and Jeanie. Sometimes Pepe Rodríguez and José Guevara stayed with him (the women were not permitted to sleep in the fields). They rose before sunrise, when the air was cool, and crouched down in the faint light, looking for stem rust. Always they found it. The number of plants shrank as summer progressed.

Shrank almost to zero, in fact. *Puccinia graminis* destroyed each and every one of the eight thousand varieties from the Bajío, as well as the six hundred sent by the governments of Mexico and the United States. The second, smaller field—planted with ninety-nine wheat samples from Stakman and the Texas breeder—fared slightly better. Four types did not succumb to rust, two from each man. The four spindly rows were all that remained of the 110,000 plants put in the fields by Borlaug and his assistants. Once again, the Rockefeller experts had failed.

In the months he spent yanking out rust-afflicted plants, Borlaug had had a long time to think. No record survives of his rumination. But by following his actions, reading later interviews, and poring through the papers of his co-workers it is possible, perhaps, to trace the outline of his thoughts.

What he was thinking (I believe) was that Harrar had it wrong. So did Stakman, the entire Rockefeller hierarchy, and Sauer, too. What he was thinking was that the Rockefeller project wasn't going to succeed. It would have too little impact and take too much time.

The Rockefeller Foundation had been charged with boosting harvests in Mexico by working with farmers in the Bajío. But these were some of the nation's most deprived people on some of its most degraded land. Increasing their productivity would benefit them—a great thing. But the land was so marginal that even tripling its output would do next to nothing for the nation as a whole. (Three times a small number is another small number.) On top of that, the nation's poor infrastructure ensured that any extra grain from the Bajío could not be sent from the highlands to other places. It would be like trying to solve the problems of the United States as a whole by assisting the farmers of Saude, with their bad soil and lack of access to railroads. Bajío farms could boom and Mexico would still need to import maize and wheat.

The better way, he decided, was to raise yields all over the nation—to target Mexico as a whole, rather than only the Bajío. As Vietmeyer put it, Borlaug thought the objective should be to "feed *everyone*; not just the hungry. Opt to feed the whole populace." Produce enough not only to feed every man and woman in Mexico but also to export to other food-short nations.

Alas, that goal seemed next to impossible. With its mountains, deserts, and wet valleys, Mexico was ecologically diverse. To breed wheat suited to each climatic and soil zone, Borlaug would have to set up programs in many different places. The project didn't have the staff or funds to do that. But even if it did, Borlaug thought, the process would be too slow. As a rule of thumb, wheat breeders needed ten to fifteen harvests to select, test, and propagate a new variety. The process couldn't be hurried; farmers could grow only one crop of winter or spring wheat a year. But the Rockefeller Foundation wasn't going to wait fifteen years. And the farmers needed help *now*.

In the fields, Borlaug came up with a solution. As he told Harrar, he wanted to combine two breeding methods. One was difficult but conventional; the other was equally difficult, but not conventional at all. The first, conventional method was high-volume crossbreeding—breeding together many different varieties in the hope of producing favorable new variants. Genetically speaking, high-volume crossbreeding is equivalent to throwing a huge number of darts, in the belief that chance will eventually produce a bull's-eye. High-volume crossbreeding was typically the province of big laboratories with large staffs. Borlaug and his small team would have to sow and raise thousands upon thousands of wheat plants, collect each plant's pollen individually, hand-pollinate the blossoms, harvest the resultant grain plant by plant, and then grow that grain to discover the results of their crossbreeding.

Harrar thought that this procedure was possible, though he worried about whether Borlaug's team had sufficient resources and personnel. But he vehemently opposed Borlaug's second, unconventional method: *shuttle breeding*, as it has come to be called.

Shuttle breeding was intended to speed up crossbreeding by taking advantage of Mexico's diverse terrain, which stretches two thousand miles from the country's semitropical south to its semiarid north. This vast area held three main wheat-growing regions: the Bajío, in central Mexico, where the Rockefeller Foundation was working; the Pacific

Coastal Plain in the state of Sonora, along the Gulf of California; and a smaller area called La Laguna, north of the Bajío. Borlaug's idea was to split the breeding program between two of these zones: the Bajío (and nearby Chapingo) and Sonora, seven hundred miles to the northwest.

In November, after the harvest, Borlaug would take his four surviving varieties to Sonora, where he would breed them with each other and many other cultivars in an effort to produce new cultivars that both resisted stem rust (as the four survivors did) and produced a lot of grain (as the other strains would if they didn't succumb to rust). In April he would harvest the seed from the best plants and take it to the Bajío, where he would perform a second round of crossbreeding. Because summer in the Bajío was wet, the area was like an incubator for plant diseases. Borlaug could use the second generation as a screen to check susceptibility to diseases other than *P. graminis*: viruses, bacteria, different types of fungi. In October he would harvest the most resistant seed from the second round—and take it to Sonora for a third round, where he could scan for plants that produced the most grain. The fourth generation, in the Bajío, would be examined to ensure that these high-producing plants still maintained their resistance to multiple diseases. By the sixth generation, Borlaug believed, he could have varieties ready for farmers to field-test. By ricocheting back and forth between Sonora and the Bajío, new, rust-resistant varieties could be ready for wide use in five years, half the usual time. (Left unnoted in this description: Margaret would be left to raise their daughter by herself for half the year.)

Borlaug had not chosen Sonora at random. Blessed with plenty of sun and good soil, coastal Sonora was the site of several big new irrigation projects. Borlaug had wangled a ticket that summer to Ciudad Obregón, Sonora's second-biggest city, thirty miles from the Gulf of California. Between the city and the gulf were about 145,000 acres of irrigated rice and wheat. The land was in the delta of the Yaqui River, enriched over the centuries by flooding. Like the Bajío, the region was afflicted with stem rust. But if its farmers could get rust-resistant wheat, Borlaug thought, they could produce more wheat than would ever be possible in the Bajío. Sonora could become a breadbasket for all of Mexico.

Harrar wouldn't hear of it. The Mexican government had been clear about restricting Rockefeller to the Bajío. But even if the foundation had been permitted to work in Sonora, Borlaug's scheme violated a cardi-

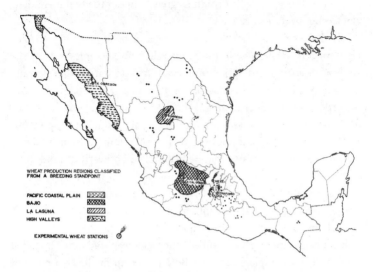

Mexico's wheat-growing areas, as shown in a map from the Mexican Agricultural Program in the early 1950s

nal rule of plant breeding: breeders had to develop new varieties in the environment in which they would be grown, so that they would be well adapted for that region. Winter wheat thus could not be bred in a place where farmers grew spring wheat, and vice versa. It was worse to try this in Sonora and the Bajío, where the climates were so different.

Borlaug stubbornly insisted. The irrigation projects in the Yaqui Valley were opening up farmland so good that it would be folly for him to ignore it. And he promised to do the work entirely on his own; the foundation wouldn't have to pay anything extra. As for his idea's workability—to Borlaug, the virtue of being able to move faster would be worth whatever problems came from not developing these new wheat varieties in the place where they were to be grown.

Prizing loyalty, Harrar expected his subordinates to follow him, and in return he would back them completely. Borlaug's intransigence infuriated him, but his position was weak. Rockefeller's maize program was faltering. Firing Borlaug would put the wheat program in jeopardy, too. In the end, Harrar grudgingly agreed to let Borlaug try Sonora for a season, but only if he spent no program funds and concealed his work from the authorities. Still, a coldness grew between the two men; their relationship never recovered.

Late that year a ramshackle, six-passenger plane took Borlaug to Ciudad Obregón, at the edge of the Yaqui Valley. Twelve miles outside the city was an abandoned agricultural experimental station that Borlaug planned to use as his base. Because he had no car, he hitchhiked from the airport to the experimental station. In his luggage were his clothing, many cans of beans and stew, and the total product of the previous year's labor: a couple thousand seeds from the four resistant varieties and samples of new strains that he had picked up nearly at random—seeds that he had taken from an abandoned field in his first visit to Sonora, for example.

The experimental station, built in 1938 to work with cattle and pigs, was a shambles: "Windows broken, roofs beat to hell, the machinery all broken up, and the livestock and whatnot had been sold off or had died or both. Just a complete disaster." Borlaug set up his bed—a piece of canvas stretched from poles—in a loft above a dilapidated storage shed. There was no electricity, telephone, or running water; insects, rodents, and rain came in freely through the broken windows. The station had no staff except a part-time caretaker. Borlaug built a fire from maize cobs, heated a can of beans, and went to sleep.

In the morning he inspected the station. The 250 acres of former experimental fields and pastures were overgrown with weeds and shrubs. He would need equipment to clear and plow the land. Walking from farm to farm along the road, he knocked on doors and asked in his bad Spanish if he could borrow a tractor for a couple of days. The neighboring farmers, puzzled and suspicious, did not want to lend their valuable machinery to a random *norteamericano* who claimed to be a researcher but dressed like a laborer. Borlaug came back to the station that afternoon angry, sunburned, and empty-handed.

The next day he prowled through the ruined buildings, looking for equipment. He found broken spades, rusted rakes—and an ancient wooden cultivator, the kind meant to be pulled by a mule. When the caretaker came to work, Borlaug brought him to the field, communicating mostly by gesture. Borlaug strapped the harness to his body and pointed to the plow. Then he began dragging the cultivator through the earth, the old caretaker unsteadily guiding the blade. In Chapingo, Borlaug had been able to swap positions with the other members of his crew.

Now it was entirely on him; the caretaker was too old to pull the plow. People passing by stopped to watch the two men as they worked. Borlaug ignored them, leaning into the straps. By noon he was exhausted. He put away the plow, raked weeds from the soil he had turned, and planted wheat until sunset. At night he built a fire, opened a can of beans, and staggered to his cot.

On the third day of hand-plowing he was approached by the owner of the farm next door. Borlaug stopped dragging the cultivator, wondering why the man was in his Sunday best. Then he realized it *was* Sunday—his visitor was returning from church. Why are you doing this? the man asked. Why are you working like a mule on the Lord's day? Borlaug tried to explain. The neighboring farmer listened uncomprehendingly to this pathetic foreigner with his filthy clothing and dreadful Spanish. Finally he told Borlaug that he could borrow a tractor on weekends.

Having the tractor made a difference, even if it was only part-time. So did meeting a local U.S. expatriate who offered him rides into town to buy groceries. By Christmas Borlaug had planted five acres—about 140,000 plants, he later estimated. He flew to Mexico City and his family. The first thing he heard was that his son had died in the Baltimore hospital. Shoving his grief to the back of his mind, he spent as much time with Margaret and Jeanie as possible. He also tried to make amends with the project office. Harrar had hired more staff, including a forester, Joe Rupert, a Stakman student whom Borlaug had known slightly at Minnesota. He became a friend and soon moved into a spare room in the Borlaugs' apartment, helping Margaret when her husband returned to Sonora.

Back at the research station, Borlaug found his seeds sprouted and the plants ready for cross-pollination. Wheat flowers, known as florets, grow in little bunches on the spikelets. Like most flowers, each floret is bisexual, with both male and female reproductive organs. Rising up on thin stalks from the center of the floret are the stamens, the male parts of the plant, which contain the pollen in little pods at the tips. Below these are the small, delicate filaments of the female part of the flower, the stigma, with the ovary below. When the stigma has developed enough to be capable of reproduction, a biochemical signal causes the stamen tips to burst, releasing spore-like grains of pollen in tiny golden puffs. In flowering plants like wheat, each pollen grain contains a generative cell and a vegetative cell. The former produces two sperm cells; the latter swings into action when the pollen settles on the fronds of the stigma. There it produces a germ or pollen tube, a cylindrical extension of the pollen grain that carries the actual sperm cells. Within as little as an hour of landing on the stigma, the pollen grain's germ tube has penetrated the ovary beneath the stigma. Inside the ovary is the ovule, which contains the egg cell. Male and female mechanisms join and begin creating a seed, the grain that the farmer will harvest.

Because both sperm and egg come from the same plant, the new seed will have the same genes as its parent. To create new varieties, plant breeders must stop wheat from fertilizing itself. In practice, this boiled down to Borlaug sitting on a little homemade stool in the sun, opening up every floret on every spikelet, carefully plucking out the pollen-containing stamens with tweezers, and discarding them. Every stamen in every plant had to be removed to ensure control. Now the entire field of wheat was entirely female—the plants had been, so to speak, emasculated. Only pollen from other plants could fertilize them.

The next step was to place a flattened paper cylinder over the heads of the now-female plants. Once it was in place, Borlaug folded over the top of the cylinder, sealing it with a paper clip. Now no pollen could enter. Each bagged plant had to be labeled and logged. And all of this had to be done in the few days between when the egg cell became fertile and when the pollen was released by the stamen.

A few days later, when the plants had recovered from surgery, Borlaug snipped off the florets from another wheat variety he hadn't cut up. Opening the paper cylinders on the emasculated grain heads, he inserted the florets from the second variety, twirled them to release their pollen, then resealed the paper. Each spikelet on the plant had to be pollinated by the same pollen; no other pollen could enter. After a few days he removed the paper and let the fertilized florets produce grain. When the grain matured, he harvested it, separately packaging and labeling the seeds from every plant, and transported the lot to the Bajío, where he did it all over again.

The possibilities were almost endless, but they were slow to appear. The first generation of crosses usually would end up looking halfway between the two parents. But if breeders combined the offspring, the second generation would have many individuals with two identical copies of significant genes—and these would begin to look different from the original plants. Important positive or negative traits could show up. These would be amplified by the third or fourth generation of crosses. If one of these plants had especially desirable qualities, it would need to be studied and bred on a large scale before field trials could begin, which also took time. Borlaug's shuttle breeding was intended to turn the wheel faster.

Hearteningly, the initial breeding at Sonora was successful. Seed from the four rust-resistant varieties produced plants that survived the annual onslaught of *Puccinia graminis*. By the fall Borlaug had five acres of tall wheat plants bending under the load of their grain-filled spikelets. Next Borlaug would have to harvest the grain and transport it to the Bajío for the next round of crosses. The tiny planes that flew between Ciudad Obregón and Mexico City could not carry hundreds of pounds of grain. In April 1946 Borlaug flew to Mexico City, where Harrar reluctantly allowed him to borrow the Rockefeller program's sole vehicle, a Ford pickup truck.

In those days no paved roads crossed the mountains between Mexico City and Sonora. From the capital, drivers had to take a two-lane highway northwest three hundred miles to Guadalajara, then go six hundred miles north on a path through the brush, crossing deserts and fording rivers, carrying their own gasoline for most of the route. The prospect was so daunting that Borlaug decided to go the long way around: drive from Mexico City to El Paso, Texas; cross most of Arizona; re-enter Mexico at Nogales; then forge two hundred miles south through scrublands to Ciudad Obregón. To take shifts at the wheel, he recruited Joe Rupert, his apartment mate, and one of Harrar's new hires, a Chapingo graduate named Teodoro Enciso.

The three men loaded equipment into the truck—two small, battered threshing machines, four spare tires, and a pile of gunny sacks to use as mattresses—and headed north. When they reached the border at El Paso, three days of rough road later, the U.S. Customs guard stopped them, claiming that Mexican government vehicles were not allowed to enter the United States. Borlaug told him that the two countries had agreed to permit official vehicles to cross the border freely. The guard responded that the truck could enter, but not its contents. To ensure that the researchers could not illegally sell equipment, they would have

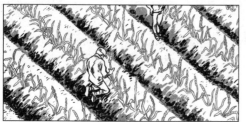

Before modern techniques, wheat breeders like Norman Borlaug had to plant thousands of different varieties, hoping that chance would produce favorable variants. Each plant had to be checked by hand for desirable characteristics—early flowering, perhaps, or disease resistance. Then breeders tried to mate them to produce plants with both good traits.

WHEAT FLORET

ANTHER ("MALE" PART OF PLANT, RELEASES POLLEN)

GLUME (HARD OUTER SHELL)

STIGMA (CATCHES POLLEN, TRIGGERING GERM TUBE THAT FERTILIZES OVARY)

OVARY ("FEMALE" PART OF PLANT, CONTAINS EGG CELL INSIDE AT BOTTOM)

To mate two wheat plants, people tweezer out the anthers (right), creating a purely "female" plant. Then they cover the now emasculated plant with a small envelope to block off errant pollen.

Meanwhile the breeder snips off the floret of the second plant, opens the envelope atop the first wheat plant, and twirls the floret—releasing the pollen and fertilizing the first wheat plant. The fertilized floret then turns into a seed—a kernel of wheat.

Each plant is harvested by itself, with the grain placed in a separate envelope and labeled with the information about its parents. These are then sown in the next season. And so the cycle begins again.

Even when breeders create a wheat variety that combines all their desired characteristics, their work is not done. Varieties must be constantly adjusted for different conditions, new strains of pests and pathogens, and new farming techniques. The work can be tedious, but it is one of the foundations of the modern world.

to unload the machines, tires, and sacks, and hire a bonded freight company to carry them to Nogales.

Borlaug was irate. Every minute of delay meant that they could miss the harvest. Even if the grain were still in good shape, he still needed to reap, winnow, dry, and pack it, measuring the weight and characteristics of every spikelet, in time to haul it to the Bajío and plant it before winter set in. The three men pushed from Nogales to the Yaqui Valley in a hot-brained hurry. To Borlaug's relief, the wheat was harvestable.

Not wanting another run-in with U.S. Customs, Borlaug, Rupert, and Enciso decided to come back through Mexico with the grain. The main peril was the rivers coursing down the flanks of the western Sierra Madre, many of which had to be driven through, a risky business in the spring rains. Bushed, battered, and bruised, the three banged across the mountains, then went to the Bajío and Chapingo to plant the second generation.

In July 1947 the wheat project was visited by Herbert K. Hayes, a revered Minnesota plant breeder who had helped develop both hybrid maize and wheat. Touring Chapingo, Hayes was shocked to learn that Borlaug, a Minnesota graduate, was trying to breed the same wheat in two different latitudes and climates, a violation of basic botanical dogma. Had Borlaug taken any courses in plant breeding, he would have found this dictum etched in the textbooks, including *Breeding Crop Plants*, a classic whose senior author was Herbert K. Hayes.

Borlaug stood his ground; Hayes was just a visitor. But he had a harder time fending off criticism that October, when Rupert and he were summoned to Harrar's office. Funds were running short, Harrar said. Although Borlaug was spending almost no money in Sonora, the Rockefeller Foundation was effectively paying other people to do Borlaug's job while he was there, pursuing a private obsession that flouted well-known scientific principles. Shuttle breeding had to stop. His tone, Borlaug said later, was icy. Borlaug again laid out his argument: conditions in the Bajío were too poor to solve Mexico's wheat problem. Possibly, Harrar said, but that is where our hosts have asked us to work. Borlaug replied, in his later recollection,

"With the water facilities and the land resources that are available, this is, in effect, tying one arm behind my back, and I can't do it. And if this is a firm decision that's been made, you better find someone else to execute this, because I will stay [only] till you find my replacement. If Joe Rupert here wants to take it over as I leave this door, why as far as I'm concerned it's to be executed." And before I got to the door, Joe stood up and said, "This goes for me, too." And we both walked out.

Fuming, Borlaug stomped into his office, Rupert in tow. There he found a stack of mail. He swept it to the floor in annoyance—then realized that one of the letters was from the farmer next door to the agricultural experimental station, the man who had loaned his tractor. To Borlaug's surprise, it was a copy of a note to Harrar. It began by congratulating the Rockefeller Foundation:

> Perhaps it is the first time in the history of Mexico that any scientist tried to help our farmers. . . . But why is it, with such a great force like the Rockefeller Foundation, that you do not give your men the tools and machinery they must have to fight with? Why does he come like a beggar to borrow the tools to grow new wheat?

On Borlaug's copy of the letter the farmer had written, "It's time somebody was helping you fight inside your own organization!"

Borlaug and Rupert were stricken with guilt—they had, after all, just quit. They stalked into a cantina and drank themselves into a near stupor. When they stumbled home that night, Margaret angrily refused to let them in. The hallway filled with their argument, Margaret shouting in the apartment, Borlaug bellowing in the hallway, Rupert sheepish and silent. A comic-opera scene, Borlaug said later. Eventually Margaret unlocked the door. Early the next morning, hungover and abashed, Borlaug slipped out of the apartment.

> I walked into the office, and here's Stakman, sitting, smoking his pipe. Have you ever seen Stakman at seven o'clock in the morning? I never have before or since. And he said, "You people act like children!"

He told Borlaug he would see what he could do. Borlaug went to Chapingo for the day. It was no consolation to see how well his wheat was growing. When he returned, he went to Harrar's office. Stakman was there. Without looking up, Harrar said, Go back to Sonora. Borlaug realized that he must have read the farmer's letter. With the ongoing problems of the maize initiative, the project couldn't ignore a project that farmers seemed to support, even if they were not the farmers Rockefeller was supposed to be working with, even if the project was not going to work.

Again Borlaug took the truck over the mountains. Landslides left him stranded in the heights for days. In Sonora he worked through the winter, mostly alone. Some of his new plants seemed to be strongly resistant to rust. That spring he organized a demonstration of his new seed for Sonora farmers. It was like a cold shower. Only a few people showed up, one of whom was the neighbor who had written a letter to Harrar. The others did not share his enthusiasm. Some loudly ridiculed Borlaug. Not one took advantage of his offer of free seed.

Green Revolution

Plant breeding in that time was a black art. A few laboratory researchers were beginning to wonder whether a molecule called DNA played a role in heredity and reproduction, but that suspicion had not filtered into the world of practical science. And even the white-coated boffins didn't know that clusters of interacting genes resided on twined DNA strands wrapped into bodies called chromosomes within the cell nucleus. Today this information is taught in elementary-school science classes, but it was unknown when Borlaug was working in Mexico. Now-familiar terms like "DNA" and "gene" do not figure in his reports or work notes.

Because Borlaug knew little about the molecular basis of heredity, he worked exclusively with plants' physical features—the thickness of their stalks, the number of leaves, the time of flowering, and so on. Thus he might give a high-yielding "female" plant pollen from a "male" disease-resistant plant (or vice versa). Or he might mate a fast-growing but rust-susceptible plant with a slower-growing but resistant plant in the knock-on-wood hope of producing offspring with the two desirable traits, not the two undesirable ones. In both cases, he couldn't know the results of these matches until the plants grew in the field.

Worse, wheat has about four times as many genes as humans, many present in multiple interacting versions. For breeders, this enormous genetic diversity is a source of both hope and frustration. On the positive side, it means that wheat has many hidden genetic treasures—valuable genes are concealed in the thickets. But on the negative side it means that these valuable genes are, needle-in-haystack-style, hard to find.

Even if Borlaug's crosses turned out well—he bred plants that resisted stem rust, say, or bore extra-plentiful grain—there was no telling if those traits could be passed on to the next generation. Most observable physical traits are associated with multiple genes; human eye color, for instance, is associated with about fifteen. If all the interacting, beneficial versions of genes were not passed on together, or if breeders unknowingly introduced other genes that canceled out the favorable effect, or if environmental conditions changed and the proper genes were not switched on, the desired characteristic could vanish. Breeders could combat this uncertainty only by performing huge numbers of crosses—throwing darts by the thousand, so to speak, praying for bull's-eyes.

Borlaug got lucky—"a case of serendipity," he later called it. Like many other plants, wheat controls its growth by a kind of biochemical clock that measures day length. Winter wheat stays dormant until the clock "observes" the days growing longer, which signals the advent of warmer spring weather. The plant then "knows" that the danger of frost has probably passed and it is safe to flower. Sensitivity to day length is called "photoperiodicity." Multiple genes direct wheat photoperiodicity, of which the most important is known as Ppd-D1. When Borlaug moved his wheat from Sonora to the Bajío, he was conducting an unknowing experiment to discover whether any of his varieties were *not* controlled by Ppd-D1.

By chance, a few of them weren't under its thumb. In a long-ago genetic accident, a snippet of DNA in one Korean wheat plant had dropped out of the gene, crippling the intricate workings of Ppd-D1 in somewhat the way that removing a single small chip from a computer can disrupt its function. Equally fortuitously, the mutated gene had been passed on through the generations—ending up, geneticists believe, in the fields of an Italian breeder, from which it passed to Kenya, where it was collected by the Texas breeder and passed on, all unknowing, to Borlaug. Wheat with malfunctioning Ppd-D1 is called "photoperiod-insensitive"—a fancy way of saying that it has a clock that can no longer tell time. Rather than waiting for the day to grow long, the plant sprouts and grows as soon as it can.

Photoperiodicity was why, as Hayes said, crops had to be developed locally. Plants bred in one area had clocks adapted for that area, and moving them too far away would confuse them, so to speak. When Hayes and Harrar told Borlaug that his experiments with shuttle breeding would just end up with dead plants, they were correct—about normal wheat. But it turned out that not all of Borlaug's wheat was normal. Stubborn ignorance allowed him to discover that the offspring of some of his crosses with the Kenyan wheat were photoperiod-insensitive, and would grow whenever he planted them. He had taken advantage of that lucky break to breed wheat twice as fast as anyone else—wheat that could withstand a much wider range of conditions than other varieties.

He would soon need more luck. By crossbreeding the four wheat types that had survived stem rust with hundreds of others, he succeeded in producing five new varieties that were photoperiod-insensitive, rust-resistant, and highly productive. Despite their doubts, a few Sonora farmers tried them out—and nearly doubled their wheat harvests. Others quickly jumped on the bandwagon. Farmers in Sonora had already been benefiting from increased irrigation, which allowed them to plant more land. Production soared further when they adopted Borlaug's rust-resistant seeds.

Borlaug was achieving his goal, but he couldn't enjoy it. He spent the entire time worrying that the triumph would be short-lived. As more and more farmers planted the new varieties, two problems had appeared.

The first was *lodging*. To achieve their maximum potential yield, his new varieties needed more nutrients than Mexico's poor soils could provide. When Sonora farmers added chemical fertilizer, the new Kenya-based varieties responded avidly, bearing huge heads of grain. Too avidly, in fact: the heads in Borlaug's wheat—unlike those in the traditional types it was replacing—were so large that in high winds the top-heavy plants fell over easily, one knocking down the next, domino-style. Lodging leads to disaster, because the crimped, bent-over stems can't deliver the water and nutrients the juvenile grain needs to mature. The new varieties were wildly productive but vulnerable to wind and rain in a way that the traditional varieties never had been. Entire fields could be destroyed by a single hard gust.

To prevent lodging, Borlaug would have to modify his wheat by breeding it with wheat varieties that had stalks—"straw," in the jargon—short and rigid enough to support the increased grain. Breeding for shortness, in fact, might be a double win. It would protect the grain that his varieties were already producing from lodging. But it could also mean that the plants would spend less energy producing inedible straw. Short plants would take in just as much energy from the sun as tall plants, but they could channel more of that energy into creating grain. A field of short plants might feed more people than a field of tall plants.

Unfortunately, viable short plants would not be easy to find. Taller plants are more likely to see the sun, and thus more likely to grow and reproduce. Genes for shortness, conversely, put plants at an evolutionary disadvantage, and so are discouraged. In breeding for shortness, Borlaug would be fighting natural selection.

Compounding the difficulty was the second problem: a powerful new variety of *P. graminis* had shown up in the United States. Since the First World War, Stakman had cultivated a network of farmers who sent him samples of the rust growing in their fields. Much as medical researchers today monitor the development of new strains of influenza, Stakman was looking for the onset of new strains of stem rust. His Minnesota students would test each sample against a cluster of twelve standard wheat varieties, comparing its effects to those from other known varieties. In 1938 a farmer had sent Stakman a rust sample from a barberry bush in New York that had escaped the anti-barberry crusade. In tests, the New York rust rapidly killed all twelve standard wheat varieties. U.S. Department of Agriculture (USDA) researchers conducted their own experiments and "established beyond a doubt that every commercial variety of wheat then grown in the United States and Canada was susceptible." So were all of Borlaug's varieties in Mexico.

Luckily for farmers, the new rust strain—15B in Stakman's catalogue of stem-rust "races"—didn't seem to spread easily. It was deadly, but not especially contagious. Stakman watched 15B for more than a decade as it confined itself to a few areas around remnant barberry bushes. He worried that it would suddenly become more virulent. Exactly this came to pass in 1950; 15B exploded across the entire U.S. wheat belt in a matter of weeks. Nobody could account for the change. A mutation, a buildup of critical mass, favorable weather conditions—whatever the reason, decades of progress against stem rust were about to be wiped out.

Alarmed, Stakman and his USDA counterparts convened the first-ever international stem-rust conference in November. Borlaug attended and gave a report on Mexico. The attendees decided to establish in seven nations rust-monitoring programs that in years to come would prove a valuable tool for rust-fighting. But that tool would do nothing in the short term. Even as the researchers met, 15B was racing south on the Puccinia highway, crossing the Rio Grande. Borlaug knew that the Bajío, already plagued by rust, would soon be hit harder; Sonora, now thriving, would be flattened. In the affluent United States, 15B would lead to reduced harvests and hardship for farmers; in poor rural Mexico, it would lead to malnutrition and ruin.

In April 1951 race 15B was detected in Mexico for the first time; by summer it was charging through the Bajío. Seeking to avoid calamity, Borlaug set up four large nurseries in central Mexico and tested 15B against sixty-six thousand wheat varieties: sixty thousand from his own work, five thousand from a USDA collection of world wheats, and a thousand other lines from various parts of the Americas. This was a huge task, though he now had more help; over the years, the Mexican Agricultural Program had trained more than five hundred young researchers, of whom it had hired more than a dozen. Despite the assistance, the results were disturbing. Of the mass of wheat varieties, only four survived 15B, most of them relatives of the Kenyan types Borlaug was already using. And none of the four helped him with his other problem—they weren't short. A few of the sixty-six thousand varieties had short straw, but the rest of the plant was small, too, including the cluster of grain-bearing spikelets at the top. Miniature plants, miniature harvests. And in any case all the short plants were susceptible to rust.

The following summer Borlaug both examined the candidates from the previous year more carefully and tested all of the thirty thousand varieties in the USDA wheat collection. The task was even larger, but it met with even less success. Every single one of the USDA wheats fell to 15B, including all the varieties with short straw; at one growth stage or another, so did almost all of the potential rust-resisters from the previous year. Borlaug's team had now exposed tens of thousands of lines of wheat to 15B. Only two varieties had survived, both descended from the four original survivors: Kentana 48 and Lerma 50 (the numbers represent the year each variety was first produced). Lerma 50 was highly resistant but made terrible bread. Until it could be crossed with wheat with better baking characteristics, Mexican farmers would have to plant Kentana 48—what choice did they have? It would keep the rust away, keep the harvests growing. But a single variety for an entire nation—it was not a situation that could endure for long.

Astonishing in retrospect, Harrar and the Rockefeller Foundation leaders in Manhattan chose this moment to expand the Mexican Agricultural Program. Wellhausen, the maize breeder, had resuscitated Rockefeller's maize venture by establishing high-tech facilities to breed U.S.-style hybrid maize for Mexican conditions—exactly the sort of effort Carl Sauer had warned about, exactly the sort of effort Mexican researchers and politicians had wanted. But the project flagship had become Borlaug's wheat. Harrar, once skeptical, had become an enthusiast. Ignoring the new varieties' susceptibility to rust, Harrar seized on the early signs of success to argue that the Mexican Agricultural Program should be expanded to other nations, even other continents. The foundation agreed and promoted him to program director.

To begin the expansion, Harrar went to India; Joe Rupert, Borlaug's former housemate, traveled to Colombia. Borlaug was sent to Argentina. Accompanying him was USDA researcher Burton B. Bayles. Arriving in November 1952, the two men discovered, quickly and unhappily, that rust afflicted Argentina, too. Borlaug was in a grim mood. Bad enough that his tests had generated just two potential candidates that resisted 15B, he told Bayles. But a *second* new strain of *P. graminis*—race 49 in Stakman's stem-rust catalogue—had shown up in the Bajío that spring. By August it had become clear that both Kentana 48 and Lerma 50 were susceptible to 49, Kentana perhaps a little less than Lerma. Now Borlaug did not have a single variety that could withstand attack. And he still had not found a single plant with both a short stem and a normal cluster of spikelets. Despite more than a decade of drudgery, Borlaug's program was tottering.

A prerequisite for a successful scientific career is an enthusiastic willingness to pore through the minutiae of subjects that 99.9 percent of Earth's population find screamingly dull. Bayles and Borlaug ruminated about the quirks of *Puccinia graminis* as they toured the pampas, running through one possibility after another. The basic problem had been understood for decades: wheat that fended off one rust race could always be susceptible to another. Borlaug asked if it would be possible to create a "composite" wheat—a variety with many types of resistance. What if he found a line of wheat that was immune to 15B and crossed it with a line immune to 49 and then crossed the result with other lines immune to other races? Bayles told him that the standard answer was that wheat was simply too complex. Each time he crossed one line with another, there was a chance that the genetic reshuffling would bury the desirable traits he had elicited in previous crosses. The likelihood skyrocketed if the breeder was crossing many lines at once. Every throw of the darts threatened to undo the results of the previous throws. Borlaug asked if the odds could be overcome by running truly massive trials—tens of thousands of crosses, year after year, a staggering amount of work— with high selectivity. "You might just get away with it," Bayles said, in Borlaug's recollection.

Along the way, Bayles told Borlaug that a colleague, Orville Vogel of Washington State University, was also experimenting with a variety of short wheat. Known as Norin 10, it had been sent to Vogel by a U.S. agricultural researcher in postwar Japan. Ordinary wheat was almost the height of a tall man; Norin 10 plants barely reached to the knee, dwarfs "so damn short," the researcher wrote to Vogel, "they're pretty much underground!" After returning from Argentina, Borlaug wrote to Vogel. Cheerful and generous, a tinkerer known as much for his farm-equipment inventions as his plant breeding, Vogel was happy to share this odd Japanese wheat. A carefully packed envelope arrived in Mexico the following summer. Inside were eighty seeds: sixty of the original Norin 10, twenty from Vogel's own experimental crosses.

In November 1953 Borlaug planted the Norin 10 at Sonora. As Vogel had said, Norin 10 had short straw and normal-sized cluster of spike-lets, exactly the architecture he was looking for. But the dwarfs proved quivering vulnerable to 15B. The new race of *P. graminis* consumed them like so many sticks of kindling in a fire, felling each and every plant before it produced a single kernel of wheat. Borlaug's supply of Norin 10 vanished, and with it his only source of genes for the right kind of shortness. Meanwhile, none of the thousands of other crosses he had performed grew up short.

Outside the experimental station, Kentana was still keeping 15B at bay. In an inexplicable bit of luck, race 49 had been mostly quiescent that year. Up north, 15B was causing the worst catastrophe to hit midwestern wheat farmers since the Dust Bowl. Sooner or later, both would arrive in Sonora. Borlaug had limited time to get ready, and he had just lost a breeding season—and all of his Norin 10—with nothing to show for it.

Borlaug returned to Mexico City and his family the following spring. Dourly he went to Chapingo to make the next big round of crosses. By this time the Rockefeller Foundation had built a new building for the work; the tarpaper shack, still in use, slumped against it. When he went inside the shack, he came across Vogel's envelope of seeds, thumbtacked to a wall with scores of others. To Borlaug's surprise, eight kernels of Norin 10 were still inside.

Eight seeds! He still had a chance to grow plants with short straw—if he could keep them away from rust. Working with exquisite care, he sowed the seeds in individual pots under grow lights in the basement of the new building. Each pot was wrapped in fine gauze that would block rust spores. Visiting his eight plants daily, he watched them sprout and grow—but only to two feet. By controlling the lights, he was able to induce the dwarfs to flower at the same time as his other varieties. He crossed them with the newest versions of his most resistant wheat varieties, Kentana and Lerma, which he had also grown in gauze-wrapped pots in the basement. At the end of the summer, he harvested about a thousand seeds from the crosses.

Packed in bags, the seed went with him to Sonora, where he planted it in the open. No grow lights, no gauze wrapping, just a thousand plants, crosses of Kentana and Lerma with Norin 10, on a plot tucked away in the rear of the experimental station. Most of the plants were short—the dwarfing gene was dominant, as geneticists say. But rust quickly killed hundreds of them. Worse, many of the survivors were sterile. Still, about fifty of the plants fended off 15B, barely reached Borlaug's thighs, and produced grain. And these plants looked good in so many other ways. Each plant not only produced more grain heads than typical varieties— they "tillered profusely," as farmers say—but each head had more and bigger spikelets than typical varieties.

After so much failure, things were abruptly going right. Chance, finally, had favored him: he had hit a clutch of bull's-eyes at once, suc-

The difference between the dwarf and ordinary wheat was striking, as shown in this photograph, taken in Sonora in 1957, of two plots of wheat planted side by side at the same time.

cess quick and unpredictable as failure. The fifty short plants provided a few thousand seeds—just enough for the next generation of crosses. Wary of setbacks, Borlaug said nothing to his superiors about what he was doing. He went to Chapingo, planted these seeds, and bred them to other plants with multiple strains of resistance to *P. graminis*—15B, 49, and every other weapon in rust's arsenal. Then back to Sonora to do it again. Now test for milling quality. With the predictable perversity of nature, the dwarfs—so superior in so many ways—had soft, shriveled, low-protein seed that made terrible flour. So he had to breed in milling quality, too. Then taste, and maybe color. The process took another five years, but the plants stayed short, and they kept tillering profusely and not lodging, and the grain was plentiful, and beginning to be usable.

By 1960 Borlaug and his team were ready to show off the new varieties. Every April, just before harvest, the Sonora station had a field day, where local farmers could see what the scientists were doing. Tractors pulled wagonloads of visitors through the fields, stopping at a dozen stations where researchers and staff standing by chartboards would explain their work. The dwarf varieties, grown in plots that couldn't be seen by visitors from the road, would be a surprise. Borlaug knew that farmers plagued by lodging would understand instantly the implications of dwarf wheat. But he didn't want them simply to grab the seed—the new varieties still didn't produce grain that could be easily milled. To keep farmers out of the field, Alfredo García, the young Mexican agronomist at that station, instructed the tractor drivers not to let visitors off the wagons when they stopped by. On field day, Borlaug later recalled,

> I stood off at a distance and watched what happened when these first wagons pulled up. As the first one pulled up—there were about six of them—[García] motioned to the tractor driver to move that one up further so he could bring another three in, and then he was going to tell them not to get off the wagons. But after he had finished talking to the tractor driver, he looked around, and I'll bet there were 50 farmers in these plots [of dwarf wheat], pulling out the heads and showing this and showing that. From then on, it was chaos. [García] lost his temper. He lost control. He started to curse. It was a real show. And then some of these wheats got away right there.

The seed, Borlaug knew, would be impossible to retrieve. It would be planted the next season and taken to the mill. If the mill owners rejected the grain because it didn't make good flour, that would be a disaster for the Mexican Agricultural Program. He was pretty sure that he was just a year or two away from wheat that would make excellent flour. The key, he thought, was to convince the mills to take the grain, even if it was not yet ready. By this time, Borlaug had gained some credibility and had some support in the capital. He had learned to speak rough Spanish and had a forceful manner when needed—he was not above jabbing a finger into a man's chest and leaning close to make a point. He called on the mills and told their owners that next harvest they would be getting batches of poor-quality grain from Sonora and they should— they *would*—buy it anyway, because it would be good for the country, absorbing the losses was their patriotic duty, and it would be only one or two seasons because he would fix the dwarf strains in the interim. (He was right; the better varieties were released in 1962.)

At the same time, he was exultant—García's bellowing, impotent fury on field day was the soundtrack of success. In the ass-end of nowhere Borlaug and his Mexican team had created something new to the world: an all-purpose wheat. Short, fecund, and disease-resistant, it could be sown in soil rich or poor anywhere in Mexico and produce well. As long as farmers provided water and fertilizer, the plant would thrive and the harvest would be large. The fertilizer could be cow manure or bird guano or bags of chemicals made in factories. The water could be from rain or concrete irrigation channels. It didn't matter: pile on the inputs and the grain would grow in quantities greater than ever before, he thought, cascades of wheat that would banish hunger for millions.

The new wheat varieties were not a permanent solution to farmers' problems. Driven by the inexorable forces of evolution, new variants of stem rust and other pests would emerge. Wheat breeders would have to create new varieties to combat them. Hidden inside the genes of the new varieties were now-unknown negative traits that sooner or later would show up in the field (scientists call this "residual heterozygosity"). Breeders would have to figure out how to eliminate their effects. And there were places where the new varieties wouldn't prosper: farms in unusually hot places, unusually wet places, or places with soil contaminated by metal or salt. Breeders would need to develop special strains for them. Still, Borlaug and his team had created something new.

Later Borlaug would think of these new, high-yielding seeds as part of a package, the other pieces being adequate nutrients (which meant, mostly, plentiful fertilization) and water management (which meant, mostly, careful irrigation). The package—seeds, fertilizer, water—was like one of the antibiotic pills that had, astonishingly, come into doctors' offices after the war: an entity, the product of faraway scientists, which could work its magic anywhere, at any time, as good at wiping out bacteria in Ireland as it was in Indonesia. Borlaug could take the wheat package to any part of the globe. The seeds and fertilizers and water practices would have to be adjusted to local conditions, much as pharmaceutical companies packaged their universal antibiotics into shots or pills or liquids or nasal sprays according to local preferences. But the idea was that the guts of the package would work anywhere.

The package was a turnkey, ready for use. Switch it on, and yields would skyrocket. No longer did farmers have to worry much about local varieties or particular soil conditions or (if they had irrigation) even the weather. Just follow the instructions, same here as anywhere else; the package was good for everyone. *It freed farmers from the land*—or, anyway, took a giant step toward that goal.

Farmers in the U.S. Middle West always had an advantage: uniform, flat terrain with deep topsoil and little climatic variation. The same wheat or maize could be grown anywhere within a thousand miles. Mile after mile after mile could be sown with the same crop, identical stalks waving to the horizon and beyond. Now, in effect, the same would be true for Mexico, or any other nation. A central breeding facility could develop a crop suitable for every region. It would make all the world an Iowa.

In 1968, the year a U.S. aid official coined the term "Green Revolution" to describe the Rockefeller package, Borlaug gave a victory-lap speech at a wheat meeting in Australia. Twenty years before, he said, Mexican farmers had reaped about 760 pounds of wheat from every acre planted. Now the figure had risen to almost 2,500 pounds per acre—triple the harvest from the same land. The same thing was happening in India, he said. The first Green Revolution wheat had been tested there in the 1964–65 growing season. It had been so successful that the government had tested it on seven thousand acres the next year. Now it was covering almost seven *million* acres. The same thing was happening in Pakistan. And this didn't count Green Revolution rice—also short and disease-resistant—which was spreading across Asia.

It was not all open skies and sunlight. By 1968 the Green Revolution was being criticized as environmentally, culturally, and socially destructive. But Borlaug brushed aside the complaints, as did governments in Asia and Latin America. In Mexico, the Rockefeller program had been revamped into a permanent research agency: the International Maize and Wheat Improvement Center, known as CIMMYT after its Spanish acronym. Wellhausen was its first director-general. CIMMYT joined the International Rice Research Institute in Manila, funded by the Ford Foundation but modeled after Borlaug's work. Now there are fifteen such centers, linked in an association (CGIAR, an acronym of its former name, the Consultative Group for International Agricultural Research) that is as profoundly important to global agriculture as it is little known.

Borlaug, too, remained little known, even after he won the Nobel Peace Prize in 1970. But he and the Green Revolution had become exemplary to a certain sort of scientist, journalist, and environmentalist. The Borlaug package has become an emblem of the view that the road through humankind's environmental difficulties lies through the groves of scientifically guided productivity. "Ours is the first civilization based on science and technology," Borlaug said that day in 1968. "In order to assure continued progress we scientists . . . must recognize and meet the changing needs and demands of our fellow men." The future of the world depends on science, he said, and on politicians guided by scientists. This vision of a scientific elite was like Vogt's vision, in its way, except that Borlaug was possessed by the hope of more, rather than a call for less.

To the end of his life, he kept his head down and worked ferociously hard. He always believed that hard rational work would lead him to the goal in the end. It was impossible for him to understand that there were people who didn't want to go there.

FOUR ELEMENTS

FOUR

Earth: Food

"A World Population 50 or 60 Times the Present One"

One of William Vogt's readers was a mathematician named Warren Weaver, director of the Division of Natural Sciences at the Rockefeller Foundation. Ambitious and multi-talented, Weaver was convinced that science and technology, carefully used, could improve the lot of humanity—which was why he left academia to join the foundation in 1932. Presciently, he believed that the life sciences were about to take the giant strides summed up by the words "molecular biology"—a term that Weaver himself coined. At Rockefeller, he was like the producer for the movie of molecular biology: the man who chose the scientists and funded the research that led to the main discoveries of DNA and RNA. Between 1954 and 1965, eighteen scientists received Nobel Prizes for molecular biology; fifteen were funded by Weaver at Rockefeller.

Equally remarkable was his work in field biology. When U.S. vice president Henry Wallace asked the Rockefeller Foundation to improve Mexican agriculture, Weaver urged his superiors to take on the task. As a reward for his advocacy, Weaver was asked to supervise the Mexican Agricultural Program. A small part of that duty was, in theory, overseeing Norman Borlaug.

In July 1948 the foundation acquired a new president: Chester Barnard, a retired telecommunications executive who had written classic books on management. A few weeks after he came to Rockefeller, *Road to Survival* appeared. Barnard quickly read it. He scoffed at Vogt's "vituperative" attacks on "everything from private property to the Pope and Communism." But he found himself unable to dismiss Vogt's claim that humankind was overwhelming Earth's carrying capacity. Could Rockefeller's efforts to improve health and food supply actually be counterproductive, because they would increase human numbers, hastening the ecological day of reckoning? Fearing that *Road* could "stir up" what he dryly called "blasphemous criticisms" of the foundation, Bernard asked Weaver to look into the matter. In particular, he asked, "How do we justify the Mexican agricultural program, considering [Vogt's] strictures?"

Weaver was a busy man. Even as he was setting up the molecular biology revolution and preparing for the Green Revolution, he was inventing the key concepts of machine translation; co-writing, with Claude Shannon, *The Mathematical Theory of Communication*, the founding document of information theory; establishing the basic ideas of what is now known as complexity theory; and, as a passionate hobby, researching the composition of *Alice in Wonderland*. Months later, he finally went through Vogt's book, along with Fairfield Osborn's *Our Plundered Planet*. He summed up his reaction in a confidential, seventeen-page report in July 1949. Informal, even slapdash, it reads more like a long email than a careful corporate memorandum. Nevertheless, it was one of the first—perhaps *the* first—modern statements of the Wizardly credo.

Vogt's warnings in *Road to Survival*, Weaver said, were "exaggerations"—"inaccurate, even if timely." They were mired in "the traditional patterns of the past." Environmental questions had to be thought about in a new way, he said. And he provided one. It was based on physics and chemistry, rather than biology.

To survive, Weaver said, humans have a single basic need: "usable energy." That energy comes in two forms: energy for the body (food and water, in other words), and energy for daily existence (that is, fuel to power vehicles, heat and cool buildings, and make essential materials like cement and steel). "In the United States," Weaver estimated, "each person uses, on the average, 3,000 calories per day for food, [and] 125,000 calories per day for heat and power."

Ultimately, those 128,000 calories had but one source: "nuclear disintegration." By this Weaver meant both the nuclear reactions inside the sun that create sunlight and those in atomic power plants. The latter was intriguing, but in 1949 nuclear technology was still so new and secret that Weaver believed "it is at this time not feasible to make any realistic estimates about atomic energy." For that reason, he ignored nuclear plants, at least for the moment, describing them simply as "potentially important."

Weaver *could* say something about the sun, though. In principle, the sun pours onto Earth enough energy— vastly more than enough—to provide all humanity with the necessary 128,000 calories a day. "If solar energy could be utilized with full efficiency,

Warren Weaver, 1963

the United States alone could sustain, energy-wise, a population over 40 times the present total population of the planet." The global population then being about 2 billion, Weaver was suggesting that in terms of energy the theoretical carrying capacity of the United States was about 80 billion people.

The limit of 80 billion would never be reached, because nobody would want to live in such a jam-packed country. But thinking in these terms was clarifying, Weaver thought. It showed that viewing the human dilemma in terms of an ecological carrying capacity was a mistake. The planet's actual, physical carrying capacity was so large—scores of billions of people—as to be irrelevant. The true problem was not that humankind risked surpassing natural limits, but that our species didn't know how to tap more than a fraction of the energy provided by nature.

Harnessing these energy sources would require new technology. But once people learned how to make "direct use" of the sun (or nuclear power), all human needs for heat, air-conditioning, transportation, electricity, steel, cement, and everything else would be satisfied for eons to come. In this respect, Weaver thought, Vogt was flat-out wrong.

Food energy, the second kind of energy, was a different matter: more complicated, harder to resolve. Here, Weaver conceded, Vogt's warnings might be borne out. Food energy derives from plants, either directly (when people eat them) or indirectly (when people eat animals that have eaten them). And the energy in plants comes from the sun, captured by photosynthesis.

Nobody in 1949 knew how photosynthesis worked. Almost two hundred years before, Jan IngenHousz (or Ingen-Housz), a Dutch doctor and biologist working in England, had established that sunlight, water, and carbon dioxide somehow went into a plant and were transformed into roots, leaves, and stems. But after IngenHousz the scientific project had come almost to a stop. For decade after decade, photosynthesis remained a black box. Sunlight, water, and carbon dioxide went in, plant growth came out. What happened inside was unknown.

By measuring the solar energy falling onto a plant and its associated growth, scientists had roughly calculated how much of that energy the plant actually used. Not very much, was the answer. Photosynthesis, Weaver said, "has an over-all efficiency surely less than 0.00025%"—one-quarter of one-thousandth of one percent! The inefficiency was mind-boggling.

The bright side was the potential for improvement. In theory, Weaver argued, photosynthetic efficiency could be increased

by a factor of something like 400,000. . . . If we had a more efficient way of turning solar energy into food—and let us now say, to be more reasonable, a way that had efficiency of only 1 percent—then an area the size of 1/100 the state of Texas would produce food enough to give 3,000 calories per day to a world population 50 or 60 times the present one.

As Weaver knew, researchers were far from being able to revamp photosynthesis. So he proposed other measures that, while difficult, seemed closer to reality: developing "the food potentialities of the sea," controlling rainfall on farmland, selecting and modifying bacteria (our "tiny servants," he called them) "to form the molecular aggregates that man needs as food." But these ideas, in his view, were placeholders for the real paths to the future: tapping new energy supplies, solar or nuclear, and hacking photosynthesis to grow more food.

Weaver never published his ideas. His memorandum lay unnoticed in the archives of the foundation, now stored underground on one of the Rockefeller estates. And his dream of reworking photosynthesis would be almost forgotten for sixty years, until it was revived by the descendants of the molecular biologists whom Weaver had funded and the successor to Rockefeller as the world's biggest charitable foundation.

When the idea did return, it would be entangled in an argument between Wizards and Prophets over how to feed tomorrow's world. Because that world will be (almost certainly) more numerous and (probably) more affluent, it is commonly stated that harvests will have to double by 2050. Some researchers believe this figure is overstated—a 50 percent increase would do the trick. In either case, though, how can it be done?

Wizards see an essential part of the answer in a new technology: genetic engineering. Hacking photosynthesis exemplifies its potential—reaching into the heart of life to ensure a better existence for millions of our fellows. By contrast, Prophets view reworking photosynthesis as embodying an ecologically foolish mania for growth and accumulation that will lead to destruction. At bottom, in their view, genetic engineering has the same fundamental fault as Weaver's memo: imagining that the world in all its complexity can be boiled down to a small number of physical parts that can be freely measured and manipulated.

"Reductionism" is the term of art for this idea, and agriculture is but one focus of a broader disagreement over its place. In this section of the book, I look at how Vogtians and Borlaugians view four great, oncoming challenges—food, water, energy supply, and climate change—each represented as one of Plato's four elements. The subjects greatly differ, but in every one Wizards and Prophets have taken up old quarrels and transformed them. In agriculture, for instance, the fight over genetic engineering represents the extension of a dispute, surprisingly heated and now almost a century old, over a seemingly arcane question: the proper manner of providing nutrients, especially nitrogen, to plants. And this, in turn, is related to an even older struggle—a quarrel over the nature of life itself.

The Story of N (Natural Version)

Then, what is life?

So reads the last line of "The Triumph of Life," the last poem that Percy Bysshe Shelley wrote before his death in 1822. Because Shelley died before completing it, we cannot know whether the surviving draft embodies his final intentions. But the version we have seems both elegantly assembled and intellectually muddled—life is at the same time a numinous, uplifting spirit and a crushing destroyer of that spirit. As a college student, assigned the poem in class, I was baffled by its inconsistency. Later I realized that Shelley's confusion was widely shared. He was writing just as scientists had begun a tussle over the definition of life.

In ancient times, life was typically viewed as a principle or essence: *qi* in China, *ase* in Nigeria, *mana* in Polynesia, *manitou* in the Algonkian cultures of North America, *pneuma* to the Greeks, the Force in a galaxy far, far away. Living creatures and non-living things are both made of matter, long-ago thinkers said. But the former eat, reproduce, act with intent, and do a hundred other things that seem beyond the capacities of the non-living. It was easy for the ancients to explain the gulf between life and non-life by imagining that a special kind of immaterial energy flows through and sustains living tissues. Without this essence, a live body would be a mere mechanism, not an organism.

For Aristotle, plants were a special case. Intrigued by how they grow despite lacking any visible mechanism for taking in food, he proposed that they obtain their nourishment by drawing in humus—decayed plant and animal matter—through their roots. Humus could nourish plants because it, like them, was charged with *pneuma*. Although Aristotle's hypotheses had what seem today like obvious problems—humus doesn't disappear into roots, for one—his ideas continued to be accepted in the West well into the eighteenth century, where they were promulgated by Johan Gottschalk Wallerius, a Swedish chemist who became the founder of agricultural chemistry. In his *Agriculturae fundamenta chemica* (1761), the first important treatise on the subject, Wallerius proclaimed that living creatures are driven by an internal energy unique to life. He called this energy the "spirit of the world." Other thinkers used different names: vivifying fire, active principle, vital force (or *vis vitalis*—the term was often left in Latin). Whatever the title, it was also characteristic of humus, which had once been alive, and still retained this animating drive.

Big breaks with the past are rare, but one was initiated by Carl Sprengel. Born in Germany in 1787, Sprengel was something of a prodigy, beginning his agronomy studies at the age of fifteen. As a professor at the University of Göttingen, he analyzed the chemical constituents of humus in a series of careful experiments in the 1820s. He concluded that everyone from Aristotle to Wallerius had it wrong: plants fed on the individual nutrients in humus, not on humus as a whole. Which meant that humus wasn't imbued with some unique life-spirit. It was just a stockpile of minerals and salts, some of them essential for plant growth. To promote these discoveries Sprengel wrote five textbooks in the 1830s.

At the time, the most prominent chemist in Germany was Justus von Liebig (1803–1873). Ambitious, charismatic, and contentious, Liebig was an educational innovator, a scientific visionary, and a conniving fraud—he faked his doctoral dissertation and, apparently, entire lines of experiments. Nonetheless, his reputation was such that in 1837 the British Association for the Advancement of Science commissioned Liebig, rather than any British scientist, to write a report on the state of organic chemistry. Three years later the great man returned with a report on a somewhat different topic: agricultural chemistry, a subject he had never previously researched. His conclusions were strikingly similar to Sprengel's, though he mentioned the other man's earlier research only in passing, dismissively. Sprengel complained, but Liebig's celebrity ensured that he received the credit for Sprengel's ideas. This turn of events was particularly unjust for what has become known, incorrectly, as Liebig's Law of the Minimum: plants need many nutrients, but their growth rate is limited by the one least present in the soil.

In most cases, that nutrient is nitrogen. At first blush, the notion of nitrogen being a limit seems odd; there is more nitrogen in the world than carbon, oxygen, phosphorus, and sulfur combined. Unfortunately, more than 99 percent of that nitrogen is nitrogen gas. Nitrogen gas—N_2 in chemical notation—consists of two nitrogen atoms bound together so tightly that plants cannot split them apart for use. Instead, plants are able to absorb nitrogen only when it is in chemical combinations— "fixed," as scientists say—that are easier to break up.

In the soil, nitrogen is mainly fixed by microorganisms. Some break down organic matter, making its nitrogen available again; others, such as the symbiotic bacteria that live in and around the roots of beans, clover, lentils, and other legumes, directly fix nitrogen gas into compounds plants can take in. (A small amount is fixed by lightning, which zaps apart nitrogen molecules in the air, after which they combine with oxygen into compounds that dissolve in rainwater.) When farmers put additives like

Justus von Liebig in a portrait from 1846

ashes, blood, urine, compost, and animal feces in their fields, they are providing fodder for nitrogen-fixing soil microorganisms; when they grow legumes, it increases the supply of nitrogen-fixing bacteria. The implication of Liebig's work was that dumping artificially created nitrogen compounds—chemical fertilizers—into fields would do the same thing.

Smarmy but far-sighted, Liebig envisioned a new kind of farming: agriculture as a branch of chemistry and physics. In this scheme, soil was just a base with the physical attributes necessary to hold roots. What mattered to agriculture were the chemical nutrients on which plant growth depended: nitrogen, potassium, phosphorus, calcium, and so on. If farmers wished, they could plant seeds in soil with zero humus (expanses of sand, perhaps), sprinkle the seeds with the doses of water and chemicals prescribed by experts, and the seeds would germinate and grow. A farm would be an organic machine, in the phrase of the historian Richard White; the new agriculture, precise as a clock, would be a harnessed flow of energy and matter. *Vis vitalis,* living humus, and *pneuma*—not only were they irrelevant, they did not exist. Crops and soil were brute physical matter, collections of molecules to be optimized by chemical recipes, rather than flowing, energy-charged wholes. In today's terms, Liebig was taking the first steps toward industrial agriculture regulated by farm chemicals—an early version of Wizardly thought.*

---

* Liebig sought to profit from his ideas by launching a fertilizer company. His celebrity attracted backers and they set up a factory in 1845. Bizarrely, he initially refused to put nitrogen in its products, maintaining that plants received plenty of nitrogen in the form of ammonia (NH_3) released by decaying roots and leaves in the soil. Supposedly that ammonia rose up into the air as a gas, dissolved into rainwater, and fell back plentifully to Earth. For nitrogen, Liebig said, manure and other traditional fertilizers were "unnecessary" and "superfluous"; instead they supplied different limiting nutrients: potassium and phosphorus. Alas, Liebig's refusal to put nitrogen in his products meant that they were ineffective; his cocksure refusal to test them meant that he didn't discover their ineffectiveness until after his customers did.

At the time, the biggest known fertilizer source was Peruvian guano. As demand drove up prices and reduced supplies, attention turned to sodium nitrate. Sodium nitrate ($NaNO_3$) consists of a sodium atom, a nitrogen atom, and three oxygen atoms, all bound together loosely enough for plants to assimilate the nitrogen. The world's biggest nitrate deposits are in the high desert of northern Chile. Although it almost never rains there, the area is constantly bathed in a fine spray from the Pacific Ocean. The spray is very thin—less than an inch per year—but it contains the nutrients from the Humboldt Current that feed the anchovetas that feed the cormorants. Other nutrients fall from the sky as dust or well up from groundwater. With next to no rainfall to wash away the residue, the deposits build up over time. The result: a layer of naturally deposited fertilizer four hundred miles long, twelve miles wide, and up to nine feet deep. It was eagerly exploited. Nitrates from Chile became a principal ingredient in packaged fertilizers—and, alas, bombs. By the beginning of the twentieth century, as Vaclav Smil of the University of Manitoba has written in his history of nitrogen use, almost half the nitrates shipped to the United States were used to make explosives.

In 1898 the British chemist William Crookes rang an alarm: the nitrogen would run out. Crookes was the new president of the British Association for the Advancement of Science, the group that had commissioned Liebig's report. In his inaugural address, Crookes focused, quite literally, on Europe's daily bread. The "bread-eaters of the world," as he called them, were increasing by more than 6 million a year. To feed these new bread eaters, farmers either would have to expand into unused land or produce more from their existing land by fertilizing it more heavily. Neither course was possible, Crookes thought. Most suitable land was already under the plow. And increasing the demand for fertilizer would exhaust the supply of Peruvian guano and Chilean nitrates in "a few years." By the 1930s, Crookes predicted, the world's wheat supply "will fall so far short of the demand as to constitute general scarcity." Science, Crookes hoped, would somehow save the bread eaters.

He got his wish. Science did save the day—at least for a while.

By the time he admitted that nitrogen was the key factor, he and his backers had lost a lot of money. Tempers can't have been improved by his subsequent claim that he had always promoted nitrogen's central role.

The Story of N (Synthetic Version)

Six years after Crookes issued his warning, an Austrian chemical company asked the German chemist Fritz Haber to look into synthetic fertilizer. More precisely, the Austrians asked Haber to look into synthetic *ammonia*. For decades researchers had believed that if they could manufacture ammonia it could be used as the basis for a synthetic fertilizer—something made in a factory instead of being dug out of the ground and shipped across the ocean. Chemically speaking, ammonia (NH_3) is simple: three hydrogen atoms, one nitrogen atom, arranged in a rough pyramid. Both hydrogen and nitrogen normally exist as gases, H_2 and N_2. In theory, one should be able to split gaseous nitrogen and hydrogen into single N and H atoms, then put together the separate atoms into NH_3 like so many building blocks. Scientists had figured out how to split the hydrogen. But, like plants, they had been unable to pry apart N_2. Every attempt to make synthetic ammonia had failed.

"Failed" in this case meant "failed to come up with something that industry could use." Chemists actually *had* synthesized ammonia, but only at ultra-high temperatures and pressures in costly laboratory experiments. And even in these extreme circumstances the reaction needed a *catalyst*, a substance that facilitates a chemical reaction but is not itself affected by it. Catalysts are like jaywalking pedestrians who cause car accidents but walk away from them without being affected. But unlike the disruptive pedestrians, catalysts are essential to the smooth functioning of thousands of chemical processes.

Several metals served as catalysts for ammonia. In the right conditions, the metal adsorbs hydrogen and nitrogen gases, dissociating them into separate hydrogen and nitrogen atoms. Now unattached, the nitrogen atoms can easily bond with hydrogen atoms, creating ammonia molecules. Some of the energy from the newly formed N-H bonds helps the ammonia molecule leave ("desorb," in the jargon) the metal surface and float into the air. The metal is left unchanged.

Haber tried this himself. Like his predecessors, he found that blowing hot, high-pressure nitrogen and hydrogen gas over metals like iron, manganese, and nickel produced tiny but measurable amounts of ammonia—it converted about one-hundredth of 1 percent of the original gases. By repeatedly recirculating the hydrogen and nitrogen, Haber could very slowly fix most of the nitrogen into ammonia. But, as he told

Fritz Haber (right) supervises a laboratory assistant in a newspaper photograph from 1918, the year he won his Nobel Prize.

the Austrians, the process was too arduous to justify the cost. It would be like spending millions of dollars to build an orange-juice factory that produced a teaspoon of juice a day.

Soon after, luck entered the picture in the form of Walther Nernst, a brilliant but caustic physicist. Nernst's experiments on the effects of heat in chemical reactions led him to conclude in 1907 that Haber's estimates of ammonia production were much too high. Even though Haber thought his figures were pessimistic, they overestimated ammonia production by almost a factor of four. Haber repeated his earlier tests and this time obtained results that were close to Nernst's claims. Chagrined, Haber acknowledged his error. In a fit of pettiness, Nernst publicly derided Haber's "strongly inaccurate" work, which Nernst said had led him to believe "that it might be possible to synthesize ammonia from nitrogen and hydrogen." As Smil, the nitrogen historian, has pointed out, this was nonsense: Haber had concluded that fixing nitrogen was *not* feasible, the opposite of Nernst's suggestion. Still, the incident was humiliating.

Now determined to redeem himself, Haber returned to ammonia.

But now he had acquired a new ally: Badische Anilin- und Soda-Fabrik (BASF), the world's biggest chemical firm. The basic problem was that ammonia was best synthesized when the hydrogen and nitrogen were at high temperature and high pressure, but those very conditions pushed up the procedure's cost and difficulty. BASF helped Haber build a better high-pressure apparatus and search for better catalysts. A breakthrough came on July 2, 1909, when, for five hours straight, Haber pumped hot gas into the apparatus and "produced continuously liquid ammonia."

Haber's experimental model was just two and a half feet tall, too small for commercial production. And his catalysts, osmium and uranium, were commercially unsuitable: the total global supply of osmium was less than 250 pounds and uranium was dangerous—not just because it is radioactive, but because it reacts explosively with oxygen and water. Nonetheless, he had demonstrated that it was possible to synthesize ammonia at high volume.

BASF put a chemical engineer named Carl Bosch in charge of scaling up Haber's process and finding more affordable catalysts. Bosch, too, had spent years trying to fix nitrogen. When he learned that Haber had beaten him to the punch, he told the company without regret that it should immediately develop his rival's design. Building the requisite high-pressure tanks proved to be especially difficult, because—an unhappy surprise to Bosch—the hydrogen diffusing into the walls combined with the carbon in the steel, weakening the metal. Meanwhile, Bosch set up a team that tested thousands of compounds to find a better catalyst. The best proved to be iron with a little aluminum, calcium, and magnesium. By 1913 BASF's first big ammonia plant was running.

Five years later Haber received a Nobel Prize for synthesizing ammonia; Bosch and his main assistant received a Nobel in 1931, for developing "chemical high pressure methods." Ammonia synthesis remained so costly that artificial fertilizers did not truly become common until the 1930s. Nonetheless, the Nobels were richly deserved; the Haber-Bosch process, as it is called, was arguably the most consequential technological development of the twentieth century, and one of the more important human discoveries of any time. The Haber-Bosch process has literally changed the land and sky, reshaped the oceans, and powerfully affected the fortunes of humanity. The German physicist Max von Laue put it neatly: Haber and Bosch made it possible to "win bread from air."

Carl Bosch (left) receives the 1931 Nobel Prize in Chemistry from Crown Prince Gustav of Sweden.

Today the Haber-Bosch process is responsible for almost all of the world's synthetic fertilizer. A little more than 1 percent of the world's industrial energy is devoted to it, as the futurist Ramez Naam has noted. Remarkable fact: "That 1 percent," Naam says, "roughly doubles the amount of food the world can grow." Between 1960 and 2000 global synthetic fertilizer use rose by about 800 percent. About half of that production was devoted to just three crops: wheat, rice, and maize. One way to look at this figure is to say that the accomplishment of Borlaug and his associates was to create strains of wheat, rice, and maize that could use what Haber and Bosch had provided.

Increasing the food supply has led to a concomitant increase in human numbers. Vaclav Smil has calculated that fertilizer from the Haber-Bosch process was responsible for "the prevailing diets of nearly 45% of the world's population." Roughly speaking, this is equivalent to feeding about 3.25 billion people. More than 3 billion men, women, and children—an incomprehensibly vast cloud of dreams, fears, and explorations—owe their existence to two early-twentieth-century German chemists.

The magnitude of the change wrought by artificially fixed nitrogen is hard to grasp. Think of the deaths from hunger that have been averted, the opportunities granted to people who would otherwise not have had a chance to thrive, the great works of art and science created by those who would have had to devote their lives to wringing sustenance from the earth. Particle accelerators in Japan, Switzerland, and Illinois; *One Hundred Years of Solitude* and *Things Fall Apart*; vaccines, computers, and antibiotics; the Sydney Opera House and Stephen Holl's Chapel of St. Ignatius—how many are owed, indirectly, to Haber and Bosch? How many would exist if this Wizardly triumph had not produced the nitrogen that filled their creators' childhood plates?

Hard on the heels of the gains were the losses. About 40 percent of the fertilizer applied in the last sixty years wasn't assimilated by plants; instead, it washed away into rivers or seeped into the air in the form of nitrous oxide. Fertilizer flushed into rivers, lakes, and oceans is still fertilizer: it boosts the growth of algae, weeds, and other aquatic organisms. When these die, they rain to the ocean floor, where they are consumed by microbes. So rapidly do the microbes grow on the increased food supply that their respiration drains the oxygen from the lower depths, killing off most life. Where agricultural runoff flows, dead zones flourish. Nitrogen from Middle Western farms flows down the Mississippi to the Gulf of Mexico every summer, creating an oxygen desert that in 2016 covered almost 7,000 square miles. The next year a still larger dead zone—23,000 square miles—was mapped in the Bay of Bengal.

Fueling the fire are automobile engines, which as a by-product of combustion convert nitrogen gas to various types of nitrous oxides (NO_x, in the language of chemists). Rising into the stratosphere, nitrous oxides combine chemically with the planet's protective ozone, which guards life on the surface by blocking harmful ultraviolet rays. Mixing with nitrous oxide pulls ozone off duty. Down below, NO_x leads to pollution. The total cost of unwanted nitrogen has been estimated at hundreds of billions of dollars a year. Were it not for climate change, suggests the science writer Oliver Morton, the spread of nitrogen's empire would be our biggest ecological worry.*

---

* Confusingly, nitrous oxide (N_2O) is not one of the nitrous oxides (NO_x). Because the extra nitrogen in N_2O makes it behave differently than NO_x, chemists place it in a separate category.

Law of Return

Action brings reaction, every *yes* from one followed by another's *no*. Justus von Liebig's proto-Wizardly vision of industrial agriculture, the farm as organic machine, generated a proto-Prophetic attempt to bring back the living spirit he had banished. The counterforce saw each achievement touted by the modernizers as a deficit, every new landmark as a ruin. Had they been able to read Warren Weaver's manifesto, they would have rejected it from the first page: agriculture was about more than "usable energy." In their view, the modernizers had forgotten about the living humus—*pneuma*, so to speak. On many levels this was a dreadful mistake, the counterforce believed, one that would reverberate through time and space.

"It is notoriously difficult to identify precisely the beginning of a cultural movement," wrote the University of Leicester historian Philip Conford. In his history of organic farming, Conford argued that the most appropriate date for the beginning of the pushback against Liebig-style agriculture was the 1920s, when Haber-Bosch fertilizer was beginning to spread across the world. Resistance awoke in Africa, Germany, Great Britain, and the United States. But the most important source was South Asia, where challengers found inspiration in the small, traditional farms that Norman Borlaug would later try to modernize.

Among the first naysayers was Robert McCarrison, later Major General Sir Robert McCarrison, C.I.E., F.R.C.P. Raised in Northern Ireland, McCarrison joined the military as a surgeon right after obtaining his medical degree. In 1901 he came to what is now Pakistan and was then the northern tip of British India. Twenty-three years old, he had no training in epidemiology, public health, or environmental science. Nonetheless he made contributions to all of these, discovering the environmental causes of diseases (bacteria-carrying insects, vitamin deficiencies) and methods to prevent them. Eventually he became the colonial director of nutritional research, a post he retained until his retirement in 1935.

In his work as a physician, McCarrison traveled through the high reaches of northern Pakistan, where he encountered the Hunza Valley, inhabited by an Ismaili Muslim people "whose sole food consists to this day of grains, vegetables, and fruits, with a certain amount of milk and butter, and goat's meat only on feast days." The Hunza were superb

Although he was a proud servant of the British Empire, Robert McCarrison (third from left, during a tour of his workshop in India in 1926) became convinced that Asian farming methods were superior to those in Europe.

physical specimens: "unsurpassed in perfection of physique and in freedom from disease in general." In seven years of visits McCarrison "never saw a case of asthenic dyspepsia [chronic stomach distress], or gastric or duodenal ulcer, or appendicitis, of mucous colitis [irritable bowel syndrome], of cancer." Education and affluence were not the cause of their "extraordinarily long" lifespans; the Hunza were illiterate and so impoverished that most could not afford to keep dogs. McCarrison came to believe that their good health was due to their diet.

McCarrison's research, published in 1921, was an early example of a new scientific genre: the study of poor people in isolated places who live long, vigorous lives. Missionaries on the Ogowe River in Gabon; anthropologists in the U.S. Southwest; gold-mine officials in South Africa; doctors in Inuit villages—all encountered groups in remote areas who rarely experienced cancer, cardiovascular disease, diabetes, and the other chronic conditions that have come to be called "diseases of civilization." Despite poverty, poor sanitation, and lack of medical care, the Hunza were healthier than their British counterparts. The reason, McCarrison concluded, was that British working-class diets, with their tinned meat, sugary tea, and poofy white bread, were much poorer in vitamins than Hunza diets.

In the early 1920s McCarrison met two Indian agricultural chemists, Bhagavatula Viswanath and M. Suryanarayana. Based at a research institute in Coimbatore, in southern India, they had been comparing synthetic fertilizers and manure.* The former had been backed by the revered Liebig, whereas the latter had been used to replenish soil in South Asia for millennia. Liebig believed that grain produced with synthetic fertilizer and grain produced with manure should be identical, as long as the additives had the same nutrients. In practice, he thought, the synthetic should be better, because scientists could create mixes that targeted individual soils, giving the farmer more control. To test the great man's ideas, Viswanath and Suryanarayana grew wheat and millet in identical plots with manure and synthetic fertilizer, then chemically analyzed the resultant grain.

Intrigued, McCarrison suggested that the real question was not the different grains' chemical makeups, but their quality as *food*. Pushing his way into the enterprise, he commandeered a laboratory and fed the two men's experimental wheat and millet to rats and pigeons. Animals fed on the manure version grew robustly; those given chemically fertilized wheat and millet suffered from malnourishment. There was something in the manure-fed humus, he concluded, a factor that Liebig hadn't known about. McCarrison published this finding in 1926, assigning himself the byline and the lion's share of the credit.

Then he went further—beyond the ideas of his reluctant collaborators, Viswanath and Suryanarayana. A logical next step could have been to pin down the identity of the special nutrients in manure, then add them to synthetic fertilizers. But instead McCarrison had a conversion experience: one of those *aha!* moments that transform lives. He decided that the underlying issue was Liebig's reductionist view of the soil as a passive reservoir of chemical nutrients.

---

* "Manure" is commonly thought of as animal excrement, usually from cows, horses, or humans. But there are actually two forms: brown manure (animal feces) and green manure (crop remains or cover crops that are plowed back into the land). When I use the term, I refer to both.

In a set of lectures in 1936, McCarrison laid out what would become the ideology of the counterforce. "Perfectly constituted food," McCarrison said, was the single biggest determinant of good health. The most important part of this perfectly constituted food was plants: fruits, vegetables, and whole grains. In turn, he said, the nutritive qualities of those plants depended on how they were cultivated. And here Liebig was simply incorrect; chemistry was not the whole story. To grow the best, most nutritious crops, farmers needed (to put it in modern terms) to view their land not as a store of chemicals to be managed efficiently but as a complexly interacting living system to be cherished and maintained. Every part of the system contributed to the whole, but one predominated: the soil.

Impoverishment of the soil leads to a whole train of evils: pasture of poor quality; poor quality of the [live]stock raised upon it; poor quality of the foodstuffs they provide for man; poor quality of the vegetable foods that he cultivates for himself; and, faulty nutrition with resultant disease in both man and beast. Out of the earth are we and the plants and animals that feed us created, and to the earth we must return the things whereof we and they are made if it is to yield again foods of a quality suited to our needs.

The soil! The soil! When fed by plant remains and animal excrement, it became a vibrant, circulatory network that nourished the plants and animals which fed it. Rather than trying to replicate this system in the laboratory—an attempt doomed to fail—farmers should simply let the soil ecosystem create it naturally from humus, as Asian farmers had for millennia.

McCarrison's ideas overlapped with those of Albert Howard, another expatriate Briton in India. Born in 1873, Howard grew up on a farm on the western edge of England. Early experiences with plow and scythe led to unyielding skepticism of the microscope-wielding lab dwellers who never had dirt beneath their nails. At the same time Howard himself had impeccable academic credentials: a first-class degree in chemistry at the Royal College of Science, top of his class in agricultural science at Cambridge.

In 1905 Howard was lured to the new Agricultural Research Institute in Pusa, in northeast India. Accompanying Albert was his new wife, Gabrielle, a Cambridge-educated plant physiologist. The Howards became partners in the laboratory as in life, though he received more recognition—inevitable, one assumes, given the time. Individually and together, they bred new varieties of wheat and tobacco, studied root distribution, developed novel types of plow, and tested the results of providing oxen with a super-healthy diet and living quarters but not vaccinating them against disease. From the beginning, Gabrielle had urged Albert to think holistically, to see connections between different fields of inquiry. By 1918 their ideas were clear: "the health of the soil, plant, and animal were linked to each other, that fertile soil held the key to increased crop yield, and that manure was the key to soil fertility." (I am quoting Conford, the University of Leicester historian, whose work I am following here.)

On a practical level, the Howards' most important contribution may have been developing what is called the Indore process of composting, Indore being the region of central India where their work took place. McCarrison had focused on soil additives, especially manure; the Howards looked at composting, in which bacteria and fungi break down agricultural and household waste, fixing its nitrogen into usable ammonia and nitrates. In the Indore process, waste was inoculated with bacteria and fungi and mixed with ash in a five-to-one ratio; the material was periodically turned to expose it to oxygen, maximizing nitrogen fixation. The Howards' methods, slightly modified, remain in use today for large-scale composting.

Others had previously focused on composting, including, famously, the celebrated French novelist Victor Hugo. Violating every rule of narrative, Hugo interrupts the climax of his great novel *Les Misérables* (1862) to hector the reader for fifteen pages about the Parisian sewer system. The city's sewers discharged vast quantities of excrement into rivers, which carried it to the sea. That excrement, Hugo proclaimed, should instead be applied to farmers' fields: "the most fertilizing and effective of manures is that of man." The guano trade—shipping bags of fertilizer across the ocean!—was an intercontinental folly.

We fit out convoys of ships, at great expense, to gather up at the south pole the droppings of petrels and penguins [Hugo wrote, inexactly], and the incalculable element of wealth which we have under our own hand, we send to sea. . . . To employ the city to enrich the plain would be a sure success. If our gold is filth, our filth is gold.

Hugo's call for greater use of compost and manure was echoed by later writers, though none with his rhetorical exuberance. But neither Hugo nor his successors had much effect—the sewers kept flushing golden filth into the water. And Howard's work might have received equally little notice if his wife, Gabrielle, had not died unexpectedly in 1930.

Bereft, Howard resigned and returned to Britain to putter sadly about his garden. He believed his career was over. Instead it entered an active new phase. Soon after coming to England, he married Gabrielle's equally remarkable younger sister Louise, a Cambridge-educated pacifist and suffragist who had taught classics, become a character in a Virginia Woolf novel (in life, she had been an editorial assistant to Woolf's husband, Leonard), and chaired the agricultural division of the International Labor Organization in Switzerland. Louise did not want Howard to abandon her sister's work. Howard found himself traveling the world, promulgating the Indore process—and more.

"The basis of all Nature's farming," he said, was the Law of Return: "the faithful return to the soil of all available vegetable, animal, and human wastes." When bacteria, bugs, and birds die, their bodies return to the soil and provide nutrition for other life. The same occurs for their wastes. Humans, too, must return the residues of their existence to the earth. Civilizations fall because societies forget this simple rule. We depend on plants, plants depend on soil, soil depends on us. The Law of Return embodies an insight: everything affects everything else.

McCarrison had been promulgating much the same message, but Howard had a knack for coining a phrase and a willingness to promote his views with a kind of cheerful, over-the-top viciousness—"amiable brutality," Louise Howard called it. His *An Agricultural Testament*, published in 1943, is often called the founding document of the organic movement. In its pages Howard didn't just extol composting, he took after "the NPK mentality"—named after the chemical symbols for nitrogen, phosphorus, and potassium, the key ingredients in synthetic fertilizer. He didn't simply criticize the scientist who rarely ventured into fields, he called out this wretched specimen as a "laboratory her-

Organic pioneers (from left) Sir Albert Howard, Lady Eve Balfour, and Lord Northbourne

mit," "all intent on learning more and more about less and less" in the bowels of an "obsolete research organization." He didn't merely limit himself to decrying the overuse of synthetic fertilizers, he contended that they were actively toxic: "The slow poisoning of the life of the soil by artificial manures is one of the greatest calamities which has befallen agriculture and mankind."

Howard became the nucleus of a small but influential movement, the influence due partly to the fact that many of its members were aristocratic Christians who saw industrial agriculture as a threat to both the social and divine orders. Boldfaced names like Lord Bledisloe (governor of New Zealand), the Duke of Bedford (founder of the humus-promoting, anti-Semitic British People's Party), and Lord Northbourne (author of *Look to the Land*) were among its most prominent exponents. Howard himself was knighted in 1934. Indeed, toffs were so heavily represented in the Soil Association, Britain's leading farm-reform organization, that its early meetings were like house parties at Downton Abbey, except that the discussions over sherry were about manure and earthworms.

Exemplary in these respects was Howard's leading convert, Soil Association founder and president Lady Eve Balfour, author of *The Living Soil* (1943). Balfour's background was a mix of money, power, and mysticism: one of her grandfathers was viceroy of India; the other, a socially prominent occultist and writer. Her father was the chief secretary of Ireland at the same time that her uncle was the British prime minister. Inspired by a deep but idiosyncratic Christian faith, Lady Balfour sought a "spiritual and moral revival" in which taking care of the land would play a central part. Through "service to our God, service to the soil, service to each other," she said, poor humankind would ascend into "the next evolutionary stage," creating "the Kingdom of God on Earth."

Across the Atlantic sprang up a similar movement. It, too, owed inspiration to Howard and McCarrison and had many adherents who were inspired by Christianity. But it was in no way aristocratic. Instead, its central figure was Jerome I. Rodale, a hardscrabble entrepreneur, publisher, playwright, gardening theorist, food experimenter, and anti-vaccine advocate. Born in a Jewish ghetto in New York City in 1898, Rodale was a sickly youth, prone to headaches and colds; his family was plagued by congenital heart problems. After working as an accountant and tax auditor, he launched a successful electrical-equipment firm with his brother. When the Depression hit, the Rodale siblings cut costs by moving the plant to rural Pennsylvania. Once relocated in the quiet countryside, Jerome had time on his hands. Almost as a hobby, he set up a publishing company, issuing pamphlets about manners, humor, and health.

In 1941 he read a magazine article about Howard. Rodale was still beset by headaches and colds and feared he had a weak heart. Diet, maybe, was the answer. Scientists might scoff, but the idea made sense to him. Intrigued, he read *An Agricultural Testament*. He had been primed, years earlier, by hearing McCarrison speak. Now Howard's book went through his eyes to his heart and burned there like a flame. He contacted Howard. He bought a sixty-acre farm nearby and began working it according to the Law of Return. He read more of Howard and Howard's circle, like Lady Balfour, Lord Northbourne, Lionel Picton (editor of *The Compost News Letter*), and McCarrison. Indeed, Rodale liked McCarrison's writing so much that Rodale, too, wrote a boosterish book about the Hunza Valley even though he had never set foot outside the United States. *The Healthy Hunza* was published by Rodale Press in April 1948, four months before *Road to Survival*. By that time Rodale was hailing his bumper crops and his improved health.

Rodale died in 1971—bizarrely, on a television talk show, suffering a heart attack minutes after declaring "I never felt better in my life!" and offering the host his special asparagus boiled in urine. Naturally, this attracted ridicule. But he left a mighty legacy—and he lived twenty years longer than his siblings, all of whom, like him, had heart conditions.

Soon after reading Howard's book, Rodale had repackaged his ideas in an article—"Present Day Crops Unfit for Human Consumption!"—in his small magazine, *Fact Digest*. It caused such a stir that Rodale shuttered *Fact Digest* and opened a new magazine, *Organic Farming and Gardening*. Later he realized that the number of people who might buy organic food was much greater than the number who would be interested in growing it themselves. In 1950 he launched *Prevention* magazine to "build the organic movement stronger and stronger" by promoting organic food to consumers.

J. I. Rodale

Lord Northbourne had coined the term "organic farming" in 1940, almost as an aside. By splashing "organic" on a magazine cover, Rodale transformed it from a neutral word that meant relating to or derived from living creatures to a special label for the "life-giving" food that came from abjuring factory-made chemicals. It was a cudgel with which to beat industrial agriculture. In 1948 *Organic Farming and Gardening* had ninety thousand paid subscribers. By the time of Rodale's death the number had grown to half a million. *Prevention* had more than a million. Despite basing himself in the middle of nowhere and refusing most mainstream advertising, Rodale had built up an empire of belief.*

"Muck and Magic"

Such effusions naturally generated a pushback of their own. Agriculture officials, farming associations, chemical companies, and university researchers denounced Howard and Rodale for decades. "Half truths, pseudo science, and emotion," said the dean of a Kansas farm school in a

---

* Parallel to the Howard-inspired soil movements in North America and Europe was a second, independent soil movement in Germany. Its central figure was Rudolf Steiner (1861–1925), an Austrian philosopher/social reformer/Christian mystic who founded a movement known as anthroposophy. One component of anthroposophy was a spiritually driven form of soil restoration, separately derived but strikingly similar to Howard's. Steiner's movement spread across the world but in the end had little deep influence over the organic movement outside of Germany.

widely cited attack. Rodale's followers were a "cult of misguided people" who had turned their backs on science. "Alarmists," sneered a California farm official about the organic movement. "Are chemical fertilizers harmful? The answer is no!" Nor would the industry allow pesticides to be deprecated. "A Poison for Every Bug," proclaimed *The Country Gentleman*, the biggest U.S. agricultural magazine. Food faddist! Charlatan! Crackpot!—Rodale and his followers received every insult in the thesaurus. "I was touching off a powder keg," he said, proudly.

Some of the criticism was justified. Laboratory-hermit researchers reproached Howard, often accurately, for overstating his case. When Howard asserted that "artificial manures lead inevitably to artificial nutrition, artificial food, artificial animals, and finally to artificial men and women"—well, ordinary scientists rolled their eyes. *Inevitably?* Howard had *proof*? And what, exactly, are "artificial animals" and "artificial men and women"?

Compounding the sin, in critics' eyes, the organic movement blithely ignored costs. To make food for millions with compost, wrote the fertilizer chemist Donald Hopkins, would require "a truly colossal effort in terms of labor, transport, and planning." The outlays would drive up the price of food—terrible for people with limited incomes. Maybe food produced by "the disciples of Liebig" was somehow "artificial," but it made life easier for the vulnerable poor.

Most of all, opponents charged that Howard and his followers based their ideas on spirituality, rather than science; ideology, rather than empirical data. In their eyes, Howard's "extremist views" were bringing back the long-discredited, flat-Earth belief that humus was imbued with a special life-force. The purported agricultural reform, critics scoffed, was little more than a species of right-wing mysticism—"muck and magic," they called it.

Again, some of the criticism was merited. Lady Balfour indeed approached Aristotle and *pneuma* when she talked about "the living principle in nature . . . the ingredient of life itself, which permeates each individual cell of all the countless millions that go to make up the plant or animal's body." And it was hard for secular researchers to contain themselves when Howard rhapsodized over decaying plant and animal matter ("glorious forest humus") and called it "the very beginning of vegetable life and therefore of plant life and of our own being."

In other ways, though, the conflict seems absurd. Howard was in-spired by a religious faith in a natural order with limits that could not be exceeded with impunity. But when he lauded the living nature of humus, he was referring to the community of soil organisms, the dynamic rela-tions between plant roots and the earth around them, and the physical structure of humus (humus stickily binds together soil particles into airy crumbs that hold water instead of letting it run through)—all of which were very real, and all of which were unknown when Justus von Liebig formed the basic ideas behind chemical agriculture. Howard's argument that the industrialization of farming was depopulating the countryside and disrupting an older way of life was accurate, too, though his oppo-nents disagreed with him about whether this was a bad thing. And his fears about soil seem prescient—a landmark study from the U.N. Food and Agricultural Organization concluded in 2011 that up to a third of the world's cropland is degraded.

Industrial and organic boosters both agreed that soil needed to fur-nish nutrients to plants, especially nitrogen. The difference was that the Liebigs believed that synthetic fertilizer could deliver them—a nitrogen atom from a factory, they said, was identical to a nitrogen atom from the rear end of a cow—and the Howards believed that they were better provided via the Law of Return as part of a natural system. At the begin-ning it might have been possible to reconcile the two points of view. One can imagine industrial advocates considering humus, humus advocates willing to use chemicals as a supplement to good soil practice. But that didn't happen. Hurling insults, the two sides moved ever further apart.

As early as 1940, Northbourne had described the conflict between organic and conventional farming as a war that would last for "genera-tions of concentrated effort." Howard was a general in that war—"the warrior at the apex of the phalanx," one disciple called him. "Chemicalist versus organiculturalist," Rodale characterized the fight; he was happy to enlist. "The Revolution has begun," he announced in the first issue of *Organic Gardening,* capitalizing the *R* to show that he was serious. The magazine lost money for years. But Rodale was in it for the long haul. McCarrison, Howard, Balfour, Northbourne, and so many others stood at his back. All set in motion a battle that not only continued into the twenty-first century, but with the onset of genetically modified crops became ever more intense.

Slow

Picture William Vogt on his guano island in 1940: a footnote in the history of nitrogen. The Haber-Bosch process has been known for thirty years; chemical companies like BASF are capitalizing on it. But natural fertilizer is still important enough that Peru has hired a foreign biologist to protect it. Strikingly, neither scientists nor corporations nor organic advocates understand *why* it is so important to provide nitrogen to the soil. Not until years after Vogt left Peru did researchers learn the answer: nitrogen is critical to photosynthesis.

Photosynthesis is hard to describe without sounding like a hand-waving mystic. By blending water from below with sunlight and carbon dioxide from above, photosynthesis links Earth to the sky. The crops in every farmer's field are air and sunlight in cold storage. So are the trees around the field and the algae in nearby ponds. Every dot of green on the landscape is a ceaselessly active photosynthetic factory. If this furious microscopic churning stopped, Oliver Morton, the science writer, has remarked, "so would everything else that you care about." The planet would survive. But it would no longer be green.

Plants need nitrogen chiefly to make a substance called rubisco, a prima donna in the dance of interactions that is photosynthesis. Rubisco is an enzyme, which means that it is a biological catalyst. Like the iron in the Haber-Bosch process, enzymes cause biochemical reactions to occur but are left unchanged by those reactions. Tens of thousands are known—the online enzyme database BRENDA alone has data on eighty-three thousand. Essential to every cell of every living creature, invisible to the eye but feverishly in motion, enzymes typically catalyze thousands of reactions per second. Some accelerate them a billionfold or more.

Rubisco is the essential catalyst for photosynthesis. Like military recruiters who induct volunteers into the army and then return to their work, rubisco molecules take carbon dioxide from the air, insert it into the maelstrom of photosynthesis, then go back for more. The name "rubisco" was coined, jokingly, in 1979, to sound like a breakfast cereal; it is a sorta-kinda acronym for the compound's scientific name, *ribu*lose-1,5-*bis*phosphate *c*arboxylase/*o*xygenase. Rubisco's catalytic actions are the limiting step in photosynthesis, which means the rate at which rubisco functions determines the rate of the entire process. Photosynthesis walks at the speed of rubisco.

Alas, rubisco is, by biological standards, a sluggard, a lazybones, a couch potato. It causes reactions to occur, but very slowly. Whereas typical enzymes catalyze thousands of reactions a second, rubisco deigns to involve itself with just two or three per second. It is one of the pokiest enzymes known. When Warren Weaver bewailed the inefficiency of photosynthesis, he was unknowingly bewailing the torpor of rubisco. Years ago I talked to biologists about photosynthesis for a magazine article. Not one had a good word to say about rubisco. "Nearly the world's worst, most incompetent enzyme," said one researcher. "Not one of evolution's finest efforts," said another.

Not only is rubisco slow, it is inept. Carbon dioxide (CO_2) consists of a carbon atom (C) flanked by two oxygen atoms (O), the whole in a straight line. Oxygen gas (O_2) consists of two oxygen atoms. Like the atoms in carbon dioxide, the oxygen atoms are bound together linearly. Schematically, they look like this:

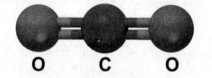 

Carbon dioxide (CO_2, left) and oxygen (O_2)

Rubisco is constantly searching, so to speak, for a linear molecule with two oxygen atoms at either end. But as much as two out of every five times, rubisco fails to pick up carbon dioxide, fumblingly grabs oxygen instead, and tries to shove the oxygen into a chemical reaction that can't use it. To get rid of the unneeded oxygen, plants have evolved an entire secondary process that pumps it out of the cell and re-primes the rubisco to try again for carbon dioxide.

The mistakes waste energy. Rubisco's penchant for oxygen reduces the maximum efficiency of photosynthesis by almost half. Because rubisco gets worse at distinguishing between oxygen and carbon dioxide as temperatures rise, the problem is worse in the tropics than in cooler zones. But even in cool climates, wheat harvests would rise by a fifth and soybean harvests by a third if rubisco could distinguish oxygen from carbon dioxide.

To overcome rubisco's lassitude and maladroitness, plants make a lot of it. As much as half of the protein in many plant leaves, by weight, is rubisco—it is often said to be the world's most abundant protein. One estimate is that plants and microorganisms contain more than eleven pounds of rubisco for every person on Earth. The biological chain seems clear: more nitrogen ⇒ more rubisco ⇒ more photosynthesis ⇒ more plant growth ⇒ more food from farms.

Researchers discovered the import of rubisco in the early 1950s. Almost immediately came a follow-up question: Could scientists develop plants with better rubisco? Was that the way to create faster-growing, more productive, less-fertilizer-intensive wheat, rice, and maize for the coming world of 10 billion? Botanists set to work.

Here's the short version of what they found out:

No—you can't improve rubisco by any means known then or now.

Here's the long version:

Rubisco is as old as photosynthesis itself. Photosynthesis apparently evolved about 3.5 billion years ago, in the ancestors of today's cyanobacteria, blue-green, single-celled creatures (the name comes from *kyanos,* a Greek word for "blue"). For more than a billion years, cyanobacteria proliferated without incident. Then one was engulfed by some microscopic organism, likely a protozoan. Usually this would be routine: the protozoan would be consuming the cyanobacterium as food. But on this occasion the protozoan allowed the cyanobacterium to remain more or less intact—how is unclear—bobbing inside its cell walls. More than that, the protozoan eventually learned how to harness—"enslave" is sometimes used—the cyanobacterium's photosynthetic abilities for its own benefit. When the altered protozoan reproduced, creating a daughter cell, the cyanobacterium reproduced, too; the two creatures were in a long-term symbiotic relationship.

This symbiosis was fantastically improbable. In 3.5 billion years of history and trillions of trillions of interactions between protozoa and cyanobacteria it seems to have happened exactly once. But this single incident had huge effects—it is responsible for the existence of plants. Over the eons the cyanobacterium shed many of its original characteristics, and became a chloroplast: the free-floating body in plant cells in which photosynthesis occurs. Plant cells today can have hundreds of chloroplasts, each a descendant of that long-ago cyanobacterium.

Guffaws greeted this scenario when it was first proposed by Russian biologists in the early twentieth century. Billion-year-old symbiotic entities hidden in most plant cells? The notion seemed like bad science fiction. In the 1950s and 1960s researchers slowly discovered that chloroplasts have their own, separate DNA; their own, separate genes; their own, separate process for creating proteins. They were like tiny alien beings with a history and purpose of their own. Suddenly the old idea seemed less crazy, at least to Lynn Margulis, who assembled the evidence for ancient symbiosis in a powerful article in 1967. Not only were chloroplasts the result of a long-ago symbiotic event, she said, but so were other objects in cell protoplasm, notably the mitochondria, the minuscule entities that regulate energy flow. In fact, some of the symbiotic protozoan-cyanobacteria associations had themselves been engulfed by other, larger creatures, forming new symbiotic associations. These symbiotic acts were rare, but they had shaped the course of life on Earth. Fifteen journals rejected Margulis's paper before it was accepted by the *Journal of Theoretical Biology*. Today it is regarded as a classic.✓

At first glance the skepticism seems merited: cyanobacteria typically have several thousand genes that encode the full panoply of molecules necessary for life; chloroplasts have fewer than 250 genes and cannot survive on their own. How could they be connected? The answer is that over the eons most of the cyanobacterial genes have migrated from chloroplasts to the cell nucleus. Among these are some of the genes for making rubisco itself. Rubisco consists of two big subunits, one bigger than the other. The big subunit is encoded by genes in the chloroplast, the small by genes in the nucleus.*

This constant genetic shuffle has allowed rubisco to evolve in many ways. Today it exists in at least four main forms, each with several sub-forms, as well as "rubisco-like proteins" that look like rubisco but have

---

* Many researchers viewed the idea of gene migration with skepticism until it was directly observed in tobacco in 2003. In the furious shuttling of DNA that creates tobacco pollen, about one out of every sixteen thousand pollen grains ends up with bits of chloroplast DNA mixed into its nuclear DNA. Usually the snippet of chloroplast DNA doesn't contain an entire gene, but sometimes it does, and when that occurs there is a good chance the offspring created from the pollen will have that gene in their nuclear DNA. By comparing modern cyanobacteria DNA to the DNA in the cell nuclei of the plant *Arabidopsis thaliana*, a team of German researchers concluded in 2002 that about one-fifth of the Arabidopsis nuclear genome originated in its chloroplasts.

changed to do something else. But despite all the tinkering, no version of rubisco is better at avoiding oxygen than the original. In this respect, 3.5 billion years of evolution has accomplished nothing.

What evolution seems to be saying, explains Jane Langdale, is that "there is an inescapable trade-off between precision and speed." If rubisco could better distinguish between carbon dioxide and oxygen, it would be even slower; if it catalyzed more reactions per second, it would make more mistakes. "It looks like there's a balance—one that hasn't changed fundamentally for a few billion years."

Langdale is a molecular geneticist at Oxford University's Department of Plant Sciences. When I spoke with her, she had recently been placed in charge of an enormous effort to make an end run around rubisco. Scientists in eight nations were collaborating in an effort to change the way photosynthesis works in rice: perhaps the biggest-ever project in plant sciences. In its effort to hack photosynthesis, the C4 Rice Consortium is an attempt to realize Warren Weaver's vision of the future—the logical extension of von Liebig's dream. To Langdale, it is the kind of effort that will be necessary to feed everyone in tomorrow's crowded, affluent world. It is an attempt to fashion a second Green Revolution. But another way of saying this is that the C4 Rice Consortium is everything Howard and Rodale didn't want.

Special Rice

The Green Revolution had two main branches. One, directly derived from Borlaug's work, was centered in Mexico at CIMMYT, the research agency descended from the Mexican Agricultural Program. The other was inspired by his work and headquartered in the Philippines, at the International Rice Research Institute. Known as IRRI, the rice institute was initially conceived in the early 1950s, when Warren Weaver and George Harrar traveled through Asia to find out if Asian nations would support a version of the Mexican wheat project for rice. Half a dozen Asian nations promised they would back the new program—but only if it was based on their territory. "That reaction eliminated any hope of creating a research center financed by multicountry contributions," sighed Robert F. Chandler, IRRI's first director. Rockefeller was unwilling to pay for the whole project by itself and shelved it. To revive the idea, Harrar and Chandler met with the Ford Foundation. After automobile pioneers Henry and Edsel Ford died, their wills bequeathed so much cash to the foundation that it superseded Rockefeller as the world's richest charity. The two foundations—one with money, one with expertise—jointly built IRRI on land donated by the University of the Philippines. The campus, sprawling and modernist, was dedicated in 1962.

At the time at least half of Asia lived in hunger and want; farm yields in many places were stagnant or falling. Governments that had only recently thrown off colonialism were battling Communist insurgencies, most notably in Vietnam. U.S. leaders believed the appeal of Communism lay in its promise of a better future. In consequence, Washington wanted to demonstrate that development could occur best under capitalism. With IRRI, the hope was that top research teams would transform East and South Asia by rapidly introducing modern rice agriculture—"a Manhattan Project for food," in the historian Nick Cullather's phrase.

From the beginning, IRRI's main plant breeders, Peter R. Jennings and Te-Tzu Chang, focused on developing a rice version of Borlaug's wheat: fertilizer-responsive, photoperiod-insensitive, disease-resistant, short-straw rice. The task was as daunting for them as it had been for Borlaug, but they had the advantage of being second. Thanks to Borlaug, they knew what the target was and that it had already once been reached. In addition, as it turned out, they were favored by chance—"sheer luck," Jennings called it.

Chang, who was from Taiwan, brought to the Philippines three types of Taiwanese rice, curiosities that had short straw but were unproductive and susceptible to disease. Jennings crossbred them with tall tropical varieties, hoping for favorable mixes. It was a comparatively tiny effort: thirty-eight crosses. And the results, Jennings recalled later, "looked terrible." Combinations of their parents' worst features, they were all tall, unproductive (most were sterile), and blasted by disease. The best of this unpromising lot produced 130 grains of rice, which Jennings put in the ground with the rest of the seed. In a few of the offspring, the shortness reappeared; some of them also were less susceptible to fungus. These were crossed and harvested; the seed was planted. In the next generation, a single plant in row 233 seemed perfect. Code-named IR8–233–3,

According to its architect, Ralph T. Walker, the modernist IRRI campus, constructed entirely of imported materials, symbolized "a new type of imperialism" of "specialized knowledge generously given to backwards peoples."

grain from that fortunate rice plant was multiplied and planted in test farms all over South and East Asia. It was staggeringly successful. Borlaug's wheat had doubled or tripled harvests, but the new rice did even better—one trial in Pakistan yielded ten times more than the average of the day. Under the brand of IR-8, the new variety was released to the public in early 1966.*

IR-8 was the foundation of the rice wing of the Green Revolution. It was embraced on all levels, by nations big and small, capitalist and Communist, and even by both sides in the Vietnam War. U.S. president Lyndon Johnson visited an IR-8 rice field at IRRI in the fall of 1966 and theatrically crumbled the soil between his fingers while promising to "escalate the war on hunger." Hoping that "miracle rice" would win Vietnamese hearts and minds by leading peasants to consumer prosperity, his administration set up IR-8 demonstrations throughout South

* The gene variants ("alleles," in the jargon) that produce short-strawed plants in Green Revolution rice and wheat are related in their functions. A recessive mutation in a rice gene causes the plants to produce lower-than-normal levels of the key growth hormone gibberellin. A dominant mutation in wheat leads the plants to respond less to gibberellin, even though they produce it at normal levels. In both cases, the plant is always slamming on the brakes—constantly repressing growth.

Vietnam. U.S. helicopters dropped pro-IR-8 propaganda leaflets on Vietnamese villages; officials in Saigon, a visitor wrote, were "running around waving IR-8 pamphlets like Red Guards with the Mao books." North Vietnam fought back by spreading rumors that IR-8 was a U.S. plot to poison villagers. But after the North won the war, miracle rice became the centerpiece of the new government's brutal rural rebuilding program. By 1980 about 40 percent of the rice grown in East and Southeast Asia was from IRRI; twenty years later, the figure had risen to 80 percent.

Like Borlaug's wheat, IR-8 was an essential part of a "package" that included irrigation and artificial fertilizer. Between 1961 and 2003, Asian irrigation more than doubled, from 182 million acres to 407 million acres; fertilizer use went up by a factor of twenty, from 4.2 to 85 million tons. The consequences were drained aquifers, fertilizer runoff, aquatic dead zones, waterlogged soils, social upheaval—and a near tripling of rice production in Asia. Even though the continent's population soared, Asians had an average of 30 percent more calories in their diet. Millions upon millions of families had more food, better clothing, money for school. Seoul and Shanghai, Jaipur and Jakarta; shining skyscrapers, pricey hotels, traffic-choked streets ablaze with neon—all are built atop a foundation of laboratory-bred rice.

By 2050, researchers believe, it will have to happen all over again—a second Green Revolution for the world of 10 billion. As a journalist, I have been reporting about population and agriculture, off and on, since the early 1990s. In that time I can't recall meeting an agricultural researcher who wasn't worried about what lies ahead. "What is unsure," IRRI researcher Paul Quick told me not long ago, "is whether that additional demand can be met, and whether it can be met without undue environmental or economic cost."

How much will harvests have to rise? Typical projections claim that the world will have to lift food output by 50 to 100 percent by 2050. But in truth nobody really knows, because nobody knows how wealthy the world will be, and what the world's new middle-class people will want to eat. The biggest part of that uncertainty is how much of their diets will consist of animal products—cheese, dairy, fish, and, especially, meat. In the past, increasing affluence has always led people to eat more meat and less grain and legumes (though there is some evidence that at extreme levels of affluence meat consumption declines). At the beginning of the twentieth century, according to Smil, the environmental researcher, barely 10 percent of the world's grain harvest went to animals, mostly horses, mules, and oxen used as farm labor. By the beginning of the twenty-first century, the figure had risen considerably, though by exactly how much is difficult to calculate: perhaps 40 percent, Smil estimates, the great majority of it destined for dairy and meat animals.

How much grain is required to produce a pound of beef, pork, or chicken? The small farm down the street from me as I write provides one answer: zero. It has fifteen cows and a dozen pigs, all fed by grazing fallow land and eating scraps (uneaten or damaged produce, pulled weeds, spent pea and bean vines, and so on). Industrial farms, the source of the vast majority of meat on grocery-store shelves, provide a different—and, alas, more complicated—answer. Beef farmers in the Midwest buy 650-pound steer that have been raised on pasture and feed them leavings in the form of silage (mowed grass and clover, wheat, and maize plants that are cut after harvest) and distillers' grains (maize, rice, or barley from which the starches have been stripped to make products like beer, ethanol, or high-fructose corn syrup). The feedyards where steer are fattened before slaughter surround ethanol and corn-syrup plants like moons around a planet. As a result, the animals form a critical component of the overall grain industry—but actually eat less grain themselves than one might think.

Every uptick in meat consumption is associated with a bump in grain production. But the precise amount of the increase is not straightforward. For beef, it is affected by a host of other factors, including the subsidies for ethanol, the price of corn syrup (and the sugar for which it is a substitute), and the demand for leather, bone, fat (an ingredient in airplane lubricants), keratin (extracted from hooves and used in fire-extinguishing foam), and other meat by-products. Matters are just as complex for pork, chicken, and farm-raised fish. Almost no matter what the scenario, though, if tomorrow's newly affluent billions are as carnivorous as Westerners today, the task facing tomorrow's farmers will be huge. Between 1961 and 2014, the world's meat production more than quadrupled. Simply reproducing that jump could easily require doubling the world's grain harvest.

To double grain output by 2050, arithmetic indicates, harvests would have to rise by an average of 2.4 percent per year. Unluckily, yields are nowhere close to keeping up. A widely cited study from 2013 demonstrated that average global increases in wheat, rice, and maize production have been between 0.9 percent and 1.6 percent per year, about half of what is needed. And in some areas harvests aren't increasing at all. ✓ "Basically, the breeders have been pulling rabbits out of their hats for fifty years," Kenneth G. Cassmann, a yield specialist at the University of Nebraska, told me years ago. "Well, they're starting to run out of rabbits."

Logically speaking, only two paths to increasing harvests exist. One is to lift *actual* yields—the yields produced by farmers, some of whom are better at their work than others. If they are provided better equipment, materials, and technical advice, farmers can bring their harvests closer to the theoretical maximum. The other is to increase the *potential* yield—the theoretical maximum—which should bring up the actual yield with it.*

Both approaches were employed in the Green Revolution. By planting more land, deploying more irrigation, and pumping in more synthetic fertilizer, farmers increased their actual yields. At the same time, Borlaug and his successors at IRRI and CIMMYT dramatically increased the potential yield of wheat and rice by breeding high-yielding dwarf varieties. Channeling the energy of photosynthesis and the nutrition provided by fertilizer into grain, these varieties had a "harvest index"— the percentage of the plant's mass that is grain—of about 50 percent, almost twice the previous figure. (For maize, dwarfing didn't work, ✓ because the shorter plants shaded themselves too much. Instead scientists bred plants that could tolerate being packed closer together.) The sum of the two methods was the Green Revolution.

The situation is different today. Farmers can't plant much more land;

---

* I am oversimplifying here—there's a third alternative. On a global level, a quarter or more of the food produced for human consumption is lost or squandered—left in the field, ruined by poor storage, wrecked in packaging, spoiled in transport, rejected in markets, or simply thrown away by consumers. The exact amount depends on the definition of "waste," which varies dramatically in different studies, and how it is measured, which also varies. In wealthy places, most of the waste comes from people not eating food they have bought. By contrast, the losses in poor nations are concentrated in the field, storage, and transport. Cutting waste obviously would reduce the need to increase harvests. Unfortunately, it will not be easy. In poor nations it would require significant improvements to agricultural infrastructure— costly investments that are difficult for cash-strapped nations to make. Reducing losses in rich nations would involve changing the behavior of huge numbers of busy people, an equally challenging endeavor. Both should be attempted, but progress is likely to be slow and modest.

in Asia, almost every acre of arable soil is already in use. Indeed, as cities expand into the countryside the supply of farmland may be *decreasing*. Nor can fertilizer be increased; it is already being overused everywhere (except some parts of Africa). Irrigation, too, cannot readily be expanded. Most land that can be irrigated is already irrigated. Some increase in actual yield is certainly possible. But most scientists believe they must raise the potential yield—which brings us back to rubisco.

Nature, as one recalls, has not been able to develop more-efficient rubisco. But evolution has produced a work-around: C4 photosynthesis. Named, prosaically, after a molecule with four carbon atoms that is involved in the process, C4 involves a wholesale reorganization of leaf anatomy. The change is almost invisible to the naked eye, but has profound consequences for the plant.

Ordinary photosynthesis is a cycle with two main stages. In the first stage, chloroplasts trap solar energy and use it to break apart water molecules into hydrogen and oxygen atoms. This sequence is known as the "light" reactions because it makes use of sunlight. The hydrogen is plugged into the second stage; the oxygen filters out of the cell and into the air.* In the second stage—the "dark" reactions—the hydrogen from the first stage combines with the carbon from the carbon dioxide grabbed by rubisco. The result is a compound called G3P that other cellular mechanisms break down and rebuild into the sugars, starches, and cellulose that make up plants. In ordinary photosynthesis, both stages—the light and dark reactions—take place in a layer of cells right below the surface of the leaf. The excess gases and the sugars, starches, and cellulose produced in this layer of photosynthetic cells pass into the interior of the leaf. The gases filter up through spaces in the cells to small holes in the leaf surface while the other materials are passed down into interior cells and then into veins and the rest of the plant.

By contrast, C4 plants split photosynthesis in half. The light reactions—the reactions in which chloroplasts use captured solar energy to break apart water molecules—take place near the leaf surface, as in ordinary photosynthesis. But something different occurs with the dark reactions, those that incorporate carbon dioxide. When carbon dioxide

---

* The release of oxygen in the first stage is a necessary part of photosynthesis. It is not the oxygen release due to rubisco grabbing the wrong molecule, which occurs later.

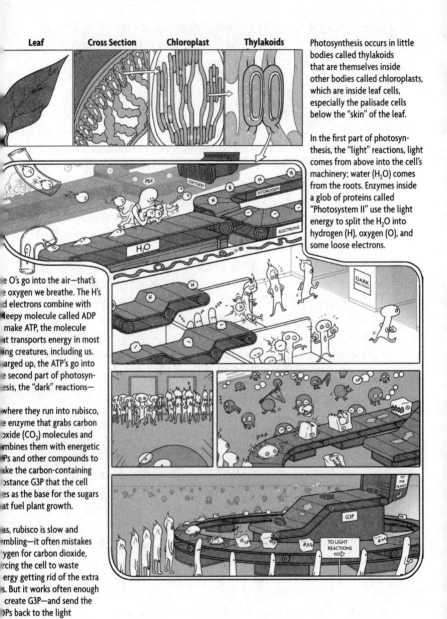

Photosynthesis occurs in little bodies called thylakoids that are themselves inside other bodies called chloroplasts, which are inside leaf cells, especially the palisade cells below the "skin" of the leaf.

In the first part of photosynthesis, the "light" reactions, light comes from above into the cell's machinery; water (H_2O) comes from the roots. Enzymes inside a glob of proteins called "Photosystem II" use the light energy to split the H_2O into hydrogen (H), oxygen (O), and some loose electrons.

e O's go into the air—that's e oxygen we breathe. The H's d electrons combine with leepy molecule called ADP make ATP, the molecule at transports energy in most ing creatures, including us. arged up, the ATP's go into e second part of photosynesis, the "dark" reactions—

where they run into rubisco, e enzyme that grabs carbon oxide (CO_2) molecules and mbines them with energetic Ps and other compounds to ake the carbon-containing ostance G3P that the cell es as the base for the sugars at fuel plant growth.

as, rubisco is slow and mbling—it often mistakes ygen for carbon dioxide, rcing the cell to waste ergy getting rid of the extra s. But it works often enough create G3P—and send the Ps back to the light actions, to begin the cycle ew.

comes into a C4 leaf, it is grabbed not by rubisco but by a different enzyme that uses it to form a compound known as malate (this is the molecule with the four carbon atoms). The malate is then pumped into special cells in the interior of the leaf called "bundle sheath" cells.

Bundle sheath cells are deep inside the leaf, wrapped in a living layer around the veins. In C4 photosynthesis, bundle sheath cells are where rubisco acts. Because they are deep in the leaf, oxygen from the air doesn't easily slip into them. Meanwhile, carbon dioxide is released from the imported malate. Almost without oxygen and boosted with carbon dioxide, each bundle sheath cell is a microscopic replica of the ancient atmosphere in which photosynthesis evolved. More than 3 billion years ago the atmosphere had a hundred times as much carbon dioxide as it does now and almost no oxygen. Rubisco's inability to distinguish carbon dioxide and oxygen was not a problem, because oxygen was rare. Bundle sheath cells have little oxygen, and rubisco is denied the opportunity to mistake it for carbon dioxide. C4 photosynthesis is thus much more efficient. Barely 3 percent of the flowering plants are C4, but they are responsible for about a quarter of all the photosynthesis on land.

The impact of C4 is evident to anyone who has looked at a recently mowed lawn. Within a few days of mowing, the crabgrass in the lawn springs up, towering over the rest of the lawn (typically bluegrass or fescue in cool areas). Fast-growing crabgrass is C4; lawn grass is ordinary photosynthesis. The same is true for wheat and maize. Plant them on the same day in the same place and soon the maize will overshadow the wheat—maize is C4, wheat is not. In addition to growing faster, C4 plants also need less water and fertilizer, because they don't waste water on reactions that lead to excess oxygen, and because they don't have to make as much rubisco. And they better tolerate high temperatures—C4 is especially common in the tropics.*

Remarkably, C4 photosynthesis has arisen independently more than sixty times. Maize, tumbleweed, crabgrass, sugarcane, and Bermuda grass—all these very different plants evolved C4 photosynthesis on their

---

* An academic note: In its madcap way, evolution has created a *second* rubisco work-around, called crassulacean acid metabolism (CAM), which splits light and dark reactions in a different way. CAM occurs mainly in dryland plants like cacti and pineapple and is of little import here.

own. When many different species develop the same traits, the implication is that a lot of plants are "pre-adapted" to create that trait. Somewhere in their DNA, very likely, are genetic switches that promote it.

Further evidence for this idea is that a few species are intermediate—some parts of the plant use ordinary photosynthesis, some use C4 photosynthesis. One of these in-between species is maize: its main leaves are C4, whereas the leaves around the cob are a mix of C4 and ordinary photosynthesis. If two forms of photosynthesis can be encoded from the same genome, they cannot be that far apart. Which in turn implies that people equipped with the tools of molecular biology might be able to transform one into another. ✔

In the botanical equivalent of a moonshot, an international consortium of almost a hundred agricultural scientists is working to convert rice into a C4 plant—a rice that could grow faster, require less water and fertilizer, withstand higher temperatures, and produce more grain. Funded largely by the Bill & Melinda Gates Foundation, the C4 Rice Consortium is the world's biggest genetic-engineering project. But the term "genetic engineering" does not capture the ambition of the project. What the researchers are trying to develop bears the same resemblance to typical genetically modified organisms that a Boeing 787 does to a paper airplane.

The genetic engineering that appears in news reports largely involves big companies like Monsanto sticking individual packets of genetic material, often taken from other species, into crops. The paradigmatic example is Monsanto's Roundup Ready soybean, in which DNA from a bacterium found in a California waste pond is inserted into soybeans, making them assemble a chemical compound in their leaves and stems that resists Roundup, Monsanto's widely used herbicide. The foreign gene lets farmers spray Roundup over their fields, killing weeds but leaving soy plants unharmed. Except for making this one tasteless, odorless, nontoxic substance—a protein with the unwieldy moniker of 5-enolpyruvylshikimate-3-phosphate synthase—Roundup Ready soy plants are, in theory, wholly identical to ordinary soy plants.

The C4 Rice Consortium is trying something different in scale and process. Rather than companies tinkering with individual genes to sell branded goop, the consortium is trying to refashion the most fundamental process of life—with the intent of giving away the result. And

Although plant breeding has advanced since Borlaug's day, it remains a long, labor-intensive process. At IRRI, rice seedlings are sprouted in climate-controlled tanks (bottom, left), then manually sown in greenhouses (top). Much as Borlaug and his team did at Sonora and Chapingo, IRRI staffers still sort through the harvested grain by hand (bottom, right).

instead of slipping genes from other species into rice, the initiative is hoping to switch on chunks of the DNA already in rice to create, in effect, a new, more productive species—common rice, *Oryza sativa*, will become something else, *Oryza nova*. (Or, possibly, the team may use genes from related species that are similar to rice genes but are for technical reasons easier to manipulate.)

When I visited IRRI, scores of people were doing what science does best: breaking a problem into individual pieces, then attacking the pieces. Some were sprouting rice in petri dishes. Others were trying to find chance variations in existing rice strains that might be helpful. Still others were studying a model organism, a C4 species called *Setaria viridis*. Faster growing than rice and not needing to be raised in paddies, Setaria is easier to work with in the lab. There were experiments to measure variations in photosynthetic chemicals, in the rate of growth of different varieties, in the transmission of biochemical markers. Twelve women in white coats were sorting rice seeds on a big table, grain by grain. More workers were in fields outside, tending experimental rice paddies. All the appurtenances of contemporary biology were in evidence: flat-screen monitors, humming refrigerators and freezers, tables full of beakers of recombinant goo, Dilbert and XKCD cartoons taped to whiteboards, a United Nations of graduate students a-gossip in the cafeteria, air-conditioners whooshing in a row outside the windows.

The cell's photosynthetic machinery is programmed by scores or even hundreds of genes. As the project began, it was plausible to doubt that so much DNA could be altered in a controlled fashion. Multiple techniques for genetic engineering have been invented, but at the time the most common for cereals like rice, wheat, and maize involved shotgunning thousands of microscopic particles of gold or tungsten at plant embryos (the first precursor cells for leaf, root, and stem, which in this case have been pulled out of the seed and grown in petri dishes). The particles are coated with snippets of DNA that contain desirable genes. In a process that astonished biologists when they discovered it could occur, a few of these particles slam through the cell walls and hit the cell nucleus in just the right way that the nucleus—or, more exactly, the DNA in the nucleus—incorporates the new bits of DNA. And every now and then the DNA from the particles is transferred in a way that allows the new genes to be switched on. The method was clumsy; because it inserted the DNA randomly, its effects were unpredictable. And it could insert only one gene at a time. Nobody knew whether it was possible to alter multiple groups of genes in this way and end up with a coherent result. In 2012 scientists at Harvard and Berkeley unveiled a new method of gene editing called CRISPR that promised more precise control. The C4 Rice Consortium went on alert.

The project has two main goals: (1) locating and switching on the precursor genes that will create the physical structures of C4 photosynthesis (the bundle sheath cells as well as a network of extra veins for them to wrap around), and (2) locating and switching on the precursor genes that create the substances involved in C4 photosynthesis (the malate-producing enzyme and other molecules that are involved in the reaction). In a sense, they want to create both the arena and the players in the arena. Initial research suggests that about a dozen genes play a major part in the leaf structure; another ten genes, perhaps, have an equivalent role in the biochemistry. Alas, properly altering them to create a C4 organism, difficult as that is, would be only the first step. The next, possibly more arduous, would be breeding varieties that can channel the extra growth provided by photosynthesis into grain, rather than roots or stalk. All the while, the new varieties must be disease-resistant, easy to grow, and palatable to their intended audience of several billion Asians.

"I think it can happen, but it might not," Langdale, the project leader, told me. She was quick to point out that even if C4 rice runs into insurmountable obstacles, it is not the only biological moonshot. Nitrogen-fixing maize, wheat that can grow in saltwater, enhanced soil microbial ecosystems—the list of possibilities is as long as imagination allows.* The odds that any one of them will succeed may be small. But the odds that all of them will fail are equally small.

---

* There may even be other ways of improving photosynthesis. Plants protect themselves from intense sunlight by dissipating some of the excess energy as heat, which means that energy is lost to photosynthesis. Plants switch off the dissipating mechanism when the light is dimmed by clouds, dusk, or shade, including the shade cast on one plant leaf by another, or on one crop by its neighbor. But the adjustment is slow enough and occurs frequently enough to reduce the total photosynthesis in wheat by about a fifth; maize may lose as much as a tenth. In 2016, a research team based at the University of Illinois demonstrated that it is possible in principle to speed up the reaction, possibly making up for some of the loss.

Kantian Interlude

The protests were a surprise. Not many activists come to Brentwood, a small town about thirty miles east of San Francisco. Nobody expected that they would slip into the strawberry patch at night and rip out 2,200 seedlings. The company that had planted the strawberries was Advanced Genetic Sciences, of Oakland, California. Its researchers discovered that most of the vandalized plants were still alive and replanted them. The next night, guards watched the plot from a white van. Protesters snuck up on the van and slashed its tires. The next day, April 24, 1987, technicians in moon suits sprayed the replanted strawberries with bacteria.

The bacterium was *Pseudomonas syringae*. In ordinary circumstances *P. syringae* sits on plant leaves and obtains nutrients from dust and rainfall. Like all bacteria, it has a protective coating on its surface. The coating contains a protein that interacts with water in a fashion most unfortunate for farmers. Liquid water doesn't turn easily into solid ice; it must be cooled well below freezing to crystallize spontaneously. But water molecules will transform quickly into ice if they have an object—a *nucleus*, scientists say—to crystallize around. The protein on the outer coat of *P. syringae* is just the right size and shape for an ice nucleus. As a result, bacteria-coated plants freeze more readily than those without the bacteria. Estimates of the cost to U.S. farmers of *P. syringae*–induced freezes at the time ranged up to $1.5 billion a year. Researchers at the University of California at Berkeley used genetic-engineering techniques to incapacitate the gene that produces the offending surface protein. Advanced Genetic Sciences, based in Oakland, wanted to turn these altered bacteria into a product that farmers could use to protect their fields. In 1983 it announced plans to test its "ice-minus" bacteria by spraying them on plots of strawberries and potatoes in rural California. The hope was that the engineered bugs would crowd out their natural, ice-causing cousins, protecting their host plants from frost. It would be the first release of genetically modified organisms into the wild—and the opening skirmish in a battle that continues to the present day.

The experiment had been approved by the Recombinant DNA Advisory Committee, a semi-official body at the National Institutes of Health formed after a scientific conference in 1975. The conference was itself the culmination of a series of meetings that had begun when scientists realized that they were now able to manipulate DNA to transfer genes between species. Held at the Asilomar center on California's Monterey Peninsula, the conference was attended by 145 people from thirteen nations, along with sixteen journalists who had agreed to defer publication until the meeting ended. In three days of debate, the group did what it could to assess risks and then set out measures to avoid them. Before closing, the conference issued a declaration that, as conference organizer Paul Berg put it, could be summed by a single sentence: "With reservations, some form of experiments should proceed; some, however, should not." Later a longer, official Asilomar statement was published. Universities and governments around the world used it as a basis for biotechnology regulation. The United States, for instance, formed the Recombinant DNA Advisory Committee in response.

Except for the journalists, two lawyers, and a historian of science, everyone at Asilomar was a molecular biologist. Not invited were people with expertise in ethics, public health, the human sciences, or even other branches of biology, like ecology or agricultural science. Politicians and civic groups were not present, nor were any members of the public. In an opening address, the conference co-chair, David Baltimore of MIT, ruled out "topics peripheral to the meeting." Among these, he said, were "the complicated question of what's right and what's wrong" and "complicated questions of political motivation." From the scientists' point of view, the absence of these subjects had no effect on the outcome. Under the Recombinant DNA Advisory Committee, research continued almost without restriction; more than 250 biotech firms were founded, many of them by scientists, to take advantage of the new findings. Prizes were awarded for this work—Baltimore won a Nobel in 1975. In all this activity, scientists pointed out, genetic engineering didn't hurt a soul. Hardly anyone even seemed to lift an eyebrow.

An activist named Jeremy Rifkin, along with three environmental groups, sued to stop the ice-minus test in November 1983. Because the Asilomar conference—and the Recombinant DNA Advisory Committee it spawned—consisted almost exclusively of biologists, he said, it didn't have the "interdisciplinary capability" to judge environmental hazards. Nor could it evaluate the ethics of imposing risks on the public. The experiment should be stopped—now.

It was easy to mock Rifkin. A social activist since the Vietnam War, he liked to attack "the Boys"—Isaac Newton, Karl Marx, Adam Smith, Francis Bacon, René Descartes, and Charles Darwin, among others—who had, to Rifkin's way of thinking, created a worldview that valued efficiency rather than empathy and the spirit. By "reducing all life from matter to energy to information," he told a reporter at the time, disciples of the Boys were seeking to "create the perfectly efficient living utility, be it a microbe, a plant, an animal species, or perhaps even human beings." Without scientific expertise himself, Rifkin was notorious for stoking alarms that researchers found absurd. In his ice-minus complaint, for example, he claimed that because "the naturally occurring ice-nucleating bacteria blown by the wind into the upper atmosphere may play a role in the global climate," the "recombinant DNA mutant bacteria" could alter weather patterns across the world. This was nonsense—*P. syringae* can't live in the air, isn't blown into the upper atmosphere, and doesn't affect the climate.

Nonetheless, Rifkin was on to something. For ice-minus, the court "emphatically" agreed with Rifkin: the expert panel hadn't considered "the possibility of various environmental effects." The company went back and performed an environmental assessment, measuring wind-dispersal patterns during spraying; the Environmental Protection Agency conducted its own review. At Berkeley, researchers examined the effects of ice-minus on sixty-seven local plants, including every important crop grown in the proposed test region. Four years after its initial approval, the experiment was again allowed to proceed. (It was successful, but the product was never marketed.)

None of the evaluations persuaded Rifkin that ice-minus was safe—he fought it until the end. Rifkin didn't think that *any* product of gene-splicing could be adequately tested or should be released in the wild. Much of the public was equally unmoved. Local officials and environmental groups in Brentwood joined Rifkin in protesting ice-minus. Vandals tried to destroy the strawberry plants and, later, the potato plants in another test. In effect, all were saying: *We don't trust you white-coats—there have been too many examples of unintended consequences.* National and international organizations have followed Rifkin's lead. Genetically modified crops have been banned in parts of Europe, Asia, Latin America, and Africa.

To counter the perception of risk, scientific groups have repeatedly issued statements supporting the safety of genetically modified foods. The tally of pro-gene-splicers is like a Who's Who of global research: the American Association for the Advancement of Science board of directors (2012); the World Health Organization (2005); the German Academy of Sciences (2006); the European Union (2010); the American Medical Association (2012); the British government (2003); the Australian and New Zealand governments (2005); three British molecular biologists (2008); three U.S. agronomists (2013); the U.S. National Academies of Sciences, Engineering, and Medicine (2016). And so on.

These efforts have been entirely unsuccessful. A Gallup poll in 1999 found that a little more than a quarter of U.S. citizens thought that food from genetically modified organisms (GMOs, to activists) was unsafe. Sixteen years and a dozen scientific reports later, the Pew Research Center found that the fears had actually increased: 57 percent of the U.S. public now thought that GMOs were dangerous; 67 percent believed that scientists didn't understand the health risks. Levels of distrust are even higher in Europe. "The most ironic aspect of this long-running and unfinished controversy is that the brilliant minds that figured out this world-transforming technology in the first place have yet to figure out a way to ease public fears about it," remarked the journalist Stephen S. Hall at the time of the ice-minus fight.

Researchers find this infuriating. Why don't people care that the people who best understand the technology believe it to be useful and safe? After all, scientists eat the food, too. But they are looking at the question as if it were one of *risk*, whereas men and women on the street also think in terms of *fairness*—equity, to use the two-dollar word. In the laboratory, scientists ask: Is it feasible? In the world outside the laboratory, people ask: Is it right?

Contrast the behavior of the European and North American consumers who fear putting GMOs into their bodies as food with that of the European and North American patients who willingly put GMOs into their bodies as medicine. Genetically modified *E. coli* creates synthetic insulin for diabetics; genetically modified baker's yeast produces hepatitis B vaccine; genetically modified mammal cells make blood factor VIII for hemophiliacs and tissue plasminogen activator for heart-attack victims. Although activists have occasionally campaigned against these drugs, their efforts have not caught fire. The divergent reactions are not because people are foolish but because the two circumstances have different ethical benefit-cost calculations. In both cases scientists assure non-scientists that the likelihood of negative side effects is small. But the diabetics who use synthetic insulin personally benefit from it, compensating for any risk. The same is true for the hemophiliacs and cardiac patients. Meanwhile, the Californians who lived around the strawberry and potato patches would receive no benefit whatsoever from the ice-minus test. From their point of view, they were being asked to expose themselves to an unknown peril for the benefit of some rich venture capitalists in a city hundreds of miles way. Imposing *any* risk, however tiny, would make them worse off. They were being used purely as a means to somebody else's end—something that philosophers have regarded as unethical since the days of Kant.

As a rule, GMOs have made life easier and cheaper for large-scale farmers in developed nations by reducing the costs of chemicals, labor, or storage. But they have provided few tangible gains for the middle-class folk in those nations who buy the farmers' products in supermarkets. The food doesn't look or smell or taste better; it doesn't seem less expensive. Why should they accept *any* risk, no matter how small the white-coats claim them to be?

Matters are more complex in poor places, where consumers are more likely to be farmers, too. Reducing costs for farmers there means reducing costs to consumers. Assume for the moment that C4 rice becomes everything that scientists like Langdale hope—productive, resilient, efficient, tasty. It could directly benefit, say, a village of Cambodian subsistence farmers, especially if (a big if) every villager can obtain the necessary water, fertilizer, and credit. In those circumstances, C4 rice could be a potent weapon against hunger and malnutrition. In affluent California, though, its virtues seem less evident. Farmers in the Central Valley would have bigger harvests, but would likely export the surplus. Local food prices would be little changed. The biggest benefit to Californians would be increased income-tax revenue—nothing to scoff at, but also nothing that has ever motivated large numbers of people to action.

The conundrum is that poor nations are less likely to accept an innovation if it is rejected by their richer neighbors—it can become stigmatized. The stigma turns into outright economic harm if the rich nations ban the innovation; if C4 rice cannot be exported, farmers seeking extra cash are less likely to grow it. The action by which middle-class people refuse to take risks on behalf of rich companies becomes a way of blocking aspirations of the distant poor. Weighing the relative pluses and minuses is an exercise in morality that is outside the realm of science.

Coupled with debate about what is right is a second about what is good. Wizards, Borlaug loudly among them, have repeatedly claimed that GMOs are essential to feeding tomorrow's world, which they identify with large-scale industrial agriculture. Prophets, who believe that large-scale industrial agriculture endangers tomorrow's world, naturally resist any innovation that is said to be central to perpetuating it. In this way GMOs became a focus for a larger disquiet, a synecdoche for a larger anxiety about being an insignificant part of a vast economic complex that did not have the citizen's best interests at heart. Prophets saw giant farms, almost empty but for huge machines, producing endless streams of protein and carbohydrates that were sent to factories for processing—and they recoiled at the image of life this conjured. They didn't want something as intimate as breakfast to be out of any possibility of control. To be so far removed from any identifiable human touch. To be (as some thought) under the thumb of corporate capitalism. All the scientific reports in the world wouldn't address the source of the foreboding. The fundamental disquiet would still be there.

Trees and Tubers

For a long time Lloyd Nichols worked in ramp services at O'Hare Airport, outside Chicago. He worked for one airline that got bought by another and then that airline merged with a third. His co-workers were constantly being laid off. It seemed clear that his future in the airline industry was limited. In retrospect, he told me, this might have had something to do with how much attention he paid to his garden.

Nichols had always liked taking care of plants and animals. As time passed, the gardens got bigger. He was hoeing in the morning a couple of hours before going to the airport and weeding until dark after he came home. When he heard about a new variety or a new crop, he stuck it in the ground to see what happened. Wanting more space, Lloyd dragged his wife, Doreen, and their kids from the suburbs in 1977 to a ten-acre parcel in Marengo, in northwest Illinois. Four acres went for the garden. He was thirty-three.

For a couple years he commuted, an hour back and forth five times a week. All the while he thought about quitting his job and being self-sufficient. Going into the garden, picking up lettuce or zucchini or tomatoes, having them fresh on the dinner table an hour later. Lloyd and Doreen planted a little orchard, apples and apricots. Soon he was producing more than his family could eat, so he loaded up the surplus on the back of a truck and sold it at one of the new farmers' markets that were opening around Chicago. He bought neighboring parcels of land and planted them. Before long some of his customers were chefs at fancy restaurants in the city. When the tumult in the airline industry claimed his job, he thought, well, maybe I can make a go of this.

The family split up on weekends to take produce to different farmers' markets. As sales rose, Lloyd ran out of family and had to hire people to help. Then he had to hire more people to help the help. When I visited, Nichols Farm had expanded to 527 acres and eight outbuildings with refrigerators powered by solar panels and a wind turbine. It had eleven full-time, year-round workers and thirty or more seasonal employees. They grew a thousand different crop varieties, give or take. On one wall were handwritten charts that listed them all, together with current growing notes—a spreadsheet in non-digital form.

Nichols drove me around the operation on a kind of golf cart. It was evident that he was one of those people born with a limitless store of energy and enthusiasm. He spoke rapidly as I tried to take notes in the jouncing vehicle. He said he had known about J. I. Rodale and Albert Howard from back when he was a kid reading his father's *Prevention* magazines and more or less farmed according to their ideas because he liked the idea of having his crops spring up from the richest possible soil. But he refused to call his farm organic or be certified as an organic grower because he didn't relish some standards committee telling him what to do. "If I need to use some chemical that is not on their list, I'm going to do it if it's best for my plants."

Spread over the gently undulating contours of his land was a tribute to the breeder's art: eleven types of cauliflower, twelve types of broccoli, thirteen types of lettuce, fourteen types of melon. A mosaic of salad greens, lustrous in the sun. Pumpkins the size of a mastiff. Calendula, canna, and carrot; celosia, coleus, and collard. For a while, Nichols told me, he hadn't grown potatoes, because he thought people bought them in anonymous, fifty-pound bags. Now he planted twenty-three varieties. Pride of place went to his apple orchard: more than three hundred varieties. He said he really liked apples. "Even the spitters."

I was seduced. Nichols and his farm were a pleasure to be around. So were the similar people and places that I had visited in California, Louisiana, Massachusetts, and North Carolina—and the farms like them, most of them newly established, that I had come across in India, Thailand, and Brazil. The enthusiasm of their owners was palpable and infectious.

Less clear was their comparative role in the food system, today and tomorrow. Nichols's concern for maintaining the richness of his soil and the health of the farm ecosystem it supported would have heartened William Vogt. But at the same time (as Nichols was quick to point out), his food was more expensive than supermarket food from industrial farms. Like a custom-furniture maker, he was making beautiful things for a restricted clientele. Nothing wrong with that, Wizards say. But don't imagine that this kind of operation can play a big role in feeding the world of 10 billion. At a time when we should be thinking about doubling global harvests, it's a mistake to focus on boutique farms, no matter how charming. That's the line, anyway.

"Organic" may be the wrong term for these enterprises—like the Nichols farm, not all of them follow the official rules for certification. More important, it implies that they are somehow "natural," not constructed landscapes based on scientific information. To run these sorts of farm requires integrating multiple forms of knowledge—botanical, biochemical, pedological, economic, legal, cultural—that are constantly interacting and changing. The results are exotic cosmopolitan objects, as thoroughly driven by contemporary technology as the products of the pharmaceutical plants that they (mostly) abjure.

Nichols took me to visit his neighbor, Harold Heinberg. Heinberg had a bit more than 1,200 acres in wheat and maize. Good-looking plants, at least to my untrained eye. While I walked down the rows of maize, Nichols and Heinberg stayed in the shade cast by a barn, talking amiably. The two men were friendly neighbors but their farms might as well have been in different countries. The Nichols farm had forty or so workers and a thousand different crop varieties. The Heinberg farm had one and a half workers—Heinberg's son pitched in when not at his day job as a trucker—and two crop varieties. What did the work was a million-plus dollars' worth of farm machinery and a stream of chemical treatments, each applied on a precise schedule. Heinberg let me walk about the shed where he kept his tractors, harvesters, threshers, and so on. The intricate machines were a model of the Wizard's art. Some had tires the size of a tall man. A workbench had a dusty laptop, open to a spreadsheet—a database not all that different in form from the one in Nichols's barn.

Which of the two farms is more productive? Wizards and Prophets would disagree about the answer, because they disagree about the question. To Wizards, the question means: Which farm creates more calories—more usable energy, in Weaver's terms—per acre? Scores of research teams have tried to appraise the relative contributions of organic and conventional agriculture. These inquiries in turn have been gathered together and assessed, a procedure that is also fraught with difficulty (researchers use different definitions of "organic," compare different kinds of farms, and include different costs in their analyses). Nonetheless, *every* attempt to sum up the data that I know of has shown that in side-by-side comparisons, Howard-style farms grow less food per acre overall than Liebig-style farms—sometimes a little bit less, sometimes quite a lot. The implications for the world of 2050 are obvious, Wizards say. If farmers must grow twice as much food to feed the 10 billion, following *An Agricultural Testament* ties their hands.

Prophets smite their brows in exasperation at this logic. To their minds, evaluating farming systems wholly in terms of calories produced—in terms of usable energy—is a perfect example of the flaws of reductive thinking. It does not include the costs of overfertilization, habitat loss, watershed degradation, soil erosion and compaction, and pesticide and antibiotic overuse; it doesn't account for the destruction of rural communities; it doesn't consider whether the food is tasty and nutritious. It's like evaluating automobiles entirely by their gasoline mileage, without taking into account safety, comfort, reliability, emissions, or any of the other factors that people consider when buying cars.

The difficulty is that both arguments are correct on their own terms. At bottom, the disagreement is about the nature of agriculture—and, with it, the best form of society. To Borlaugians, farming is a species of useful drudgery that should be eased and reduced as much as possible to maximize individual liberty. Borlaug's life is an example: when his family mechanized their farm by buying a tractor, it freed him to go to school, and from there to change the world. The farm is a springboard, essential as a base, but also a trap. Howard-style farms may mimic natural ecosystems, but they are also ensnarled in them, unable to rise above their limits.

To Vogtians, by contrast, agriculture is about maintaining a set of communities, ecological and human, which have cradled human life since the first of Gause's inflection points, ten-thousand-plus years ago. It can be drudgery, but it is also work that reinforces the human connection to the earth. The two arguments are like skew lines, not on the same plane.

Wait a minute, Wizards in effect say. Calories per acre *is* the fundamental measure! People need food before they need anything else! To feed the world of 10 billion, ordinary agriculture maize will not be enough. Industrial operations—farms like the Heinberg spread—are pre-adapted for change. Swap out the old seeds for the new, add some machinery, and they are ready to go. Fears about ecological consequences are mistaken—new technology can fix or avoid them. Worries about community and connection are secondary. (Bertolt Brecht, succinctly, in *The Threepenny Opera*: "First comes food, then comes right from wrong.")

Prophets insist that this is not true: Howard-style agriculture can respond to these pressures. Organic farmers, too, have radical alternatives: domesticating new cereal species, hybridizing ordinary crops and their wild relatives, or even switching to entirely new crops.

Wheat, rice, maize, oats, barley, rye, and the other common cereals are *annual* crops, planted anew every year. By contrast, the wild grasses that used to fill the prairies of the Midwest, Australia, and central Eurasia are *perennial*: plants that come back summer after summer, for as much as a decade. Because perennial grasses build up root systems that reach deep into the ground, they better hold on to the soil and are less dependent on surface rainwater and nutrients than annual grasses. Not needing to build up new roots in the spring, perennials emerge from the soil earlier and faster than annuals. And because they don't die in the winter, they keep photosynthesizing in the fall, when annuals stop. Effectively, they

After rinsing the dirt from its roots, Land Institute researcher Jerry Glover holds an intermediate wheatgrass plant (A), demonstrating the power of perennial plants to hold the soil. By contrast, the root network of the adjacent bread wheat (B) is much less extensive.

have a longer growing season. At the same time, perennials have drawbacks. Because they devote more photosynthetic energy to building up their roots, they expend less on reproduction: their seeds are few and small. Annual crops, by contrast, expend less energy on roots and focus on grain—exactly what farmers want.

In an echo of the Rockefeller Foundation program's initial collection of wheat varieties, the Rodale Institute, the research arm of the Rodale imperium, gathered three hundred samples of intermediate wheatgrass (*Thinopyrum intermedium*) from around North America in the early 1980s. A perennial cousin to bread wheat, wheatgrass is native to central Eurasia. It was introduced to the Western Hemisphere in the 1920s as fodder for farm animals. Working with U.S. Department of Agriculture researchers, Rodale's Peggy Wagoner planted the samples, measured their yields, and crossed the best performers with each other. The work was slow; because wheatgrass is a perennial, it must be evaluated over years, rather than a single season—shuttle breeding is not an option. Rodale and Wagoner passed the baton in 2002 to the Land Institute, in Salina, Kansas, a nonprofit agricultural-research center dedicated to replacing conventional agriculture with processes that mimic natural ecosystems. The Land Institute, collaborating with other researchers, has been working on wheatgrass ever since. A trade name emerged for the new crop: Kernza.

Like engineering C4 rice, domesticating wheatgrass is a decades-long quest that may not fulfill its originators' hopes. As of now, the tools of molecular biology would be of little help even if wanted—the task is too complicated. Wheatgrass, as one Land Institute researcher put it to me, "is irredeemably old-school." Wheatgrass kernels are a quarter the size of wheat kernels, sometimes less, and have a thicker layer of bran. Unlike wheat, wheatgrass grows into a dark, dense mass of foliage that covers the field; the thick layer of vegetation protects the soil and keeps out weeds, but it also reduces productivity. To make wheatgrass useful to farmers, breeders will have to increase seed size and alter the architecture of the plant. As they do, they will also have to avoid its penchant for lodging. (Wild plants, used to fighting for every scrap of sun, tend to lodge in the bright light of a farmer's field.) The Land Institute hopes to have field-ready wheatgrass with kernels that are half the size of wheat in the 2020s, but nothing is guaranteed.

Domesticating wheatgrass is the long game. Others have been trying for a shortcut: creating a hybrid of wheatgrass and bread wheat, hoping to mix the former's large, plentiful grain and the latter's perenniality and disease resistance. The two species produce viable offspring just often enough that Soviet-era biologists tried for decades to breed useful hybrids. They eventually gave up in the 1960s; smaller programs in North America and Germany also failed. Bolstered by new developments in biology, researchers at the Land Institute, Australia, and the Pacific Northwest began anew at the beginning of this century. When I visited Stephen S. Jones, of Washington State University, he and his colleagues had just suggested a scientific name for the new hybrid: *Tritipyrum aaseae* (the species name honors the pioneering cereal geneticist Hannah Aase). Much work remains; Jones told me that he hoped bread from *T. aaseae* would be ready for my daughter's children. "The problem isn't going away," he said. The world was moving, slowly but implacably, toward a demographic cliff. Jones and his colleagues were working with imperfect tools to build a safety net.

African and Latin American researchers scratch their heads in bafflement when they hear about these projects—some of them, anyway. Perennial grains are the hard way to do this, Edwige Botoni told me. Botoni is a researcher at the Permanent Interstate Committee for Drought Control in the Sahel, in Burkina Faso. Traveling through the edge of the Sahara, she had given a lot of thought to the problem of feeding people on marginal land. One part of the answer, she told me, would be to emulate the tropical places like Nigeria and Brazil. Whereas farmers in the temperate zone focus on cereals, tropical agriculture is centered on tubers and trees, both often more productive than cereals.

Consider cassava, the big tuber also known as manioc, mogo, and yuca. The tenth-most-important crop in the world, it is grown in wide swathes of Africa, Latin America, and Asia. Nigeria is the world's biggest producer. Because cassava is a tuber, not a cereal, the edible part grows underground; no matter how big the tuber, the plant will never lodge. On a per-acre basis, cassava harvests far outstrip those of wheat and other cereals. In optimal conditions, cassava farmers have pulled 160,000 pounds per acre from the ground—more than fifty times the average for wheat. The comparison is unfair, because cassava tubers contain more water than wheat kernels. But even when this is taken into account, cassava produces many more calories per acre than wheat. "I don't know why this alternative is not considered," Botoni said. "It seems easier than breeding entirely new species."*

Much the same is true for tree crops. Nichols grows more than a hundred different types of apple. A mature McIntosh apple tree can grow between 350 and 550 pounds of apples per year. Orchard growers commonly plant 200 to 250 trees per acre. In good years this can work out to 35 to 65 tons of fruit per acre. The equivalent figure for wheat, by contrast, is about a ton and a half. Again, apples contain more water than wheat—but not *that* much more water. Papaya is even more productive. So are some nuts.

---

* Potato is a northern equivalent. The average 2016 U.S. potato yield was 43,700 pounds per acre, more than ten times the equivalent figure for wheat.

Am I arguing that farmers around the world should replace their plots of wheat, rice, and maize with fields of cassava, potato, and sweet potato and orchards of bananas, apples, and chestnuts? No. The argument is rather that Vogtians have multiple ways to meet tomorrow's needs. These alternative paths are difficult, but so is the Borlaugian path exemplified in C4 rice. The greatest obstacle for Vogtians is something else: labor. Lloyd Nichols's operation requires a lot of workers, and so does every farm like it.

Heinberg, Nichols's neighbor, was able to take advantage of a host of incentives and subsidies provided by the state of Illinois and the federal government: land-tax incentives, depreciation allowances, crop subsidies, and so on. Nichols couldn't, because almost everything he planted was not on the official state list of eligible crops. And he didn't devote sufficient acreage to those on the list that he did plant to qualify as a grower. On the official, regulatory level, it was as if his farm didn't exist. "I've never got a subsidy check in forty years of doing this," he said to me. As he pointed out, many of the supports were intended to promote the acquisition of machinery, rather than labor. He might get a special low-interest loan to buy a combine, but not to hire a human being.

These policies are not accidental. Beginning with the end of the Second World War, most national governments have intentionally directed labor away from the land (Communist China was long an exception). Farmwork was seen as "stagnant" and "unproductive." The goal was to consolidate and mechanize farms, which would increase harvests and reduce costs, especially labor costs. Farmworkers, no longer needed, would migrate to the cities, where they could get better-paying work in factories. Ideally, both the remaining farm owners and the factory workers would earn more, the former by growing more and better crops, the latter by obtaining better-paying jobs in industry. The nation as a whole would benefit: increased exports from industry and agriculture, cheaper food in the cities, a plentiful labor supply.

There were downsides—cities in developing nations acquired entire slums full of displaced families. But in many places, including most of the developed world, the countryside was emptied. In the United States, for example, the proportion of the workforce employed in agriculture went from 21.5 percent in 1930 to 1.9 percent in 2000. Meanwhile, the number of farms fell by almost two-thirds. The average size of the surviving farms increased to match; their owners increasingly focused on exports to the world market. Because the rules that encourage large-scale, industrial production for export remain in effect, farmers like Lloyd Nichols are swimming against the tide.

To Vogtians, the best agriculture takes care of the soil first and foremost, a goal that is difficult to accomplish when growing large swathes of a single crop. But tending multiple crops, as Nichols does, unavoidably requires more human attention. Nichols pays for the labor by selling his food to affluent foodies. Truly extending this type of agriculture would require bringing back at least some of the people whose parents and grandparents left the countryside. Providing these workers with a decent living would further drive up costs. Some labor-sparing mechanization is possible, but no small farmer whom I have spoken with thinks that it would be possible to shrink the labor force to the level seen in big industrial operations. The whole system can grow only with some kind of wall-to-wall rewrite of the legal system that encourages the use of labor. Such large shifts in social arrangements are not easily accomplished.

Even then, everything could be thrown off by water.

FIVE

Water: Freshwater

Tomatoes

In the early 1980s an editor took a chance on an inexperienced writer and gave me an assignment to write an article about the tomato-processing industry. I knew nothing about the subject but neglected to emphasize this to the editor. About nine-tenths of the processed-tomato acreage in the United States was in California. I went there and met up with a photographer named Peter Menzel. Lucky for me, Peter knew quite a bit about the tomato industry. We drove his truck to the Central Valley, an agricultural wonderland that I had never seen or maybe even heard of before the magazine called.

The editor had it in mind that the humble tomato had become the subject of a gigantic technological enterprise. He was correct. We saw vast storage and processing facilities in which layers of crisscrossing conveyor belts carried red rivers of tomatoes past advanced sensors. A manager demonstrated how breeders had created tomatoes with extra-thick, machinery-proof skins by dropping one from chest height onto a concrete floor. Another manager took us to high-tech laboratories in which masked women tested tomato juice and loud salsa music played. All the workers are illegals from Mexico, the manager said, explaining the music. In the fields Peter and I gawped at great combines like swaying ships that sucked up tomatoes as they passed. Conveyor belts ran down each side and teams of men and women picked through the tomatoes as they coursed through the mechanism. At one point a small plane flew overhead and the workers abruptly abandoned the combines and fled into the cover of some nearby trees. They're afraid it's the immigration police, a supervisor said. Labor is a big issue here.

Tomatoes are essentially little balls of flavored water. During my visit the weather was hot and dry—100-degree days, cloudless skies. Outside the farms the landscape was sere. It occurred to me to wonder where the water for the tomatoes came from. Peter asked if I had seen the movie *Chinatown*. That stuff about people killing for water in California? It's all true.

California produces more fruits, vegetables, and nuts than anywhere else in North America, and most of them are grown in the Central Valley. The valley runs for about 450 miles between the Coastal Mountains on the west and the Sierra Nevada on the east. Its floor is a trough of impermeable rock. Over the eons, the mountain ranges have eroded and filled the trough with bands of silt, gravel, sand, and clay several thousand feet deep. Water from melting snow in the heights runs into the deposits and is trapped there by the impermeable rock. Some of the water eventually spills out at the edge of the valley into streams. The rest is stored far underground. Early in the twentieth century, deep-well drills were invented. Suddenly farmers could draw from the ground as much water for irrigation as they wanted. Within a few decades they had sucked out so much water that many parts of the Central Valley were drained and some were sinking like foundering ships into the earth. Here and there the water table declined by more than three hundred feet.

Farmers begged for help. The California state legislature responded in 1933 with the biggest infrastructure project since the Great Wall of China. In the next forty years the Central Valley Project captured and channeled two-thirds of the runoff in the state. It retooled two big river systems with a thousand miles of giant canals and aqueducts, more than twenty big dams and new reservoirs, and a score of huge pumping plants. Naturally, the state was not able to cover the bills. Two years after the Central Valley Project's inception, Washington took over. That left California free in the 1960s to pursue an additional, overlapping State Water Project that was almost as large—twenty-one dams and more than seven hundred miles of canals that funnel water from the far north of the state down the west side of the Central Valley to within fifty miles of the Mexican border.

Peter told me all of this as we drove through the valley in blazing heat.

I asked what California did with all the millions upon millions of gallons of water that it shuffled around. He pointed outside. We were passing through a land of rice paddies. From one end of the horizon to the other were shallow rectangular pools with brilliant green strands of rice waving above. As I remember it, I could almost see the water steaming off the surface and flooding the sky.

I was startled. They spent all that money to send all this water here from hundreds of miles away and then they just let it *evaporate*?

He nodded.

Is that crazy?

Not if you're a rice farmer, he said.

The 170-Mile Sphere

A trope in science fiction novels is the approach to Earth by first-time visitors from other stars. In their spaceships the aliens pass by the outer planets without much interest—ice giants like Neptune and Uranus and gas giants like Saturn and Jupiter are interstellar commonplaces. So are barren rocks like Mars and the asteroids. Then the visitors see Earth and are thunderstruck: it is sheathed in water.

About three-quarters of our world's surface is covered by water, either liquid or ice. Above it is yet more water in the form of clouds. There is lively scientific debate about where all the water came from and why it isn't seen on other planets. What isn't in doubt is that water—H_2O—is one of the most common molecules on Earth, perhaps the most common. Which makes the idea of water scarcity seem odd. How could people run short of something so abundant?

The reason is that 97.5 percent of the world's water is saltwater—undrinkable, corrosive, even toxic. More than two-thirds of the remainder is locked into polar ice caps and glaciers, the great majority of it in Antarctica. The rest—all of the planet's lakes, rivers, swamps, and groundwater—is less than 1 percent of the total. That is the theoretically available freshwater supply. Put together, it would form a sphere about 170 miles in diameter. In fact, though, this is a wild overestimate, because more than nine-tenths of that water is groundwater, and most groundwater is unusable or inaccessible.

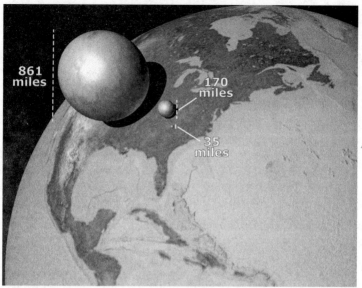

Although water is common, the total global supply (large sphere) is less than one might imagine. Smaller still is the total supply of freshwater (medium sphere), and most of that is locked up in glacial ice or buried too far underground to reach. The total supply of available, usable freshwater—all the water for the world of 10 billion—is shown in the tiny sphere.

Exactly how much of that sphere people already use is not readily ascertained, because it is difficult to measure water flows, and because it is difficult to define water use. One often-cited study from 1996 claimed that humankind already used almost a third of the world's supply of renewable freshwater. For the year 2000, estimated the environmental historian J. R. McNeill, the figure was almost 40 percent. That year a respected Russian researcher, I. A. Shiklomanov, put out a lower number: 12 percent. Whatever the tally, it is a lot—the rest of the water, after all, has to nourish all the other millions of species as they provide our air, break down our wastes, and produce our food.

In any case, global figures are almost beside the point, according to Peter Gleick, founder of the Pacific Institute, a water-research agency in Oakland, California. "If you think about how much water is available, the total amount is still more than what we use," he told me. "The real problem is that water is incredibly short in the western United States and the Middle East and parts of Africa and China, whereas water in Canada and Norway is not scarce at all." Brazil, which has one-sixth as many people as India, has more than four times as much water. The total supply is enough for both nations, but there is no way to distribute it from one to the other.

Children learn in school that freshwater goes through the hydrologic cycle. The cycle begins when water evaporates from seas and lakes. As water vapor rises into the air, it cools and condenses to form clouds. The clouds produce rain and snow, which fall to the surface. There the water either evaporates back into the atmosphere, runs into rivers and streams and then to the sea, or penetrates into the land and becomes groundwater.

River and surface water moves through the cycle quickly, in weeks or months; groundwater moves through it slowly, in years or decades. Either way, economists describe freshwater as a *flow*: a current with a value that is measured by time (gallons per day, for rivers). The sunlight streaming in a window, the electricity issuing from hydroelectric dams, the wheat growing in a field, the wind passing over the wheat—each in its own way is a flow. By contrast, resources like marble and gold and coal are *stocks*—they exist in a fixed amount. Turn on a tap and fill a bucket with water. The water gushing from the spigot is a flow; the water in the bucket is a stock.

The difference may seem academic, but it has repercussions. Every use of a stock reduces the supply of that stock. Take a ton of marble from a quarry, and the next day the quarry has less marble. Keep mining the marble, and eventually the stock runs low. When that happens, the cost of extraction goes up, and with it, typically, the price. People respond by searching for additional supplies (new marble quarries), finding substitutes (home builders use granite instead of marble for kitchen counters), or inventing cheaper methods of using the material (mass-manufacturing marble counters, for instance, thus reducing their cost). Problems have a tendency, however imperfect and slow, to self-correct.

Flows are different. Some flows, like sunlight or wind, cannot be affected by human action. No matter how many solar panels I put on my roof to absorb sunlight, they will have no effect on what the sun does tomorrow. But other flows—"critical-zone resources," in the jargon—

can be exploited to exhaustion. Consider an archetypical critical-zone flow: the run of salmon swimming upstream to spawn. Drop a net across the watercourse and the fish will swim right into it. As long as the number of fish taken from the river every year doesn't exceed the number of survivors from that year's crop of newborns, fishing can continue indefinitely—the supply won't go down, no matter how many years people put in nets. But leave the net in too long one year and it will take every single salmon and there will be no more fishing after that. Catching the last fish is just as easy as catching the first—laying the net across the stream doesn't get more costly as the supply diminishes. With critical-zone flows, things typically go fine until they suddenly don't.

Flows can be wrecked in ways that are uncommon for stocks. It is hard work to ruin an iron-ore deposit. But anyone who has ever seen a toxic chemical spill knows that a river can be contaminated in a careless instant. Interruptions of water flows can be particularly severe because water, unusually, has no substitute. If a salmon run stops, there are other salmon streams and other species of fish. Or people can go without. But everyone has to drink water every day. Coca-Cola or Chianti, apple juice or aquavit, all are just flavored water. Every month millions of householders look at their water bill. Not one of them says, "Oh, this is too much cost and bother—I guess I'll stop drinking water this month."

Groundwater flows take longer to destroy than rivers but are just as vulnerable. Most important groundwater sources are aquifers: underground layers of permeable, water-holding rock. The bands of silt and sand in California's Central Valley form an aquifer. Similar sediments comprise the Ogallala Aquifer, among the world's biggest, which runs from South Dakota to Texas. Aquifers can also be made of porous, sponge-like stone like limestone and dolomite. Water seeps slowly into aquifers and moves through them with equal torpor. The flow through the northern part of the Ogallala Aquifer, to give one example, is on the order of fifty to a hundred feet a day. Drill a few wells and the current is unaffected; the water will pass through as before. But take more than the flow allows and bad things happen—the particles in the porous sand and silt, which had been held apart by water pressure, suddenly compact and become impermeable. The flow is interrupted and often cannot resume.

Contamination is a still more potent worry. Farmers spray pesticides and herbicides on their crops; the residues dissolve in rainfall; the rainfall seeps into groundwater, carrying the chemicals; the chemicals turn up, toxic additives, in people's wells. According to the European Environment Agency, nitrates, heavy metals, or harmful microorganisms contaminate groundwater in nearly every European country and former Soviet republic. Some of these will filter out of groundwater over time, but all too often the damage is permanent. When people pump too much water from coastal aquifers, saltwater can rush in. Thick with salt and minerals, seawater is denser than freshwater; once in an aquifer, there is no known way of flushing it out. Coastal aquifers are imperiled from Maine to Florida; on the Arabian coast; in the suburbs of Jakarta (metropolitan population, more than 10 million); throughout the Mediterranean; and in a host of other places.

Journalists sometimes describe unsexy subjects as MEGO: My Eyes Glaze Over. Alas, water quality is the essence of MEGO. Nonetheless, the stakes—human and environmental—are high. Today, according to the International Water Management Institute, a Sri Lanka–based cousin to IRRI and CIMMYT, one person out of every three on the planet lacks reliable access to freshwater, whether because the water is unsafe, unaffordable, or unavailable. The problems are not restricted to poor nations. By 2025, the institute predicts, all of Africa and the Middle East, almost all of South and Central America and Asia, and much of North America will either be running out of water or unable to afford its cost. As many as 4.5 billion people could be short of water.

Typically such reports focus on urban water supplies. The emphasis is understandable: most people live in metropolitan areas and water from their taps is what will make them sick if contaminated. But most freshwater is actually used by agriculture—almost 70 percent, according to the U.N. Food and Agricultural Organization. Just 12 percent goes to direct human consumption: drinking, cooking, washing, and so on. (Industry takes the rest.) For most of human history, agriculture's overwhelming thirst didn't matter; water was plentiful enough for all. But now populations have risen enough that the requirements of families and the requirements of agriculture are colliding.

The water problems of cities and agriculture are both difficult, but the latter are possibly more consequential, probably more expensive, and certainly more intractable. Domestic water services involve smaller amounts of water. And because people are concentrated in cities, the infrastructure is less expensive on a per-capita basis. By contrast, farms need more water and spread it over bigger areas. Urban water is delivered to homes and businesses where the surplus and waste can be collected for reuse and treatment. Farm water goes into fields; because any excess sinks into the ground or evaporates into the sky, it is not easy to gather for reuse.

Agricultural losses are costly to prevent. Most irrigation is deployed through canals. They lose water because it seeps through the bottom, evaporates during transmission, and spills out at junctions; a rule of thumb is that almost two-thirds of the water is lost, and often much more. (The figures are imprecise, because some of the "lost" water flows usefully into neighboring fields or percolates back into rivers.) Reducing such losses for the Central Valley Project would involve relining and roofing over more than a thousand miles of large canal—and would do nothing for the losses in fields. Farmers could not possibly pay for the measure; costs would be passed to others, either through taxes or higher food prices.

If global affluence continues to rise, more people will want dishwashers, washing machines, and other water-using appliances. Meanwhile, the same rising prosperity indicates (as I have discussed) that food production will have to increase, possibly even double. Unavoidably, more food means more water to grow crops, especially if people eat more meat. In the world of 10 billion, water experts project, the demand for water could be 50 percent higher than it is now. Where will it all come from? New supplies will not be easy to find. Few lakes and rivers are unexploited, and aquifers are being depleted. Equally difficult would be stretching existing water supplies by reducing waste and encouraging thrifty use. Adding to the pressure, climate change is shrinking glaciers and drying streams.

As with food, the disciples of Borlaug tend to react in one way to these worries; those of Vogt, in another. These have been called the paths of hard and soft water, and the choice between them will resonate in the lives of generations to come. The debate between hard and soft is occurring in many places, but can be seen with especial clarity in the Middle East and California. With its rapidly growing population and febrile political tensions, the former bids fair to have the world's most severe water problem. If the latter were a separate nation, it would have one of the world's ten biggest economies. Its water problem may be the biggest in scale. √

Fertile Crescent

The battered Buick drove down the new Mussolini Highway through Tunisia and Libya to Egypt. At the wheel was Walter Clay Lowdermilk, associate chief of the U.S. Soil Conservation Service. Accompanying him were his wife, Inez, a Methodist missionary and social activist; his son, fifteen, and daughter, eleven; a teenaged niece, brought as a babysitter; a personal assistant; and an unruly dog, bought at an Algerian market to keep the children company. Amazingly, this was the full roster of an official U.S. scientific expedition.

The Lowdermilks arrived in Cairo in January 1939, intending to cross the Sinai Peninsula to Palestine, then ruled by Great Britain. The family was advised not to follow this plan. For three years Palestinians had been in revolt against the British authorities. Although London had bloodily suppressed the insurgency, no travelers had crossed the Sinai for six months. Desert villages had been destroyed; nomadic people, hunted by both sides, were robbing survivors. Some had concealed homemade mines—improvised explosive devices—beneath rocks in the road. Lowdermilk decided to go anyway. His family would be safe, he explained, because the Buick could go faster than the men on camels who would be chasing it. And they would keep a lookout for mines—detouring around rocks on the road, for instance, rather than moving them. Soldiers at the guard post of Beersheba in Palestine were astonished to see the Buick coming in from the desert. We are here, Lowdermilk informed them, to see the Holy Land.*

Walter Lowdermilk had made a great career from failing to look before he leaped. As a boy he left his family in Arizona and worked his

---

* Terminology in this region is contested. I use "Palestine" to refer to historic Palestine—the area under British rule from 1922 to 1948—and "Palestinian territories" for the non-Jewish areas established in 1948 by U.N. General Assembly Resolution 181. That resolution split Palestine into two separate entities, Israel and the Palestinian territories, neither of which accepted its proposed borders.

The Lowdermilk expedition, stuck in the mud in Syria

way through high school in Missouri. A friend there told him about a new scholarship program named after the British tycoon Cecil Rhodes that paid for foreign students to attend Oxford. In an instant Lowdermilk decided the Rhodes scholarship was his future. He quit studying science and devoted himself to the required Latin and Greek. He won the Rhodes and went to Oxford, where he abandoned classics to study forestry. On his return to Arizona he took a job with the Forest Service, becoming friends with Aldo Leopold. Just before Lowdermilk left for Oxford he had met a young woman at church, Inez Marks, a pastor's daughter. She became a missionary and social reformer in China, building schools for girls and campaigning against foot-binding. When Inez returned briefly to the United States, Lowdermilk drove to California and saw her for the first time in eleven years. Forty-eight hours later, he proposed marriage. Inez was returning to China; Walter quit the Forest Service to join her. He learned Chinese and took a two-thousand-mile solo trip up the Yellow River. Along the way inspiration filled him and he understood the path of history and the rise and fall of civilizations.

Erosion was the key, and its cause was water. To grow food, societies needed to harness rainfall or deploy irrigation. But they failed in both. Rainfall rushed down slopes and carried topsoil into rivers and flushed it into the sea. Or water evaporated in irrigation channels and left behind salts that poisoned the land. Or water wasn't saved when rain fell and the fields withered. Later Lowdermilk would also point to the role of overgrazing—especially by goats, animals he came to detest—but for the most part the devastation was wrought by water. Human incompetence with water management had destroyed countless societies over the millennia.

Civil war forced the Lowdermilks to flee China in 1927. Walter barely escaped with his life, but his zeal to preserve soil and water was undimmed. With a fellow anti-erosion advocate, Hugh Hammond Bennett, he helped establish the Soil Conservation Service, possibly the world's first national anti-erosion agency. Bennett became its head. The two men did not get along. In 1938, Secretary of Agriculture Henry Wallace suggested that the soil service could better plan for the future if it better understood soil problems of the past. Embracing the chance to work away from Bennett, Lowdermilk decided to survey soils in Europe, North Africa, and, especially the Middle East. Both Inez and Walter were devoted to their Christian faith. Visiting the Fertile Crescent—the Promised Land of Abraham, from the eastern Mediterranean shore to Iraq's Tigris and Euphrates rivers—was a long-held dream. Now they had the chance to realize it.

Lowdermilk was ready to see what Moses had seen on the hills east of the Jordan River: "a good land, a land of brooks of water, of fountains and springs that flow out of valleys and hills." He was ready to see the lush cedar groves of Lebanon, evoked in the Bible as "full of sap" ("His [the Lord's] countenance is like Lebanon, excellent as the cedars"). He was ready to see Babylon, "fairest of cities," where the Hanging Gardens had been one of the seven wonders of the ancient world and King Nebuchadnezzar II had built "great canals" that "brought abundant waters to all the people" and erected "magnificent palaces and temples" with "huge cedars from Mount Lebanon" and overlaid with "radiant gold."

Instead Lowdermilk found an almost treeless waste: exhausted soil, untended ruins, scattered and scrubby vegetation, impoverished goatherds in "poverty, ignorance, and squalor," "the Hanging Gardens of Babylon now heaps and piles in a salty desolation." The Fertile Crescent was no longer fertile; the land of milk and honey had neither. Where once the mighty forests of Lebanon had provided wood for ships and cities across the Holy Land, only three small cedar groves remained, the largest with about four hundred trees. Because there were no roads, the family followed oil pipelines across Syria to Baghdad. The biggest city in Mesopotamia, it was a "dirty place," Lowdermilk thought, "a heap and a pile." And this was the heir to glorious Babylon! "What a decline in wealth, in buildings, in population, in attainments, in glory!" he wrote in a Soil Conservation Service report on his journey. "Is this the best that mankind can do? Is this our end after 7,000 years of civilization?"

Lowdermilk knew just what had happened. Prosperity in the Fertile Crescent, he said, depended on commanding the waters of its major rivers: the great Tigris and Euphrates in the east, the smaller Jordan and Litani near the Mediterranean. Around 5000 B.C., he said, ancient Mesopotamian societies began to construct canals for irrigation, which promoted agriculture and allowed these societies to flourish. By the days of Nebuchadnezzar, the canal network extended many miles into the dry land. Unfortunately, the rivers came from the mountains and filled with silt as they cut their way downstream. When the fast-moving rivers were channeled into irrigation canals, they lost velocity; the suspended silt precipitated to the bottom of the canals and eventually choked them. Maintaining the network required scraping out the silt. "A standing army of slaves for this task was required to toil without ceasing on this endless removal of silt from the canals," Lowdermilk explained. These were the Israelites of the Old Testament. Roman and then Byzantine invaders took over but continued to dredge the canals. Then came Arab nomads with their new religion of Islam. "Despising the tilling of the soil and hating trees, the nomads sought to live off herds and off the plunder of settled areas." As the water infrastructure fell into disuse, levees eroded; floods washed away topsoil. The canals dried up and left salt in the soils that glittered at night. Goats ate what was left. Nobody tried to restore the land—inactivity Lowdermilk attributed to "Moslem fanaticism, with its fatalistic belief that what happens is the 'will of Allah.'"

Modern archaeologists think most of this is wrong. The irrigation systems were built up more slowly than Lowdermilk thought, not reaching their apex until about the birth of Christ. By then erosion was already epidemic. Societies throughout the Fertile Crescent had cut down forests to build cities and, especially, feed the forges that made bronze and iron. Without tree cover, the hills could not retain water; floods destroyed canals downstream. The same biblical peoples who had created the great city of Babylon set in motion its destruction. Islam and goats had little impact. And the failure to restore the land was due not to nomadic peoples but to the wholly sedentary Ottoman Empire, based in faraway Istanbul, which ruled the area from the fifteenth century until the end of the First World War. In its bureaucratic way, the empire extracted wealth from the area while refusing to invest in it. Still, Lowdermilk got one big thing right: the Fertile Crescent had become a desert and a major cause was human incompetence with water.

Escaping from fascism, European Jews had poured into Palestine—more than sixty thousand in 1935 alone. Arab residents reacted angrily to the flood of immigrants. The British government was convinced that the hostility was due, in part, to the region's lack of resources; the immigrants were exceeding Palestine's "absorptive capacity" (that is, its carrying capacity). The limit to absorptive capacity was water—British experts argued that regional supplies couldn't sustain a big influx of immigrants. In this arid, eroded landscape, the supply of well-watered farmland was so small that incoming Jews who used their superior financial resources to acquire it would necessarily create "a considerable landless Arab population." Zionist groups sent out water testers, who proclaimed that they had found much more water than Britain allowed. London ignored the reports and in 1939 restricted Jewish immigration to fifteen thousand a year.

No! Lowdermilk protested. Britain had it backward! The new Jewish settlements were the only bright spots he had seen in the entire dismal region! In the midst of the desolation were Zionist village cooperatives where jointly owned farms grew newly bred crop varieties that thrived in the dry heat. The farms were investing their profits to buy advanced well-boring equipment and create small industries—carpentry and printing shops, food-processing facilities, factories for building material. Most important to Lowdermilk were the irrigation and soil-retention programs—"the most remarkable" he had encountered "in twenty-four countries." If the British increased immigration, rather than restricted it, he said, Palestine would be able to support "at least four million Jewish refugees from Europe."

Lowdermilk's years at the Soil Conservation Service had been a series of battles with agronomists ("plant men," he called them) who thought that bad land should be revived mainly by revegetation—covering the soil with a mosaic of ground-hugging plants that would shield it with an absorptive mulch. To Lowdermilk, a proto-Wizard, engineering had to come first. Dams, pumps, and pipelines were needed to divert water from places where it was plentiful to places where it was not. The land that would receive it had to be prepared by reshaping its contours to avoid runoff and erosion. Only then could planting play a role. And all of the work should be on the biggest scale possible—entire regions guided by a precisely hewn plan—rather than working plot by plot, in small-scale endeavors.

Palestine, Lowdermilk proclaimed, offered "a splendid opportunity" for a water-and-power project of transformative scope. The region's north had water—almost forty inches of rain per year; the Sea of Galilee, fed by the Jordan River—but little arable land. In the south, 130 miles away, the Negev Desert had arable land but little water—no lakes, less than four inches of rain per year. Redirecting water from north to south, Lowdermilk said, would allow farmers to irrigate more than 300,000 acres of good soil.

The Sea of Galilee drains through the Jordan Valley into the Dead Sea, a salty lake 1,412 feet below sea level. If much of the Sea of Galilee's water were diverted to the south, other water would need to feed the Dead Sea. An obvious solution: pumping desalinated seawater from the Mediterranean into the Jordan Valley. The water coursing into the deep valley could be used to drive turbines, Lowdermilk said, generating electricity for "well over a million" people. Water and power would be supplemented by a program of range management and reforestation; the project would also extract minerals from the Dead Sea.

The Second World War broke out as Lowdermilk was completing his trip. Swept up in the war effort and beset by health problems, he didn't publish his ideas about water for the Holy Land until 1944. The book, *Palestine: Land of Promise*, appeared just as the U.S. public was learning of the plight of German Jews. By an accident of timing, the book offered good news to offset the bad news of the concentration camps. There is "hope in Palestine," the book declared. In a region of "utter decline," its Jewish settlements were "one of the most remarkable phenomena of our day." They proved that with proper technology even ruined land can bloom. "Water was the main problem in ancient times as it is today," Lowdermilk proclaimed. "But in this machine age we have more perfect instruments for our purposes." The dams, conduits, "great reservoirs or artificial lakes," tunnels "boring through mountains" in his Palestine plan—all would demonstrate how "modern engineering" and "scientific agriculture" could "transform waste lands into fields, orchards, and gardens supporting populous and thriving communities."

Lowdermilk's ideas had been anticipated years before by Zionist dreamers like Theodor Herzl, Levi Eshkol, Aaron Aaronson, and Simcha Blass. But *Palestine: Land of Promise* drew widespread support for them in Western nations—the book supposedly lay open on President Franklin D. Roosevelt's desk when he died, in the spring of 1945. Nonetheless, Britain rejected the Lowdermilk plan as too costly and impossible to administer. After becoming a nation, Israel built it. The nation's leaders believed they had no choice—three-quarters of a million refugees had come to Israel in its first five years. Debates about absorptive capacity had been rendered irrelevant. First came a kind of practice pipeline from Tel Aviv along the Mediterranean coast to the Negev, near the Sinai Peninsula. In 1956 Israel committed to building a Lowdermilk-style, north-to-south project: the National Water Carrier.

The National Water Carrier was and is a Wizardly demonstration of technological prowess. Thousands of workers carved an underground pumping station 250 feet long and 60 feet high at the edge of the Sea of Galilee. Working twenty-four hours a day, its three huge pumps push millions of tons of water almost a thousand feet through the surrounding hills to the newly constructed Jordan Canal. Ten feet deep and forty feet wide and about ten miles long, the canal conducts the water through a series of reservoirs, canals, and pumps into a nine-foot pipe that runs for more than fifty miles to southern Israel, where the water is spread through a specially built irrigation network. The cost was enormous: on a real per-capita basis, the nation spent more on the National Water Carrier than the United States did to build the Panama Canal.

Lowdermilk was invited to its inauguration in 1964. He ended up spending six years in Israel. He was sure that more and greater projects would come. Israeli water technology would transform the Fertile Crescent. Already Israel was talking about using nuclear power to pump water from the Red Sea to replenish the Dead Sea. Monitored by advanced sensors and controlled by computers, a region-wide network

The Israeli National Water Carrier

of dams, reservoirs, canals, pipelines, desalination plants, and pumping stations would transfer huge quantities of water from areas of surplus to areas of want. The high-tech water web would bind neighboring states together, calming political conflicts.

It didn't happen that way. Counterforces emerged.

Water in the Garden City

Justus von Liebig's connection to Israeli sewage-treatment policies is insufficiently appreciated. The great chemist became famous for portraying agriculture as, at bottom, a process of funneling chemicals—nitrogen, phosphorus, potassium, and so on—to plants. But Liebig also observed that farmers effectively transferred the applied chemicals to the city in the form of harvested grain, vegetables, and fruit. Urbanites excreted the nutrients, which then accumulated as pollution or were dumped into rivers. To return those chemicals to the soil, Liebig wrote, "all the proprietors of the soil in every great country, should form a society for the establishment of reservoirs where the excreta of men and animals might be collected."

One of Liebig's more attentive readers was Karl Marx. Liebig's shipment of nutrients from country to city, Marx said, amounted to nothing less than a "rift" between people and the land, causing urbanites to live in toxic filth while robbing farm landscapes of their fertility. City and country must become one! Marx said. Agriculture and industry must join to preserve the soil! Marx blamed capitalism for the rift, rather than seeing it as a side effect of industrialization. Nevertheless, he was on to something. His interpretation of Liebig was taken up by writers like the designer-activist William Morris, the anarchist geographer Peter Kropotkin, and the socialist science-fictioneer Edward Bellamy, who in turn passed them to a young English clerk named Ebenezer Howard.

The son of a baker, Howard (no relation to Albert Howard, the organic campaigner) immigrated to the United States, failed as a farmer in Nebraska, and returned to Britain five years after leaving. He became a parliamentary clerk in London and spent his off-hours with anarchists, socialists, and other freethinkers. Gradually Howard became convinced that reknitting the seams between city and country was key to improving the human condition. In 1898 he released *To-Morrow: A Peaceful Path to Real Reform*. It was his first published work. Four years later a revised edition came out under a new title: *Garden Cities of To-Morrow*.

The *To-Morrow* books transformed the idea of the city. They were to urban planning what *Road to Survival* was to environmentalism—a summation and extension of others' thoughts that created a movement. Like William Vogt, Ebenezer Howard put together a complex of beliefs now so conventional that it is surprising to discover that it had an origin. And, like Vogt, Howard has mostly been forgotten, even by the urban planners whose visual language of open spaces and connectivity was largely his creation.

Distilling Liebig, Marx, and their successors into a few succinct pages, Howard's books offered concrete plans for integrating urban and rural life, simultaneously preserving and balancing human relationships with society and nature. No longer would the landscape be divided between

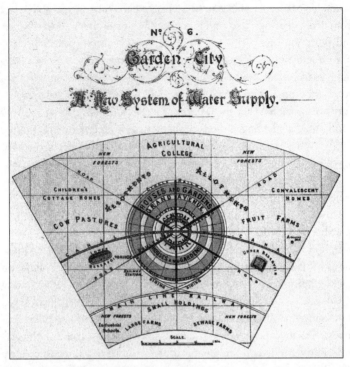

In Ebenezer Howard's garden cities, stormwater and wastewater are collected in reservoirs and pumped with windmills into canals that run through the city, where they can be used for fountains, parks, and agriculture.

crowded, dirty, culturally rich cities and the depopulated, lonely, intellectually barren countryside. Instead both would be woven together in a mesh of communities—the garden cities of his title—with green belts between. People could live in open countryside, small towns, or big cities, and all would be close at hand. "Town and country *must be married*," he wrote in excited italics, "and out of this joyous union will spring a new hope, a new life, a new civilization." Connoisseurs of irony will note that this proto-environmentalist laid the intellectual underpinnings for what his successors would lambaste as "suburban sprawl."

European cities' water problems, Howard believed, were evidence of an "essential mistake at the very root of our social life." Accordingly, his garden cities would use water efficiently, cheaply, and cleanly. They would have three parallel water systems: one for carrying potable water from nearby springs; a second for collecting stormwater and wastewater and pumping them with windmills into reservoirs for reuse by agriculture and public works; and a third for gathering sewage for recycling, as Liebig had wanted. The idea of separate water systems was not new—the Roman emperor Augustus built a special aqueduct to transport non-potable water for watering gardens and filling a big artificial pond on which he staged mock naval battles for entertainment. Rome's parallel water network was intended to avoid drawing down the supply of good water for the frivolous purpose of watching fake wars. Taking the idea further, Howard sought to harness wastewater to improve society.

Howard created an association to bring his vision to fruition. It quickly discovered that building brand-new cities was expensive. Moreover, few places had the right combination of idealistic beliefs and platoons of new inhabitants without housing. Among them was Jaffa, on the Mediterranean shore, one of the first destinations for Jews migrating to Palestine. An ancient port, Jaffa had been inhabited for at least seven thousand years. Its long history was manifest in its tangle of dark, narrow streets and open sewers. The newcomers had not come all the way from Europe to live in the squalor of the past. They wanted to create a homeland that was clean and modern, filled with sunlight and healthful air. Zionist leaders in Germany supported this ambition. They sent *Garden Cities of To-Morrow*.

Half a dozen new Jewish settlements around Jaffa were modeled on Howard's ideas. Over the years so many Jews moved to them that what had originally been suburban villages swelled into the city of Tel Aviv, which swallowed Jaffa. All the while, other new Jewish settlements were built according to Howard's precepts. And in 1956, the municipalities that made up greater Tel Aviv committed to recycling wastewater and sewage water, much as Howard had proposed.

Contemporary sewage treatment occurs in three steps, in an ascending ladder of squeamishness. The first, primary treatment occurs when the sewage is placed in large tanks where the most disagreeable solids settle to the floor. This sludge is removed and buried in landfills (in a few exceptional places, it is converted to fertilizer). Secondary treatment involves adding bacteria to the still-vile water to consume remaining organic matter. After eating their fill, the bacteria, too, sink to the bottom, and are scraped away. Tertiary treatment usually means disinfecting the wastewater with chlorine or ultraviolet radiation, after which it is safe to discharge into rivers or seas. All of these are costly, which is why governments typically have resisted them until forced to adopt them by public pressure.

The municipalities around Tel Aviv—seven at first, and later, as the project grew, eighteen—agreed to conventional primary and secondary treatment. But rather than deploy tertiary treatment they decided to pipe the water from secondary treatment to a sand-dune region a few miles away. Layers of fine, packed sand sit there atop a coastal aquifer. (Or, more precisely, atop a section of the large sandstone aquifer that runs along most of the Israeli coast and down to the mouth of the Nile.) The wastewater would be channeled into newly created ponds on the dunes. In six months to a year, the water would filter slowly through the sand and then through a layer of sandstone. When it reached a hundred feet or more beneath the surface, it would recharge the aquifer. Water from the aquifer could then be pumped through a fifty-mile pipe into a reservoir in the Negev, where it would be disbursed for irrigation. Some of the irrigation water would in turn percolate into the ground, recharging another portion of the coastal aquifer.

After five years of testing, the first water went into an experimental recharging pond in 1977. Political opposition delayed the project's expansion for more than a decade. Only in 1989 did treated wastewater go to the Negev. By then, the idea was being replicated across Israel. Laws directed every town to transform its sewage into irrigation water—a parallel, Howard-style water infrastructure. The advantage was clear: unlike the flow of rain, the flow of sewage is as constant as the North Star. But because farmers' water requirements change with the seasons, the parallel infrastructure included reservoirs to store the treated water until needed. To keep the two networks separate, the new one had to be built from scratch. Today, about 85 percent of Israel's wastewater—more than 100 million gallons a year—is used for irrigation, according to Seth M. Siegel, the author of *Let There Be Water* (2015), a study of Israeli water use that I am following here.

Reusing sewage water is an example of what the natural-resource economist David B. Brooks, of the International Development Research

Centre in Ottawa, called the "soft path" of water management.* The soft path is the Prophet's path. The contrasting "hard path," the Wizard's path, the path of Lowdermilk and Israel's National Water Carrier, has long been the conventional way: centralized infrastructure to capture, deliver, and treat water supplies. It is the path of huge concrete structures, vast engines, top-down planning, landscape changes on a geographic scale. The hard path asks: How can we get more water? Focused on increasing supply, first and foremost, it stems from the belief that the demand for clean freshwater is inexhaustible. Its logical outcome, according to Peter Gleick of the Pacific Institute: "ever-larger numbers of dams, reservoirs, and aqueducts to capture, store, and move ever-larger fractions of freshwater runoff."

Enabled by the development of cheap cement and cheap fossil fuels, the hard path has produced drinking and irrigation water for huge numbers of people. Like the invention of synthetic fertilizer, the reshaping of water systems has profoundly affected the contours of everyday life, allowing the inhabitants of today's megacities to live at a level of cleanliness, health, and comfort that would astonish our ancestors. But it has also sucked water from rivers and lakes, degrading their ecosystems, and made water tables sink in every corner of the world.

The soft path, by contrast, is something new. Decentralization, efficiency, and education are its hallmarks. It "draws all 'new' water from better use of existing supplies and changing habits and attitudes," Brooks and a co-author explained in 2007. It eliminates waste and squeezes inefficiencies from the system. It says it is usually easier and cheaper to be smart about current water uses than to build big new projects.

Hard-path Wizards ask: How should we get more water? Soft-path Prophets ask: Why use water to do this at all? In arid areas like the U.S. Southwest, middle-class households use as much as three-quarters of their water on the lawn. Hard-path boosters seek supplies to maintain them. If droughts occur, they endorse water restrictions—more-efficient sprinklers, no-watering days—but regard them as temporary measures, not long-term fixes. Soft-path believers see the goal as attractive landscaping and argue for replacing lawns permanently with dryland plant-

---

* Brooks borrowed the term "soft path" from the energy advocate Amory Lovins, whom I discuss in Chapter 6.

ings that need little water. Replacing the lawn is an exemplary soft-path solution: hyper-efficient, locally oriented, bottom-up, low-tech. In saving water, it seeks to transform the featureless, universal landscape of the lawn with area-specific plantings that embody the essential qualities of a place.

Like Howard's vision for agriculture, the soft path is about limits and values. It is, Brooks has said, "a human vision toward a sustainable future." At one level, it is about reforming institutions; at another, about changing habits. Ultimately, though, it is a vision of the human place in nature. Hard-path supporters see technology placing humanity in charge: we can move H_2O molecules wherever we want to satisfy our wishes. Soft-path people think this level of control is illusory—cooperation and adjustment, not command and control, is the way to live.

The problem is, as Gleick says, "We cannot follow both paths." As a practical matter, saving water cuts into the justification for providing more, and vice versa. Implementing both paths at once requires attention and funding that are hard to come by in a world of clamoring needs.

Exactly that conflict occurred in Israel when water authorities adopted soft-path methods in the 1980s and 1990s. In addition to mandating wastewater recycling, regulators successfully pushed farmers to stop growing cotton, a water-intensive crop. They mandated low-volume showers and dual-flush toilets (one button for big loads, one for small). They raised water prices. They launched water-education programs in schools that taught children to value water—classroom posters exhorted children "not to waste a drop." ("If my shower goes too long, my kids yell, 'Dad! You're draining the Sea of Galilee!'" Noam Weisbrod of Israel's Zuckerman Institute for Water Research told me.) Utilities fought leaks by equipping every meter with a cell phone that reported unusual flows.

Notably, Israel provided incentives for farmers to switch to drip irrigation, in which pipes with tiny holes provide small, precisely adjusted flows of water. Ideally, drip irrigation provides water at just the rate at which it can be absorbed by plant roots. Invented by the Israeli engineer Simcha Blass, it can use half or less of the water used in ordinary irrigation to nourish the same number of plants. At the same time, drip irrigation is one of those ideas that sounds simple but is hard to accomplish in practice. To exude water in regular amounts from the holes, the

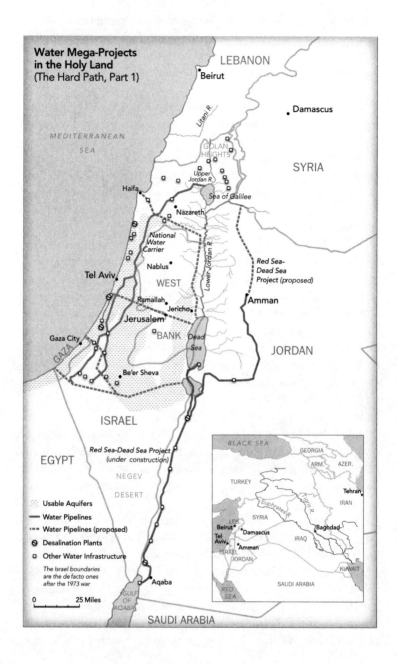

**Water Mega-Projects
in the Holy Land**
(The Hard Path, Part 1)

MEDITERRANEAN
SEA

LEBANON

Beirut

Damascus

Litani R.

GOLAN
HEIGHTS

SYRIA

Upper
Jordan R.

Haifa

Sea of Galilee

Nazareth

National
Water
Carrier

Lower Jordan R.

Red Sea-
Dead Sea
Project (proposed)

Nablus

Tel Aviv

WEST

Ramallah

Amman

Jericho

Jerusalem

BANK

Dead
Sea

Gaza City

JORDAN

GAZA

Be'er Sheva

ISRAEL

Red Sea-Dead Sea Project
(under construction)

EGYPT

NEGEV

DESERT

:::: Usable Aquifers

—— Water Pipelines

-■-■- Water Pipelines (proposed)

◉ Desalination Plants

□ Other Water Infrastructure

*The Israel boundaries
are the de facto ones
after the 1973 war*

0   25 Miles

Aqaba

GULF
OF
AQABA

SAUDI ARABIA

BLACK SEA

GEORGIA

TURKEY

ARM.   AZER.

Tehran

IRAN

Euphrates R.

Tigris R.

Beirut   LEB.
SYRIA

Tel
Aviv   Damascus

Baghdad

ISRAEL

Amman

IRAQ

JORDAN

KUWAIT

SAUDI ARABIA

RED
SEA

water pressure must be exactly the same down the entire length of the pipe, an engineering challenge; if the pipe is underground, which maximizes contact with roots, the holes must not be susceptible to clogging by dirt or being infiltrated by water-seeking roots. The first Israeli drip-irrigation firm was established in 1966; by the 1990s, the method was being used in about half of Israel's farms.

Hard-path boosters scoffed at these efforts as Band-Aids that treated symptoms instead of the actual disease: scarcity. Scarcity occurred in two forms: natural and political. The natural scarcity referred to the region's recurring droughts and meager stores of groundwater. Political scarcity referred to the likelihood that some of Israel's water would be seized by its hostile Arab neighbors. These neighbors had objected vehemently to the National Water Carrier, which they viewed as an illegitimate regime's scheme to steal the region's water. In the 1950s Israeli and Syrian forces shot at each other over the former's efforts to implement it. The opening of the National Water Carrier prompted Egypt to convene the first Arab Summit, which in turn led to the creation of the Palestine Liberation Organization. The target of its first attack, in December 1964, was the National Water Carrier. Since then, Jordan has diverted more water upstream from Israel, as has Syria. Palestinians have repeatedly demanded a greater share. Diplomats have suggested that Israel trade water for peace.

All of this, hard-path backers say, is why Israel will eventually need to come up with more water. Dual-flush toilets and school posters are not enough. Agreeing, Israel built five huge desalination plants on the Mediterranean between 2005 and 2015. They produced so much drinking water—about 80 percent of the nation's needs—that Israel has discussed replumbing the National Water Carrier to send any surplus to the Sea of Galilee. Meanwhile, Jordan, Israel, and the Palestinians announced in 2013 a vast project to link the Red Sea to the Dead Sea. For its first phase, to be completed in 2021, a desalination plant on Jordan's Red Sea shore would provide water to southern Israel and the Palestinian territories. In a swap, Israel would send desalinated water from its Mediterranean facilities to Amman, the water-short Jordanian capital. The leftover brine from Red Sea desalination, too salty to be dumped safely into the environmentally fragile Red Sea, would be pumped north into the Dead Sea, making up for some of the Jordan River water lost to dams and the National Water Carrier. The project was scaled back after scientists decried the ecological risks of linking those two seas for the first time ever. Nonetheless officials from all three governments told me they were excited. They saw themselves as changing the game, not simply fiddling with the rules.

Economic Interlude

Urban planning is easier in a dictatorship than in a democracy. Since its founding, in the thirteenth century, Shanghai has occupied the west bank of the Huangpu River. On the east bank were the farms and gardens that fed the big city. In 1993 the Chinese government decided that this arrangement was unsatisfactory. It bulldozed the farms and gardens, created a "special economic zone" in their place, and in a hot-brained hurry threw up a brand-new city, Pudong ("East of the Pu"). Pudong now has the world's second-, ninth-, and twenty-third-tallest buildings, and more than 5 million inhabitants, most of them middle class, housed in faux-Mediterranean villettes as deracinated as so many Orange County strip malls.

By chance, I visited Shanghai a year after construction had begun. Out of curiosity, I crossed the river one evening to see what was going on. I walked to the edge of the then-small construction zone and looked to the east, toward what would become the rest of the city: an endless expanse of rice paddies and truck gardens. Few of the inhabitants seemed to have electricity; only a handful of lights were visible in the darkness. Fifteen years later I returned to the same spot and saw a city almost twice the size of Chicago. Its light covered the stars.

Rapid urbanization is a hallmark of our age. In 1950 fewer than one out of three of the world's people lived in cities. By 2050, according to United Nations projections, the figure will be almost two out of three. Meanwhile the world's population will have more than tripled. In 1950, 750 million people lived in urban areas; by 2050, demographers project, 6.3 billion will—more than eight times as many. For the most part, farmers have kept up with the increase in urban numbers, growing more food and distributing it in the newly expanding cities. Water has a poorer record. Cairo, Buenos Aires, and San Antonio; Dhaka, Istanbul, and Port-au-Prince; Miami, Manila, Monrovia, Mumbai, and Mexico

City—all have greatly expanded, and all have failed to keep up with the demand for clean, plentiful water.

Providing water to cities entails four basic functions: purifying the water that goes into the system; delivering it to households and businesses; cleaning up the water that leaves those homes and businesses; and maintaining the network of pipes, pumps, and plants. Simple to describe, these tasks are hair-pullingly complex on the ground. The cost and technical challenge of building and operating a water system that can supply the daily onslaught of morning flushes and showers while not overwhelming users at light times is the sort of thing that keeps engineers in heavy demand. Challenges increase if the city is growing rapidly, like Pudong. Because demand is constantly increasing, new capacity must be built at top speed while administering the old.

But cost and technical difficulty are not the primary reason so many modern cities have been unable to provide water to their inhabitants. Again and again, the biggest obstacle has been what social scientists call governmentality and what everybody else calls corruption, inefficiency, incompetence, and indifference. French cities lose a fifth of their water supply to leaks; Pennsylvania's cities lose almost a quarter; cities in KwaZulu-Natal, the South African province, lose more than a third. So much of India's urban water supply is contaminated that the lost productivity from the resultant disease costs fully 5 percent of the nation's gross domestic product. More than thirty North American cities improperly test for lead in their water, including, famously, Flint, Michigan, where bungling local, state, and federal officials have forced residents to drink bottled water for years. The Mountain Aquifer that straddles the border between Israel and the Palestinian territories is the most important source of groundwater for cities in both nations. In an atypical act of collaboration, both societies are polluting it. And so on.

So long is the litany of government failure in urban water systems that in the 1990s organizations like the World Bank and the International Monetary Fund began to argue for turning over water management to the private sector. One of the biggest examples was China, where in 2002 the Shanghai government contracted the task of expanding and operating Pudong's new water system to a French company named Veolia. In return for $243 million and a fifty-year, 50 percent stake in Pudong's water utility, Veolia became very, very busy. In its first five years in Pudong, Veolia laid almost 900 miles of large-diameter pipe, hooked up

300,000 new structures to the growing water system, built sewage and water-treatment plants, and hired 7,000 local workers. Along the way it built a new office tower and on its ninth floor created a customer-service call-in center—a novelty in China—staffed around the clock by young women in powder-blue Veolia uniforms. During my visit, one proudly showed off a war room with a twelve-foot screen displaying the real-time status of every water connection in Pudong.

Veolia has a longer, stranger history than one would expect from its dull corporate moniker. It was founded in 1853 by Napoleon III, France's last emperor, along with many French nobles, and financed by the bankers Baron de Rothschild and Charles Lafitte. The Compagnie Générale des Eaux, as it was then called, became an essential part of the emperor's plan to modernize his country. Signing decades-long contracts to expand, modernize, and operate the water systems in France's biggest cities, CGE became integral to the national infrastructure—a private enterprise holding a public commission, and entrusted with the water supply even of the glorious capital.

In the 1980s the firm abruptly woke up to the possibilities of modern financial markets, and realized that it could use the revenue flow from its millions of water customers to acquire other companies, most of them in industries more glamorous than plumbing. After buying publishers, software firms, music companies, television networks, and movie studios, the company's CEO celebrated by moving to a $17.5 million duplex apartment on Park Avenue. The unwieldy enterprise soon foundered. The CEO was forced out and in 2011 was convicted of embezzlement. Meanwhile lawyers and financiers feasted like vultures on the parts, tearing off corporate pieces and gulping them down. Out of the morass emerged Veolia, the biggest private water-service operator in the world. It runs water systems in nineteen nations, including China.

Veolia's smooth operation in Pudong was a testament to the power of private enterprise. It takes formidable organization to deliver water to so much new construction—though even here, incredible to Westerners, the water must still be boiled before drinking. (Habituated to bad water, Chinese people don't expect to be able to drink it straight from the tap.) To let Veolia recoup its costs, Shanghai gradually raised water prices—not enough to make Pudong's nouveaux riches blink, but enough to ensure that they will not leave the taps on all day. "This model is a unique combination of private efficiency within overall public ownership," Veolia's Shanghai director said to me. He said he had a hard time understanding why anyone would be upset at the thought of privatizing water. So did his boss, Antoine Frérot, now head of Veolia. "Private companies can better manufacture cars" than governments, Frérot told me, and water "is exactly the same."

Veolia and other water multinationals—Big Water, let me call them—make an argument straight out of Economics 101. I heard a version of it from John Briscoe, who was for ten years the senior adviser on water at the World Bank. Briscoe kept a doggerel poem by the economist Kenneth Boulding on his wall that read, in part:

> Water is politics, water's religion,
> Water is just about anyone's pigeon. . . .
> Water is tragical, water is comical,
> Water is far from the pure economical.

Boulding had captured something, Briscoe said. Many of the world's water problems arise because the sacred aura around water induces governments to treat it "as common property—it's free to use, no matter what you do with it and how much you use." In consequence, huge quantities are wasted. Equally bad, the fact that water is free means that governments can't recoup the cost of extending water networks—so they don't. Utilities don't fix leaky pipes for the same reason. All over the world, Briscoe said, "you have these hugely underfunded, very inefficient services producing very bad service." He said, "They don't have enough to operate the system properly, so the existing system rations water, and of course it's the elite that gets to the front of the queue."

The best way to deliver water to people's homes efficiently, Big Water argues, is to put the process in the hands of the market. If water is scarce, then raise the price—let the law of supply and demand take over! If people want water that is not only plentiful but actually clean, then raise the price again. The market will find the balance between what consumers want and what they can afford. And if the water company does not make good on its promises, it can be ejected in favor of another firm. The threat of competition will force utilities to be accountable.

Implicit in the free-market scenario, however, is the assumption that families can actually afford to pay for their water—and are willing to accept the price. I saw how that assumption can play out when I visited the city of Liuzhou, a small city of 1.5 million in southwestern China. In its rapid industrialization, Liuzhou has surrounded itself with more and more factories, simultaneously increasing living standards, attracting more residents, and poisoning the Liu River, formerly its water supply. The city government realized at the turn of the century that it would have to create a city water network, including sewage-treatment plants. Unable to afford the cost, Liuzhou borrowed $100 million from the World Bank in 2005. Then it signed a thirty-year contract with Veolia. The company sent out construction crews. Naturally, it needed to recoup its investment. It raised the cost of water.

The historic center of Liuzhou occupies a promontory in a big oxbow of the Liu. Toward the northern tip of the promontory is a beat-up public square near an area that city maps label Maojin Chang—Towel Factory. The name comes from a factory that went out of business before the current wave of modernization. Most of the neighborhood's inhabitants still lived in the shuttered factory's dormitories, or so they told me when I came by. Many were pensioners; some were farmers who were thrown off their land to make way for the factories outside town. They were not wealthy people, and they were keenly aware of water prices. In a few hours of wandering around Towel Factory Square I found half a dozen men and women who were paying a quarter or more of their income just for water.

Economics 101 does not readily apply to neighborhoods like these. A few blocks away from Towel Factory Square, an elderly man named Wei Wenfang waved at me. He had heard me asking about water and was eager to contribute to the discussion. (Like other residents, I suspect, he was also tickled by the opportunity to talk to Josh, my friend and translator—tall, blue-eyed Mandarin speakers were a local rarity.) Until 1975, Wei said, city water was free: you simply dipped a bucket into the Liu. You could literally see the bottom of the river back then! he said with an emphatic flourish of the hands. But now Wei paid almost $10 a month, more than a quarter of his pension, for water that was nowhere near as good. When I asked if he could save money by conserving water, he barked with laughter. Each water meter in the neighborhood, he said, covered sixty to seventy apartments, many of which are occupied by more than one family. The total bill was divided equally among the residents. "There's no way to save," he told me. "Your efforts are just lost in the mass. That's the way it is all over the city."

The contract created a company, Liuzhou Water Services, its ownership split between Veolia (49 percent) and the city (51 percent). To judge by my conversations, people felt uncertain about the implications. Some said 51 percent ownership meant that the government was actually in control. Others fretted that the entire scheme was a way for officials to avoid responsibility—they had hired a foreign company to build a plant rather than stop the politically powerful factory owners who were polluting the river. It was a small instance of the human race's perennial inability to govern itself. "There's plenty of water to drink," Wei said. "Everyone knows *that* is not the problem."

Three R's

The facility was big, white, and noisy. Sandwiched between an interstate highway and the Pacific Ocean, it stood in the shadow of a nearby power plant. The complex had three parts, each a bland metal-and-glass box in the style of Suburban Car Dealership. Inside the largest of the three parts was a hall the size of a football field with a thirty-foot ceiling. It was filled with the throbbing sound of large pumps. Workers wore hard hats and noise-canceling headphones. Reaching to the ceiling were endless racks of eight-foot-long gray tubes connected to fat white pipes by blue hoses. The pipes were connected to even larger pipes and then to a ten-mile-long underground conduit. The facility was located in Carlsbad, in southern California, and named after a beloved former mayor. Opened in December 2015 after seventeen years of legal and political squabbles, it provides freshwater to half a million people. It is the Western Hemisphere's biggest water-desalination plant. Every day, it removes the salt from about 50 million gallons of water.

Carlsbad first proposed desalination in 1998 because its officials feared that the region would run out of water as its population and economy grew. The city is at the end of the pipeline that brings water from the Colorado River to southern California. If supplies became limited, the city administration worried, those who were last in line would be stuck. The Carlsbad project would be equivalent to sticking a straw into a boundless, ever-present supply of water: the Pacific Ocean. It would be a test of the Wizard's way out of problems: surpassing local ecological constraints, eliminating water anxieties once and for all.

The Carlsbad desalination plant, the biggest in the Americas, opened in 2015.

About 3.5 percent of the weight of seawater consists of dissolved salts, most of it table salt. The most common way of removing the salt is known as "reverse osmosis." Simple in principle but complex in practice, reverse osmosis involves forcing seawater through a membrane with extremely fine holes—so fine that they allow water molecules to pass through but block the slightly larger salt molecules. One sort of complexity stems from the need to make membranes that are both strong enough to withstand continual pressure but fine enough to allow water to pass through. Another stems from the cost of fueling the motors that pump millions of gallons through the membrane. The Carlsbad plant provides 10 percent of the region's water, but accounts for 25 percent of its cost. The price of building it was about $1 billion.

The reverse-osmosis machinery in Carlsbad was designed by IDE Technologies. IDE was founded by the Israeli government in 1960 under the name Israel Desalination Engineering. A special project by Israel's first prime minister, David Ben-Gurion, it was created in the hope of eliminating the nation's dependence on the Jordan River, contested by Jordan and Syria. After their initial hope of freezing freshwater out of seawater proved unworkably expensive, the IDE engineers tried out one scheme for desalination after another. They achieved enough success that they were contracted to build several desalination plants in areas that could not readily be provided with water in any other way—the Canary Islands, for instance, and several remote air bases in Iran. But practical desalination remained a distant prospect.

In 1966 Israel invited a California scientist, Sidney Loeb, to spend a year at Ben-Gurion University in Beer-Sheva (the new Hebrew name for Beersheba). Loeb had worked for industry after taking an undergraduate degree in engineering in 1941. Feeling restless, he quit his job at the age of forty and went to graduate school at the University of California at Los Angeles. Like the researchers in Israel, scientists at UCLA had been seeking practical desalination methods. Loeb joined the quest with another student, a Canadian named Srinivasa Sourirajan. They developed the first successful reverse-osmosis process in 1960—"successful" in the sense that it worked in a laboratory, not that it could be deployed in the real world. Sourirajan soon ran into visa problems and Loeb continued alone, constantly tweaking the all-important membrane. By 1965 the technology had advanced enough that Loeb was able to build a commercial reverse-osmosis plant—the first in the Americas—in Coalinga, a town of about six thousand in the San Joaquin Valley. So thick with salts was its groundwater that residents had always brought in potable water by tanker cars. The apparatus, small enough to fit into the village firehouse, provided about a third of the town's drinking water. It was operated by Coalinga's lone fireman.

Loeb gained little from his success—Coalinga's water was costly, and most of the rest of North America had had cheaper sources. But the Coalinga plant resonated in Israel, where he was asked to teach a course on reverse osmosis and set up a reverse-osmosis plant at a kibbutz at the southern edge of the Negev. Afraid that the water was full of chemicals, kibbutz residents initially refused to drink the water. The university sent a doctor to explain that the kibbutz had already pumped out so much groundwater that the minerals in the remainder were building up to toxic levels. The kibbutzniks reconsidered and dipped in their cups. The modern desalination industry was born. Today more than eighteen thousand desalination plants are operating around the world. The field is growing—but it is also contentious. Some of the biggest disputes are in California.

California has always experienced droughts, but they have been more

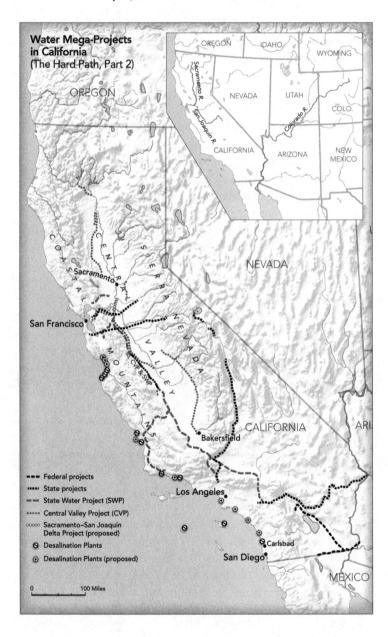

Water Mega-Projects in California (The Hard Path, Part 2)

Federal projects
State projects
State Water Project (SWP)
Central Valley Project (CVP)
Sacramento–San Joaquin Delta Project (proposed)
Desalination Plants
Desalination Plants (proposed)

frequent in the twenty-first century. Every year but one between 2007 and 2017 was a <u>drought year</u>; the years between 2011 and 2014 were the driest period in California for more than a millennium. Wizards and Prophets agree that something must be done to waterproof the state, but they disagree on everything else.

The former point fingers at the state's dam reservoirs, which are surprisingly small; unable to store California's occasional heavy rains, they have often been forced to dump water in dry periods. Expanding the dams is a Wizardly option; so is digging a reservoir almost twenty miles long to store water from the Sacramento River. Long in the making is a multibillion-dollar plan to build two thirty-five-mile-long underground tunnels to funnel water from the same river to the Central Valley Project and the State Water Project. Perhaps most important, California localities have proposed more than twenty major desalination plants (not all are actively being pursued). The desalination plants must be built now, Wizards argue, so people can develop the technology for later, when it will become critical; costs will come down with experience, as has been the case with, for example, solar power.

Finding more water is especially important for agriculture, Wizards say. If the world needs to produce a lot more food—maybe twice as much—California, with its good soil and warm climate, is one of the places where it will happen. If that food is provided by C4 rice or some other super-productive grain, California will need water to grow it. Already <u>farms use four-fifths of the state's water</u>. Greatly increasing agricultural productivity means increasing the supply, which in the long run means desalination. Ultimately, Wizards argue, it is the only way for people to surpass local limits.

Prophets resist these claims. Desalination plants kill marine life, they argue, pollute the seas with their discharges, and increase utility rates—all because big businesses feel crimped by the soft path. Prophets  instead point to water recycling, stormwater capture, lawn and garden watering rules, leak tracking, graywater reuse, appliance- and fixture-efficiency standards, well controls (drilling is almost unregulated, leading to groundwater depletion)—an array of small-scale changes that mostly involve nudging people and businesses to change their habits and become more efficient. Encapsulating this approach is the mantra of the Three R's: reduce, reuse, and recycle.

Many of these measures apply to cities, but Prophets believe the soft path also could apply to agriculture. More than half of California farms use flood irrigation—covering their fields a few inches deep in water—as did the farmers of the ancient Fertile Crescent. Converting these to sprinklers would save huge amounts of water; more still could be conserved by simply not using irrigation when the soil is already moist. (Incredibly, irrigation still continues even when it is raining.) Further savings could be made by adjusting what farmers grow, as Israel has done with cotton. Most of the almonds in the United States, for example, are grown by about four hundred large operations in the San Joaquin Valley, which use about 10 percent of the state's water supply. And the state remains the nation's biggest producer of alfalfa, used for cattle feed. Most of that feed is sent to cattle in other states that could grow their own; some is exported across the Pacific. Meanwhile, more water is used to grow alfalfa than is consumed by all the households in California. And so on.

The dispute in California will be echoed in different ways across the globe. The hard path creates universal Wizardly solutions that do not depend on local conditions or knowledge. It leads quite naturally to broad fields of waving grain—visions of concentrated productivity. Societies that adopt the soft path will lead toward networks of smaller farms with drip irrigation and multiple crops—the inhabited, networked spaces preferred by Prophets. One values a kind of liberty; the other, a kind of community. One sees nature instrumentally, as a set of raw materials freely available for use; the other believes each ecosystem has an inner integrity and meaning that should be preserved, even if it constrains human actions. The choices lead to radically different pictures of how to live. What looks like a dispute over practical matters is an argument of the heart.

SIX

Fire: Energy

Pithole

First the derricks, then the bars and brothels. After that, the wasteland.

In 1859 the first successful oil well in the United States was drilled in Titusville, Pennsylvania. Six years later, in January 1865, more oil—a *lot* more oil—was found where I was standing, eight miles away in the almost uninhabited slopes near Pithole Creek. Within weeks new wells were being dug by the score and spewing petroleum across the snowy slopes that rose from the river. Wagons carrying tons of oil in improvised barrels jolted one after another on the muddy tracks out of town. When one wagon got trapped in the mire, those behind it could be stuck for days. Some outfits turned to ferrying the oil down shallow, rocky Pithole Creek. They dammed the stream, piled barrels onto rafts, and then broke the dam, letting the rafts surf downstream on the wave. Vessels capsized so often that people made a profitable business of skimming crude from the riverbank.

In every direction the landscape was stripped of trees to build oil silos, oil barrels, oil roads—and a new oil city, population fifteen thousand. Conjured out of nothing, it was the world's first petroleum boomtown. It had no legal existence or official name or town charter or anything but petroleum, so much petroleum that it seeped to the surface and covered every horizontal surface in a foot-thick impasto of oily mud mixed with snow and excrement (the city had no sewers). Most people

Pithole's main street (above) was a muddy wreck during the town's brief heyday. Today (opposite, in a photograph taken from the same location) almost nothing remains of the world's first oil boomtown.

called the new settlement Pithole. Newspapers called it Oil-Dorado or Petrolia. Whatever the name, it was a frenzy of extraction. Entire buildings were thrown up in a few days, then caught fire and were rebuilt. Oil seeped into ordinary water wells, a fact discovered when firefighters who had been dumping water onto one of Pithole's many fires realized that they were instead feeding the conflagration. Seizing the opportunity, an inventor developed a wheeled firefighting dredge that scooped up mud, hundreds of pounds at a time, and catapulted it into the flames. While demonstrating his invention at a fire, its creator fell into the machine and was thrown into the blaze.

By August 1865, seven months after the first Pithole strike, more than three hundred derricks were operating and hundreds of other wells were under way. People bought and sold drilling land in a fever, waving bricks of cash. The atmosphere was filled with smoke and ash and the baying of human beings in chase of money. So many sex workers flooded into the town that they marched down First Street in a daily parade. Hooker and john alike were sure they were at the beginning of something that would last forever and change the world.

That same month a big well stopped flowing. Others followed—the oil was running out. Brothel owners, sensitive to customers' moods, quickly vacated their establishments; other, less perceptive business-people followed later. By the spring of 1866 scores of buildings stood empty. Pithole was barely a year old but already in breakdown. In 1870 only 281 people lived there. Eight years later somebody bought the entire town for $4.37.

Today not one of Pithole's original structures remains. During my visit I strolled down paths that had once been streets past vacant land that had once been real estate. Not another soul was in sight. Some-body had planted trees along the paths. Hand-stenciled signs identified the locations of vanished buildings: hotels, law offices, banks. A small museum with irregular hours stood atop a hill. The age of petroleum seemed to have alighted there and left.

A brief, tawdry flowering, followed by collapse—surely Pithole's inhabitants had not imagined this as their future (most of them, any-way). Walking through the city's ruins, I found it hard not to wonder whether our industrial era was not simply Pithole writ large: an evanes-cent surge of wealth, much of it squandered, doomed to end when the world's fuel supply was consumed.

Stuffed in my briefcase as I drove to Pithole were reports from scientists, oil companies, and international agencies, a paper blizzard of charts and graphs projecting how much energy the world would want tomorrow, and the day after tomorrow. Still more estimates clogged the hard drive on my computer. A 37 percent rise between 2013 and 2035. A 37 percent rise between 2014 and 2040. A 61 percent rise by 2050. A 100 percent rise by 2050. The numbers differed from one forecast to the next, but in every single one the demand for energy went up— sometimes fast, sometimes faster.

What would happen if the requisite supply failed to appear? If, instead, the world of 10 billion abruptly ran short? The answer is easy to picture: industrial civilization, imploding in an awful smash. Pithole's citizens, wannabe wildcatters all, had been certain they were creating a prosperous, long-lasting tomorrow. Centuries from now, will our descendants look back in scorn at our equally feckless view of the future?

Strange Forests

Fossil fuels are ancient light. Three hundred million years ago, in the Carboniferous epoch, strange forests covered the world. Many were ruled by giant, shaggy lepidodendrons: scaly, hundred-foot-high poles topped with grass-like leaves. Others were dominated by horsetails the size of trucks and ferns as tall as an apartment building. Although these creatures resembled no trees on Earth in our time, they were, like modern trees, the product of photosynthesis, which is to say that they were organic batteries, storing energy from the sun. During the Carboniferous, the world's landmasses were shifting to form a single vast supercontinent cut by mountain ranges and huge, boggy basins that ran alongside them. Entire forests fell into the basins' airless muck. When plants die today, fungi decompose them, releasing their trapped solar energy. In the Carboniferous, most fungi apparently had not yet evolved the ability to break down lignin, the tough compound that gives plant stems their strength and bulk. Buried in almost oxygen-free sludge, attacked only slowly or not at all by fungi, the lepidodendrons, horsetails, and giant ferns decayed at an infinitesimal rate, creating layers of peat. Over the eons, crushed and heated by the slow churning of the earth, the peat became coal. All the while, in a parallel process, the earth was crushing and heating ocean-floor layers of dead plankton, algae, and other marine organisms to form the sticky gumbo of oil, gas, and other compounds known collectively as petroleum. In these smashed jungles and seabeds, glossy and black, solar energy waited, frozen in time, ready to be tapped.

The first known human use of fossil fuels—burning coal for heating and cooking—occurred in China, probably around 3400 B.C. Coal didn't catch on quickly. People found it easier to cut down nearby forests for fuel, and even burn grass and dung, than to dig for coal in faraway mines. Because Britain was among the first areas to be thoroughly deforested and had shallow, easily accessible coal deposits, Britons were early coal adopters. Records show that the black stuff has been powering iron foundries, lime kilns, and brewery boilers since at least the days of Henry III, who ruled in the thirteenth century. The coal, mostly low-quality and rich with impurities, released so much toxic smoke that Henry's queen, Eleanor of Provence, fled coal-crazy Nottingham, unable to tolerate the fumes. Despite the pollution, Britain and the rest of northern Europe kept using fossil fuels; having little wood, they had little alternative. The choice paid off in the eighteenth and nineteenth centuries, when the invention of the steam engine, the blast furnace, and the cement kiln vastly increased the demand for energy—new coal beds to begin with, then oil and natural gas.

The impact of fossil fuels exhausts hyperbole. Energy has any number of sources (solar, wind, hydroelectric, geothermal), but for all of the modern era the overwhelming majority has been derived from fossil fuels (coal, oil, natural gas), and it was fossil fuels that transformed daily existence. Take any variable of human well-being—longevity, nutrition, income, mortality, overall population—and draw a graph of its value over time. In almost every case it skitters along at a low level for thousands of years, then rises abruptly in the eighteenth and nineteenth centuries, as humans learn to wield the trapped solar power in coal, oil, and natural gas. "The average person in the world of 1800 was no better off than the average person of 100,000 B.C.," writes the economic historian Gregory Clark of the University of California at Davis. "Indeed in 1800 the bulk of the world's population was *poorer* than their distant ancestors." The Industrial Revolution, driven by fossil fuels, changed that, possibly until the end of days.

Before fossil fuels, even the wealthiest houses were cold when temperatures dropped. A visitor to the palace of Versailles observed in February 1695 that guests wore furs to dinner with the king; at the king's table, the royal water glasses were filmed with ice. A century later, Thomas Jefferson had a magnificent home (Monticello), the nation's finest wine collection, and one of the world's great private libraries, which would become the foundation for the Library of Congress. But Monticello was so frigid in winter (12°F indoors!) that Jefferson's ink froze in his inkwell, preventing him from writing to complain about the cold. ✓

A century after Jefferson's death, these fundamental aspects of life were being transformed, at least in the upper- and middle-class West. For the first time in history people in large numbers could heat their entire residence, bedrooms included; for the first time they could, if they wished, illuminate every room in the house. Central plumbing suddenly became more feasible, because the temperature inside buildings was less likely to be below freezing, and pipes were less likely to burst. On a larger scale, fossil fuels lighted city streets, drove railroads and steamships, and allowed for the mass production of steel and cement, the physical underpinning of every industrial society. "Coal is a portable climate," Ralph Waldo Emerson marveled in 1860. "Every basket is power and civilization."

None of this was hidden from view. Educated nineteenth-century Westerners like Emerson knew that they were living in a time of unparalleled prosperity. Their twenty-first-century descendants are richer than any dream of Solomon. To ward off cold, people used to chop down trees, then stack the wood in enormous piles; today billions of people can flick a switch and feel hot air gush into the room. The average American car engine is, unthinkably, more than two hundred horsepower—as if every suburban Mom and Dad had two hundred ponies at their disposal, but without the need to feed the animals, take them to the veterinarian, or shovel their manure.

Those educated Westerners also understood that their wealth and well-being was tied to the lavish use of fossil fuels—which is why Western politicians and businesspeople have worried for more than a century about whether the supply would last. The apprehension came out in the open as early as 1886, when Pennsylvania state geologist J. P. Lesley declared in a widely publicized speech that the "amazing exhibition of oil and gas" begun at Pithole was "a temporary and vanishing phenomenon." Within a few years, he proclaimed, "our children will merely, and with difficulty, drain the dregs." One of the first and most enduring products of the age of fossil fuels was the fear that the age would rapidly end.

The last two chapters discussed two related subjects, food and water, showing how Borlaugian Wizards and Vogtian Prophets approach the task of providing them for a world of 10 billion. This chapter and the next also treat two related subjects, but the relationship between them is different. The first of the two chapters concerns *energy supply*: Will there be enough energy in the world of 10 billion to provide everyone with the comforts of modern existence? The next chapter is about what might be called *energy by-products*—the environmental effects of using large amounts of energy. By far the biggest of these, at least in potential, is climate change. The reason for splitting the discussion in this way is that the world looks different depending on whether one focuses primarily on energy supply or energy by-products.

Wizards and Prophets disagree about energy, as they do about food and water. Wizards support big, high-tech, centralized power plants based on concentrated energy sources (coal, oil, natural gas, uranium); Prophets place their hopes in small-scale, distributed, low-impact, neighborhood- and household-level facilities that harness diffuse forms of energy (sunlight, wind, geothermal heat). Prophets have proclaimed their bottom-up vision for a century and a half, and enthusiastically envisioned its triumph. Nonetheless, big, Wizard-style utilities have been so economically advantageous that until recently the other view has never had a chance. With few exceptions, distributed sun and wind power gains viability only in situations where people consider the by-products of massive energy consumption.

More than 80 percent of the world's energy now comes from fossil fuels, and every bit of it is mined from the earth.* That is another way of saying that all the fossil fuels humankind will ever have are already here,

---

* Nuclear plants provide a bit more than 5 percent of the global energy supply; renewable-energy facilities, a bit under 5 percent. The remainder is from biofuels: wood, charcoal, ethanol (from corn and sugarcane), biodiesel (usually made from vegetable oils), and so on. The most important types of renewable energy are solar, wind, and hydroelectric. I focus on solar power, rather than wind power, because it has, rightly or wrongly, drawn the most interest from Prophets, and because many of the same arguments apply to wind. I don't discuss hydroelectric power, because the lack of suitable untapped rivers means that it is unlikely to expand greatly.

waiting to be extracted from the ground—in contrast to food, which is grown every season from the soil, and freshwater, which is drawn in constant but limited amounts from rivers, lakes, and aquifers. In theory one could mine every ounce of coal the world will ever use and put it in a huge warehouse; the same applies for oil and natural gas. Nobody could do that for food—imagine trying to grow a hundred-year supply of food in a single season. And because the great majority of the world's freshwater is either in the atmosphere as water vapor or being filtered and stored by Earth, it also can't be mined—not, at least, without wrecking the natural systems that sustain it, which would make life impossible.

In economic terms, as I said in the last chapter, food and water can be thought of as a *flow*—or, more precisely, a critical-zone flow, a current with a volume that must be maintained. By contrast, fossil fuels are like a *stock*, a fixed amount of a good. Few dispute that the flow of food and water could be interrupted, with terrible effects. But people have disagreed for a century and a half—since the days of Pithole—about whether the world has an adequate stock of fossil fuels.

Today the notion that the stock of fossil fuels will run out is called "peak oil," after the idea that global petroleum output will soon peak, then fall. Coursing through history like waves of panic, the conviction that civilization was hurtling toward an energy disaster has become embedded in the culture. Time after time, decade after decade, presidents, prime ministers, and politicians of every party have predicted that the world will soon run out of oil and gas. Time after time, decade after decade, new supplies have been found and old reservoirs extended. People forgot their apprehensions until the next alarm, the next prophecies of catastrophe.

None of this would matter if the fears had no cost. But that is not the case. Fear of running out has been a malign presence for more than a century, driving imperialist forays, stoking hatred among nations, fueling war and rebellion. It has cost countless lives. Equally problematic, peak oil helped establish a set of wholly mistaken beliefs about natural systems—beliefs that have repeatedly impeded environmental progress. It laid out a narrative that has led activists astray for years. Far too often, we have been told that the future will be wracked by crises of energy scarcity, when the problems our children will face will be due to its abundance.

Fear of Oil

If Andrew Carnegie didn't think of himself as the smartest person in the room, he certainly acted as if he did. Canny and ruthless, a cross-grain mix of avarice and generosity, Carnegie prided himself on his ability to see ahead farther than other people. In his later years he would become one of the richest people who ever lived. But he was merely a successful twenty-six-year-old railroad executive when he became one of the first to envision the consequences of peak oil.

In 1862 Carnegie toured the Pennsylvania oil patch and was taken aback. This frenzy, he in effect said, cannot possibly last. With a friend, Carnegie decided to set up a company that would profit from the coming collapse. As Carnegie recalled in his autobiography, his partner "proposed to make a lake of oil by excavating a pool sufficient to hold a hundred thousand barrels . . . and to hold it for the not far distant day when, as we then expected, the oil supply would cease." When that happened, Carnegie and his friend would be sitting pretty.

The two men raised $40,000 (about a million dollars in today's money) to lease an oil field, dug a pit the size of six Olympic swimming pools, filled it with their oil, and waited for the apocalypse. Meanwhile, the reservoir leaked—a lot. Carnegie and his partner realized that if they waited for the end of oil it would be the end of *their* oil. They were forced to sell. Contrary to their expectations, the wells in their field kept producing oil, which they sold at a high profit. The two men made several million dollars from their $40,000 investment. It was, Carnegie said later, the best investment he ever made.

Undeterred by Carnegie's blunder, other oil entrepreneurs continued to expect the day of reckoning. At the time, Pennsylvania contained the Western world's only big, proven oil field. Geologists at Standard Oil, the largest firm in the industry, reported to headquarters that the odds of finding another like it were a hundred to one. The looming end of easy oil became common wisdom at energy firms. Told in 1885 that oil might be found in Oklahoma, Standard's John D. Archbold, one of the first U.S. petroleum refiners, scoffed, "Are you crazy?"

Standard's beliefs were both prescient and misguided. Pennsylvania oil indeed hit a peak in 1890 and thereafter fell, though the wells never quite ran dry. But new fields were emerging in Indiana, Ohio, Okla-

Two years after the first oil strike at Beaumont, the landscape had been transformed from a sparsely populated mix of cattle ranches and rice farms into a forest of closely packed oil derricks.

homa, and, especially, Texas. In 1901 a crew in East Texas, near the Gulf of Mexico, struck black gold. Oil shot 150 feet in the air at a rate of 100,000 barrels a day, a gusher bigger than any seen in Pennsylvania. Flailing about in the surreal black rain, workers took nine days to control the spout, by which time a new Pithole—Beaumont, Texas—was already forming. Unlike Pithole, Beaumont produced oil for decades.

Each new discovery was bigger than the last, but each seemed only to enhance the feeling of vulnerability. Even as oil poured out of Texas, President Theodore Roosevelt in 1908 invited all forty-six U.S. governors to the White House to decry the "imminent exhaustion" of fossil fuels and other natural resources—"the weightiest problem now before the nation." Afterward Roosevelt asked the U.S. Geological Survey to assay domestic oil reserves, the first such analysis ever undertaken. Its conclusions, released in 1909, were emphatic: if the nation continued "the present rate of increase in production," a "marked decline" would begin "within a very few years." Output would hit zero about 1935—a prophecy the survey repeated, annual report after annual report, for almost twenty years.

The survey didn't know it, but its geologists were echoing admonishments from across the Atlantic. Great Britain, the first nation to industrialize, was also the first to realize its dependence on fossil fuel—coal in this case—and dread its depletion. As far back as 1789, when the country had only a few hundred coal-fired steam engines, the Welsh engineer John Williams warned that the coal supply would soon come to an end—and, with it, "the prosperity and glory of this flourishing and fortunate island."

Williams's dire projections set off a decades-long dispute. On one side were wide-eyed optimists, most of them scientists. Prominent among them was Robert Bakewell, one of Britain's best-known geologists, who claimed in 1828 that the nation's coal deposits would last for two thousand years. On the other side were pessimists, most of them economists, none gloomier than the young British savant William Stanley Jevons, who devoted the 380 pages of *The Coal Question* (1865) to detailing why "we cannot long maintain our present rate of increase of consumption" of coal.

Bakewell had argued that companies were using coal with ever-increasing efficiency, which would make the national coal supply last longer. Wrong, Jevons said. In what is now called the "Jevons paradox," he reasoned that improvements in efficiency would reduce the cost of energy from coal. Lower cost would encourage people to use more, draining Britain's reserves faster. Luminaries from the philosopher John Stuart Mill to future prime minister William Gladstone endorsed these cheerless views and called for conserving coal. The Jevons paradox was true, Lord Kelvin, the great British physicist, announced in 1891: the "coal-stores of the world are becoming exhausted surely, and not slowly."

Britain's coal output peaked in 1913, as Jevons had warned, but global reserves continued to climb, and no shortages occurred. The pleasing outcome made little difference. London was again beset by fossil-fuel nightmares—but about oil, rather than coal. Sounding the alarm was the First Lord of the Admiralty, Winston Leonard Spencer-Churchill. Appointed in 1911, the preternaturally vigorous Churchill set about modernizing the Royal Navy. Britain had just converted its entire fleet from the unsteady power of wind to the constant force provided by coal.

Now, Churchill declared, Britain had to transform its navy a second time. Burning a pound of fuel oil produces about twice as much energy as burning a pound of coal. An oil-fueled ship could thus travel roughly twice as far as a coal-fueled ship of similar size. Oil's greater *energy density* meant that it, rather than coal, was the fossil fuel of choice.

Because Britain had little oil, British officials worried that converting would make the fleet dependent on foreign entities, a frightful prospect. The obvious solution, Churchill told Parliament in 1913, was for Britons to become "the owners, or at any rate, the controllers at the source of at least a proportion of the supply of natural oil which we require." The government soon bought 51 percent of what is now British Petroleum, which had rights to oil "at the source": Iran (then known as Persia).

The initial oil concession with Iran, negotiated in 1901, had been on terms so favorable to London that the Iranians showed signs of seller's remorse. To forestall protests, Britain temporarily seized control of the Iranian government. An attempt in 1919 to make the arrangement permanent led to uprisings. Two years later Britain coordinated a coup d'état that led to the installation of a new shah. He swore publicly to protect Iran from foreign influence while privately assuring the same foreigners he would never interrupt the flow of oil.

Iran was not the only focus of oil fear. During the First World War, Britain, France, Italy, and Russia made plans to carve up the Ottoman Empire, which had allied against them with Germany and Austria-Hungary. Other than the strategically located Ottoman capital, Istanbul (then Constantinople), the most valuable spoils were the petroleum zones in what are now Iraq, Kuwait, Bahrain, and Saudi Arabia. These were parceled out in a series of covert meetings, but the United States rejected the deal—it would, for example, have awarded Istanbul to Moscow. In the middle of the bickering, Greece invaded the Ottoman Empire, unwittingly igniting the revolution that created modern Turkey. Not willing to interfere, the Europeans gave up their designs on the Ottoman heartland—today's Turkey—and focused on oil regions, too far away for the revolutionary army to defend. Only in 1928 did the parties agree how to divvy up drilling rights, with Britain winning the proverbial lion's share.

From today's perspective, the frenzy over Middle East oil seems bizarrely disconnected from reality. At the time, two nations dominated petroleum production: the United States, responsible for about two-thirds of it, and the Soviet Union, which pumped an additional fifth. Both were finding petroleum at ever-rising rates. Between 1920 and 1929, U.S. crude-oil reserves nearly doubled, despite constantly increasing consumption. Meanwhile, the Russian oil industry, which had crashed after the revolution of 1917, returned with a roar in the early Soviet period; production almost quadrupled in the 1920s. And new sources were coming online. Venezuela, for instance, went from pumping almost nothing in 1920 to 500,000 barrels a day in 1929. Petroleum was flooding the world.

Nonetheless, politicians throughout the West continued to invoke the phantasm of an impending petroleum drought. When I searched through an archive of 1920s newspapers, I turned up more than a thousand articles prophesying an inevitable "oil crisis," "oil famine," or "oil shortage." Some of those articles mentioned that oil executives were baffled by the cries of doom. But the overall tone was ominous. "The United States is face to face with a near shortage in petroleum supplies so serious it threatens the very economic fabric of the nation," cried the *Los Angeles Times* in 1923. A year later, the *Houston Post-Dispatch* forecast "oil famine within two years." "Oil exhaustion in fifteen or twenty years," said the *Brooklyn Daily Eagle* in 1925. A special twelve-part wire-service investigation in 1928 flatly decreed, "There is no possible excuse for assuming an adequate future supply of oil."

The drumbeat of negative forecasts had its effect: the United States and the European powers rushed to control every drop of oil in the Middle East, Latin America, and Africa. In light of the last eighty years of history in these regions, it is hard to view these moves as enduring successes. Coups and attempted coups in Iran, Venezuela, and Nigeria; oil shocks in 1973 and 1979; failed programs for "energy independence"; wars in Iraq, Kuwait, and Syria—this cancerous relationship, a mix of wrath and dependence, has continued with little change for nine decades. Driven by the recurrent panic of peak oil, it sometimes seems as fundamental to the structure of global relations as the law of gravitation is to the rotation of Earth around the sun.

Although many other factors, religion notable among them, have had their hand in this state of affairs, it is easy to wish that peak oil had never been invented. But this fantasy may be unreasonable. Could the doomsayers have been correct, but rung the alarm a little too early? After all, Earth is finite, so the amount of energy it contains must also be finite. Isn't it wholly rational to expect fossil fuels to run out?

"A Giant Lampshade, Reversed"

In June 1866 a high school mathematics teacher in Tours, in central France, attached a toy steam engine to a small metal container full of water. Working with a mechanic friend, the teacher placed the device in front of a curved mirror. Shaped like a shallow trough, the mirror focused the sun's rays on the container. After an hour, the water began to boil. Steam gushed out, driving the steam engine—"a success that surpassed my expectations," the teacher crowed. It was the first true example of solar power: converting energy from the sun into mechanical force that could accomplish useful tasks.

Today the math teacher is a historical footnote, but it seems likely that tomorrow he will have a place in the main text. His name was Augustin-Bernard Mouchot. He was born in 1825 in a village southeast of Paris, the youngest of the six children of a poor locksmith. After scrambling his way through school, Mouchot became a teacher, drifting from one position to another in the boondocks. Along the way he gradually acquired professional degrees in mathematics and physics. He was teaching in northwest France in 1860 when a powerful insight set him on a decades-long crusade.

Like their counterparts in Britain, the French upper and middle classes understood that their prosperity depended on energy from coal. But France, unlike Britain, had little coal and thus was forced to import much of its supply at high cost. Many French people, Mouchot among them, feared that foreigners would stop selling coal to France—or, worse, that the foreign deposits would run out. If that unhappy day arrived, Mouchot warned in a manifesto, French industry would no longer have "the resources that are part of the cause of its prodigious expansion. What will it do then?" In a flash he realized where the solution lay: "The sun! that is to say, a powerful hearth ready to provide its heat for mechanical applications."

Augustin Mouchot was far from the first to realize that the sun's energy could be tapped. For more than two thousand years Chinese architects had been aligning windows and doors with the southern sky to let sunlight flood into rooms during winter, heating cold interiors. Thousands of miles away, Greek savants expounded the same architectural principles to their disciples. So, later, did the Romans, according to the solar-energy chronicler John Perlin, whose work I am drawing upon here. To heat the rooms in public baths, Romans built giant south-facing windows—those in Pompeii's *caldarium* were 6'7" x 9'10".

Augustin-Bernard Mouchot

Historians used to call the European era after the fall of Rome the Dark Ages. Now we know that scholarship and the arts continued and flourished. Still, use of the sun nearly ceased. Rich people stopped placing glass windows on the south side of their villas and mansions; poor people didn't orient their shacks to take advantage of sunlight. (In this respect the Dark Ages actually were dark.) Not until the Renaissance did Europeans again collect solar heat, installing glass walls in greenhouses and conservatories. And not until the eighteenth century did natural scientists try to understand why, exactly, "a room, a carriage, or any other place is hotter when the rays of the sun pass through glass." The quotation is from the Swiss scientist Horace-Bénédict de Saussure, who in 1784 built the first "hot box"—a small wooden box, insulated with cork and topped by sheets of glass. De Saussure put a container of water in his box and took it outside on a sunny summer day. Impressively, the water quickly began to boil.

Working almost a century later, Mouchot put his own spin on de Saussure's idea: focusing the sun's rays with a mirror. To be sure, people had concentrated sunlight with mirrors before—Chinese farmers carried small mirrors to set fires as long as three thousand years ago. But Mouchot was the first to use sunlight and mirrors to boil water and then employ the steam to drive engines.

Mouchot's initial efforts attracted enough attention that he was given access to a prestigious military workshop. After three years of sporadic tinkering, he had a working model—the one he tested with a toy steam engine. Elated, he presented his invention to Emperor Napoleon III in September 1866. Soon after, Mouchot began writing the inevitable self-promotional book.

By then he had moved to a more prestigious high school in Tours, negotiating a reduced teaching load so he could devote more time to his mirrors. Unburdened by family or friends, he spent every moment in his studio laboratory. In 1870 he erected a seven-foot sun engine in the Jardin des Tuileries, in the center of Paris. Onlookers marveled to see an engine without any visible fuel source. A motor that ran on sunbeams! Little wonder the crowds were excited.

Unluckily for Mouchot, France declared war on Prussia a few months after he put the machine on display. After a series of military disasters, German forces rampaged through Paris and Emperor Napoleon III fled into exile. In the chaos, the solar engine disappeared forever.

Mouchot, ever tenacious, assembled another solar engine, mounting it in front of the Tours library in 1874. It had a conical mirror that surrounded the boiler, focusing heat on every side. "A giant lampshade, reversed," one enthusiastic journalist called it, "turning its concavity towards the sky." A subsidiary mechanism allowed the mirror to track the sun as it moved through the sky. On a hot, clear day, the apparatus could boil five quarts of water an hour, enough to drive a half-horsepower motor. It was an enormous popular success, attracting crowds of gawkers. But Mouchot was learning the limitations of solar power.

Sunlight is plentiful and free, but it comes as an intermittent flow, not a reliable stock. Mouchot's engines were useless at night or on cloudy days—and French skies were often cloudy. Even when the sun shone, the mirrors were costly. One skeptical engineer noted in a review of Mouchot's work that running a typical one-horsepower steam engine required "about two kilograms [4.4 pounds] of coal." To drive the same engine with the sun, Mouchot would need a mirror of about 320 square feet. Operating factory-scale machinery would require hundreds of giant mirrors—a huge expense. Meanwhile, French industry was not running out of fuel, as so many had predicted it would. Paris had signed a trade agreement with London, and the nation was awash with British coal.

In an 1882 demonstration in Paris, Mouchot's assistant used his solar engine to drive a printing press.

Desperate to save his work, Mouchot came up with a new justification for solar power: as a tool of imperialism. In the 1870s France was conquering Algeria, dispatching thousands of colonists to brand-new villages along the coast. The occupation was hobbled by energy problems; not only did the colony have to import almost all of its coal from across the Mediterranean, it had no railroads to transport it from the ports to those new French villages. Solar power, Mouchot promised, would transform Algeria into a productive adjunct of the French imperium. Winning a government grant, he traveled across the colony, testing solar irrigation pumps and solar distillation plants. In the desert an infection left him nearly blind. A second bout of fever left him mostly deaf. He ignored his afflictions and wrote reports on his demonstration projects that so excited colonial authorities that they asked him to represent Algeria with a sun engine at the 1878 Universal Exhibition in Paris. Featuring what Mouchot modestly described as "the largest mirror ever built in the world," the device astounded visitors by running a freezer. Using the sun's heat to make ice! Mouchot won a gold medal at the exhibition; barely able to see and hear, he became a chevalier in the French Legion of Honor.

Two years later he gave up his crusade. Blindness and deafness did not defeat him. Coal did. Historians estimate that in 1800 all of the steam engines in Britain could generate perhaps 50,000 horsepower. By 1870 the figure had soared to more than 1.3 million horsepower, a twenty-six-fold increase. Nobody was going to wait for solar enthusiasts to fiddle with mirrors that didn't work on rainy days. Mouchot was trying to persuade society to switch from a stable stock of coal to an inconstant flow of sunlight. And society was not terribly interested.*

Others, though, took up the solar cause, most notably John Ericsson, a Swedish-American engineer famed for his design of the *Monitor*, the U.S. Navy's first commissioned iron-clad warship. In 1868, four years after Mouchot's initial demonstration, Ericsson revealed the solution to the coming "exhaustion of our coal fields": "the concentrated heat of the solar rays." Eight years later, in his own self-promotional book, Ericsson proclaimed that he had invented seven types of solar engines, though he hadn't shown them to anyone. They were, he said, the world's first true "sun motor"; Mouchot's device, he jeered, was a "mere toy."

There is reason to question whether Ericsson—a gifted engineer, but also a pioneer in vaporware—actually made a single solar machine. Zealously secretive, he wouldn't allow visitors to his laboratory, often refused to allow his inventions to be examined even by their financial backers, and repeatedly promised new breakthroughs that never appeared. In 1888, soon after another premature announcement, he had a heart attack and died. The engine "occupied his thoughts up to his last hour," one obituary reported. "While he could hardly speak above a whisper, he drew his chief engineer's face close to his own, gave him final instructions for continuing the work on the machine, and exacted a promise that the work should go on."

Ericsson failed as a solar inventor but nonetheless had a true vision for the future, which he broadcast in manifestos and articles. Tomorrow, he avowed, would be as clean and luminous as sunlight itself. It would be a world without smokestacks or toxic furnaces or lightless coal mines. Buoyed on a refulgent tide of free solar power, communities everywhere would provide themselves with heat and light from millions of local solar engines. A new era of universal prosperity!—all from harnessing the inexhaustible light of the sun. It was the first articulation of what

---

* Mouchot's last years were bleak. Retired on a modest pension, he lost track of his finances; creditors seized his possessions. He died in lonely squalor in 1912.

John Ericsson insisted that his sun motor design (shown here in an 1876 drawing) owed nothing to Mouchot's design.

would, in the 1970s, become a rallying cry for Prophets: clean, cheap, distributed power, a global tapestry of light and energy generated and distributed on the level of the neighborhood, the farm, the workshop.

Today's version of Ericsson's Prophetic vision was first put together by Amory Lovins, an environmental activist with no formal degree. In 1976 Lovins published an article in the magazine *Foreign Affairs* that introduced the "soft path" for energy—the ideas that inspired water advocates twenty years later to invoke a "soft path" for water. The hard path, Lovins said, consists of distributing ever-increasing amounts of energy from big, integrated facilities: giant power plants, giant pipelines, giant tankers. All are massive, brittle, and ecologically destructive; all require control from repressive, technocratic bureaucracies. The soft path, by contrast, consists of bottom-up power generation from networks of renewable sources. It is small-scale, flexible, and respectful of environmental limits; it fosters community control and democracy. Lovins, needless to say, was a soft-path guy.

These ideas attracted enormous attention—and vehement condemnation from shocked energy executives. But neither Lovins nor Ericsson was offering anything new. Their vision of the soft path was simply an extension of the millennia-old, pre-modern power system, with individual stacks of wood for each home replaced by individual assemblages of mirrors (Ericsson) or windmills and solar panels (Lovins). What *was* novel in the long term was the opposing vision of the Wizards, initially formulated in the early nineteenth century by a man named Frederick Winsor.

Half genius, half fraud, Winsor helped create an institution so fundamental to modern life that it is almost invisible: the power utility. Born in Braunschweig, Germany, he saw an experimental gas lamp in Paris in 1802 and was immediately hooked. (The gas was "coal gas," a flammable mix of methane, hydrogen, and carbon monoxide made by heating coal in an oxygen-free oven.) Leaving behind a mass of unpaid debts, he immigrated to England, changed his name from the overly Teutonic Friedrich Albrecht Winzer of his birth to the eminently English Frederick Albert Winsor, and began a frenzy of "secret" experiments that he publicized at every opportunity. His work, he said, was creating "the most prolific source for the wealth of nations, that ever was recorded in the history of the world." A font of blarney, bunk, and braggadocio, Winsor constantly fell out with his partners and was so careless with money that at the moment of his greatest success he had to flee abroad to escape his creditors. Nonetheless, he changed the world.

Winsor was the first to realize that energy, like water, could flow through pipes from a central location, the equivalent of a well. By building a network of pipes that pumped gas from big central plants, he could charge for energy, monitoring customers' usage and cutting them off if they failed to pay their bills. After multiple legal and financial battles Winsor's Gas Light and Coke Company opened its doors in 1812. It was the beginning of the hard path. Within eight years the firm was feeding gas along 120 miles of London street mains to about thirty thousand lamps. Competitors arose: by 1825, every big city in England had at least one gas-lamp firm. Similar enterprises quickly appeared in other nations; the first U.S. gas-light company, for instance, was established in Baltimore in 1816. Decades later, when electricity became common, its inventors followed Winsor's model, creating high-tech, consolidated distribution companies that fed power through wires—"pipes," in effect—to faraway customers.

Father Himalaya's Pyrheliophoro, seen here in a 1904 photograph, marked both the apex and the end of the first solar-power movement.

Because energy is critical to modern life, these utilities, as we now call them, became so politically important that many governments seized them as essential tools of the state; other nations contented themselves with heavy regulation. Either way, utilities have become a prominent feature of the contemporary landscape. Economically speaking, the advantages of Wizard-style, hard-path centralization and scale were so overwhelming that until recently efforts to promulgate Prophet-style distributed power systems almost vanished.

To be sure, the solar dream staggered along for a while. An Ericsson-inspired inventor provided solar power in 1901 to an ostrich farm in Pasadena, California. A Massachusetts schoolteacher wrote a solar-power textbook, probably the world's first, in 1903. A Philadelphia engineer built a solar irrigation plant in Cairo in 1906. But the last and greatest descendant of Mouchot and Ericsson was an uncompromising Portuguese priest, Manuel António Gomes, known as Father Himalaya for his extreme height.

Born into a poor family in 1868, Father Himalaya went into the priesthood mainly for the paycheck. A polymath who taught himself chemistry, biology, and optics, he invented an explosive, himalayite, said to be safer and more powerful than dynamite, and the first rotary steam engine. He was critical of other solar pioneers, because their mirrors did not track the sun with sufficient precision: their inexact alignments meant that the central boiler cast a shadow on the mirror. By contrast, Father Himalaya's first solar engine, which avoided both problems, generated temperatures hot enough to melt iron. Indeed, its very effectiveness was a problem. His second prototype, incorrectly operated by an assistant, melted itself down before a delighted mob. Undaunted, Father Himalaya decided to build a new machine that would track the sun automatically, without needing unreliable human operators.

The Pyrheliophoro was unveiled at the 1904 World's Fair in St. Louis. Shaped like a sail, its reflector consisted of 6,117 hand-sized mirrors attached to a steel frame forty-two feet high. The Pyrheliophoro focused light so intensely that an awed *New York Times* reporter noted that the reflected heat killed birds forty feet above it. It could produce temperatures up to 7000°F, then the highest ever seen on Earth. Fascinated businesspeople thronged Father Himalaya, who firmly rejected their blandishments. "The sun machine is not yet in an industrial stage," he said. "I cannot lie to form a company, and it is necessary to lie [to do so]."

Instead he took the Pyrheliophoro back to Portugal. There he found out that his backers had pulled the plug. Portugal, like France, had too much coal. A British firm had obtained a monopoly on power distribution and was building giant coal plants. Nobody paid attention when Father Himalaya proclaimed his Ericsson-style vision of free neighborhood electricity powered by mirrors. Solar research had been the product of anxiety about fossil fuels. When the anxiety faded, so did the interest.

Hubbert's Pique

Marion King Hubbert, an idealist through and through, believed in the power of Science to guide the human enterprise. A geophysicist at Columbia University in the early 1930s, he was one of the half-dozen cofounders of Technocracy Incorporated, a crusading effort to establish a government of all-knowing, hyper-logical engineers and scientists—

men rather like Hubbert himself, as it happened. (Hubbert had impeccable academic credentials: undergraduate, master's, and doctoral degrees from the University of Chicago.) Technocracy adherents believed that the world was controlled by flows of energy and mineral resources, and that society should be based on this understanding. Rather than allowing economies to dance to the senseless, febrile beat of supply and demand, Technocrats wanted to organize them on the basis of a quantity controlled by the eternal laws of physics: energy.

Politically unbiased experts in red-and-gray Technocracy uniforms would assay each nation's yearly energy output, then divide it fairly among the citizenry, each person receiving an allocation of so many joules or kilowatt-hours per month. If people wanted to buy, say, shirts, they would look up the price on a table of energy equivalents calculated by objective Technocratic savants. The leader of the system, the Great Engineer, would oversee a new nation, the North American Technate, a merger of North America, Central America, Greenland, and the northern bits of South America. No more would self-interested businesspeople and short-sighted politicians run rampant; the North American Technate would be smooth, efficient, and rational. Practical skills and the hard sciences would be prized; law, politics, the liberal arts, and the so-called "life of the mind" would be relegated to the cellar. Hubbert spent more than a decade laboring to transform this vision into reality.

Hubbert was a poor boy from central Texas who scrabbled his way to the top. Even before he finished his dissertation he was invited to lead Columbia's new geophysics program. These early accomplishments, real and impressive, gave Hubbert an equally real and impressive estimation of his own abilities. He came to believe that he was destined to exert an impact upon society. In this he was wholly successful. Hubbert, one of the nation's most important petroleum scientists, built much of the intellectual framework for the environmental movement. He was a Wizard who became a Prophet.

In New York Hubbert fell under the spell of a charismatic Greenwich Village layabout named Howard Scott. According to Mason Inman, Hubbert's biographer, Scott claimed to be, variously, the son of the head of the Berlin-Baghdad Railway; the scion of an aristocratic family that had lost a fortune in Constantinople; an honors graduate of Berlin's select Technische Hochschule; the manager of a huge nitrate plant

M. King Hubbert

during the war; and the creator of a revolutionary Theory of Energy Determinants. A newspaper exposé proved that none of these claims was true, except possibly the last. It didn't matter, either to Hubbert or the thousands of others who were mesmerized by Scott's apocalyptic predictions of the capitalist system's unavoidable collapse by 1942. Guided by the Theory of Energy Determinants, Scott proclaimed, Technocracy would then step in to create a Utopia in the Technate. Alas, working out the theory took Scott so long that when Hubbert met him he had been jobless for years and was about to be ejected from his apartment. Hubbert paid his debts and moved in with him.

Scott never managed to put his theory on paper. Moved by a disciple's fervor, Hubbert set aside his other work and devoted 1934 to writing a definitive, 250-page statement of Technocracy dogma. The *Technocracy Study Course*, as it was called, dismissed the world's businesspeople, social scientists, lawyers, and teachers as charlatans. Rather than being products of economics, psychology, culture, and history, the *Course* said, societies were ruled by the kind of immutable natural laws that the biologists Raymond Pearl and Georgii Gause had discovered when experimenting with fruit flies in bottles and protozoa in petri dishes.

Pearl had placed a breeding pair of fruit flies in a bottle with a food supply that was replenished at a constant level. He found that the fruit-fly population increased in a way that could be described by an *S*-shaped curve—an initial rise followed by a leveling off. The leveling off was because the fruit-fly population hit the limit of its food supply. (The curve for Gause's protozoans was almost identical, except that the protozoa had a finite food supply, so they died out after exhausting it.) Scott's great insight—Technocracy's central dogma—was that humans behaved exactly like fruit flies. Hubbert laid this out in the *Technocracy Study Course*:

Should the fruit flies continue to multiply at their initial compound interest rate, it can be shown by computation that in a relatively few weeks the number would be considerably greater than the capacity of the bottle. This being so, it is a very simple matter to see why there is a definite limit to the number of fruit flies that can live in the bottle. Once the number is reached, the death rate is equal to the birth rate, and population growth ceases. Very little thought and examination of the facts should suffice to convince one that in the case of the production of coal, pig iron, or automobiles, circumstances are not essentially different.

Politicians and economists who argued for perpetual economic growth were deluded, Hubbert said. The population of the United States would hit a maximum "of probably not more than 135,000,000 people" in the 1950s, and after that the nation simply would not contain enough new consumers to need more consumer products. Hoodwinked by the fantasy of continuing growth, the ruling class had lost sight of these basic scientific realities. They were rushing toward inevitable disaster—after which they would be replaced, thank Heaven, by an elite corps of eco-engineering mandarins with the technical know-how to "operate the entire physical equipment of the North American Continent." In other words, Technocracy.

Surprising to Hubbert, Technocracy was mocked rather than embraced. The group split into factions, and Hubbert slowly became disenchanted. According to Inman, his biographer, a "visibly drunk" Hubbert showed up in 1949 at Technocracy's Manhattan headquarters. He demanded to know if Scott had predicted that capitalism would fall apart by 1942. No, Hubbert was told. This was a lie. Scott had long predicted just that. Now it was seven years past the deadline, and capitalism was still there. For Hubbert, the failure of Scott's predictions was empirical proof that the Theory of Energy Determinants was wrong. "Hubbert never attended another Technocracy meeting," Inman wrote. It was the end of his career as a Wizard.

That same year Hubbert visited a friend who was attending a big natural-resource conference sponsored by the new United Nations. At the conference Hubbert was startled to hear a prominent geologist assert that the world still had 1.5 *trillion* barrels of obtainable oil, enough to last centuries. "I nearly fell out of my seat," Hubbert recalled later. "I was up here, relaxed, visiting with my friend—and good God Almighty! And nobody said boo." A trillion-and-a-half barrels was "just an utterly preposterous amount of oil." Annoyed, Hubbert raised his hand at the end of the session. The geologist's claims, he said, were "an exercise in metaphysics." The dispute grew heated and did not end in agreement.

Lacking a real theory of petroleum formation, early petroleum geologists had assumed that oil and gas deposits must be located in zones similar to those where oil and gas had been found before. They looked, so to speak, for more Pitholes. Because few such areas were known, researchers believed that petroleum deposits therefore must be rare. In reality, new oil was found repeatedly—by wildcatters who, unaware of expert opinion, searched for it in all the wrong places. After many such stories, scientists had become persuaded that petroleum could be found in some form almost anywhere. The main obstacle to finding oil, the famed petroleum geologist Wallace Pratt wrote in 1952, was the conviction that it wasn't there: "Where oil is first found, in the final analysis, is the minds of men."

To Hubbert, this kind of thinking was sheer mysticism. Earth, being finite, contains a finite number of hydrocarbon molecules in a finite set of locations. Supplies are therefore limited—a fact that Hubbert had been pondering since his undergraduate days, when he speculated that the world might run out of the most desirable types of coal "within fifty years." Now, spurred by his annoyance with the 1.5 trillion figure, he developed the first formal model of petroleum peak output. Between the first Pennsylvania wells and 1947 the world had produced 57.7 billion barrels, Hubbert estimated. "Of this, one half has been produced and consumed since 1937"—that is, in the previous ten years. "One cannot refrain from asking, 'How long can we keep it up? Where is it taking us?'" The answer, in his view, was obvious: "the production curve of any given species of fossil fuel will rise, pass through one or several maxima, and then decline asymptotically to zero."

Decline asymptotically to zero! The potential consequences were vast. Hubbert believed that the fossil-fuel explosion had created the population explosion—that consuming coal, oil, and gas had provided the impetus to drive our species up Gause's S-shaped curve. Because the amount of the world's oil was by definition finite, the supply would

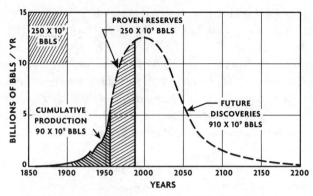

Hubbert's original 1956 diagram of the rise and fall of global oil production

fall to zero after too much use. As a result, we were bound, so to speak, to hit the edge of the petri dish. Hubbert drew Gause-like graphs showing the simultaneous rise in energy use and population—and the inescapable future peak in both.

Hubbert's views echoed those in *The Road to Survival*, published the year before, except that he thought in terms of physical limits, rather than biological limits. Still, he ended up in the same place: capitalist-style economic growth was not only unsustainable, it was actively driving humankind beyond its limits to disaster. "The future of our civilization largely depends," he wrote, on whether humanity will be able "to evolve a culture more nearly in conformity with the limitations imposed on us by the basic properties of matter and energy."

These ideas might have been expected to draw fire from Hubbert's employer—he had become second-in-command at a big Shell Oil research center in Houston. But they attracted little notice until 1956, when Hubbert explained his thinking at a meeting of the American Petroleum Institute in San Antonio. Just before Hubbert gave his talk, he later alleged, he was telephoned by an appalled Shell public-relations executive. "Couldn't you tone it down a bit?" he recalled the man asking. "Couldn't you take the sensational parts out?" Hubbert, rarely in doubt about his own abilities, refused to back down. Between 1965 and 1970, he told the audience, crude-oil yield in the continental United States would peak. Global production would hit its maximum by the beginning of the twenty-first century.

In 1964 Hubbert left Shell to work for the U.S. Geological Survey. As the University of Iowa historian Tyler Priest has written, Hubbert didn't have it easy at USGS: his boss, USGS director Vincent E. McKelvey, became his most rabid critic. Like Hubbert, McKelvey saw himself as a grand thinker with wisdom to impart about society at large. But unlike Hubbert, his vision was sunny and optimistic, Borlaugian to the core. Human ingenuity and technical prowess, this Wizard believed, were the sturdy vehicles that would carry us into a future of unbounded affluence.

Unsurprisingly, the two men clashed. McKelvey's USGS sent out a flood of cheery projections of the country's oil reserves, as did the oil industry. All the while, Hubbert broadcast jeremiads about imminent exhaustion, none of them published by USGS. The dispute soon grew personal: Hubbert accused McKelvey of stealing his papers; McKelvey accused Hubbert of withholding information; the two men wrote dueling reports for different branches of the government. In a fit of pique, McKelvey snatched away Hubbert's secretary, a low blow in the pre-computer era. According to Priest, the historian, Hubbert struck back by blackballing McKelvey when he was nominated for the National Academy of Sciences and the American Academy of Arts and Sciences.

In a setback to McKelvey, Hubbert's estimate proved to be correct: U.S. crude-oil production hit a peak in 1970 and then slowly fell. As the output declined, former interior secretary Stewart Udall—"a Hubbert man," he called himself—ridiculed McKelvey's work as "an enormous energy balloon of inflated promises and boundless optimism [that] had long since lost touch with any mainland reality." In 1977 President Jimmy Carter forced McKelvey to resign—the first such ouster, Priest reported, "in the survey's ninety-eight-year history."

McKelvey's fate may have been sealed by the Arab oil shock, which resonated with Hubbert's message of limits. During the 1973 Arab-Israeli war, several Arab nations decided to slap the United States for supporting Israel. They cut oil production for four months. Huge public alarm ensued. Passions boiled over as people waited for hours in gas lines; line-jumpers got into fistfights. To Hubbert, the oil shocks presaged "the end of the Oil Age."

Today most historians and economists instead view the oil shock as a product of mistaken government policies. Arab petro-states could not target individual nations, the energy analyst Michael Lynch told me, because national oil companies sell oil and gas to what is, in effect, a single worldwide pool controlled by middlemen. Any embargo thus could only raise prices equally across the planet, rather than striking at a single nation. Or, rather, the Arabs couldn't have targeted a single nation if President Richard Nixon had not imposed price caps on U.S. oil and gas two years before as an inflation-fighting measure. The embargo cut global oil output by about a quarter, pushing up petroleum prices worldwide. Middlemen could take advantage of the higher prices only if they sold their oil to nations other than the United States, with its price caps. Doing just that transformed a modest global shortfall into a U.S.-specific oil drought.

That was not how events were understood at the time. One year previously, an MIT-based research team had created an international furor with *The Limits to Growth,* a book-length study that used computer models to predict that unless radical steps were taken the world would soon run out of resources, precipitating civilizational collapse. Hubbert's name does not appear in *Limits.* Nor does Vogt's. Nonetheless their fingerprints are all over it—indeed, the *Limits* authors, influenced by Hubbert's peak-oil theory, had futilely begged him to collaborate with them. The final result was as if the MIT team had plugged Hubbert's equations into a computer and applied them beyond oil to resources like coal, iron, natural gas, and aluminum. Graph after graph depicted a Hubbertian race to a peak of production, followed by a ruinous decline. Like Hubbert, the *Limits* writers saw a direct connection between economic growth and calamity. As the Yale historian Paul Sabin has written, the oil shock "seemingly confirmed the thesis of *The Limits to Growth*." The fistfights at the gas pump were viewed as a harbinger of a coming crisis caused by overconsumption. The Vogtian vision of inescapable limits to carrying capacity had become an organizing principle of environmental thought.

Propelled by the oil shock, fears of scarcity wafted across the nation like a bad smell. Rumors of shortages in any number of goods—gasoline, salmon, cheese, onions, raisins—caused brief, unwarranted episodes of anxiety, some of them about commodities one would never imagine could run out. The Great Toilet Paper Panic of 1973 occurred after talk-show host Johnny Carson joked about a shortage, causing frightened consumers to buy out stores. Carson's jest ricocheted to Japan, which imported almost all of its paper from the United States; toilet-paper supplies disappeared from Hokkaido to Kyushu. The next elected president, Jimmy Carter, was a Hubbertian. Soon after his inauguration, he gave a nationwide address warning that the planet's petroleum could be gone "by the end of the next decade"—i.e., by 1989. Hubbert himself thought the disaster would occur a bit later, in about 1995.

Perversely, the most enduring consequence of the 1970s belief that energy supplies were running out was not to use less, but to look for more. In this quest, Jimmy Carter, arguably the most ecologically minded president in U.S. history, endorsed policies that today seem like environmental folly. Notably, his administration sought to offset the approaching decline of oil and gas by tripling the use of coal, a much dirtier fuel. Just as peak oil had provided justification for foreign-policy misadventures in the 1920s and 1930s, it proved a friend to Big Coal in the 1970s and 1980s. Meanwhile, oil firms found so much crude that by the end of the 1990s real prices had fallen to half—sometimes a fifth—of what they were during Carter's day.

Central to the misunderstanding was the concept of a "reserve." Both Hubbertians and McKelveyans agree that an oil reserve is a physical stock: a finite pool of hydrocarbon molecules. To Hubbertians, the implication is clear: pump out too much and you will eventually empty it. How long you can pump depends primarily on the size of the pool. To McKelveyans, though, what matters most is not the size of the pool, but the capacity of the pump.

The reason for this apparently counterintuitive belief is that a petroleum reserve is not, in fact, a subterranean pool, like the underground lake where Voldemort conceals part of his soul in the Harry Potter books, but rather an imprecisely defined zone of permeable, spongelike rock that has petroleum in its pores. (A reserve can also occur in thin sheets between layers of shale.) Nor is petroleum a uniform substance, a black liquid like the inky water in Voldemort's lake. Instead, it is a crazy stew of different compounds: oil of various grades mixed with ethane, propane, methane, and other hydrocarbons. These range from the purely gaseous (methane, or natural gas) to syrupy liquids (crude oil) to semi-solids (the petroleum precursors sometimes called tar sands, for example). Squashed into stone deep underground, this jumble of glop, goo, and gas is usually under great pressure. Layers of impermeable rock prevent it from seeping to the surface. When drilling bores through the caps, the pressurized liquids and gases shoot up in orthodox gusher fashion.

How much can be extracted depends on how deep the drilling operation can probe, the composition of the regions it can reach, which of the different compounds in that area it can handle, and—a key variable—whether the current price justifies the required effort. If a company's engineers develop new equipment that can pump out more petroleum at a lower cost, the effective size of the reservoir increases. Not the *actual* size—its physical dimensions—but the *effective* size, the amount of oil and gas that can be extracted in the foreseeable future.

An often-cited example is the Kern River field, north of Los Angeles. From the day of its discovery in 1899, it was evident that Kern River was rich with oil. Pithole-style, wildcatters poured into the area, throwing up derricks and boring wells. In 1949, after fifty years of drilling, analysts estimated that just 47 million barrels remained in reserves—a triviality, a finger snap, a rounding error in the oil business. Kern River, it seemed, was nearly played out. In reality, the field was still full of oil. But what remained was so thick and heavy that it almost didn't float on water. No method existed for sucking such dense stuff out of the ground.

Petroleum engineers in the 1970s figured out how to extract it: shoot hot steam down wells to soften thick oil and force it from stone. At first, the process was hideously inefficient: boiling the water to produce the steam required up to 40 percent of the oil that came out of the wells. Because companies were making steam by burning unrefined crude oil at the wellhead, they released torrents of toxic chemicals into the air. But this wasteful process squeezed out petroleum that had seemed impossible to reach. Over time engineers learned how to use steam with less waste and pollution. By 1989 they had taken out another *945 million* barrels from the Kern River field. That year analysts again estimated Kern reserves: 697 million barrels. Technology continued to improve. By 2009 Kern had produced more than 1.3 billion additional barrels, and reserves were estimated to be almost 600 million barrels.

Meanwhile, the industry was learning how to tunnel deeper into the earth, opening up previously inaccessible deposits. In 1998 an oil rig at a field adjacent to Kern River bored thousands of feet beyond any previous well in the area. At 17,657 feet, it blew out in a classic gusher. Oil and gas shot three hundred feet in the air, caught fire, and destroyed the well. Energy firms guessed that the blowout indicated the presence of undiscovered oil and gas deposits far underground. Investors rushed in and began to drill. They indeed found millions of barrels of oil at great depths, but it was mixed with so much water that the wells flooded. Within a few years, almost all the new rigs ceased operation. The reserve vanished, but the oil remained.

Stories like that of Kern River have occurred all over the world for decades. After hearing them over and over again from geologists, I realized Hubbert and *Limits* were going about matters the wrong way. An oil reservoir in the earth is a stock. If it becomes too costly or difficult to extract, people will either find new reservoirs, new techniques to extract more from old reservoirs, or new methods to use less to accomplish the same goal. All of this means that the situation constantly changes, which in turn means that we can see only a limited distance ahead.

"It is commonly asked, when will the world's supply of oil be exhausted?" wrote the MIT economist Morris Adelman. "The best one-word answer: never." On its face, this seems ridiculous—how could a finite stock be inexhaustible, when a constantly renewed flow can run out? But more than a century of experience has shown it to be true. As a practical matter, we know only that there is more than enough for the foreseeable future. That is, fossil-fuel supplies have no known bounds. In strict technical terms, this means they are infinite. Hardly anyone who is not an economist believes this, though.

"Not a Commodity to Be Bought or Sold"

It was a time when a new electronic communications network had suddenly made it possible to transmit data around the world at almost the speed of light. Fashions spread from one corner of the earth to another, then vanished. A bold new breed of super-rich entrepreneurs was launching enormous technological enterprises. Media empires were rising and falling.

This was the 1870s; the electronic network was the telegraph. The "Victorian Internet," as the writer Tom Standage has called it, "revolutionized business practice, gave rise to new forms of crime, and inundated its users with information." The super-rich entrepreneurs were laying submarine cables across the English Channel, the Mediterranean, and the Atlantic. Completion of the first transatlantic cable had been greeted by a jubilant parade in New York City. Its insulation failed rapidly. Similar problems plagued the other undersea cables. The future was being held back. Innovations in cable technology were required.

One of the engineers developing the next generation of cables was a Briton named Willoughby Smith. Smith was supposed to test the cable as it was being laid. Looking for the right material for his tests, he tried selenium, a gray, metal-like element. Irritatingly, he couldn't ascertain exactly how much electricity selenium let through. Sometimes it blocked the current like so much rubber, sometimes it allowed it to flow almost as freely as copper. "Pieces of high resistance at night would be only half the resistance in the morning," Smith recalled. He eventually realized the difference was due to light. In sunlight, selenium conducted electricity; in darkness, it did not. Smith was baffled; nothing in physics said this could be possible.

Taking up the puzzle, a King's College physicist named William Grylls Adams found something more surprising still. If Adams put a strip of Smith's selenium in a dark room and then lighted a candle, he wrote in 1876, it was "possible to *start a current in the selenium merely by the action of light.*" The excited italics expressed Adams's amazement. In all of human history, people had generated power either by burning something or by letting water or air or muscle turn a crank. Adams had created electricity by shining light at a lump of stuff.

In retrospect it seems clear that many of Adams's colleagues didn't believe it. Even when the New York inventor Charles Fritts actually built functioning solar panels—he spread a layer of selenium over a layer of copper and placed the assembly on his roof, thereby generating electricity—most researchers still dismissed them. "They appeared to generate power without consuming fuel and without dissipating heat," wrote the solar historian John Perlin. Fritts's panels "seemed to counter all of what science believed at that time." They sounded like perpetual-motion machines. How could Adams's "photoelectric effect" possibly be real?

Only in 1905 was the panels' puzzling behavior explained—by Albert Einstein, a newly minted Ph.D. with a day job in the Swiss patent office. In what may have been the greatest intellectual sprint for any physicist in history, Einstein completed four major articles in the spring of that year. One described a new way to measure the size of molecules, a second gave a new explanation for the movement of small particles in liquids, and a third introduced special relativity, which revamped science's understanding of space and time. The fourth explained the photoelectric effect. ✓

Physicists had always described light as a kind of wave. In his photoelectric paper, Einstein posited that light could also be viewed as a packet or particle—a photon, to use today's term. Waves spread their energy across a region; particles, like bullets, concentrate it at a point. The photoelectric effect occurred when these particles of light slammed into atoms and knocked free some of their electrons. In Fritts's panels, photons from sunlight ejected electrons from the thin layer of selenium into the copper. The copper acted like a wire and transmitted the stream of electrons: an electric current.

Einstein received the Nobel Prize in 1921 for explaining the photoelectric effect. But Fritts's invention remained a laboratory curiosity. Photovoltaic panels, as they are known today, were fascinating but useless. Converting only a tiny fraction of the sun's energy into electricity, they were much too inefficient for any practical use. Decades of sporadic research into photovoltaics brought little improvement—something that Warren Weaver, Borlaug's supervisor at the Rockefeller Foundation, lamented in 1949. The lack of progress was demonstrated four years later, when the Bell Laboratories physicist Daryl Chapin tested an array of selenium panels. No matter what he did, they converted less than 1 percent of the incoming solar energy into electricity. Then two of Chapin's colleagues presented him with a surprise.

The two researchers, Calvin Fuller and Gerald Pearson, were members of the team that transformed the transistor, invented at Bell in 1947, from a finicky laboratory prototype into the mass-produced foundation of the computer industry. At the center of this work was silicon, common and inexpensive, a principal constituent of beach sand. Silicon forms crystals, each atom linked to four neighbors in a pattern identical to that formed by carbon atoms in diamonds. As students learn in high school chemistry, the atoms bond to each other by sharing their outer electrons. Silicon crystals can be adulterated—"doped," in the jargon—by replacing a few of the silicon atoms with atoms of boron, arsenic, phosphorus, or other elements. If the added "dopant" atoms have more shareable electrons than the silicon atoms they replaced, the crystal as a whole ends up with extra electrons. Because electrons have a negative charge, the crystal becomes negatively charged. Similarly, if the dopant atoms have fewer shareable electrons, the doped crystal has, in effect, some electron-sized "holes"—it is positively charged. Like the electrons in the crystal, the "holes" are shared, meaning that their locations can move about somewhat in the way that physical electrons can move about.

Fuller and Pearson placed a thin layer of the first type of doped silicon (extra electrons) atop a layer of the second type (extra holes). The two Bell researchers attached the little assembly to a circuit—a loop of wire, in effect—and an ammeter, a device that measures electric currents. When they turned on a desk light, the ammeter showed the two-layer silicon suddenly generating an electric current. The same thing happened with sunlight. Fuller and Pearson realized that the photons were penetrating the top layer with enough force to knock electrons into the bottom layer, creating a flow of electrons that moved into the wire: a current. The two men had accidentally created a new type of solar panel.

When Chapin tested these novel photovoltaics, they converted about five times more solar energy to electricity than the older selenium panels. But they were still terribly inefficient. Chapin estimated the cost of silicon panels that could supply electricity for a typical middle-class home at $1.43 million (about $13 million in today's dollars). It would be cheaper to cover the entire roof in gold leaf.

Daunted by the economics, most researchers gave up on photovoltaics until the oil shocks of the 1970s revived peak-oil fears—and hopes that the sun might be the way to escape them. The numbers seemed so overwhelming, so alluring. Every second, the sun bathes Earth with 172,500 terawatts of energy. (A terawatt—a trillion watts—is the biggest energy unit in common use.) About a third of this prodigious flow is promptly reflected into space, mainly by clouds. The leftover—roughly 113,000 terawatts, depending on cloud cover—is available for capture. All human enterprises together now use a bit less than 18 terawatts. In other words, the sun furnishes more than six thousand times the energy produced today by all of our power plants, engines, factories, furnaces, and fires combined. Its light won't run out for billions of years. Who cares about oil in the Middle East?

Hubbert was a solar champion, but the noisiest advocates were in the new counterculture, which swarmed about solar installations like bumblebees around flowering thyme. Publications like *The Whole Earth Catalog* and *CoEvolution Quarterly* extolled sun power in Vogtian terms as "soft technology"—the route to a small-scale, decentralized, individual-empowering future with technology that was "alive, resilient, adaptive, maybe even lovable." Solar hot-water heaters! Solar heat exchangers! Solar shutters! Solar dryers! Solar buildings!—all untouched by corporate greed. (By contrast, Borlaugian-style "hard tech," exemplified by smoke-belching coal plants, was alienating, wasteful, environmentally ruinous, and above all old-fashioned.) Solar power, proclaimed the eco-activist Barry Commoner, is *inherently* liberating. "No giant monopoly can control its supply or dictate its use. . . . Unlike oil or uranium, sunlight is not a commodity to be bought or sold; it cannot be possessed."

Sun power's image as the province of baling-wire hippies was at odds with reality. Today's multibillion-dollar photovoltaic industry owes its existence mainly to the Pentagon and Big Oil. The first wide-scale use of solar panels had come in the 1960s: powering military satellites, which couldn't use fossil fuels (too bulky to lift into space) or batteries (impossible to recharge in orbit). By the 1970s photovoltaics were cheaper, but the industry had acquired only one major new user: the petroleum industry. Some 70 percent of the solar modules sold in the United States were bought to run offshore drilling platforms.

Realizing that solar had become essential to oil production, petroleum firms set up their own photovoltaic subsidiaries. Exxon became, in 1973, the first commercial manufacturer of solar panels; the second, a year later, was a joint venture with the oil giant Mobil. (Exxon and Mobil merged in 1999.) The Atlantic Richfield Company (ARCO), another oil colossus, ran the world's biggest solar company until it was acquired by Royal Dutch Shell, the oil and gas multinational. Later the title of world's biggest solar company passed to British Petroleum (now known as BP). By 1980 petroleum firms owned six of the ten biggest U.S. solar firms, representing most of the world's photovoltaic manufacturing capacity.

Why did Big Oil keep investing in a technology with such a slow potential payoff? One reason was a new wave of peak-oil anxieties. After falling in the 1980s, worries slowly built up in the 1990s until they were publicly detonated by the British geologist Colin Campbell and the French petroleum engineer Jean Laherrère, who predicted in a widely read *Scientific American* article in 1998 that the oil party was almost over. "Before 2010," the two men proclaimed, global petroleum output would permanently decline. "Spending more money on oil exploration will not change this situation. . . . There is only so much crude oil in the world, and the industry has found about 90 percent of it." Humankind was not running out of oil per se, they stressed. What was vanishing was "the abundant and cheap oil on which all industrial nations depend."

As in previous oil panics, the fears spread widely. "The supply of oil is limited," President George W. Bush told world leaders in Switzerland. Saudi petroleum is in "irreversible decline," proclaimed the peak-oil pundit Matt Simmons in 2005. Oil baron/corporate raider T. Boone Pickens agreed; the world is "halfway through the hydrocarbon era," he said that year. "Slowly at first and [then] at an accelerating rate," the best-selling peak enthusiast James Howard Kunstler predicted at about the same time, "world oil production will decline, world economies and markets will exhibit increased instability . . . and we will enter a new age of previously unimaginable austerity." Warnings flooded from the presses: *Hubbert's Peak* (2001), *Powerdown* (2004), *Twilight in the Desert* (2005), *The Long Emergency* (2005), *Peak Oil Prep* (2006), *The Post-Petroleum Survival Guide and Cookbook* (2006), *Confronting Collapse: The Crisis of Energy and Money in a Post-Peak-Oil World* (2009).

"The price of oil was an index to the Western world's anxiety," the novelist Don DeLillo had suggested. "It told us how bad we felt at a given time." If so, people were feeling rather bad; petroleum panic had taken hold as never before. The University of Maryland's Program of International Policy Attitudes surveyed fifteen thousand people in sixteen countries: 78 percent believed that we were running out of oil. Another poll: 83 percent of Britons thought that oil and gas could become unaffordable. Another: three-quarters of Americans believed that a petroleum drought was coming. "I don't see why people are so worried about global warming destroying the planet," Simmons said in 2008. "Peak oil will take care of that." Seeming to echo his admonition, the price of oil soared that year to its all-time high: $147.27 per barrel.

As the fears mounted that oil would run out, nations around the world built solar parks of ever-increasing scale. The biggest in Asia (as I write, anyway) is the Charanka solar park, an Ozymandiac installation located on a wasteland in the coastal Indian state of Gujarat, a hundred miles from Ahmedabad, its biggest city. When I visited Ahmedabad not

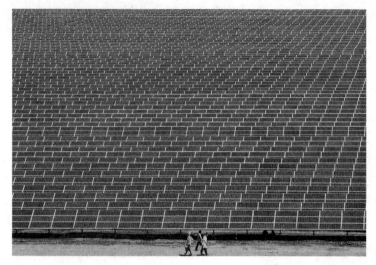

The Charanka solar park: a sea of photovoltaics

long ago, I could see Charanka gleaming from my airplane window: dozens of rectangular photovoltaic arrays, regular as midwestern wheat fields, scattered in a broad U about three miles on a side. By squinting a little I could talk myself into thinking I saw power lines spiderwebbing from the arrays: hundreds of megawatts from the desert. Twenty miles from the airport was a metallic ribbon half a mile long and a hundred feet wide: a solar park built atop an irrigation canal. Southeast of the city was another, even longer one. As the plane approached the city, solar panels stood like sentinels atop buildings everywhere—one of the world's most important efforts to bring into being a solar future.

The center of India's oldest civilizations, Gujarat is at once a cradle of Hindu identity and a busily cosmopolitan place, full of traders from across Asia. It is also the homeland of Narendra Modi, elected prime minister of India in 2014, the principal author of its solar program. Modi was born in 1950, the son of an impoverished tea-stall owner in a remote Gujarat village. A political activist since adolescence, he joined, in 1987, the Bharatiya Janata (Indian People's) Party (BJP), a pro-Hindu, nationalist party tied to nativist organizations known for attacking Christians, Muslims, and other non-Hindus. He rose steadily and won election as chief minister of Gujarat in October 2001. A few months later a Gujarat train loaded with Hindu pilgrims and activists caught fire, killing dozens of passengers. Angered by rumors that the blaze had been set by Muslims, club-wielding Hindu thugs murdered a thousand or more people, most of them Muslim. Human rights groups and political rivals charged that Modi and the BJP had encouraged the attacks. Inquiries found no evidence for the charge, but the riots stained his reputation; in 2005, Modi became the only person ever denied a U.S. visa for "severe violations of religious freedom."

Alarmed by the fallout, Modi shifted gears, refashioning himself as a nattily dressed, tech-friendly progressive who lured major companies, foreign and Indian alike, to invest in Gujarat. He also became one of the world's most prominent advocates for solar power. In a "green auto-biography" published in 2011, Modi promised to transform hot, dry Gujarat, with its 55 million people, into an emblem of sustainable development, simultaneously increasing irrigation and recharging aquifers, converting thousands of cars from gasoline to natural gas, and turning the state capital, Gandhinagar, into a "solar city." He created Asia's first ministry of climate change and led a pioneering program to install solar panels atop irrigation canals, shielding the canals from evaporation and generating power without covering scarce farmland. "I saw more than glittering panels," said then U.N. secretary-general Ban Ki-moon, inaugurating a canal-top project in 2015. "I saw the future of India and the future of our world."

On one side of the Charanka solar park is a seven-story observation tower sheathed in glass. When I visited it, placards trumpeting advances in photovoltaic technology were mounted for perusal inside. The best modules available today, they said, convert more than 20 percent of the sun's energy to electricity. In lab tests, some solar cells reach 40 percent. (A typical coal plant converts 40–45 percent of the energy in coal to electricity.) In parallel, the cost of generating power with photovoltaics has plummeted. Exact figures are hard to nail down, but in many places the cost of building a big solar plant is now equivalent to the cost of building a big coal plant, and in all likelihood photovoltaic prices will continue to fall.*

---

* Here I am being deliberately vague. Cost estimates depend on the factors included in the assessment. For solar power, these include the location (sunshine varies from place to place), the type of photovoltaic, and the likely subsidies and taxes (almost all energy is subsidized in

I climbed to the top of the tower. The solar panels below seemed to stand at attention, a vast photovoltaic army. That day the temperature was about 110 degrees. Wind whipped dust into the air, coating the solar panels. Pipes beneath the arrays carried water to wash them. Solar parks, farms for electrons, effectively must be irrigated. Here and there the serried lines of panels wobbled, harsh conditions and land subsidence nudging them out of alignment. The panels were designed and built by engineers from temperate-zone nations; I wondered how long they would last in the heat. Energy from the sun today is responsible for about 1 percent of India's electricity; even in Gujarat, it amounts to just 5 percent. Optimistic scenarios show its share rising to 10 percent by 2022. The state Power Grid Corporation has proposed creating huge systems in Indian deserts to boost that number to 35 percent by 2050.

Atop the tower, I tried to imagine what Augustin Mouchot would have made of Charanka. Would he see such enormous installations as vindication? Or would he be dismayed by the lack of progress on the problems that had bedeviled him? Like Mouchot's mirrors, Charanka's photovoltaics generate electricity only between sunrise and sunset—6:45 a.m. to 6:45 p.m., during my visit. To provide electricity at night, energy generated in daylight must be stored in some form for later use, a practice called "load-shifting." Typical load-shifting projects heat a liquid (molten salt, say) by day; at night the stored, super-hot liquid boils water, driving a steam turbine, producing electricity in a manner Mouchot would have approved. In 2010 India announced seven solar energy-storage projects, five of them in Gujarat. Only one was under construction. The others had been abandoned when the builders learned that the state's air is so hazy that initial estimates of potential solar power were off by a quarter.

√ Germany, richer than India, has about seventy energy-storage projects, about a third of which collect the output from wind and solar plants into banks of batteries. The price of batteries, like the price of photovoltaics, has been falling. Renewable-energy enthusiasts imagine

one way or another). For coal, should one include the cost of its carbon-dioxide emissions? If so, at what price? For solar plants, how should one treat the emissions due to manufacturing, land acquisition, and installation? And so on. One prominent study argued that the true costs of solar are so high that it will always be unaffordable. Another says that coal-pollution costs are so high that coal is unaffordable. Different studies have such different results that it seems best to say that neither solar nor coal has an undisputed cost advantage.

great warehouses full of batteries, soaking up excess sun power by day, releasing it by night, keeping the lights on in the dark. But no matter how cheap the batteries are, such facilities will involve constructing a second, parallel infrastructure for energy storage, adjacent to the first, for energy production, a costly step for the foreseeable future. Today, as in Mouchot's time, free energy from the sun is surprisingly expensive.

Even this may understate the price of renewables, as I learned when I spoke to Steven Chu, the Nobel-winning physicist who was U.S. energy secretary from 2009 to 2013. Chu, who described himself as a "very strong supporter" of renewables, pointed out that the skies in places like New England or France or northern China can be cloudy for weeks on end. "There are times," he said, "when you get a week of bad weather or a week of cloudy days over hundreds of miles. There are times when the wind stops blowing across all of Washington and Oregon for two weeks. During these times—guess what?—you still need a source of reliable power."

Later Chu emailed a chart to me of the wind-energy output in January 2009 from a big wind farm run by the Bonneville Power Administration in eastern Washington State. The chart had two lines: one on top, zigging up and down every day, depicting the energy demand in the region; one on bottom, showing the contribution of the wind farm to satisfying that demand. Halfway through the month, the bottom line dropped to zero—and stayed there. The wind had ceased entirely for eleven days. As the top line showed, people's needs didn't stop with the wind. Hospitals, schools, libraries, homes, and office buildings still needed light and heat. If nations switched over wholly to renewable energy, Chu said, they would have to come up with mechanisms to supply entire regions with power during long periods of cloudy or still weather. Engineers, he said, have barely begun working on this challenge. A century and a half after Mouchot, the problems he identified with solar energy remain unsolved.

Hundreds of large renewable energy installations now dot the globe. But only one even begins to approach what advocates envision: a solar or wind installation that provides reliable, round-the-clock power to large numbers of people. Completed in 2015, the $800 million Crescent Dunes project, located in Nevada, consists of a central tower surrounded by ten thousand mirrors, each the size of a small house. Tracking the

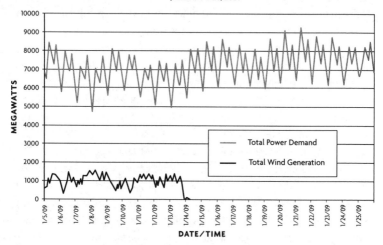

A chart emailed to the author by Steven Chu

sun, the mirrors focus sunlight on the tower, which is filled with molten salt. The hot salt boils water, driving turbine generators; the electricity is sent to Las Vegas to power streetlights, air-conditioners, and video-game consoles. Because the salt stays hot for hours after sunset, the project makes electricity at night: solar power in the dark.

Strikingly, Crescent Dunes has been fought by Prophets. As a rule, renewable-energy *leaders* see their goal as building giant, central-ized facilities like Crescent Dunes—they are Borlaugians through and through, hard-path advocates in solar guise. But many or most renewable-energy *supporters* are Prophets who view Big Solar and Big Wind with almost as much distaste as the Big Coal and Big Oil they seek to replace. From its inception, Crescent Dunes was resisted by an environmental organization, Basin and Range, because the installation, like Father Himalaya's Pyrheliophoro, kills any birds that come near it. Basin and Range also objected to the concomitant destruction of the fragile desert environment; building Crescent Dunes required bulldoz-ing about 2.5 square miles of arid land, including about 10 percent of the habitat of two rare beetles, the Crescent Dunes aegialian scarab and the Crescent Dunes serican scarab.

Similar complaints dog Big Solar in California's Mojave Desert. Litigation from the Sierra Club and the Natural Resources Defense Council in 2012 helped doom one massive solar-mirror project there. Two years later, another Mojave enterprise—the $2.2 billion Ivanpah Solar Electric Generating System, then the world's biggest solar-mirror installation— managed to begin operation. Covering more than five square miles, it knocks large numbers of bats out of the sky, which has led to near-constant attack by environmental groups.

These Prophetic anxieties are not restricted to the United States. Protests against solar farms have occurred by the score in England. Wind power has been fought in Canada, Denmark, Ireland, Italy, Mexico, and Spain. (For this century, the Oxford zoologist Clive Hambler charged, wind power is "a far greater threat to wildlife than climate change.") Even renewables-mad Germany has battled the infrastructure necessary to bring wind-generated electricity from the breezy north to the industrial south; rather than allow new high-voltage transmission lines, the Bavarian government has campaigned to return to fossil fuels.

Most objectionable, to Prophets, are these projects' scale. Some complaints, to be sure, are linked to the selfish unwillingness to sacrifice anything for the common good encapsulated in the slogan NIMBY—Not In My Back Yard. But others are rooted in a respect for limits. Prophets see the mile-long stands of photovoltaic cells in projects like Charanka as inherently destructive to communities, natural and human. Industrial giantism is the problem, in their view, not the solution. True to Ericsson's original vision, they argue instead for smaller-scale, networked energy generation: rooftop photovoltaic panels, air-source heat pumps, biological fuel cells, solar air heating, methane generated by agricultural or municipal waste, and so on.* All of this is to be accompanied by insulating buildings, installing energy- and water-saving fixtures and appliances, recirculating waste heat, and fitting sensors into buildings that automatically monitor power use, shutting off lights and temperature-control when rooms are unoccupied.

Wizard-style renewable advocates like the venture-capital-backed

---

* Air-source heat pumps take advantage of temperature differences to transfer heat from outside to inside a building, or vice versa. Biological fuel cells use bacteria that consume wastes to drive chemical reactions that produce electricity. Solar air heating involves covering buildings with thin, air-filled panels that absorb heat from the sun; the air is then piped into the building, directly or indirectly. Methane that spontaneously issues from waste— municipal dumps, say—can be tapped and burned in local power plants.

firm that built Crescent Dunes scoff at these ideas. Even in the best of circumstances, the process of replacing the present coal-and-gas grid with a new, renewable-energy grid—all the while keeping the old grid running—would be long, expensive, and risky even if it weren't being sabotaged by the people who are supposed to support it. Insisting on using small-scale components to build in a world of 10 billion only multiplies the difficulty. Now add in the fact that fossil-fuel prices have been declining for decades. Thinking purely in terms of a reliable energy supply, one is hard-pressed to imagine why one would try to do it.

But, of course, the question is not simply about a reliable energy supply.

SEVEN

Air: Climate Change

A Quick Few Millennia

Lynn Margulis had high standards for calamity. Years ago she came into a café where I was reading a book while waiting for my daughter to finish an art lesson. I was on a jury that awarded prizes to popular books about science. One of the entries was the book in my hand: *An Inconvenient Truth*, Al Gore's invocation of the "planetary emergency of global warming." Margulis picked it up and gazed at the back cover, which showed the author standing before microphones in a rugged outdoor setting. She said nothing, but her expression was eloquent.

Feeling defensive, I asked if she thought the kind of climate change Al Gore was describing wouldn't be a catastrophe.

Sad, sure, she said, in my recollection. But a *catastrophe*—no. She paused. Oxygen, she said. Now, *that* was a catastrophe.

The oxygen she was referring to was the Great Oxidation Event, which occurred after cyanobacteria evolved photosynthesis. Photosynthetic creatures spread through the oceans, excreting oxygen all the while. The flood of oxygen permanently changed the surface of the earth, the composition of the oceans, and the functioning of the atmosphere. Most scientists believe it made the vast majority of the world's land and sea uninhabitable for the vast majority of the planet's living creatures. Margulis called the resultant slaughter an "oxygen holocaust." Over time, minerals absorbed much of the gas, stabilizing it at about 21 percent of the atmosphere—a good thing, because if the level had risen much more our air would have had so much oxygen that a single spark could have set the planet afire.

To Margulis, the Great Oxidation Event had lessons for today. The first was that people who thought that living creatures couldn't affect the climate had no idea of the power of life. The second was that the onset of climate change meant that *Homo sapiens* was getting into the biological big leagues—we were tiptoeing into the terrain of bacteria, algae, and other truly important creatures. The third was that species, like sullen teenagers, don't pick up after themselves. Cyanobacteria sprayed their oxygen garbage all over Earth without concern for the consequences—littering on an epic scale. People were doing the same with carbon dioxide.

Cyanobacteria were lucky: being inundated by their own oxygen didn't end up bothering them much. But people were going to feel the effects of carbon dioxide. That wouldn't stop us, Margulis said, unsentimental as ever. Humans were no more going to stop emitting carbon dioxide than cyanobacteria were going to stop emitting oxygen, she thought. If it is the fate of every successful species to wipe itself out, climate change looked to Margulis like a plausible candidate for the method by which *Homo sapiens* would achieve that end. The upside, she told me, was that the impacts would be relatively confined and short-lived. In a few millennia, the world would look much the same, except that people probably wouldn't be living in it.

Climate change! In the last chapter, I looked at two visions of how society should meet its needs for energy. Borlaugian Wizards favor large utilities that feed metered power to households and businesses. Vogtian Prophets like small-scale operations that harness renewable sources and are owned by the neighborhoods that use the power. Because the sunlight and wind favored by Prophets are intermittent, they have not yet been able to compete economically against the reliable power provided by fossil fuels. Indeed, the cost advantage is so extreme that there would be little reason to switch to renewables unless some new factor put its thumb on the scale. In recent decades, the prospect of devastating climate change has done just that.*

---

* As I mentioned in the Prologue, I am splitting the discussion of climate change into two pieces. Here I ask skeptics to accept—just for the moment—that climate change is a problem, so I can look at how Borlaugians and Vogtians would address it. In an appendix, I address whether one should believe in its potential impact.

Eccentrics

Was Margulis correct that we are fated by natural law to wreck our own future? History provides two ways of approaching this question. The first draws on the inspiring manner in which a group of scientific eccentrics and outsiders slowly built up today's picture of climate change just in time to use that knowledge to halt its worst effects. The second focuses on the discouraging way that political institutions have been unable to grapple with the challenge and climate change became the subject of a cultural battle over symbols and values. The second approach leads to the conclusion that Margulis was correct: indecision and political tensions will give the opportunity for our wastes to destroy us. Only the first approach leads us to do something about climate change, following the path either of Wizards or of Prophets.

The first approach begins, like most scientific tales, with someone asking a question. That someone was Jean-Baptiste-Joseph Fourier (1768–1830), the nineteenth child of a tailor in Burgundy. The tailor exerted his main impact on Fourier's life by dying when he was eight. The boy and his many siblings ended up in an orphanage. Fortuitously for Fourier, a local rich person suggested that the town bishop enroll him in an academy run by Benedictine monks. Initially Fourier wanted to become a mathematics teacher at another Benedictine school. This required him to become a Benedictine. In October 1789 the French Revolution began. Fourier had not yet taken his monastic vows. Among the first actions of the new, anti-clerical government was to forbid any French person from taking monastic vows. Fourier, disappointed, returned to Burgundy, where he was arrested for insufficient revolutionary enthusiasm and sentenced to death. The revolution's leaders were executed in 1794, while Fourier was still on death row. He was released and became a math professor in Paris, intending to work on equations for the rest of his life. Instead he and two hundred other savants were drafted by General Napoleon Bonaparte to accompany his invasion of Egypt. The scientists were to study the conquered land and identify objects worth stealing. In Cairo Napoleon appointed Fourier to his new Institut d'Égypte. Unlucky for him, Napoleon (1) noticed that he was an able administrator and (2) had in the interim staged a coup and become France's sole ruler. When Fourier came home in 1801 the dictator made

Joseph Fourier, 1823

him head of a province in southeast France.

The story goes that Egypt's hot climate somehow damaged Fourier's internal thermostat. Back in France, he was perpetually cold; late in life, he wore heavy overcoats in summer and refused to leave the fire in winter. His constant chill may account for his interest in the physical question of how heat spreads, which he studied between bouts of administration. His ambition was to become the Newton of heat, the man who set down the "simple and constant laws" that explained how "heat penetrates, like gravity, all objects and all of space." He struggled with the work until the end of his life, his body so weak from neurotic cold that he often worked in a custom-built padded wooden box with holes for his head and arms.

Around 1820 Fourier asked: Why doesn't sunlight keep heating up Earth until it becomes as hot as the sun? Earth, he knew, reflects some heat back into space. But why isn't *all* of it reflected? What keeps our planet cozily warm, Goldilocks-style, and not too hot or too cold?

Fourier wrote up his conclusions four years later. Three different mechanisms account for Earth's temperature, he said. First, the sun shines on the surface, heating it. Second, the ground is warmed by the planet's molten core—leftover fire from Earth's creation, in Fourier's phrase. And third, heat could come in from outer space, which Fourier noted is irradiated by the light from "innumerable stars." Fourier thought that the starlit warmth of outer space acted as a kind of Goldilocks cap, surrounding our planet and preventing it from radiating away too much of the sun's heat.

Much of this is wrong. The temperature of outer space is not, as Fourier thought, "a little below what would be observed in polar regions." (It is actually a few degrees above absolute zero.) And Earth's core has almost no effect on the surface. Still, Fourier got the basic idea right: Earth's climate is a balance of constantly interacting forces.*

Clearer light was provided by an Irish researcher named John Tyndall. Like Fourier, he had a modest background: Tyndall, born in 1820, was the son of a cobbler. Despite having no money he was able to attend school and become a surveyor. In school he contracted the science virus. The infection worsened in adulthood until he quit his surveying job at twenty-eight and moved to Germany to study physics at the prestigious University of Marburg. There he acquired a new disease: mountain climbing. When he returned to Britain he became a professor at the Royal Institution in London. Combining his passions, he investigated glacial ice.

In northern Europe and America, geologists had uncovered gigantic, strangely placed boulders, inexplicable gravel ridges, and rock strata that had been scoured by something heavy. Slowly scientists had come to believe that these were signs of the growth and recession, eons ago, of continent-sized glaciers. To create such enormous amounts of ice, temperatures must have plunged across the globe for millennia. Nobody understood how this could have happened.

Tyndall came to suspect that the atmosphere somehow must have been involved. By this time, physicists had established that the temperature of any substance is a measure of the average energy with which its constituent atoms or molecules move, vibrate, and spin. The faster they zip and whirl, the hotter the substance. Scientists also had learned that atoms and molecules can absorb or emit light, and that this increases or decreases the energy with which they move, which in turn increases or decreases their temperature. One type of light—infrared light, invisible to the eye—is especially associated with temperature; all warm or hot objects radiate it. (The see-in-the-dark goggles in old spy movies use infrared light.) Tyndall hypothesized that the atmosphere absorbed infrared light, that this absorption governed global temperatures, and

---

* It is often said that Fourier discovered the "greenhouse effect." This is wrong for two reasons. First, Fourier never said the atmosphere acts like a greenhouse—the term *serre* (greenhouse) doesn't appear in his articles. Second, the atmosphere *doesn't* act like a greenhouse. The atmosphere is warm because it absorbs heat radiation from the surface. A greenhouse is warm because the glass physically blocks hot air from wafting away, a different process. Because many scientists whom I spoke with regard "greenhouse effect" as misleading, I avoid it in this book.

John Tyndall, ca. 1880

that in some fashion it was responsible for the ice ages.

Fourier had treated air as if it were a uniform material. Tyndall looked at its constituent gases individually, suspecting that one of them was the agent that took in infrared light and warmed the atmosphere. To answer this question, Tyndall filled long tubes with each gas, one per tube—a tube for nitrogen, a tube for oxygen, and so on. At one end of each tube he placed a hot metal box. At the other end he put a thermopile, a device that converts heat into electricity. The metal box radiated infrared light through the tube to the thermopile. Along the way, some of the infrared was absorbed by the gas in the tube; the thermopile converted the rest of the energy to electricity. The more infrared taken in by the gas, the smaller the electric current. In this way Tyndall could find which gas heated the atmosphere.

It didn't work. No matter what Tyndall did, infrared shot through the tube as if the nitrogen or oxygen weren't there. Together nitrogen and oxygen make up more than 99 percent of the atmosphere—everything else is less than 1 percent. Tyndall's apparatus was telling him that more than 99 percent of the atmosphere couldn't take in infrared radiation. If that were the case, most of the infrared radiation from Earth's surface would shoot into space like a bullet through tissue paper. Our world would be a frigid snowball, almost as cold as the moon. The question would be not why ice ages occurred, but why the world would ever be warm enough for life.

After weeks of bafflement Tyndall was about to give up when, in a what-the-hey moment, he tried coal gas, which by this time was piped throughout London and used for lighting. To his surprise, it soaked up "about 81 percent" of the cube's infrared radiation. Coal gas was colorless; like a clear window, it allowed visible light to stream through. But for infrared light, coal gas was like a frosted bathroom window, blocking most of it.

Excited, Tyndall tried a variety of other gases, including ether, perfume, alcohol vapor, carbon dioxide, and water vapor. All happily sponged up infrared radiation. Tyndall was particularly interested in the last. When he removed the water vapor from air, it absorbed about one unit of infrared radiation. But when he added just a small amount of water vapor back to the air, it "produced an absorption of about 15." This finding, announced in 1861, allowed Tyndall to assemble the first roughly correct picture of atmospheric physics. ✓

As schoolchildren learn, the sun washes Earth with every imaginable type of light wave—X-rays, ultraviolet light, visible light, infrared radiation, microwaves, radio waves, you name it. About a third of the total is reflected from clouds. Another sixth is taken in by airborne water vapor. That leaves roughly half of the incoming light—most of which is visible light, as it happens—to pass through the atmosphere. Almost all of that half is absorbed by the land, oceans, and vegetation on the surface. (A little is reflected.)

Having taken in all this solar energy, the ground, water, and plants naturally warm up, which makes them emit infrared light, radiating it into the air. Most of this secondary infrared is absorbed by airborne water vapor, heating it up. Usually water vapor comprises between 1 and 4 percent of the atmosphere by weight. (The exact number changes with temperature, wind, and surface conditions.) But this relatively small quantity—1 to 4 percent—packs a big punch.

When water vapor molecules take in infrared light, the extra energy kicks them into an "excited" state—their electrons go into a new, higher-energy configuration. (Here I am using modern language that Tyndall wouldn't have recognized; electrons weren't discovered for another three decades.) Left to their own devices, water molecules typically release this energy back into space in a few thousandths of a second in the form of infrared light. Then, like middle-aged suburbanites recovering from a fling, they settle into more stable, lower-energy states. If this was all that occurred, the atmosphere as a whole would absorb no infrared light and there would be no effect on temperature. But that isn't what happens in the air—molecules there *aren't* left to their own devices. Instead, a typical molecule in the atmosphere collides with one of its neighbors about 10 million times a second. In the few milliseconds before water molecules would release their extra energy, they may collide with nitro-

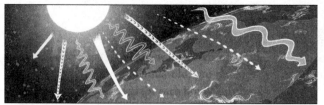

Every second of the day, the Sun emits almost every form of light—radio waves, ultra-violet light, visible light, X rays, gamma rays, microwaves, you name it. Most of it is visible infrared light, though.

Almost a third of the light that comes to the Earth is reflected by clouds, dust and the surface. Some is absorbed, notably by the ozone layer that soaks up dangerous ultraviolet light.

The rest is absorbed by land, water, and vegetation. All heat up and release most of the light energy as infrared light—a good thing, as otherwise the planet would become unbearably hot.

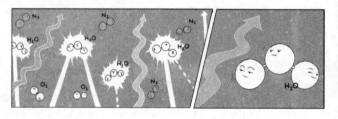

The nitrogen (N_2) and oxygen (O_2) that make up 99% of the atmosphere don't react with infrared light. But water vapor (H_2O) absorbs it—a good thing, as otherwise the infrared energy would shoot out of the atmosphere and the planet would get unbearably cold. Water vapor doesn't absorb *all* the infrared energy. Instead it lets a few frequencies pass by—that lets just enough energy escape to keep the air from getting unbearably hot.

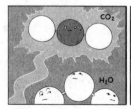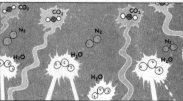

Unfortunately, carbon dioxide (CO_2) absorbs just those frequencies—a bad thing, because it closes the escape valve. Even though there is very little CO_2 in the air, there is just enough to slowly warm the Earth.

gen and oxygen molecules thousands of times. In those collisions, they transfer the extra energy they have picked up from the infrared radiation to the nitrogen and oxygen molecules, and thus increase their temperature. Another way of putting this is to say that a water-vapor molecule is like a machine that feeds infrared energy indirectly into nitrogen and oxygen molecules that can't take it in directly.

Some water-vapor molecules *do* emit infrared light, the waves bouncing down to Earth or up to the sky, only to be re-absorbed and re-admitted, or re-absorbed and the energy transferred, and re-re-absorbed and re-re-admitted, and so on. And nitrogen and oxygen molecules do dissipate their heat slowly into the cold upper atmosphere in another ramosely complicated snarl of interactions. Adding up all the pieces, the whole system ends up with just enough infrared energy being stored in the atmosphere, second by second, to keep Earth tolerably warm—and just enough leaking into outer space to prevent Earth from getting too hot. And this, Tyndall realized, was the answer to Fourier's question. Water vapor is the master switch that controls the climate. The ice ages, he thought, must somehow have been set off by changes in water vapor. These were, in Tyndall's excited, italicized words, the "true causes" of "*all the mutations of climate which the researches of geologists reveal.*"

Tyndall paid next to no attention to carbon dioxide, because so little of it was in the air. At the time, carbon dioxide comprised about .03 percent of the atmosphere by volume (the level has risen slightly since then). If somebody collected ten thousand scuba tanks of air, the carbon dioxide in them would be enough to fill up three tanks. It was hard to credit that anything so tiny could be important—as if a child's toy bulldozer could knock down a skyscraper. Carbon dioxide, Tyndall thought, was too inconsequential to have any real effect.[*]

For the rest of the nineteenth century researchers followed Tyndall's lead. Few thought airborne carbon dioxide to be of any interest. Among those few, though, were two Swedish scientists, Arvid Högbom and Svante Arrhenius. Both born in 1857, they both studied at the Univer-

---

[*] Three years before Tyndall, a U.S. scientist published a two-page paper that described how different gases absorb solar energy. The scientist was Eunice Foote, a suffragist from upstate New York. Little is known about Foote; she published just one other article, on a different subject, and no further trace of her work has been found. Foote's research was similar to but less comprehensive than Tyndall's work. No evidence exists that Tyndall knew of it.

sity of Uppsala. In 1891 both joined the Stockholm Högskola, a private think tank that later became the University of Stockholm. But their paths to the institution were different. The charming, urbane Högbom was such a successful student at Uppsala that upon receiving his doctorate he was immediately hired as a professor. The impulsive, emotional Arrhenius produced a thesis bubbling with so many novel but undeveloped ideas that his exasperated supervisors almost flunked him. Convinced that his low marks had put an end to his career, Arrhenius spent the next two years bewailing his fate and sleeping on his parents' couch. Eventually he picked himself up and took temporary jobs in German laboratories while he worked out what would become some of the fundamental ideas of physical chemistry. When these were published, Arrhenius's reputation soared; he was able to return to Sweden and take a job at the think tank with Högbom.

Högbom, a geologist, was interested in the origin of limestone. When carbon dioxide comes into contact with the ocean, it dissolves into the water. Seawater is also replete with dissolved calcium. The dissolved calcium and carbon dioxide combine to form calcium carbonate. Shellfish, coral, foraminifera (single-celled protozoa that usually live in tiny shells on the seafloor), and other aquatic organisms use this calcium carbonate to make their shells. Over time the shells pile up in drifts that are gradually compacted and turned into stone—limestone. Limestone, Högbom realized, is a storehouse for atmospheric carbon dioxide. But there were huge deposits of limestone in the earth and very little carbon dioxide in the air. Where did the carbon dioxide for limestone come from?

The biggest source of carbon dioxide that Högbom knew of was volcanic eruptions, which belch out the gas from molten limestone, coal, and petroleum. If the eruptions were really big, Högbom realized, they could jack up atmospheric carbon dioxide levels substantially. Similarly, without volcanoes carbon dioxide could become scarce in the atmosphere because it would be taken in by the sea and turned into seashells. From this he concluded that "the probability of important variations in the quantity of [carbon dioxide] must be very great." In less academic language: the air could have held much more—or much less—carbon dioxide in the past than today.

Arrhenius was intrigued. Having won a cozy academic sinecure in Stockholm, he had lost interest in slaving over test tubes in experiments.

Instead he had taken to lofty speculation about other people's data. In this
way he came up with new theories
of the formation of the solar system,
the age of the universe, and the inner
mechanics of the sun. All of these were
later proven wrong by people who
actually did experiments. Learning
of Högbom's work, Arrhenius wondered if his carbon dioxide data could
explain the ice ages. Could the glaciation have been set off by lower carbon
dioxide levels?

To answer the question, Arrhenius
decided to estimate what the effects
would be if the concentration were

Svante Arrhenius, 1909

doubled or halved. The calculation was laborious. Because of their dif-
fering latitudes and cloud covers, different parts of the world receive
different amounts of sunlight at different times of the year. Arrhenius
ended up calculating the average for each season in seven-hundred-mile
bands from the equator to the poles. A U.S. scientist had measured which
light wavelengths are absorbed by water vapor and carbon dioxide, and
which pass through. (The wavelength is the distance between succes-
sive crests of a light wave.) Different wavelengths have different energies,
which means that they affect temperatures by different amounts. Arrhenius had to include this factor, too. Snow, water, and soil don't reflect
the same amount of light; Arrhenius plugged these variances into his
calculations. Et cetera, et cetera—one complicating factor after another.

He began his "tedious calculations" on Christmas Eve of 1894. He
had just married his laboratory assistant; she left him, pregnant with
their child, while he was still bent over his desk, and went to live alone
on a remote island. Tens of thousands of calculations later, Arrhenius
finished in December 1895. "It is unbelievable that so trifling a matter
has cost me a full year," he complained to a friend. He didn't mention
his wife.

The results were worth the effort, if not, perhaps, the marriage. Arrhenius believed that he had established a remarkable fact: tiny changes in

airborne carbon dioxide could cause an ice age. Indeed, he said, halving the level of carbon dioxide—reducing it from .03 percent to .015 percent—would cool the world by about 8°F, more than enough to set off the glaciers. Högbom had pointed out that burning fossil fuels must be increasing atmospheric carbon dioxide. Arrhenius estimated that doubling carbon dioxide levels would increase Earth's average temperature by as much as 11°F, enough to turn most of the planet into a desert.

Arrhenius wasn't worried by this prospect. He thought it would take thousands of years to reach 11°F, if that ever happened. Meanwhile, he lived in frigid Sweden; rising temperatures seemed like a fine idea. In the future, Arrhenius predicted, "our descendants [will] live under a warmer sky and in a less harsh environment than we were granted."

His colleagues were even less worried, because they thought he was talking through his hat. A chief source for the disbelief was one of Arrhenius's longtime acquaintances, Knut Ångström. The son of a prominent physicist, Ångström attended Uppsala at the same time as Arrhenius and arrived at the think tank with him. But his long relationship with Arrhenius did not stop him, in 1900, from tearing into the other man's work.

Ångström focused on Arrhenius's depiction of the interactions of light and carbon dioxide. As children learn in high school chemistry, atoms are, so to speak, picky about which wavelengths of light they absorb and emit. They will interact with some wavelengths, but not others. In the nineteenth century, physicists discovered that every substance's pattern of absorption and emission identifies it as surely as a fingerprint.* Examining the patterns for carbon dioxide and water vapor, Ångström realized that they took in many of the same wavelengths. Because the atmosphere contains so much more water vapor than carbon dioxide, any absorption of those wavelengths would be almost entirely due to water vapor. Which meant that Arrhenius's whole carbon dioxide scheme was wrong—"the observations cannot be treated as Mr. Arrhe-

---

* Today we know that is because an atom's electrons surround it in complex "orbitals." An atom can absorb an incoming light wave only if it has exactly the right energy to kick the atom's electrons from one orbital to another, more energetic orbital. Any more or any less and the atom won't take it in, because the light wave's energy doesn't match the energy between orbitals. All of this baffled nineteenth-century physicists: How could atoms be so choosy? Resolving the puzzle led in the early twentieth century to the revolution of quantum mechanics.

nius did." Just as Tyndall had thought, water vapor controlled atmospheric heat and carbon dioxide was irrelevant.

With Arrhenius's carbon dioxide hypothesis seemingly disproved, scientists advanced other ideas about the origin of the ice ages. They were due to fluctuations in the brightness of the sun. To the uplift of mountain ranges. To continents moving about Earth's mantle. To volcano dust. To the solar system passing through colder bits of space. To ocean currents. To collisions with extraterrestrial icebergs. Arguments were long, bitter, inconclusive.* One of the few points of agreement was stated in 1929 by George Clarke Simpson, director of the British Meteorological Office: it was "now generally accepted that variations in carbon-dioxide in the atmosphere, even if they do occur, can have no appreciable effect on the climate."

Outsiders

Guy Callendar didn't agree. Born in 1898, Callendar was the son of Britain's leading steam engineer and grew up in an ivy-covered, twenty-two-room manse in a fashionable part of West London. He became his father's assistant and then his successor. Intrepid and curious, he was willing to delve into fields he knew nothing about, atmospheric science among them. Nobody knows why climate interested him. Possibly he simply wondered why winters had become warmer since his boyhood. Callendar himself attributed it to ordinary curiosity: "As man is now changing the composition of the atmosphere at a rate which must be very exceptional on the geological time-scale, it is natural to seek for the probable effects of such a change."

In the early 1930s Callendar began collecting measurements of the properties of gases, the structure of the atmosphere, the sunlight at different latitudes, the use of fossil fuels, the action of ocean currents, the temperature and rainfall in weather stations across the world, and a host of other factors. In effect, he was trying to redo Arrhenius's calculations. But Callendar had an advantage. Arrhenius had been forced to guess at

* Most researchers now think that the ice ages were caused by slight shifts in the earth's tilt and orbit, which changed the amount and distribution of sunlight on the surface, cooling the planet.

Guy Callendar, 1934

many values, because they had not been measured. His articles were therefore full of hedge words: "it is, therefore, justifiable to *assume*"; "I do not know if this [factor] has ever been measured, but it *probably* does not differ"; "it *seems* as if"; "I *have convinced myself* that by this mode of working no systematic error is introduced" (emphasis mine). Callendar had four more decades of data. He produced the first rough draft of the huge climate models familiar today.

Central to his work were more-precise measurements of carbon dioxide and water vapor. As the reader will remember, the ground absorbs solar energy and then sends most of it back into the air in the form of infrared radiation. The majority of this outgoing infrared energy is caught by water vapor and transferred to the rest of the atmosphere. Just enough leaks out to prevent the atmosphere from heating to unbearable levels.

Two mechanisms are responsible for the escape. The first is that the water vapor releases some of its absorbed energy as infrared radiation, and some of that released infrared beams into outer space (it is re-absorbed and re-emitted by water vapor many times along the way, but eventually passes beyond the atmosphere). The second is that water vapor doesn't absorb *all* of Earth's infrared radiation—it is effectively transparent for certain wavelengths. New measurements, Callendar learned, showed that the most important of these "windows" occurs at wavelengths around 10 micrometers—that is, water vapor lets through light waves that are about ten millionths of a meter long. Another prominent window occurs at 4 micrometers.

Callendar also learned that scientists had more precisely measured the wavelengths absorbed by carbon dioxide. Contrary to Ångström, they did *not* overlap precisely with water vapor. Quite the opposite: carbon dioxide absorbs some of the wavelengths that water vapor lets

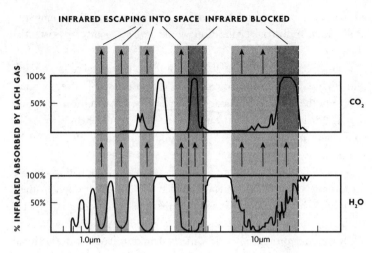

The infrared spectrum, from wavelengths of about 1 μm (1 micrometer) to those of about 10 μm. Water vapor (H_2O, bottom half of chart) in the air absorbs most of the infrared radiation from the surface. But there are gaps (light shaded areas) in the spectrum that it lets through—a safety valve that prevents Earth from overheating. Carbon dioxide (CO_2, top half of chart) absorbs only a few bands of infrared. By chance, two of them (dark shaded areas) sit in the water vapor gaps. They absorb some of the infrared radiation that water vapor lets through. The effect is to reduce—by just a bit—the amount of infrared radiation that escapes into space.

through—it shuts the windows. The more carbon dioxide in the air, the more of this radiation it can absorb.

In Callendar's scenario, the atmosphere is like a bathtub. Water pours into the tub in the form of infrared radiation. In the tub are small holes—the "windows" through which water vapor allows infrared radiation from the surface to pass. Because the outflow from the holes is approximately equal to the inflow from the spigot, the water level in the tub is constant. Now block a hole or two with chewing gum. That is like adding carbon dioxide to the air. Inevitably, the water rises.

From the human point of view, this is stupid bad luck. If the physical properties of carbon dioxide and water vapor didn't intersect in this way—if carbon dioxide didn't happen to absorb the infrared radiation that water vapor lets through—then burning fossil fuels would be of little interest to climate researchers. The carbon dioxide rise would be viewed as a dusty corner of atmospheric science, the province of pedants. Coal and oil could be burned without worry (after removing pollutants). Industrial civilization would not be facing an existential challenge.

Neither Callendar nor anyone else understood the stakes as we think of them today. That carbon dioxide blocks the outflow of infrared radiation from the earth—Callendar, like Arrhenius, thought this was a good thing. "Small increases in mean temperature" would help farmers in cold places, he said. Better yet, they would "indefinitely" postpone "the return of the deadly glaciers."

In making this argument, Callendar was effectively telling climate professionals that he, an outsider, had made a breakthrough in their discipline—one that they had wholly missed. This did not go over well. He had enough academic status to be allowed to present his ideas to the Royal Meteorological Society in 1938, and to have six professional climate scientists comment on them. But he did not have enough standing to stop the commenters from being condescending (they praised his "perseverance"). Years before, British Meteorological Office head George Clarke Simpson had stressed the consensus that carbon dioxide had "no appreciable effect on the climate." Now he was one of Callendar's commentators. The problem with Callendar's work, he sniffed, was that "non-meteorologists" simply didn't know enough about climate to be helpful.

Two criticisms from the panel were more substantial. The first was that Callendar hadn't shown why carbon dioxide from fossil fuels wouldn't be absorbed by the ocean rather than remain in the atmosphere. The second was that all measurements of atmospheric carbon dioxide were unreliable, because scientists' instruments could be affected by nearby car exhausts, factories, farms, and power plants. Although Callendar had spent years gathering data, the assembled meteorologists were "very doubtful" that the data meant anything.

Undiscouraged, Callendar kept working on climate until his death in 1964. And slowly—very slowly—climatologists began to give his ideas a hearing.

They were almost forced to, because after the Second World War a torrent of new researchers poured into atmospheric science, most of them funded by the U.S. military. During the war, the military had used unfamiliar types of radiation to great effect—infrared light for signaling and sniping, microwaves for radar detection of enemy aircraft. To develop these techniques further, the armed services wanted to know everything about unusual types of light and how they interacted with the atmosphere. Strategic Air Force commander General George C. Kenney laid out his dreams in a speech at MIT in 1947. "Below the infrareds and above the ultraviolets," he said, "there may be weapons of future warfare as devastating as the atomic bomb." For example, "An airplane equipped with a sort of super dog whistle could fly around a city for a while and upset the nervous systems of the whole population." Even more exciting to Kenney, the interaction of radiation and the atmosphere suggested the possibility of understanding and directing the weather: "The nation that first learns to plot the paths of air masses accurately and learns to control the time and place of precipitation will dominate the globe." Buoyed by visions of "climatological warfare," the military funded the computer pioneer John von Neumann's plan to create the first digital simulation of the atmosphere.

The gush of Pentagon money shifted the center of climate science from Europe to the United States—an impact that could be seen in 1957, when three California-based scientists launched two projects that answered Callendar's critics. The first project came from Hans E. Suess and Roger Revelle, of the Scripps Institution of Oceanography, in San Diego. Like Callendar, they were climate outsiders: Suess was an Austrian physicist who had worked on the fringes of the failed Nazi effort to make an atom bomb before immigrating to the United States; Revelle, an oceanographer, was director of Scripps, the most important institution in his field.

Thanks to his physics background, Suess had figured out that the recently invented technique of radiocarbon dating could be used to distinguish fossil-fuel carbon from other types of carbon.* In turn Revelle

---

* About radiocarbon dating: Earth is constantly bathed by a rain of high-energy subatomic particles from outer space. When these "cosmic rays" slam into a nitrogen atom, the violent collision can change the nitrogen into a mildly radioactive form of carbon: carbon-14 (^{14}C), as scientists call it. By happenstance, ^{14}C disintegrates into a form of nitrogen at almost exactly the same rate that it is created by cosmic rays. As a result, a small, steady percentage of the carbon in the air, sea, and land consists of ^{14}C. Plants take in ^{14}C through photosynthesis. When animals eat plants, they take it in, too. In consequence, every living cell has a steady, small level of ^{14}C—they are all slightly radioactive. When organisms die, they stop absorbing ^{14}C. Because the ^{14}C in their cells continues to disintegrate, their bodies' ^{14}C level falls in a predictable way that researchers can use to estimate when the creatures were alive. Suess realized that

realized that Suess's technique could be used to learn whether the ocean, as Callendar's critics had suspected, was taking in the carbon dioxide from fossil fuels. Working together, the two men determined that the oceans indeed were absorbing most of it. But only after they drafted a paper to this effect did they grasp that pumping carbon dioxide into the sea set off other interactions that ended up with the water quickly releasing much of the gas it had initially absorbed. In the end, the ocean *didn't* soak up the carbon dioxide emitted by coal, oil, and gas. Hastily tacking on a paragraph to the conclusion of their article, Revelle and Suess conceded that burning fossil fuels amounted to "carrying out a large-scale geophysical experiment of a kind that could not have happened in the past."

The second project, complementary to the first, was executed by Charles D. Keeling, another Scripps researcher. Born in 1928, Keeling was the son of a Chicago banker who became convinced that bankers like him had caused the Great Depression. The father quit his job and became a proselytizer for banking reform, Keeling wrote in an autobiographical essay, "thereby plunging our family into the poverty he was distressed about." The son studied chemistry at the University of Illinois, but switched to liberal arts to avoid a required course in economics—"I felt quite passionately that my exposure to economics at home had been enough." After further confusion, Keeling returned to chemistry, obtained a Ph.D., fell in love with geology, and wangled a geochemistry fellowship at the California Institute of Technology, in Pasadena. An offhand remark from his supervisor in 1956 set up the rest of his life.

Keeling's supervisor speculated that the carbon dioxide in water was in equilibrium with the carbon dioxide in the air—if you increase the amount of one, the other will quickly compensate to balance it. Keeling decided to find out if this was true. He was bored with his desk job and liked the idea of performing measurements outdoors, in the mountains near Pasadena. To his surprise, his data jumped all over the place—sometimes there was more carbon dioxide in the air, sometimes less. Keeling realized that "emissions from industry, car exhaust, and backyard incinerators" were blowing by his instruments—exactly the

---

carbon dioxide from burning petroleum would have almost no ^{14}C because the organisms that had created it had died millions of years ago. It would thus be possible to detect if fossil fuels were pumping this ^{14}C-less carbon dioxide into the environment—the overall ^{14}C level would be a hair lower than it would be otherwise.

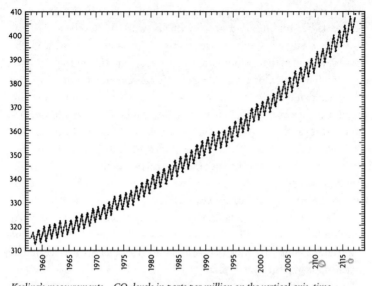

Keeling's measurements—CO₂ levels in parts per million on the vertical axis, time on the horizontal axis—were precise enough to observe slight seasonal fluctuations of carbon dioxide. As Northern Hemisphere plants grow in the summer, their increased photosynthesis sucks some of the gas from the air, regularly reducing its atmospheric concentration.

problem Callendar's critics had identified. To determine the actual level, he would need to collect long-term data in a contaminant-free place. The U.S. Weather Bureau agreed to sponsor Keeling in an effort to measure carbon dioxide on Mauna Loa, a thirteen-thousand-foot volcano in Hawaii. Because prevailing winds were from the west, the nearest carbon dioxide sources were thousands of miles away in Asia.

In the interim, Revelle had learned about Keeling's initial work and invited him to Scripps. Keeling moved there and assembled the world's first high-precision gas-analysis system. He exasperated Revelle by constantly tinkering with the design and asking for more money; Keeling, Revelle thought, was fixated on achieving a meaningless degree of precision, pointlessly piling on the decimal points. Measurement began in February 1958. Within two years his instruments showed that the world's store of airborne carbon dioxide had increased in that period from about 313 parts per million to about 315 parts per million.

Keeling worked on Mauna Loa from 1958 until his death in 2005, during which time the proportion of carbon dioxide in the air rose to 380 parts per million. Combined with the work by Revelle and Suess, Keeling's meticulous, decades-spanning measurements convinced climate researchers that carbon dioxide was accumulating in the air.

But what, if anything, would be the effects? On an absolute scale, the carbon dioxide rise was minuscule—a few parts per million. Any rise in temperature, most researchers thought, would occur only in the distant future, and could be beneficial. Revelle and Suess had concluded that humanity was conducting an experiment, but thought it unlikely to amount to much. In a few decades, they wrote in their article, enough data might accumulate to "allow a determination of the effects, *if any*, of changes in atmospheric carbon dioxide."

"The world is growing slightly warmer," *The New York Times* reported one Sunday in 1959, noting that most scientists believed that "the warming trend" is not "alarming or steep." Befitting the lack of concern, the article appeared on page 112.

Moral Interlude

In the spring of 2016, my friend Rob DeConto published an article in the scientific journal *Nature*. By chance, our families had dinner together a day after its release. Rob was both cheerful, because he had just finished a long project, and uneasy, because of the implications of that project. With a colleague, David Pollard, he had spent years examining the Antarctic sheet, by far the world's largest ice mass. Past researchers had thought that because of its size it would respond slowly to rising global temperatures. To their dismay, DeConto and Pollard had realized that Antarctica might be more vulnerable than previously thought.

Increasing temperatures would attack the ice in two ways: warmer air would melt it from above, forming pools on the surface, and warming ocean currents would eat at the underside of the sheet, creating large cracks. The pools on the surface could drain through the cracks, widening them and splitting the ice sheet into unstable pieces that would fall apart under their own weight. The remaining chunks, surrounded by warm water and air, would melt quickly, like the ice cubes in a cocktail. If the two men were correct, melting Antarctic ice could *by itself* raise the world's oceans more than three feet by 2100, enough to swamp Miami, Tokyo, Mumbai, New Orleans, and many other cities. By 2500 the rise could be as much as fifty feet.

Every few years the Intergovernmental Panel on Climate Change (IPCC), a United Nations–sponsored international scientific consortium, issues a set of lengthy, multivolume reports that attempt to portray the state of the science on climate change. In 2013 it put together a consensus projection of how sea level would be affected by a variety of factors, including the melting of glaciers in Greenland, the shrinking of the Arctic ice cap, and the fact that water expands as it gets warmer (the increase is slight, but it adds up when extended across the whole globe). Taken together, the IPCC estimated, these would push up the sea roughly twenty-eight inches by 2100. Antarctica, the biggest ice mass, would remain more or less intact, contributing only one to three inches to the total. This idea seemed borne out by the facts—in the previous few decades, Antarctica had shown little sign of melting. Indeed, parts of the ice sheet were growing. But the IPCC also thought that scientists needed to look more closely, just to be sure. Now DeConto and Pollard had performed that analysis, and they were saying that there was a chance that Antarctica could fall apart by 2100, and that therefore the seas could rise higher and faster than the IPCC had imagined possible.

"For a geophysicist, what's going on is stunning," my friend told me. "We used to believe these systems needed thousands of years to make these shifts. Instead it's happening so fast that it's terrifying. Conceivably, you could start seeing truly bad effects in *a hundred years*."

And *this* is one of the great difficulties in thinking about climate change: what seems terrifyingly fast on the geological scale is unfathomably long on the human scale. By "truly bad effects" DeConto meant flooded coasts, vanished islands, awful droughts, and, maybe, storms of unprecedented power. But even if these occur in the time he fears—even if they transpire in the geologically insignificant span of a century—they will not be seen by him or me. Quite possibly every person who reads this book will be dead before they occur, as will most of their children. How many governments make plans for such long-term contingencies? How many families do?

The most likely victims of climate change, in the short run, are people who live on oceanic islands, in very low-lying coastal settlements, in ice-bound Arctic communities, and around forests that burn after unwonted dry spells. Millions of people live in these places, but they are a small fraction of the world's billions. The greatest potential harms of climate change will be experienced by future generations—centuries in the future, or even millennia. By our actions today (burning fossil fuels), the argument is, we are dumping problems (drought, sea-level rise) on tomorrow.

On the one hand, forcing other people to clean up our mess violates basic notions of fairness. On the other hand, actually preventing climate-change problems would require societies today to make investments, some of them costly, to benefit people in the faraway future. It's like asking teenagers to save for their grandchildren's retirement. Or, maybe, for somebody else's grandchildren. Not many would do it.

Would they be wrong? How much concern *should* we have for future generations? The harder one looks at the problem, the more confusing it seems. "The problems of climate change," says the New York University philosopher Dale Jamieson, "swamp the machinery of morality." ("A perfect moral storm," says Stephen M. Gardiner, another philosopher.) The basis for arguing for action on climate change is the belief that we have a moral responsibility to people in the future. But this is asking one group of people to make wrenching changes to help a completely different set of people to whom they have no tangible connection. Indeed, this other set of people *doesn't exist*. When one tries to make plans for nonexistent people, the result is an intellectual quagmire, because there is no way to know what those hypothetical future people will want.

Today we live in a world where almost everywhere slavery is illegal, women can vote and own property, and overt obeisance to social class is frowned upon. Most decision-makers who lived three hundred years ago would have regarded these developments with horror. Had they grasped that the future could be like this, they would have sought to prevent it.

Picture Manhattan Island in the seventeenth century. Suppose its original inhabitants, the Lenape, could determine its fate, in perfect awareness of future outcomes. In this fanciful situation, the Lenape know that Manhattan could end up hosting some of the world's great storehouses of culture—the Metropolitan Museum of Art, the American Museum of Natural History, the Museum of Modern Art, the New York Public Library, the opera and symphony at Lincoln Center. All will give pleasure and instruction to countless people. But the Lenape also know that creating this cultural mecca will involve destroying a diverse and fecund ecosystem. I suspect the Lenape would have kept their rich, beautiful homeland. If so, would they have wronged the present?

Economists tend to scoff at these conundrums. Forget all of this philosophical foofaraw about the rights of hypothetical people, they say. That's just a smokescreen for "paternalistic" intellectuals and social engineers "imposing their own value judgments on the rest of the world." (I am quoting the Harvard economist Martin Weitzman.) Instead, economists suggest, one should observe what people actually *do*—and respect that. In their daily lives people care most about the next few years and don't take the distant future into much consideration—they prefer "present over future utility," in the economist's phrase.

In technical terms, this idea is expressed with the *discount rate*, which is like the antimatter version of an interest rate. Imagine that today I would pay $200,000 for a new house. How much would I pay today if I had to wait five years to receive the house? $100,000? $50,000? Today and tomorrow, the house is the same physical object. But as a rule people won't pay as much for a house they have to wait for as a house they can occupy immediately, and the longer they have to wait, the less they are willing to pay. Usually economists use 5 percent as a discount rate— for every year of waiting, the price goes down 5 percent, compounded. Doing the arithmetic, a 5 percent discount rate means that goods and services are worth roughly half as much to me in fifteen years as they are today.

The implications for climate change are both striking and, to many people, absurd. At a 5 percent discount rate, the Argentine-American economist Graciela Chichilnisky has calculated, "the present value of the earth's aggregate output discounted 200 years from now is a few hundred thousand dollars." In econospeak, "the earth's aggregate output" means "the human race and all its works." To prevent our species from being wiped out in two centuries, Chichilnisky points out, standard economics suggests that the world would pay "no more than one is willing to invest in an apartment."

Intuitively, I am hard-pressed to believe that most people would endorse the notion that the future of humankind is worth no more than a single apartment. Chichilnisky, a major figure in the IPCC, has argued that this kind of thinking about discount rates is not only ridiculous but immoral; it exalts a "dictatorship of the present" over the future. Economists could retort that people *say* they value the future, but don't *act* like it, even when the future is their own. And it is demonstrably true that many—perhaps most—men and women *don't* set aside for retirement, buy enough insurance, prepare their wills, or a hundred other precautions, even if they have sufficient resources. If people won't make long-term provisions for their own lives, why should we expect people to bother about climate change for strangers many decades from now?

Not so fast, says Samuel Scheffler, a New York University philosopher who is the author of *Death and the Afterlife* (2013). The way people feel and act about their *individual* futures is not the same as how they feel and act about our species's *collective* future. In his book, Scheffler discusses another book: *Children of Men*, a best-selling 1992 science fiction novel by P. D. James that was adapted into a movie by the filmmaker Alfonso Cuarón. The premise of both book and film is that humanity has suddenly become infertile, and our species is stumbling toward extinction. In this scenario, as Scheffler notes, nobody alive is worse off, at least in the short run. Couples are denied future children, but they lose no children they already have; nobody even loses money. The present doesn't change materially. All that is lost is a future that we would never have seen.

Both book and movie show this world as one of anomie and despair. Because our species has lost its future, life seems meaningless. Civilization wanes; aimless, violent gangs roam ruined streets. The belief that human life will continue, even if we ourselves die, is one of the underpinnings of society.

Logically speaking, the desolation in *Children of Men* is peculiar. As Scheffler points out, all people have known from childhood that they will die. As individuals, we have no long-term future. Personal extinction is guaranteed. But this tragedy—one that will be directly experienced by every single man, woman, and child—provokes no public alarm. No tabloid has ever blared the headline, "All 7.3 Billion of Us to Vanish Within Decades." Our conviction that life is worth living is "more threatened by the prospect of humanity's disappearance than by the prospect of our own deaths," Scheffler writes in *Death and the Afterlife*.

The idea is startling: the existence of hypothetical future generations matters more to people than their own continued existence—

"evidence of hitherto unsuspected reserves of altruism," as Scheffler drily comments.

What this suggests is that, contrary to economists, the discount rate accounts for only part of our relationship to the future. People *are* concerned about future generations. Even if the logic is hard to parse, they think that humanity's fate is worth more than an apartment. But trying to transform this general wish into specific deeds and plans is confounding.

Imagine a ladder of moral concern that begins with an exclusive concern for oneself and extends through concern for family to concern for culture or religion to concern for all cultures and religions and beyond that to future generations. At the far end is the Margulis position, a concern for a natural order that is so all-encompassing that it is hard to distinguish from unconcern. It is a philosophical truism that exclusively caring about oneself is not a route to a happy or satisfying life. Another philosophical truism is that a lofty concern for all of existence is the province of saints, and sainthood is not required for ordinary people to live decently and well.

In the middle, where most people spend their days, it is hard to distinguish morally between positions. It is easy to disparage people who think only of their family or neighborhood. But higher up the ladder is not necessarily better—think of the numberless instances where people, genuinely believing that they are acting for the benefit of larger entities, have ended up doing awful things. Would the world have been better off if the soldiers in the Crusades had not tried to spread the light of Christianity and instead had stayed home and improved their own villages? Or, again, consider Manhattan Island. If the level of concern is for preserving all cultures, it might make sense to bulldoze all the buildings and return the land to the Lenape. After all, there are many wonderful centers of Western culture; there is only one Lenape homeland. But removing the millions who live in Manhattan could not be done without creating terrible hardship and dislocation on the level of the community, quarter, family, and individual.

In addition, reaching for higher levels on the ladder of concern is more complex and difficult. If nothing else, the many misadventures of foreign aid have shown how difficult it is for even the best-intentioned people from one culture to know how to help other cultures. Now add in all the conundrums of working to benefit people in the future, so inherently unknowable, and the hurdles grow higher. Thinking of all the necessary actions across the world, decade upon decade—it freezes thought. All of which indicates that although people are motivated to reach for the upper rungs, they are more likely to succeed in their aspirations if they stay rooted to the lower, more local ones.

Margulis, I suspect, would have put this in biological terms. Evolution has provided the human brain with marvelous tools for detecting and resolving fast-moving, clearly visible, small-scale, near-future risks. By the same token, the brain is easily overwhelmed by slow, abstract, large, long-term problems. Patiently dribbling money, year by year, into a retirement account, calculating the insurance needed to compensate for an unlikely but real hazard, contemplating the arrangement of one's death—they boggle the mind. (Most people's minds, anyway.) Climate change is all of these and more: gradual, impalpable, world-altering, multigenerational, a situation that will not become readily tangible until irreversible lines already have been crossed. "It is not the sort of problem that Mother Nature raised us to solve or even notice," Jamieson, the philosopher, has written.

None of these considerations seem to have occurred to the early researchers on climate change. But they help to explain the reactions to their work later on.

"There Will Be Megadeaths"

The small band of climate researchers was riveted by the discovery—reported, however confusingly, by Revelle and Suess—that human-released carbon dioxide could affect the climate. Figuring out exactly what would happen, though, turned out to be vexingly difficult, because climate was a morass of feedback mechanisms.* If higher carbon diox-

---

* Feedback in this sense occurs when the output of a system gets fed back into the system, affecting the next output. An example is the way an opera audience's bravos can affect the performers. If the crowd cheers every time a soprano stops the show with a high C, the singer might, buoyed by approval, try to hit more high Cs. Enough feedback, and the performance will be nothing but flashy high notes. This is *positive* feedback: the feedback increases the change until it reaches a new state. If the audience boos, the soprano might not try any more high Cs and just sing normally—*negative* feedback, when the feedback reduces the change. The back-and-forth between singer and crowd is a *feedback loop*.

ide levels made the atmosphere grow warmer, for example, it would become more humid. On the one hand, moister air would absorb more heat, further driving up temperatures: a positive feedback loop. On the other hand, moister air would lead to more cloud cover, which would block the sun, lowering temperatures to what they were before: a negative feedback loop. Similarly, higher temperatures could melt the ice in glaciers and at the poles, leaving bare rock. Because the rock is darker than the ice, it would soak up more of the sun's heat, raising the temperature and melting more ice to expose more rock: positive feedback. But the cold meltwater from the glaciers would pour into the oceans, lowering their temperature, which would chill the air over the water: negative feedback. The permutations were endless, and adding them up was stunningly difficult.

Worse, the tangle of interacting feedback loops meant that climate was subject to the "butterfly effect." The term refers to a now-famous metaphor for how tiny changes in complex systems can have wildly disproportionate effects. It originated in 1972, when the MIT meteorologist Edward Lorenz asked a conference, "Does the flap of a butterfly's wings in Brazil set off a tornado in Texas?" His answer: Well, *yes*, actually—maybe.

By the time Lorenz asked about butterflies, he had been creating computer models of weather and climate for two decades. His biggest breakthrough had come while tinkering with a weather model in 1961. It consisted of twelve equations that expressed relationships between variables like temperature, humidity, barometric pressure, and wind speed. Although the model was of toy-like simplicity, Lorenz's computer—a clunky mass of vacuum tubes programmed with punch cards—printed out results that looked surprisingly like actual weather, with storm fronts shifting, winds blowing, and temperatures rising and falling. At one point Lorenz decided to check a calculation by repeating it. The machine made computations out to six decimal places, but for the sake of simplicity Lorenz wrote only the first three—he abbreviated, say, 0.111111 to 0.111. He fed the truncated numbers into the machine and was startled to obtain a totally different answer than the first time. The two computer runs were similar at the beginning, but then the output lines drew apart until they looked completely different. A tiny difference in initial conditions had dramatically changed the outcome.

Lorenz was baffled. Cutting off a few decimal places was a trivial change; it shouldn't have a big effect. It was as if he had repeated a three-day automobile test drive but changed the initial speed from 100 miles per hour to 100.1 miles per hour—and ended up completing the trip in twice the time on a different route. Surely he was seeing some glitch in the computer.

After a year of digging, Lorenz realized that the computer was right and his intuition had been wrong. He proved that the type of equations one would use in a model of daily weather—equations that described convections or flows of various types—were unavoidably sensitive to small initial changes. And the sensitivity applied not only to weather, but to climate as well. (Weather and climate are not the same. Weather is the daily local ups-and-downs, climate the overall atmospheric system.) People had known for centuries that entire decades could be much drier or wetter than their predecessors. But they had believed that these changes were the result of predictable, regularly occurring cycles—the increase and decrease of sunspots, perhaps, or the oscillation of ocean currents. Instead, Lorenz was saying, climate was more like a random walk, an unstable trajectory driven by trivial variations. Indeed, Lorenz couldn't "prove that there existed a 'climate' at all, in the traditional sense of a stable, long-term statistical average." The quotation comes from the historian Spencer R. Weart's 2008 book, *The Discovery of Global Warming*. At the time, Weart explained, the task of climate scientists

> was to compile statistics on past weather, in order to advise farmers what crops to grow or tell engineers how great a flood was likely over the lifetime of a bridge. . . . Yet the value of this kind of climatology to society was based on the conviction that statistics of the previous half century or so could reliably describe conditions for many decades ahead. Textbooks started out by describing the term climate as a set of weather data averaged over temporary ups and downs—it was stable *by definition.*

Now Lorenz had challenged the most basic idea in the discipline.

Climate scientists encountered Lorenz's ideas in 1965, when he gave the keynote address at a conference in Colorado called "The Causes of Climate Change," the first big scientific gathering devoted to the subject.

As he described the instability he had uncovered, his audience made the connection with carbon dioxide. Conference organizer Roger Revelle, who had been skeptical, was persuaded. If small changes in initial conditions could have enormous long-term effects, he said in a summary speech, then perhaps tiny rises and falls in atmospheric carbon dioxide could "'flip' the atmospheric circulation from one state to another." Arrhenius and Callendar had been vindicated. A scientific consensus was emerging: a tiny shift in the atmosphere's carbon dioxide load could make Earth hard to live on. And Keeling had shown that carbon dioxide levels were rising in exactly the way that might lift temperatures to new heights. Revelle was then on a panel charged by the U.S. president with writing a report about environmental pollution. He took advantage of the position to create a subpanel on carbon dioxide and write the first-ever official government report about the possibility of climate change.

That didn't end the matter. Few people care about rising temperatures per se. What matters, as Revelle, Suess, Keeling, and their colleagues realized, is their potential future influence on other things: agricultural productivity, sea levels, rainfall patterns, ocean chemistry, infectious disease. And nobody truly knew what these might be. Climate scientists had already engaged in some of the most elaborate calculations ever performed just to understand basic atmospheric physics. Now they would have to add agronomy, oceanography, disease ecology, and a host of other fields—a monumental task.

Little of this process was evident to the public or even other researchers. Climate science was a new field that relied on gigantic computer simulations of a sort unfamiliar outside the climate community. Often its practitioners had drifted into climate from other disciplines and worked not in traditional universities but at specialized institutions—Scripps, for instance, or the Joint Numerical Weather Prediction Unit (a collaboration of the Pentagon and the U.S. Weather Bureau that was disbanded in 1960)—with only loose connections to the rest of academia. Later these stand-alone climate centers appeared all over the world, including the Center for International Climate and Environmental Research in Oslo, the Wuppertal Institute for Climate, Environment, and Energy in western Germany, and the National Center for Atmospheric Research, the government-funded Colorado research organization that hosted "The Causes of Climate Change." For their part, more than a few university scientists regarded climatologists as Johnny-come-latelies who exaggerated their importance to snatch more funding for their jet-setting international conferences and glittering laboratory centers (the National Center for Atmospheric Research headquarters was designed by the star architect I. M. Pei). And they scoffed at how few climatological theories seemed to be testable in traditional scientific experiments. Climatologists retorted that a second Earth isn't available for testing climate theories. Instead, they were forced to refine ever more complex mathematical models, which grew ever harder for outsiders to understand. Later, as ever bolder claims about the effects of global warming issued from climate centers, some university-bound economists, ecologists, sociologists, and historians charged that the centers, stocked mainly with physical scientists, ignored or misrepresented the economic, social, ecological, and historical aspects of the subject, leading to a new version of the old environmental determinism.

Adding to the confusion, some climate scientists were warning of global *cooling*. In 1963 Reid Bryson, founding head of the University of Wisconsin meteorology department, flew to India. As the plane neared its destination he was surprised to see that smoke and dust entirely blocked his view below. The smoke came from farmers burning fields, the dust from wind blowing over drought-stricken ground. Adding to the soup was coal smoke from India's new factories. Bryson knew that microscopic particles of smoke, soot, smog, and dust—"aerosols," in the jargon—should scatter the sun's rays before they hit the surface. An obvious question: Could this affect the climate? At a 1968 symposium Bryson and others suggested that aerosols from human activity might cool Earth faster than carbon dioxide could heat it. Carbon dioxide researchers disagreed. In typical fashion, cooling and warming advocates split into two factions, each regarding the other as misguided.

Two scientists at the National Aeronautics and Space Administration, S. Ichtiaque Rasool and Stephen H. Schneider, tried to resolve the conflict by adding aerosols into the models used for carbon dioxide. A third NASA researcher, James E. Hansen, had developed a model to study the cloudbanks of Venus. Rasool and Schneider adapted Hansen's model to examine the smogbanks of Earth. In 1971 they published their conclusion in the journal *Science*: doubling atmospheric carbon dioxide levels would have little impact, but a sustained air-pollution increase would "trigger an ice age." Rasool went further in *The Washington Post*. If pollution kept increasing, he told a startled reporter, the next ice age could arrive in "five to ten" years—the glaciers would begin growing by 1981.

Coming as the environmental movement rose into public view, the image of humanity literally blocking the sun with its filth was apocalyptic yet accessible—and, for the next few years, irresistible. "Brace Yourself for Another Ice Age," advised *Science Digest*. "What's Happening to Our Climate?" asked *National Geographic*; it quoted two scientists warning that if pollution didn't stop, "continental snow cover would soon advance to the Equator." *Newsweek* invoked "A Cooling World." The futurist Lowell Ponte published *The Cooling* (1976), which predicted that freezing temperatures would destroy the Soviet grain harvest, setting off World War III. "There will be megadeaths," intoned George F. Will in *The Washington Post*, as he described projections of a global drop of "two or three degrees by the end of the century."

Will based his megadeaths story on a 1974 Central Intelligence Agency report about trends in population, food, and climate. The report stressed that climate-change discussions were "highly speculative" and that "experts will disagree with some or many of the implicit assumptions." Nonetheless, it provided the views of only one expert: Reid Bryson, global cooling's progenitor. So firmly did Bryson embrace cooling that he evidently neglected to tell the CIA about the other side of the debate. Like a lazy journalist, the CIA based its story on a single source. Altogether, seventy-one scientific papers devoted to the cooling/warming debate appeared between 1965 and 1979. Seven favored cooling, twenty were neutral, and forty-four argued for warming. Bryson, the CIA source, belonged to a vocal minority.

Among the warming advocates was NASA's Schneider, who had reversed his initial assessment. As a simplifying assumption, the computer model he borrowed from Hansen had treated the atmosphere as a single uniform air mass. Instead we know it to be a stack of layers, each with different characteristics: troposphere (from the surface to about ten miles up), stratosphere (from the troposphere to about thirty miles up), mesosphere (from the stratosphere to about ninety miles up), and so on. The stratosphere holds a lower level of water vapor than the troposphere, but roughly the same level of carbon dioxide. When Schneider incorporated atmospheric layers into the model, the output

For many politicians, NASA researcher James Hansen's 1988 Senate testimony that a minute additional percentage of carbon dioxide in the air was "changing our climate now" was their first, startling encounter with an issue that over the years would grow ever more contentious.

changed. Now the carbon dioxide in the stratosphere trapped an appreciable part of the infrared light from the surface that made it through the troposphere and beamed it back to the ground, amplifying the warming. By 1974 Schneider was suggesting that "a reasonable first-order estimate" for the effect of doubling carbon dioxide levels would be a rise of 3.6°F—a big change from his first answer.

Even as Schneider and others homed in on carbon dioxide, Hansen and two NASA colleagues were examining the other side: aerosols. Using a similar model, the three men tested its predictions against the results of an actual volcano: the 1963 eruption of Mount Agung in Bali, which killed more than a thousand people and shot enough junk into the air to have a measurable effect on the climate. The predictions of the model matched events closely enough that Hansen and his colleagues were able to sort out the relative contributions of aerosol cooling and carbon dioxide warming.

Comparing the impact of cooling (sharp, sudden bursts) to that of warming (slow and steady), Hansen and most other climate scientists became convinced that in standard Aesopian fashion tortoise would beat hare: warming would predominate in the long run. Not all climate scientists were persuaded—Reid Bryson, for one, went to his grave in 2008 awaiting the onset of glaciation—but the great majority were. Increasingly, research centers and government panels beat the drum for warming.

Despite their efforts, climate change attracted little public notice until June 23, 1988, when Hansen testified before the U.S. Senate. Worried about climate change, Colorado Democratic senator Tim Wirth had deliberately scheduled the hearing for what historically was the city summer's hottest day. His scheme worked beyond his wildest dreams. Hansen sat down amid a wave of bad weather that covered the entire planet. Downpours inundated parts of Africa; unseasonable cold shriveled European harvests. Droughts scorched crops in the U.S. Midwest; forests were aflame across the West. That day Washington, D.C., experienced a record 101°F—an effect amplified when Wirth shut off the air-conditioning in the hearing room. Adding to the heat was the glare from television lights. Perspiration glistened on Hansen's temples as he spoke. He said, "Earth is warmer in 1988 than in any time in the history of instrumental measurements." He said, "With 99 percent confidence we can state that the warming during this time period is a real warming trend." Carbon dioxide, he said, "is changing our climate now."

Hansen's stark words were headline news across the world. *The New York Times* put his charts on page one, and he appeared on a dozen television shows. "Changing our climate *now*" transformed parched fields, overflowing rivers, and sweltering cities from a random cluster of bad weather into a harbinger of a dystopian future. Adding to the furor, the journalist Bill McKibben published in 1989 the first popular account of climate change, *The End of Nature,* a worldwide best seller despite its ominous title. More important, scientific research took off. Before 1988 peer-reviewed journals had never published more than a score of articles in a given year that contained the terms "climate change" or "global warming." After 1988 the figure climbed: 55 in 1989; 138 in 1990; 348 in 1991. By 2000: 1,340. In 2015 it was 16,576.

By happenstance, the World Conference on the Changing Atmosphere began four days after Hansen spoke. Held in Toronto, it was the first major international gathering of scientists and politicians dedicated to global warming. Reporters primed by Hansen's testimony descended upon the attendees. They heard the conference issue a statement that called for reducing carbon dioxide emissions by 20 percent by 2005. They heard respectable politicians insisting that "it is imperative to act now." They heard, they wrote, they published. More headlines, more editorials, more predictions of doom, more calls for action.

Nonetheless, many non-climatologists—physicists, economists, politicians, geologists, even meteorologists—remained skeptical. To climatologists like Hansen, his statement felt like the culmination of 150 years of scientific debate that had in the end led to a high degree of certainty. To politicians, his ideas seemed to come out of nowhere. The notion that a colorless, odorless, nontoxic gas that formed less than 1 percent of the atmosphere might threaten civilization decades from now was so bewildering—as vague and abstract as it was vast—that they instinctively recoiled. Naturally, they sought a way to categorize it. They viewed it as the latest environmental crusade about the air. ✓

Uncivil, Uncertain

Here, at last, we join the second approach to thinking about climate change: how political institutions have grappled with it. In a sense, the answer is simple: they have viewed it as a kind of logical culmination of the environmental movement. Which is to say that it became the heir of an entire history—Vogt's history, in part.

By the late 1960s, when the modern environmental movement took off, most of the left and right agreed, however uneasily, on the need to curb pollution. Two years after winning the divisive 1968 election, President Richard Nixon, a Republican, declared that "the environment," "the great question of the '70s," was a "cause beyond party and beyond factions." The Clean Air Act of that year, which set up U.S. emissions regulations, was one of the world's first general air-quality laws, more stringent and comprehensive than any of its predecessors. Congress passed it overwhelmingly: 73–0 in the Senate, 374–1 in the House of Representatives. Business generally endorsed the legislation; the smog blanketing U.S. cities was obviously harmful and obviously in need of control. Such was the consensus that the distinguished sociologist David L. Sills could observe in 1975 that the new environmental movement "contains all shades of political opinion from the conservative right to the radical left." With little dissent, Washington passed twenty-one major environmental bills, one after another, in the 1970s.

Soon, though, business interests realized that these rules were just the beginning. Dirty air was succeeded by new alarms about the atmosphere, one after another. The ozone hole. Nuclear winter. Acid rain.* Each was more abstract than its predecessor but larger in immediate economic impact. Industry, sensing a trend, became wary. What had begun as a feel-good movement against a limited tally of palpable local harms seemed to have become an endless crusade against ever-larger and ever-remoter targets. A stream of "capitalism vs. ecology" manifestos led many in the business world to conclude that destroying the corporate world had been the aim from the beginning. Ranchers, farmers, loggers, and miners, touched by environmental laws, came to the same conclusion: a cabal of urban snobs was bent on destroying their livelihoods. Increasingly, they resisted. Reflexively, the anti-corporate left doubled down.

The result was a dance that became ever more dysfunctional, as the Emory University historian Patrick Allitt wrote in *A Climate of Crisis* (2014), his history of the environmental movement. Time and again, activists and corporate executives railed against each other. Time and again, regulatory syntheses emerged from this clash: rules for air, water, toxins. Often enough, businesspeople discovered that the new regulation was less costly than they had feared; environmentalists, meanwhile, found out that the problem was less dire than they had feared.

Rather than concluding from this history that, as Allitt put it, "environmental problems, though very real, were manageable," each side stored up bitterness, like batteries taking on charge. The process that had led, often disagreeably, to successful environmental action in the 1970s brought on political stasis by the 1990s. Environmental issues became ways for politicians to flag their clan identity to supporters—

---

* The ozone hole, discovered in 1985 by British researchers, was a severe regional drop in the stratospheric "ozone layer" that absorbs the sun's harmful ultraviolet radiation. Nuclear winter referred to the risk that a nuclear war could throw enough dust and smoke into the atmosphere to block the sun, creating a years-long winter of the sort familiar to readers of *A Song of Ice and Fire*. Acid rain was created when sulfurous air pollution from power plants combined with water vapor to form airborne sulfuric acid, which could mix with rain and fall on distant places, damaging lakes, streams, and forests.

less statements about practical problems with solutions that people could debate than symbols of identity. They signaled membership in a cause: taking back the country, either from tyrannical liberal elitism or right-wing greed.

To the growing anti-green opposition, climate change seemed like the logical extension of a long-brewing battle. Given the history, this may have been inevitable. The chain of reactions that leads from tailpipe emissions to climate change is not easy to visualize. To most people in the world, it remains an abstract peril, encountered mainly in the form of news reports about droughts and floods in faraway places or scientific studies filled with charts and graphs.

Even to scientists, climate change can be as hard to nail down as a blob of mercury. Arrhenius tried to figure out what would happen to global average temperatures if atmospheric carbon dioxide levels doubled, a measure of what today is known as "climate sensitivity." Before the widespread use of fossil fuels—before 1880, more or less— the atmospheric carbon dioxide level was about 280 parts per million. Arrhenius in effect asked what would happen if that number went up to 560 parts per million. In 1979 the U.S. National Research Council asked the same question. Its report projected that doubling atmospheric carbon dioxide would raise global temperatures by between 2.7°F and 8.1°F (1.5°C and 4.5°C). The National Research Council team had produced its estimate by averaging the results from two models and adding about 1°F on either end to account for uncertainty. The procedure— crude, but the best available at the time—was not enough to convince the Carter administration to stop boosting coal production. Since then, many other scientific groups have tried to improve climate-sensitivity estimates. Notable among them was the Intergovernmental Panel on Climate Change, which has produced five major reports on the state of climate science, the most recent in 2014. All five attempted to assess climate sensitivity. Unfortunately, as the economists Gernot Wagner and Martin L. Weitzman lament in their book *Climate Shock* (2015), the likely range of increase from doubling carbon dioxide levels foreseen in the last IPCC report—2.7°–8.1°F (1.5–4.5°C)—was exactly what it was in 1979. Four decades of additional research has not brought us closer to predicting the precise impact of dumping carbon dioxide into the air.

This is not because the researchers are lazy or incompetent. It is because global climate is a phenomenally complex problem: a system with multiple interacting parts, of which many, as Lorenz showed, can be sweepingly affected by tiny changes. Still, the uncertainty puts political leaders in a bind. A rise of 2.7°F would be tolerable, it is generally believed, whereas a rise of 8.1°F would be intolerable: enough to melt the polar ice, inundating coastlines around the world. The estimates of climate sensitivity were like announcing to politicians that something truly awful might happen—or might not. It is as if our species were running blindfolded toward a cliff. Nobody knows the cliff's precise location or height. There is a small chance that it is so low and far away as to be harmless—and a larger chance that it is high and rapidly approaching.

If the problem was nebulous, the general solution was clear. Hansen did not utter the terms "fossil fuels" or "petroleum" in his 1988 testimony. Nonetheless, everyone in the room understood that the climate change he described could not be addressed without taking on, as Arkansas Democratic senator Dale Bumpers put it at the hearing, "the industries that produce the things that we throw up into the atmosphere." It would require vast changes in the energy business, one of the most important sectors of the world economy, and probably its most politically powerful. And these changes—unlike those required for endangered species or the ozone hole—would reach into the lives of almost everyone on Earth.

Thinking about the implications, Bumpers's colleague, Senator Pete Domenici, Republican from New Mexico, recoiled. Nobody, he told the hearing, will "move in areas such as this until we either have a disaster or we have absolute concrete proof. And even when we have that, it seems that we need a game plan of some type." And he asked the panel, "Have any of you put forth a concrete proposal which I assume would involve both further investigations and a course of action?"

The answer he was clearly looking for was "yes"—followed by specified steps to reach an explicit target, which could then be evaluated for feasibility. But that was not what he got.

Coal First

A few years ago, a Chinese-speaking friend and I hired a taxi to drive around Hebei, the Chinese province that surrounds Beijing. When the capital city set up for the 2008 Olympics, the government pushed out the scores of coal-powered utilities and factories that were polluting its air. Mostly, they moved to Hebei. The province ended up with many new jobs—and China's dirtiest air. Because we were curious, we wanted to see what "China's dirtiest air" looked like. I wondered if the mental images conjured by the phrase would prove to be exaggerated.

We hired a taxi for the day in Tangshan, Hebei's biggest city. Visibility was at least a quarter mile—a good day, the driver told us. Haze gave buildings the washed-out look of an old photographic print. With the Olympics shift, formerly poor Tangshan had become China's leading steel-manufacturing center. Now the edge of town held a murderer's row of luxury-car dealerships: BMW, Jaguar, Mercedes, Lexus, Porsche. Most of the vehicles were displayed indoors. Those outside were covered with gray crud.

Rural Chinese are often reluctant to speak with foreign reporters. With good reason: local officials can be vindictive. We asked the driver to approach villagers who weren't in sight of other villagers. Serving as translator, my friend would hop out, introducing himself while I hid in the back seat.

Coal was everywhere, people told us. One truck driver told me with a kind of mocking pride that we were breathing the world's worst air. A university graduate in striped Hello Kitty socks remarked that every time she wiped her face the cloth had "black dirty stuff" in it. The stuff, she said, was PM2.5—techno-speak for aerosols that are less than 2.5 micrometers in diameter, the size most likely to lodge in the lungs. "Everybody is sick but the government would never report it," she said. My friend and I gave a ride to a steelworker who told me that Tangshan had plans to clean itself up in thirty years or so. "We are a city of industry, a city of coal," he said.

The high-speed train from Beijing to Shanghai crosses this part of Hebei on pylons thirty feet tall. Under the pylons, between power plants and factories, was a line of stunted vegetation. On this strip we saw an elderly shepherd who was urging his small flock of sheep and goats to eat the plants. The air made his animals sick, he told us. In late fall and early spring they always began to cough. He, too, got sick. Sometimes he was so sick he couldn't rise from his bed and take his animals to the vet.

Coal-stained air is not confined to obscure locations in China's flyover country. Designer anti-pollution face masks are increasingly common in great conurbations like Shanghai and Guangzhou. A company, Vogmask, sells masks on which companies can print their logos: smog as branding opportunity. Not long before my trip to Hebei, the 10 million inhabitants of the northeastern city of Harbin were enveloped by a tsunami of coal smog. Schools closed; people kept to their homes; highways shut down because drivers couldn't see the road. During my visit I picked up a Beijing newspaper with a glossy ad insert for the city's "first high-tech condominium project that realizes real-time control of PM2.5 levels." The condo's slogan: "Protect Your Lungs, We're Taking Action!"

In the past few decades, China has lifted more than half a billion people out of destitution—an astonishing accomplishment. That advance was driven by industrialization, and that industrialization was driven almost entirely by coal. More than three-quarters of China's electricity comes from coal. More coal goes to heating millions of homes, smelting steel (China produces nearly half the world's steel), and baking limestone to make cement (China is responsible for almost half the world's cement). In its frantic quest to industrialize, China burns almost as much coal as the rest of the world put together.

But that affluence has come with lethal concomitants. Outdoor air pollution in China, most of it from coal, contributes to about 1.2 million premature deaths per year, according to a major scientific study involving almost five hundred scientists in more than fifty nations. A Chinese-U.S.-Israeli research team has estimated that eliminating coal pollution in northern China would raise average life expectancy there by more than five years. (By contrast, wiping out all cancer would increase U.S. or European life expectancy by three years.) A "systematic review" in 2013 by ten Chinese researchers calculated that reducing PM2.5 to U.S. levels would cut the total death rate in big Chinese cities between 2 and 5 percent. A different way to say this is that in these places the side effects of breathing cause as many as one out of every twenty deaths.

Much the same is occurring in India. Already the world's fastest-growing economy, India will become the world's most populous nation (probably by 2022) and its biggest economy (possibly by 2048). It, too, runs on coal—with similar consequences. New Delhi, ringed by coal plants, is said to have the world's most polluted air, worse than anything in China. India's outdoor air pollution causes 645,000 premature deaths a year, according to a 2015 *Nature* study. Even in the United States, which uses less coal than other big nations, coal pollution leads to as many as 25,000 deaths per year.

Bear this all in mind when thinking about Senator Domenici's question: What is the plan? The uncertainties about climate sensitivity make the question especially confounding. Sometime before the end of this century, if nothing changes, the carbon dioxide concentration in the atmosphere will double from its pre-industrial level. If doubling carbon dioxide levels leads to a 2.7°F average temperature increase—the lower bound of the climate-sensitivity estimate—then the world has many decades to cut fossil-fuel use sharply. Societies can take their time and move carefully. But if doubling carbon dioxide levels causes a rise of 8.1°F, the transition must be much faster—a disruptive slam-down of the brakes. The two courses are completely different. What should societies do?

One way to chart a path, point out Wagner and Weitzman, the economists, is to look carefully at the climate-sensitivity estimates. When the IPCC says that the likely consequence of doubling carbon dioxide is a temperature rise between 2.7°F and 8.1°F, the scientists have a specific definition in mind for "likely." Skipping the mathematical complexities, it boils down to saying that the scientists estimate there is a roughly two-thirds chance that the temperature rise will be between these two numbers. But that means there is a one-out-of-three chance that the effect will be outside this range. Very roughly speaking, this translates into a one-out-of-six chance that nothing much will happen—and a one-out-of-six chance of complete disaster, with chunks of the planet becoming nearly uninhabitable. That small but real chance of catastrophe is the key, Wagner and Weitzman argue.*

On a personal level, people deal with this sort of risk all the time. They know that they face a small but real chance of personal calamity: a home robbery, a car accident, a cancer diagnosis. To manage the risk,

---

* Where do these numbers come from? Researchers have constructed half a dozen large climate models. These simulations are made of thousands of relations, each with measures of uncertainty—statements that if X happens, it will have an effect on Y of magnitude a, b, or c. When researchers run the simulation, the computer randomly goes through every imaginable variation—every a, b, or c of every relation—to obtain all possible outcomes. In this case, about two-thirds of the time that outcome is a temperature rise of between 2.7°F and 8.1°F.

people buy insurance. Insurance mitigates the consequences of terrible but unlikely problems. Few people are upset if they pay for fire insurance and their house does not burn down, or if they buy life insurance and fail to die. They happily invested money for security against the risk of disaster.

Managing risk is not entirely, or even primarily, a matter of hying to an insurance agency and signing a policy. Installing good locks and having a big dog are as much a form of managing the risk of burglary as buying theft insurance. Wearing a seat belt is far more effective at reducing the chance that a car crash will have catastrophic consequences than even the best auto insurance. For people in flood-prone areas, the same is true for sump pumps and flood insurance.

At the same time, people don't buy insurance against *every* potential calamity. Homes can be made almost perfectly theft-proof by turning them into fortresses, but for most householders the costs would be high enough to force them to forgo other worthwhile goals, like building a nest egg for retirement or education. Instead, people seek to ward off major threats while sacrificing as few other goods as possible—get the biggest bang for the buck, to use the operant cliché. Best of all is when the insurance measure also accomplishes some other goal. Systems analysts call this "alignment." One way of preventing break-ins is to install windows that are hard to open from the outside. These same windows seal the room better than loosely fitting windows, preventing drafts and thus both making the room more comfortable and lowering heating and cooling costs. The security goal aligns with the goal of making the building pleasant and cheap to use. Another example involves incandescent bulbs, which routinely set off fires because people drape towels and clothes over them. Replacing hot 100-watt incandescent lights with cool 15-watt LED lights cuts the risk of disaster at the same time that it cuts energy bills. In this way, LED lights are examplars of alignment, which is why architects embrace them.

Now look at the numbers for climate change. Humans produce four main types of climate-altering gases: carbon dioxide, methane, nitrous oxide, and a bunch of fluorine-containing gases (these have names like hydrofluorocarbons, perfluorocarbons, and sulfur hexafluoride). Of these, carbon dioxide is the focus of most concern. The other three types actually absorb more infrared radiation, molecule for molecule, but they don't stay in the air as long (the exception is some of the fluoride-containing gases, but they are not yet present in large quantities). Methane has as much as eighty times the effect on climate as an equivalent amount of carbon dioxide, but a typical methane molecule will only remain in the atmosphere for ten to twenty years. Carbon dioxide molecules, by contrast, will keep floating about for centuries, even millennia. They are a problem that doesn't go away. ✓

About 85 percent of the world's carbon dioxide emissions come from fossil fuels, and about 80 percent of those come from just two sources: coal (46 percent) in its various forms, including anthracite and lignite; and petroleum (33 percent) in its various forms, including oil, gasoline, and propane. Coal and petroleum are used differently. Most petroleum is consumed by individuals and small businesses as they heat their homes and offices and drive their cars. By contrast, coal is mainly burned by heavy industry: coal produces the great majority of the world's steel and cement and 40 percent of its electricity. The percentages vary from place to place, but the pattern remains. Coal provides about two-thirds of China's energy, but almost all of it is used by big industries. Coal provides less than a fifth of U.S. energy, but again almost all of it is for industry. In both places petroleum consumption is on a smaller, more individual scale.

The implications are profound. Oil and gasoline use is diffuse, scattered in the global crowd. The world has 1.3 billion vehicles and perhaps 1.5 billion households. Cutting emissions from these cars and homes means changing the daily lives of billions of people, a mind-boggling thought. Reducing global coal emissions, by contrast, means dealing with 3,300 big coal-fired power plants and several thousand big coal-driven steel and cement factories.* The task is huge, but it is at least imaginable—and it targets almost half of the world's emissions at a stroke. Fix coal, the idea is, then go, if needed, to the next thing. That's the way to insure against the small but real possibility of catastrophe.

This is not a new insight. Economists have said the same for years. Nor is it a "war on coal." Cars and trucks brought enormous benefits

---

* Because many electric plants have several power-producing facilities, the number of coal *units* is about 8,800, of which about 6,700 are bigger than thirty megawatts (an average U.S. coal plant is more than five hundred megawatts). No good census of the world's coal-fired steel mills or cement plants exists. "Several thousand" is an estimate from the World Steel Association and the Portland Cement Association. Still, the point remains: compared to oil and gasoline, coal is burned in a small number of facilities. ✓

to people, but they also had lethal concomitants: fatal crashes and air pollution. To cut these harms, governments told automakers to put seat belts and catalytic converters in their vehicles. The companies griped about the costs, but governments were not waging a "war on cars."

Focusing on coal emissions is cost-effective, because it aligns with several other problems. Even if climate change turns out to be less dangerous than activists fear, coal emissions are an urgent public health issue; controlling them could prevent several million premature deaths a year. Focusing on coal even aligns with other climate-change problems. In addition to pouring gouts of carbon dioxide into the air, coal plants release a fine soot that researchers call "black carbon." Black-carbon aerosols rise high into the air; because they are black, sunlight heats them, which in turn heats the air around them. The particles interact with clouds, augmenting their ability to trap heat. The soot lands on glaciers, covering them with a thin black film. Rather than reflecting sunlight, smoky ice absorbs it and melts. Already the dusting of black is helping to liquefy the poles and the Himalayan icepack. Meltwater from the Himalayas provides water for about 1.5 billion people. In 2013 an international team calculated that black carbon was, after carbon dioxide, the most important contributor to climate change.

Logically speaking, there are two ways to control the side effects of coal: clean up coal plants or shut them down. Which tactic people prefer is a good indication of whether they are Wizards or Prophets, hard-path Borlaugians or soft-path Vogtians. Cleaning up coal is the province of Wizards, who extol a technology known as "carbon capture and storage," or CCS. Conceptually speaking, CCS is simple: industries burn as much coal as before, but remove the pollutants. Already they filter out toxic gases. Now they extract carbon dioxide and pump it underground, where it is stored for eons.

The best-known carbon-capture technique is "amine scrubbing." It involves bubbling the exhaust from burning coal through a solution of water and monoethanolamine (MEA). MEA is unpleasant: toxic, flammable, and caustic, with an acrid smell. But it bonds to carbon dioxide, separating it from the other gases in the exhaust. The process creates a new chemical compound called, uneuphoniously, MEA carbamate. The MEA carbamate and water are pumped into a "stripper," where the solution is boiled or the pressure is raised. Heat or pressure reverses the reaction, breaking up the MEA carbamate into carbon dioxide and MEA.

Carbon dioxide gushes out, ready for burial; MEA returns to combine with the next batch of coal exhaust.

Scaling up this process is not easy. Big power plants produce huge amounts of carbon dioxide, and need big structures to capture it: multistory metal towers festooned with pipes and valves. Constantly boiling a silo's worth of toxic chemicals in a stripper requires a great deal of energy. Common estimates are that this kind of carbon capture will gobble 10 to 15 percent of a power plant's output. Given that even the most efficient coal plants translate less than 50 percent of the energy in coal into electricity, deploying CCS means that power plants will consume 20 to 30 percent more of the black stuff—at minimum. Mitigating the environmental costs of digging up and burning coal in this way means digging up and burning even more coal.

The industry jargon for these costs is "parasitic." (Sample usage, from an energy consultant: "Holy crap, the parasitics are awful.") Often parasitic costs are estimated at $90 to $100 per ton of stored carbon dioxide. A single 500-megawatt power plant emits roughly 3 million tons of carbon dioxide a year. Arithmetic suggests that sticking all that gas from the world's thousands of plants in the dirt would cost $2 trillion a year, a figure that doesn't include the billions required to build the carbon-capture facilities in the first place. This back-of-an-envelope calculation rests on implausible assumptions: coal plants of identical size, no technical progress, no economies of scale, no plant conversions to lower-emission natural gas, and so on. But the overall conclusion—that carbon capture based on present technology faces big obstacles—is all too plausible.*

The Wizardly argument for CCS boils down to: (1) the technology is new, and its cost will come down, as was the case with photovoltaics; (2) it is unwise and even unethical to assume that China, India, and other developing nations, having just built hundreds of big coal plants,

---

* The storage part of the equation is more straightforward. Engineers like to say that "nature is the proof of concept." What are petroleum fields but natural storage sites for carbon? Recall that a petroleum deposit consists of two layers of stone, a porous bottom layer beneath a nonporous cap. Carbon dioxide storage is the reverse of oil drilling: companies pump pressurized carbon dioxide through impermeable rock into permeable rock. After the rock is filled, the entrance hole is plugged forever, a reliquary for humankind's energy obsession. In principle, carbon dioxide could be tucked into such lairs until the sun explodes. In practice, it needs to be stored only for a century or so, the time required for the carbon dioxide to combine with the surrounding stone and form stable minerals. Most scientists believe this to be an achievable goal.

will tear them down and replace them—they just don't have the money. Thus the only hope for cutting coal emissions there is to deploy CCS.

CCS for existing coal plants is not enough. More than a billion people worldwide lack adequate access to electricity. Three hundred million of them live in India—a quarter of the nation's population, generally rural and poor, the world's biggest pool of the literally powerless. Most of these people use kerosene for lighting and burn wood or dung to cook food. Different groups have tried to estimate the fatalities from the fumes: between 500,000 and 1.3 million Indians a year, depending on the assumptions of the researchers. Providing power to these people is "a priority in every imaginable way—human, economic, and political," says Navroz Dubash, a senior fellow at the Centre for Policy Research in New Delhi. Partly in consequence, India's electricity demand is expected to double by 2030.

How will that power be provided? As of now, in theory, by coal. According to the World Coal Association, almost 2,400 more coal power-plant units are under construction or planned, though it is anyone's guess how many will be built.

Wizards typically believe that a big increase in coal would be a mistake. But the answer is not to leave the hard path, but to improve it by going nuclear. Nuclear plants produce no carbon dioxide at all (except the emissions released in making the cement and steel for the plant and the exhaust of plant workers' automobiles), which is why many technology-oriented environmentalists are ardent nuke boosters. Prominent examples in the English-speaking world include Steven Chu, futurist Stewart Brand, biologist Tim Flannery, planetary scientist James Lovelock, climate researchers James Hansen and Jesse Ausubel, physicist Raymond Pierrehumbert, and environmental scientist John Holdren, science adviser to the Obama administration. "If you're worried about climate change, and I am, nuclear is the greenest alternative," Brand told me.

Nuclear power plants are extremely costly to build, but Wizards argue that putting CCS in coal plants drives up construction costs to the point where a new nuclear plant and a new coal plant have about the same price tag. And once they are running, most scientists say, nuclear plants have proven more reliable, cheaper, and safer than coal plants. Because terms like "reliable," "cheaper," and "safer" can be vague, let me explain what engineers and physicists mean by them. In general, these people support nuclear power, so what follows is effectively a positive brief for it.

Reliability is measured by "capacity factor," the fraction of the time that the power plant is actually sending out electricity at its maximum rate. For U.S. coal plants, the capacity factor is less than 60 percent. Meanwhile, nuclear's capacity factor is 90 percent, higher than any other type of energy (solar photovoltaics, for instance, are rated at less than 30 percent). Decades of experience have shown that once a nuclear plant is turned on, it tends to run quite reliably, with the main downtime being maintenance breaks.

Cheapness refers to the price for a kilowatt-hour of electricity. Nuclear power is based, famously, on splitting the nucleus of uranium atoms. The nucleus breaks into two pieces that fly away from each other, releasing a lot of energy as they do. The energy is roughly a million times more than the chemical energy one would obtain by burning the same atom. Because of this greater energy density, nuclear plants need much less fuel to produce a given amount of electricity than fossil-fuel plants. As a result, nuclear facilities have proven to be less expensive to operate than any other type of power plant except hydroelectric dams. By far the biggest cost is constructing the plant. After that, actually making the electricity is cheap.

Safety usually is measured by the number of deaths in the "energy chain"—that is, how many people are killed by the entire cycle, from exploration and mining to refining and transportation to actual power generation, as well as waste treatment and disposal. Deaths from mining uranium are counted, as well as deaths from falling off roofs while installing solar panels (a surprisingly big number). Adding these together, a research team at the Paul Scherrer Institute, Switzerland's biggest research center, reckoned in 2016 that nukes thus far have caused fewer deaths than any other power source except, again, hydroelectric dams (wind power was pretty close). Coal had thirty to a hundred times the impact on human health in normal operations. Nuke boosters say, accurately, that nuclear plants kill people only in rare, awful accidents like Chernobyl (even in the frightening 2011 meltdown at Fukushima, radiation is not known to have caused a single fatality). Coal plants around the world kill millions of people every year in normal operation.

In addition, nuclear plants take up less land than other sorts of utilities. An often-cited study of U.S. plants by Nature Conservancy researchers in 2009 concluded that nuclear power uses about four times less land per unit of energy than coal and about fifteen times less than solar arrays. Other scientists have come to similar conclusions. Because different researchers adopt different assumptions, the exact numbers vary from study to study, but in every case I am aware of, researchers have concluded that nuclear power has the smallest footprint, an important factor in nations that are short of farmland or open space.

Brand cites the example of France, which constructed "fifty-six reactors providing nearly all of the nation's electricity in just twelve years." Nuclear power provides about 77 percent of French electricity, a far greater proportion than in any other nation. Today, according to World Bank figures, France emits 5.2 tons of carbon dioxide per capita. The corresponding figure for the United States is 17. France shut down its last coal-fired power plant in 2004. It is the world's biggest electricity exporter; meanwhile, French household electricity costs are among the lowest in Europe. What's not to like? Wizards ask. Why not, in this respect, make the rest of the world like France?

Wizards typically don't disparage renewable energy. They just don't see solar and wind power as playing a major role in the human comedy—not for decades, at least. Build what you can, they say, but don't expect it to make a huge difference. For the foreseeable future, renewables are unreliable (low capacity factors), pricey (high cost per unit of power, including storage), and land-wasting. Here and now, nuclear power is ready. Equip the most efficient existing coal facilities with CCS, but go nuclear for everything else.*

Unsurprisingly, Prophets disagree from top to bottom. To begin with, most regard CCS as a sham—an industry-sponsored fantasy that never

---

* Wizards and Prophets mostly agree on energy conservation—making buildings, vehicles, and machinery waste less energy. For this reason I don't discuss it, but efficiency is key to any climate strategy. The more energy you don't use, the less you need to convert from fossil fuels.

has and never will deliver on its promises. At a 2008 meeting of the Group of Eight (the world's richest nations), the assembled energy ministers lauded the "critical role of Carbon Capture and Storage" and promised to begin building "20 large-scale CCS demonstration projects" in the next two years. That didn't happen. According to the Global CCS Institute, an Australia-based association of governments and energy firms, the world had just one operational project that trapped and stored emissions from a big coal-fired power plant. Completed in 2015, the $1.1 billion Boundary Dam project, in Saskatchewan, Canada, went massively over budget but after some early stumbles has performed as advertised.

Worse, to Prophets, CCS means that humankind keeps mining coal. Most coal mining now is "open-pit" or "open-cast": companies carve a big pit in the ground, then extract an underlying coal seam. An especially large-scale form, mountaintop removal, involves the removal, as the name suggests, of entire mountaintops; giant backhoes dump the rubble into river valleys. Pioneered in the United States, mountaintop removal has occurred in about five hundred sites in Kentucky, Virginia, West Virginia, and Tennessee, permanently altering the topography and burying two thousand miles of streams. Rehabilitating these landscapes may be possible, at least in part, but so expensive that it is hard to imagine anyone actually doing it.

More distressing still, several thousand coal mines have caught fire in Australia, Britain, China, India, Indonesia, New Zealand, Russia, South Africa, and the United States; many have been burning for decades, some for centuries. An infamous example is the Jharia coalfield, in the northeastern Indian state of Jharkand. Covering 170 square miles, Jharia is India's main reservoir of coking coal, the hard coal used to make steel. It has been on fire, calamitously, since 1916; entire villages have disappeared into the smoking ground. Undermined railroad lines have fallen into the earth, followed by farms and streams. When I visited the region, toxic fumes shimmered in the air. Issuing from cracks in the earth, they wreathed the ruined buildings and black, leafless trees. In the evening, patches of smoldering red were visible, scattered like watching eyes across the charred landscape: Mordor without the Orcs. Centralia, Pennsylvania; Greenwood Springs, Colorado; Barnsley, Yorkshire; Wuda, Inner Mongolia; East Kalimantan, Indonesia—Prophets look at these smoldering places and see an insult to the future.

A woman struggles to shore up her collapsing home in the burning zone of the Jharia coalfield. Hers was one of the last remaining houses in this impoverished village; the rest had already been consumed by the fire.

As for nuclear, Prophets regard it as too expensive to be plausible. Two new plants in Georgia were so expensive that they drove Toshiba's Westinghouse nuclear division into bankruptcy; the price may soar to $21 billion before they are finished, years behind schedule. Two other Westinghouse plants in South Carolina were abandoned, half-built, in 2017, at a loss of $9 billion. Prophets agree that nukes are cheap to run, but disagree that the long-term savings in operations costs justify the massive upfront costs of construction.

And then there's the waste. Nuclear plants produce several types of waste, of which the most dangerous is "high-level" waste. Mostly spent reactor fuel and by-products from the reprocessing of spent fuel, it accounts for more than 99 percent of the radioactivity produced by nuclear waste—the other types of waste, though large in volume, emit very little. Since the beginning of the industry, according to the International Atomic Energy Agency, a United Nations–affiliated group that coordinates the peaceful use of nuclear energy, the world's four-hundred-plus nukes have produced about 300,000 tons of high-level waste. (The figure rises by 12,000 tons every year.) By volume, this is approximately 160,000 cubic yards, enough to cover a football field eighty feet deep. Given that this pileup represents nearly all of the truly dangerous material from sixty-plus years of worldwide nuclear power, Wizards do not see waste as an overwhelming problem. They argue that this material can be encased in glass and left deep underground; within a few centuries, its radioactivity level will have fallen by a factor of a million.

To Prophets, these arguments are beside the point, pragmatically and ethically. Some types of nuclear waste, like plutonium and radioactive iodine, are astonishingly deadly: radioactive for millennia, lethal in doses smaller than grains of sand. It is unrealistic, Prophets say, to imagine that something so dangerous in such small amounts can be contained for eons, when it can be carried away by the smallest draft of wind or dissolve into drops of water. Transporting relatively large amounts of such substances in accident-prone trucks or trains—mobile Chernobyls, opponents call them—is an act of folly. Most of all, Vogtians see waste deposits, even if contained, as no-go areas that will endure for what is in human terms an eternity. Leaving such noxious gifts to future generations is a moral calamity.

Instead of replacing coal with nuclear, Prophets favor almost anything else that doesn't use fossil fuels. The most detailed roadmaps to this kind of future have been issued by research teams led by Mark Z. Jacobson and Mark A. Delucchi, engineers at, respectively, Stanford University and the University of California at Berkeley. In a long study published in 2015, Jacobson, Delucchi, and eight other researchers laid out a path for taking the United States entirely to wind, water, and solar power by 2050. (Four years before, Jacobson and Delucchi wrote an outline for switching the entire world to renewables, but I have chosen the U.S. project because it was more detailed and, in my view, easier to understand.)

This version of the Prophetic vision can be summarized as seven No's and one big Yes. The No's are: no oil, no gasoline, no kerosene, no natural gas, no wood or biomass stoves, no nuclear power, no carbon capture and sequestration. The Yes is electricity, with two asterisks. The Yes is for electrifying the entire economy, including economic sectors—heating, transportation, and steel and cement manufacturing—that now run directly on coal and petroleum.

The asterisks link to two footnotes. The first note points out that the task is smaller than it might seem at first glance; the second says that it is bigger. The first footnote is that the ensemble of new renewable plants will not actually have to replace *all* of the power generated by fossil fuels. Electric motors are more efficient than engines driven by fossil fuels, because those lose a lot of energy producing heat; replacing them will therefore require less capacity than before. In addition, advocates believe, energy-efficiency measures like insulating buildings and improving appliances will further cut demand.

The second footnote is that building that smaller capacity will, paradoxically, require *more* power plants, because solar and wind power are intermittent. A solar facility might be designed to produce a megawatt of power, but if people want that megawatt day in and day out, society will have to build three or four one-megawatt plants in different places to ensure the supply.

Altogether, the Jacobson-Delucchi team estimated, the United States would need to build:

- 328,000 new onshore 5-megawatt (MW) wind turbines (providing 30.9 percent of U.S. energy for all purposes)
- 156,200 offshore 5-MW wind turbines (19.1 percent)
- 46,480 50-MW new utility-scale solar photovoltaic power plants (30.7 percent)
- 2,273 100-MW utility-scale concentrated solar power (i.e., Mouchot-style solar mirror) power plants (7.3 percent)
- 75.2 million 5-kilowatt (kW) home rooftop photovoltaic systems (3.98 percent)
- 2.75 million 100-kW commercial/government rooftop systems (3.2 percent)
- 208,100 1-MW geothermal plants (1.23 percent)
- 36,050 0.75-MW devices that harness wave power (0.37 percent)
- 8,800 1-MW tidal turbines (0.14 percent)
- 3 new hydroelectric power plants (all in Alaska, 3.01 percent).

As lagniappe, the nation also would convert all cars and trucks to run on electricity and all planes to run on supercold hydrogen—all the while building underground systems that store energy by heating up rock under most of the buildings in the United States.

Wizards criticize all of this as ridiculous. Jacobson and Delucchi propose constructing hundreds of thousands or even millions of underground heat-storage systems, for example, even though hardly any have been built. And they assume that by adding generators and turbines to existing dams they can squeeze fifteen times as much power out of them, which dam operators say is impossible. And the area that will be covered with solar and wind farms is huge—the scheme is taking us back to medieval times, when people used the landscape (in the form of forests) for power. The Prophetic reply is that the technology will improve and become cheaper. And nuclear plants and CCS are even more impractical, they say. Wizards envision building a thousand or more nuclear plants in the United States alone, each at a cost of several billion dollars. How can this be possible? Can one imagine people in developed nations approving nuclear plants in their neighborhoods? What about solar plants? Wizards reply. Already people are fighting their huge demand for land. *But wait a minute*—haven't we been here before? The bickering about practicality and costs, the endless back-and-forth argument? Is this telling us something?

To Vaclav Smil, the University of Manitoba environmental scientist, the intractability of the quarrel reflects the fact that both Wizards and Prophets are fooling themselves. "Energy transitions are *always* slow," he told me by email. Modern energy infrastructures, assembled over decades, cannot be revamped overnight. In every nation, modern electricity grids took decades to assemble. Disassembling and replacing them quickly enough to avoid the worst impacts of climate change would be an unprecedented challenge for societies that are still rapidly increasing their energy use. Worse still, in his view, there is little public appetite for beginning the process, or even appreciating the magnitude of what lies ahead. "The world has been running *into* fossil fuels, not away from them."

Smil's arguments about economic and technological rules run parallel to Margulis's arguments about biological rules. We cannot escape the laws, they are saying, and the laws will not let us escape tragedy. But there is an obvious counterargument: now, unlike in the past, humankind has a gun to its head. More generally, the impossibility of predicting the long-term future unavoidably leaves room for hope. The chance of a successful outcome cannot logically be excluded. Sweden, for example, has reduced its carbon output by two-thirds since 1970 without notice- ✓ able impact on its economic fortunes. At the same time, what if Smil is right? What if neither Wizards nor Prophets can move fast enough?

Planet-Hacking

When the next hurricane approaches New Orleans, every resident will know what to do: empty the fridge. Back in 2005, hardly anyone did that for Hurricane Katrina. Families in the city were accustomed to leaving for a couple of days during bad storms, then coming back to streets strewn with branches and trash and maybe a few shingles. When Katrina hit, the flooding was so bad that people couldn't return for weeks. This was NOLA—New Orleans, Louisiana: The weather was sunny and hot. Because of the storm, the electricity was out. Across the metropolitan area, a quarter of a million refrigerators became inadvertent experiments in the biology of putrefaction. Despite the warnings, many homeowners opened their refrigerators. Almost everyone who did realized instantly that they could never be used again.

Throughout the fall and winter, returnees duct-taped their refrigerators shut and dragged them out to the curb. White metal boxes lined the streets like gravestones. Sometimes they were spray-painted with sardonic slogans. *Feed my maggots. Caution: Breath of Satan inside. Ho ho ho NOLA*—this one decorated like a Christmas tree. Occasionally people illicitly dumped their refrigerators in faraway neighborhoods and came back home to find people from those neighborhoods had dumped refrigerators on their street.

Katrina created about 35 million cubic yards of debris in southern Louisiana—an estimate that does not include, among other things, the area's 250,000 destroyed automobiles. East of the city is the Old Gentilly landfill, shut down as a hazardous waste site. It quickly reopened and became Mount Katrina: a two-hundred-foot-tall mass of soggy armchairs, ruined mattresses, busted concrete, and moldy plywood.

By volume, refrigerators were a tiny part of this—a rounding error. Nonetheless, huge numbers came in every day. By late May the total was

Fridgelandia, December 2005

about 300,000. The refrigerators had their own staging area, separate from the stoves and dishwashers, in the foothills of Mount Katrina.

Fridgelandia was an amazing sight. Battered white boxes, stacked up hundreds of feet in every direction. Teams of workers in gas masks and crinkly hazmat suits, scooping out the writhing contents with plastic snow shovels. If people didn't shovel quickly, carnivorous dragonflies would descend on the maggots in such clouds that workers couldn't see.

Until I visited post-Katrina New Orleans I did not realize that rebuilding a flooded modern city would involve disposing of several hundred thousand refrigerators. Nor had I realized that it would involve a search for housing for relief workers there to build housing. Or criminals taking advantage of the lack of police to steal vast amounts of pipe and cable as soon as they had been replaced. Or toxic blooms of fungi new to science. Or that cities would have a hard time functioning after the sudden and immediate collapse of all local insurance bureaus.

Katrina was a relatively modest storm that overwhelmed inadequate dikes and levees. Many climate scientists believe that in days to come governments will need to get better at shoreline defense. The world has 136 big, low-lying coastal cities with a total population of about 550 million people. All are threatened by the rising seas associated with climate change. A study in *Nature* in 2013 estimated that if no preventive actions are taken annual flood costs in these cities could by 2050 reach as much as a trillion dollars. Other research teams have arrived at similarly extreme estimates. Coastal flooding could wipe out up to 9.3 percent of the world's annual output by 2100 (a Swedish-French-British team in 2015). It could create losses of up to $2.9 trillion in that year (a German-British-Dutch-Belgian team in 2014). It could put as many as a billion people at risk by 2050 (a Dutch team in 2012). Test cases occurred in 2017, when storms inundated Houston, Puerto Rico, and the Florida Keys.

Some economists argue that these figures are exaggerated; indeed, I have cited the researchers' worst-case scenarios, to emphasize the stakes. But the same economists also point out that some of the most threatened areas are irreplaceable parts of the world's cultural and natural patrimony. Venice is an obvious example, but so are places like central London, New Orleans and the Mississippi Delta, the vast ancient complex of Chan Chan in coastal Peru, and the great Sundarbans mangrove forests in India and Bangladesh.

To avoid this damage, cities would either have to shift their population to higher ground, construct networks of protective baffles, canals, dikes, and floodwalls, or both. All would be difficult and costly. Shanghai, with an average altitude of thirteen feet, is among the many Asian cities vulnerable to rising waters. Its 14.35 million inhabitants live on the low, flat delta of the Yangzi River. Because the city has withdrawn groundwater too rapidly, it has sunk more than nine feet in the last century. Meanwhile, sea levels are rising. In 1993 the city built a floodwall designed to block the surge from a once-in-a-thousand-years storm; within four years stormwater was lapping at its top. The nearest higher land is about thirty miles away, in the outskirts of the city of Hangzhou, population 2.45 million. Relocating part of Shanghai there would involve building a second or third Hangzhou atop the first.

Cities have adapted to rising waters in days gone by. Chicago, founded almost at the water level of Lake Michigan and the Chicago River, discovered that its land was too flat, low, and wet to install a sewer system. Beginning in 1856, the city installed six-foot sewage pipes in the middle of its streets, then raised the buildings around them. Some structures were jacked up as much as ten feet. Eventually the entire city was lifted into the air. In our era Venice has constructed lines of floodwalls around itself. The system began construction in 2003 and is intended to be operational in 2019.

These cities were small enough to make one wonder about their use as precedents. Chicago then had about 100,000 inhabitants; Venice and its surrounding communities have about 265,000. No one knows how much Chicago spent; Venice's defenses will cost $6.1 billion to construct, about $23,000 for every man, woman, and child in the area. Many of the at-risk cities are much bigger, especially those in Asia: Bangkok, Tianjin, Manila, Guangzhou, Jakarta, Taipei, Mumbai, Dhaka, Kolkata, Ho Chi Minh City. One report suggests that by the year 2100 a hundred-year storm in Southeast Asia could flood out 362 million men, women, and children. Protecting people in such numbers would be an unprecedented task.

What if Vaclav Smil is right, and the world's energy system cannot be remade fast enough to avoid flooded cities? Suppose that Smil's scenario has come to pass by 2050—our civilization has cut the amount of carbon it sends into the air, but not by enough. Suppose further that climate sensitivity has turned out to be on the high side. In this scenario we are racing toward an increase in global temperatures of 7°F, perhaps even 9°F. (Nothing known today rules this out.) What to do? The waters are lapping higher.

Tomorrow's leaders, whether Wizards or Prophets, would face a dance of impossibilities: the extraordinary cost and effort of rapidly replacing the world's energy infrastructure versus the extraordinary cost and effort of moving cities versus the extraordinary cost and effort of continuing on the same path. Facing this dizzying choice, who wouldn't look for an escape hatch? To save the future, some would look at the past. Two pasts, in fact—one preferred by Wizards, the other by Prophets.

The first past goes back to Mount Pinatubo, a volcanic eruption in the Philippines in 1991. The explosion killed several hundred people, covered several thousand square miles in debris, and shot gas, dust, and ash into the stratosphere. That plume of volcanic pollution contained at least 20 million tons of sulfur dioxide, a pungent, toxic gas. Water vapor in the stratosphere combined with Pinatubo's sulfur dioxide, producing shiny, microscopic droplets of sulfuric acid. Taken together, the journal-

Mount Pinatubo, June 12, 1991. Geoengineering advocates argue that buying time to stave off the worst effects of climate change could begin by pumping the rough equivalent of a Pinatubo per year into the upper atmosphere.

ist Oliver Morton has calculated, these aerosols had a surface area similar to that "of a large desert—definitely bigger than the Mojave, likely smaller than the Sahara." Like a diffuse, airborne desert of blazing white sand, the field of sulfuric acid droplets reflected sunlight into space. For two years the amount of sunlight that reached the surface dropped by more than 10 percent. Average global temperatures fell by about 1°F.

To Wizards, numbers like these are an irresistible excuse for performing some elementary arithmetic. Keeling's carbon dioxide measurements tell us that the air today contains a bit over 400 parts per million of carbon dioxide. Each part per million is equivalent to 7.8 billion tons of carbon dioxide. Four hundred parts per million times 7.8 billion comes to 3.1 trillion tons of airborne carbon dioxide. In 1880, before people began burning coal in large quantities, the carbon dioxide level was about 280 parts per million. Doing the same kind of multiplication, that is equivalent to 2.19 trillion tons of carbon dioxide in the air. Subtracting the pre-industrial number from today's number leads to the conclusion that our antic consumption of fossil fuels has added 0.91 trillion tons of carbon dioxide to the atmosphere. Call the number

1 trillion, for simplicity's sake. The result of that post-Pithole trillion has been to raise global temperatures by about 1.4°F (0.8°C), with most of that warming since 1975. A bit more back-of-the-envelope arithmetic: 650 billion tons of carbon dioxide is, roughly speaking, equivalent to 1°F of warming.

Pinatubo offset that 1°F of warming with about 20 million tons of sulfur dioxide. Doing the arithmetic again, sulfur dioxide is, molecule by molecule, more than fifty thousand times more effective at lowering temperatures than carbon dioxide is at raising them.

Actually, this understates the comparison. Here Wizardly attention shifts from arithmetic to the science of raindrops. One ton of water in a single round blob has a surface area of roughly fifty square feet. Divide that same ton of water into droplets a few ten-thousandths of an inch in diameter. The *volume* of water remains the same, but the *surface area* increases—to more than two square miles. Cut each of these little droplets into five identical but even smaller pieces. Now the surface area almost doubles—roughly four square miles of thinly spread mirror. (I have lifted this calculation from Morton's fine book *The Planet Remade* [2016].) The smaller the droplets, the bigger the mirror; the bigger the mirror, the more the reflection. Equally important, the droplets must be separated by enough space so that they don't bump into each other and merge into big droplets, which fall out of the sky faster than small ones.

The cooling from Pinatubo's 20 million tons of sulfur dioxide was geophysical happenstance; the droplets it formed were not of the optimal size. By making smaller, more effective droplets, geoengineers could achieve the same reduction by spraying just a few million tons of sulfur dioxide into the air in a year. Actually, they would probably spray sulfuric acid directly, rather than having the atmosphere convert sulfur dioxide, but the principle is the same. The most direct method to accomplish this task would be to launch specialized delivery vehicles from Earth, each with a payload of sulfuric acid.

Services already perform such tasks. They are called commercial airlines. A new Boeing 747 carries as many as 600 passengers. The average weight of a U.S. person is about 175 pounds. To make figuring easy, assume that each 175-pound passenger has 25 pounds of luggage and thus represents a unit of 200 pounds. Each 600-person flight on a 747 thus carries 120,000 pounds—60 tons—of human weight. To send aloft 2 or 3 million tons would require flying about a hundred or so flights a day. Today the world's airlines fly more than 100,000 flights a day. Ryanair, an Irish budget airline, operates 1,800 flights a day; Alaska Air, a regional U.S. airline, has almost 900. Recreating Pinatubo would require a service about a tenth the size of Ryanair or possibly a fifth of Alaska Air.

For better or worse, a fifth of an Alaska Air would not be expensive. One well-known estimate from 2012 suggested that fourteen big cargo aircraft—Boeing 747s, for example—could pull a Pinatubo for a little more than $1 billion a year. But commercial jets are not designed to fly into the stratosphere (the higher one places the sulfur, the longer it will stay aloft). Specially designed planes could be more effective and cost only $2 to $3 billion a year to operate. Either way, it is financially feasible. The cost for a decade of counteracting most of the impact of carbon dioxide, the Harvard physicist David Keith has written, "could be less than the $6 billion the Italian government is spending on dikes and movable barriers to protect a single city, Venice, from climate-change-related sea level rise."

To governments looking at rising seas, experiments with sulfur dioxide could seem worth the risk. Should carbon emissions not fall quickly enough, the idea goes, the world might dump sulfuric acid into the air for a couple of decades, buying enough time to finish the transition from fossil fuels. In theory, the injections could be focused on the skies above the poles, creating a reflective shield over the Arctic and Antarctic ice sheets. The goal would not be to eliminate all global warming, but to take the edge off, reducing it by a fifth or a fourth until it reaches the relatively safe level of 3°–4°F.

Since the 1980s such plans to alter Earth's climate deliberately have been called "geoengineering." Geoengineering fights climate change with more climate change. It is, in the jargon, a "technical fix." It replaces the idea of staying within natural limits with the goal of creating a balance on terms set by humankind. It is an audacious promise to fix the sky. It is one of the logical endpoints of the Wizards' dream of human empowerment and grandeur.

Geoengineering is an ancient idea with some old baggage. Ancient religions promised for millennia to control the weather by negotiating with heavenly powers. When the rise of science downgraded the role of priestly intercession, lunatics, imposters, and bunco artists filled the vacuum. Flotillas of phony rainmakers traveled through the nineteenth-century Middle West, taking advantage of drought fears to sell mysterious engines, bottles of vile, foamy liquids, and pamphlets filled with scientific-sounding gabble to credulous farmers. On the West Coast, Charles Mallory Hatfield, the self-proclaimed Moisture Accelerator, spent years building towers in remote locations from which he evaporated pans of chemicals. The chemicals, Hatfield claimed, used a "subtle attraction" to "woo" rain clouds. When the director of the U.S. Weather Bureau denounced Hatfield as a fake in 1905, he shrugged it off. "Censure and ridicule are the first tributes paid to scientific enlightenment by prejudiced ignorance," he said.

Perhaps the most energetic charlatan was Robert St. George Dyrenforth. An engineer, patent lawyer, and Civil War major, Dyrenforth was certain that rain was caused by thunder. In federally funded tests in West Texas in 1891, Dyrenforth and a cohort of rain-obsessed amateurs tried to simulate peals of thunder by taping explosives to balloons and kites, filling prairie-dog holes with gunpowder, emplacing ranks of improvised mortars (the barrels were sawed-off iron tubes), and festooning mesquite bushes with sticks of dynamite. All were rigged to go off at the same time. Rain fell heavily before the experiments and continued afterward. Dyrenforth reported this as proof of success and asked Congress for more money.

Legitimate experiments in "cloud seeding"—sprinkling tiny crystals of dry ice in clouds, to stimulate raindrop formation—began in the 1940s. They effectively ended the reign of the con artists, but gave rise to another breed of fraud, the overly optimistic intellectual. Promising that "[w]e will change the Earth's surface to suit us," the physicist Edward Teller proposed that atom bombs be used to shake loose recalcitrant petroleum deposits, create a second Panama Canal, and manipulate weather patterns. The most wild-eyed schemes came from Moscow, where Soviet dreamers unfurled one grandiose, loopy stratagem after another. Melting Arctic ice by bombing it with soot. Building a causeway off Newfoundland to redirect the Gulf Stream. Damming the Congo River to irrigate the Sahara. Pumping warm water from the Japanese Current into the Arctic Ocean, shrinking the ice cap. Launching thousands of rockets full of potassium dust to create Saturn-like rings around Earth that would somehow induce a "perpetual summer."

From these brainstorms came proposals to offset climate change from carbon dioxide—toss-offs at first, then more serious suggestions. In the late 1950s John von Neumann half-seriously proposed creating a planet-wide dust shroud with nuclear weapons; the haze would cool Earth, which he suspected humans were overheating with power plants and blast furnaces. Less enamored of bombs, Roger Revelle in 1965 suggested scattering millions of little floating mirrors on the sea to reflect sunlight into space. Few paid attention at the time. But in 2006, when the Nobel Prize–winning chemist Paul Crutzen resurrected the idea of geoengineering, people were ready to listen, however reluctantly.

Gone were the days of Soviet-style gusto; Crutzen's tone was anything but triumphant. "The very best would be if emissions of the [climate-changing] gases could be reduced so much the stratospheric sulfur release experiment would not need to take place," he wrote at the end of his article. "Currently, this looks like a pious wish." Other Wizards have echoed his tone. Geoengineering might be the culmination of Borlaugian dreams of power and control, but its advocates have drawn back, chastened, from the implications; their support for geoengineering is mixed with regret. A prominent geoengineering advocate, the Harvard physicist David Keith, has likened it to chemotherapy for the planet—a treatment that nobody would want to have unless forced by circumstance, because it deliberately makes the patient sick to cure a greater disease. Tinkering with the atmosphere, in the phrase of the writer Eli Kintisch, may be a bad idea whose time has come.

The potential pitfalls are many. Sulfur compounds would interact with stratospheric ozone, which protects surface-dwellers like us from the sun's dangerous ultraviolet radiation. The sulfur soon falls to the earth, contributing to lethal air pollution. (For this reason, some have suggested using particles of titanium, aluminum oxide, or calcite, which are more expensive but less likely to interact with ozone and unable to form acid.) In reflecting sunlight, sulfuric acid droplets in the stratosphere reduce the amount of energy coming into the atmospheric layers below, which affects wind and rainfall. Because wind currents are unevenly distributed around Earth, the changes in rainfall will be unevenly distributed, and temperatures along with them. Geoengineering may reduce temperatures globally, but there will still be local losers and winners—places that experience too much or too little rainfall, places subject to sudden temperature extremes. And no matter how much sulfur dioxide humankind throws into the heavens, the carbon dioxide will remain; to counteract the ever-increasing total, more sulfur must be launched into the air every year. Indeed, stopping it suddenly would be disastrous; all the hidden-away warming would emerge in a few months.

The greatest danger posed by planet-hacking comes from its greatest virtue: its low cost. Wagner and Weitzman, the economists, call it a "free-driver" problem; driving the car is so cheap, anyone can take it for a spin. Spraying sulfur is cheap and easy enough that a single rogue nation could reengineer the planet by itself. Or two countries could separately decide to alter the climate in conflicting ways. China, worried about drought, could seek to increase its monsoon; India, fearful of flooding, could try to decrease rainfall. "Both are nuclear weapons states," Keith reminds us. According to *Forbes* magazine, the world has 1,600 billionaires. Each could sponsor a course of geoengineering single-handedly. A Bill Gates could pay for it many times over by himself. "A lone Greenfinger, self-appointed protector of the planet and working with a small fraction of the Gates bank account, could force a lot of geoengineering on his own," remarked the Stanford international-relations specialist David Victor.

Strange Forests

Prophets go on tilt when they hear these ideas. Viewing climate change as a prime example of exceeding carrying capacity, they regard the idea of fighting pollution with pollution as marching precisely in the wrong direction. Not only is it crazy to begin with, Vogtians say, it's a distraction from the urgent social reforms needed for the future. Worse, geoengineering forever desacralizes Nature; it puts the final seal on the replacement of the authentic, billion-year-old natural world by a new, artificial world whose every surface bears the greasy human fingerprint. But the specter of drought, flooded cities, and ruined ecosystems is so imminent that some Prophets have begun thinking, however anxiously, about their own form of geoengineering. Like Wizard-style geoengineering, it is animated by a vision of the past. In the Prophets' case, though, the past is ancient: the end of the Carboniferous epoch.

Most modern forms of plant and animal appeared in a spasm of evolutionary creativity that began about 550 million years ago. For almost all of that time, carbon dioxide levels have been high—sometimes twenty times higher than they are now—and the world was, by today's standards, unbearably hot. Only twice during this period has the world experienced prolonged periods of lower temperatures: our own epoch—more exactly, the last 50 million years—and the end of the Carboniferous. The Carboniferous, one recalls, was the period in which large land plants emerged: lepidodendrons, horsetails, giant ferns, and a host of other now-vanished species. Forests grew in such proliferation that they sucked huge amounts of carbon from the air. Average temperatures fell from 75°–85°F to something like 50°F, lower than today's average of 55°–60°F—low enough to set off not one but two ice ages, killing huge numbers of plants and setting in motion the creation of coal.

Could natural systems (or natural-feeling systems) be harnessed to suck carbon from the air? Why not create a new Carboniferous by covering the two biggest deserts in the world—the Sahara and the Australian outback—with trees? In 2009 three researchers—two at the NASA Goddard Institute for Space Studies, one at the Mount Sinai School of Medicine, all in New York—proposed just that. At bottom, the idea is easy to understand. Very roughly speaking, humankind emits 40 billion tons of carbon dioxide a year, mostly by burning fossil fuels. About 40 percent of the total is absorbed by plants, microorganisms, and the ocean. Foresters have spent decades measuring the rates at which trees grow, which in turn is a measure of their capacity to take carbon dioxide out of the air. If one takes foresters' measurements seriously, covering all 3.8 million square miles of the Sahara with drought-tolerant *Eucalyptus grandus* would suck roughly 20 billion tons of carbon dioxide from the atmosphere every year—enough to have a substantial impact on climate change, though its exact size would depend on the reaction of the oceans and land plants. Still more carbon dioxide could be tucked away by foresting the Australian outback, which is almost two-thirds the size of the Sahara.

Tree planting, advocates say, is simpler and less risky than high-tech Wizardly schemes. Instead of building costly carbon-capture facilities and nuclear plants, people should install cheap, natural carbon-eating mechanisms—trees—in equatorial deserts. Unlike carbon-capture plants, trees in carbon farms represent a direct solution to the problem of climate change. Adding sulfur to the air, geoengineer-style, would make the world less hospitable by harming the ozone layer. Planting trees in the Sahara, the Arabian desert, the Kalahari, or the Australian outback would make these parts of Earth more habitable, even desirable. The trees would increase humidity, which in turn should increase rainfall. Land that is now sterile would become farmland for carbon, and then, possibly, just farmland.

All climate-change measures will involve people in developed nations paying a lot of money, Klaus Becker and Peter Lawrence of the University of Hohenheim, in Stuttgart, have contended. Now present those taxpayers with two alternatives, "one that requires the introduction of untried and potentially hazardous new technology on their own doorsteps, and one that involves the establishment of forests in underpopulated countries far away with possible related benefits for the local populations." Potently combining the virtues of altruism and Not In My Back Yard, carbon-farming is more politically feasible, from this perspective, than either carbon capture or nuclear power.

To get an idea of what a massive reforestation project might be like, visit the Sahel. Technically, the name "Sahel" refers to the arid zone between the Sahara Desert and the wet forests of central Africa—a broad east-west band that runs from Mauritania on the Atlantic through Burkina Faso, Niger, and Chad to Sudan on the Red Sea. Rhetorically, "Sahel" is a watchword for famine and desertification. Until the 1950s the Sahel was thinly settled. When the population boom began, people from the more crowded areas to its south shifted north, into the empty zone. Like city slickers moving into the sticks, they didn't know how to work this dry land. In the 1960s problems were masked by unusually high rainfall. Then came two waves of drought, one in the early 1970s and a second, worse episode in the early 1980s. More than 100,000 men, women, and children died in the ensuing famine—probably many more.

In Burkina Faso, an aid worker named Mathieu Ouédraogo assembled the farmers in his area to experiment with soil-restoration techniques, some of them traditions that Ouédraogo had read about in school. One of them was *cordons pierreux*: long lines of stones, each no bigger than a fist. Because the area's rare rains wash over the crusty soil, it stores too little moisture for plants to survive. Snagged by the *cordon*, the water pauses long enough for seeds to sprout and grow in this slightly richer environment. The line of stones becomes a line of grass that slows the

Yacouba Sawadogo, 2007

water further. Shrubs and then trees replace grasses, enriching the soil with falling leaves. In a few years, a minimal line of rocks can restore an entire field. As a rule, poor farmers are wary of new techniques—the penalty for failure is too high. But these people in Burkina were desperate and rocks were everywhere and cost nothing but labor. Hundreds of farmers put in *cordons*, bringing back thousands of acres of desertified land.

One of the farmers was Yacouba Sawadogo. Innovative and independent-minded, Sawadogo wanted to stay on his farm with his three wives and thirty-one children. "From my grandfather's grandfather's grandfather, we were always here," he told me. Sawadogo laid *cordons pierreux* across his fields. He also hacked thousands of foot-deep holes in his fields—*zaï*, as they are called—a technique he had heard about from his parents. Sawadogo salted each pit with manure, which attracted termites. The insects dug channels in the soil. When rain came, water trickled through the termite holes into the ground, rather than wash away. In each hole Sawadogo planted trees. "Without trees, no soil," he said. The trees thrived in the looser, wetter soil in each *zaï*. Stone by stone, hole by hole, Sawadogo turned fifty acres of desert waste into the biggest private forest for hundreds of miles.

To my untrained eye, his forest looked anything but miraculous:

Using little but rocks and shovels, farmers in broad stretches of the Sahel have restored savanna forests like this one in Burkina Faso. Comparing this area with what had been the same spot just a few years before (photo in picture) provides vivid evidence of the power of reforestation to change landscapes rapidly.

an undistinguished tangle of small trees and shrubs interspersed with waist-high grass. Then Sawadogo showed me a photograph of his land at the time of the drought: bare reddish soil, tufts of grass, a few dusty bushes. Hardly a tree was in sight. For me to think his land looked undistinguished was like looking at a functioning automobile somebody built out of junk in the basement and sneering at the paint job.

At his home Sawadogo had a list of the tree species in his forest, compiled by a botanist in Ouagadougou, the capital. Atop the list was *Jatropha curcas*, a small, shrubby tree with nuts used to make fuel oil. In 2014 German researchers dug up jatropha trees from Luxor, Egypt, and measured their carbon content. They determined that an acre of desert jatropha warehouses the carbon from 209.5 tons of carbon dioxide every year. On average, each U.S. citizen emits 18.7 tons of carbon dioxide per year; each German, 8.9; each Indian, 1.7. If jatropha carbon-storage values are typical, walking through Sawadogo's fifty-acre tree farm was like pushing through a crowd of 560 Americans, 1,175 Germans, or 6,160 Indians.

Unsurprisingly, the new techniques, uncomplicated and inexpensive, spread far and wide. The more people worked the soil, the richer it became, the more trees grew. Higher rainfall was responsible for part of the regrowth (though it never returned to the level of the 1960s). Another contributing factor, possibly, was higher atmospheric carbon-dioxide concentrations—they make rubisco's job easier. (A big study in 2016 suggested that as much as half of the world's vegetated area was becoming somewhat greener, with increased carbon-dioxide levels responsible for most of the additional growth.)* But mostly the restoration of Burkina was due to the efforts of individual men and women. Next door in Niger was even greater success, according to Mahamane Larwanou, a forester at Dioffo University in Niamey. With little or no support or direction from governments or aid agencies, local farmers used picks and shovels to reforest more than forty thousand square miles, an area the size of Virginia.

The carbon farms envisioned by Prophetic geoengineers are much bigger and located in even drier areas. Initially, they would require irrigation. In many cases the water would have to come from desalination plants on the shore. At first the plants would probably run on solar energy; after about three years, they could be driven by trimmings, leaves, and nuts from the trees. Studies suggest that trees could provide enough power to provide their water—they would be, so to speak, sustainable. After several decades, carbon farmers would harvest the trees and replace them with new, fast-growing saplings. All of this would be expensive, but all carbon remediation schemes are expensive. It is not ridiculous to imagine that the economic activity from making the Sahara habitable would offset some of the costs.

The old trees could be "pyrolized"—burned in low-oxygen environments, which turns them into charcoal. Depending on how it is produced, charcoal typically retains about two-thirds of its original carbon. The charcoal can be ground and buried, enriching the soil. Desert soils

---

* At a certain point, the benefits of higher carbon-dioxide levels to plant growth are outweighed by the negatives, including drought, heat stress, and enzyme failure. The exact line varies from species to species. For rice the limit may be quite low, because at just slightly higher temperatures than usual it can't produce fertile pollen.

tend not to hold nutrients and organic matter because they are made from types of dirt that don't bind to them chemically. Any precipitation makes them wash away. Over time, buried charcoal slowly oxidizes, providing the requisite binding sites. Nutrients and organic matter "stick" to it, providing food for the bacteria, fungi, and other microorganisms that make soil fertile. Charcoal, properly manufactured and deployed, can dramatically improve bad farmland. It also stores carbon: Johannes Lehmann, a charcoal-soil expert at Cornell University, has calculated that turning residues from agriculture and logging into charcoal could offset as much as an eighth of the world's carbon dioxide output if the gases from charcoal-making were captured and turned into fuel. The figure is higher if the climate-changing gases methane and nitrous oxide, emitted by rice paddies and fertilizer, are included. Presumably these techniques could be applied in carbon farms.

Typically Wizards react to these scenarios by pointing out their unfeasibility. The forests would destroy desert ecosystems, they say. Or they would require large numbers of people to radically change the way they live. Or it amounts to green imperialism—forcing poor people in desert areas to offset the emissions of faraway rich people. These criticisms are a distorted mirror of the Prophetic criticisms of Wizardly ideas. Prophets are appalled by the top-down nature of CCS and nuclear power, which depend on unelected technical experts. They like reforesting, which functions best when it is bottom-up, harnessing the willing participation of people like Yacouba Sawadogo. Either one must coordinate the actions of millions of people to have an impact or create processes that need so few people that they can't be controlled.

What to do, in a world brimming with fossil fuels? In climate change, all choices involve leaps into the unknown. Claims that carbon capture cannot be economically viable or that renewables will always cost too much or use too much land generally amount to saying, I prefer the unknown risks associated with this course rather than the unknown risks associated with that course because the first leads to a future that I like better. At bottom, the choices stem from private images of the good life—a life in which people are tied to the land or free to roam the skies. Only individuals can choose. The important thing is that they have choices, and we are still at the stage where, however dimly, we can imagine that Lynn Margulis was wrong.

TWO MEN

EIGHT

The Prophet

Launch

Washington, D.C., 10:00 a.m., December 27, 1947. A thousand feet, give or take, from the Lincoln Memorial. Boardroom of the U.S. National Academy of Sciences: walnut-panel walls, Persian carpet, marble fireplace, massive oil painting of a stone-faced Abraham Lincoln signing the academy charter. Pale winter light through the shades. A long oval table, gleaming in that light. Above the table a spherical glass light fixture, an electrolier, painted to resemble the map of the world drawn (or thought to have been drawn) around 1515 by Leonardo da Vinci. Bathed in its yellow glow: a huddle of men in dark suits and white shirts. High officials from the U.S. Department of the Interior, the Department of Agriculture, the National Research Council—and two civilians. The first civilian is William Vogt, head of the Conservation Division of the Pan American Union. One imagines him nervous and quietly excited to be in this room, literally in the meeting before the meeting, surrounded by people whose hands rest comfortably on the levers of power.

The second civilian was the man who had convened the gathering: Julian Huxley. Huxley was from a distinguished English family. His grandfather, T. H. Huxley, was famed for his bristling advocacy of evolution; his younger brother Aldous was the controversial author of *Brave New World*; and his younger half brother Andrew, a biophysicist, would go on to win a Nobel Prize. Julian himself had made major contributions to evolutionary theory, lectured across Europe and the United States, collaborated with H. G. Wells on best-selling books about biology, and filmed a natural-history documentary, possibly the world's first, which won an Oscar. A prominent anti-racist, he was also a prominent advocate of removing "inferior" elements from the human gene pool. He was now the founding director general of UNESCO, the United Nations Educational, Scientific, and Cultural Organization. And he was spoiling for a fight.

Urbane, domineering, and gossipy, Huxley regaled the others with detailed accounts of back-room maneuvering in Washington, London, and Paris. The maneuvering was about the shape of the postwar world. Years of conflict had laid waste to big swathes of Europe and East Asia and set off the dissolution of colonial empires. The victorious Allies were naturally intent on rebuilding. In the past, nations had been either colonizers or colonized, rulers or ruled, homelands or possessions. After global war the old hierarchies no longer seemed to apply. A different way of categorizing the world was required. In the new vision, all nations were on a single path from "underdeveloped" (like most of Africa) to "developed" (like Europe and the United States).

Before, the goal had been maximum political power for the imperial home, with colonies feeding it. Now the focus was maximum development for all: bustling industries, vibrant cities, affluent homes full of labor-saving appliances—the consumption-driven economic growth extolled by the economist John Maynard Keynes and his followers. Now the U.S. government was promising that it would "use all practicable means . . . to promote maximum employment, production, and purchasing power."

This goal should not be restricted to the United States, believed President Harry S. Truman. Everyone on Earth should be able to live like middle-class Americans! Not only was this the moral objective for Western nations in setting up the postwar order, it would win over former colonies in the struggle against Communism. And the way to reach this idyllic state was by benevolently deploying the latest science and technology—the physics, chemistry, and engineering that had created the atom bomb—to guide growth and reconstruction.

Huxley was appalled. In his view, these ideas would license big corporations to use researchers' discoveries to pillage what remained of the natural world. Like Truman, Huxley believed that scientific expertise could guide society into a more rational and prosperous form. But he thought it should do this by bringing nature and civilization into bal-

ance. Researchers could identify ecological limits, and teach governments
how to live within them. Among the
biggest obstacles to this biological
reordering of the world, in Huxley's
view, were the blindly pro-growth
policies of the U.S. government.
From the officials in the Academy
of Sciences boardroom, sympathetic
despite their positions in the Truman
administration, Huxley was seeking
advice on how to create voices for
nature in Washington. He wanted
something in addition from Vogt: to
know whether Vogt, from Huxley's

Julian Huxley, 1964

point of view a little-known but promising bureaucrat/ornithologist,
was ready to step onto the world stage.

After the meeting Huxley and Vogt talked. Surely it was an exciting
moment for Vogt. Speaking to Huxley, with his first-class Oxford degree,
his links to scientists around the world, his string of best-selling books,
was about as far from the Chincha Islands as it was possible to get. And
Huxley had sought out Vogt, had questions for him, possible plans.
No record exists of their conversation, though presumably Vogt talked
about his forthcoming book, *Road to Survival*. Whatever the course of
discussion, it is clear that Vogt satisfied Huxley. The two men kept in
touch, sometimes by letter, sometimes through their mutual acquaintance, Vogt's friendly rival Fairfield Osborn.

During the next year Huxley watched *Road* become an explosive
best seller, making Vogt—and Osborn, who had published a competing book—a prominent advocate for reducing human demands on
the world's ecosystems by reducing human numbers. Huxley and his
brother Aldous believed with equal passion in the same cause, but had
had much less success in gaining an audience. Meanwhile, Vogt was
poised to become a statesman of ecology.

Then it fell apart. Vogt was undone by the same traits that had
brought him so far—his fervent ambition, his abrasive insistence on
going his own way, his instinct for the sweeping, dramatic conclusion.

He recovered, but a few years later was again brought down. This time there was an additional, more important cause: his inability to see the human species in other than biological terms. Vogt was doggedly convinced, Margulis-style, that people were not exceptions to biological rules. And when our species hit the edge of the petri dish, it would take much else with it.

"Forty Thousand Frightened People"

At least they let him resign. He was working for the Pan American Union, a diplomatic forum for the twenty-two independent American nations. It was one thing when he criticized these nations' stewardship of their natural endowment in thickly footnoted memoranda for the union. It was an entirely different thing when he repeated those criticisms in the press for millions of readers.*

Vogt's first article, "A Continent Slides to Ruin," published in *Harper's* in June 1948, had arguably drawn more attention to the state of Latin American landscapes than all of his academic reports together. But its unsparing language had raised hackles in the Pan American Union. Vogt had attacked Chile, for instance, for not having "a single fire warden or forester in its employ," despite having lost "a quarter of a million acres" to fire two years before. Loggers—"lumber exploiters," to Vogt—were not replanting trees, leading to massive erosion and floods. In consequence, Vogt predicted, "the greater part" of Chile would become a desert "within one hundred years," perhaps less.

Unsurprisingly, the Chilean government didn't like public criticism from someone who was, technically speaking, one of its employees. The Chilean ambassador complained to the Pan American Union's governing board. This was easy: the ambassador was a board member. Vogt, the ambassador said, had to make a choice: "Continue to propagandize or leave the Union."

Union secretary general Alberto Lleras Camargo fended off the complaint. Lleras was the former president of Colombia, but he had begun

---

* In April 1948 the Pan American Union was reconstituted as the Organization of American States, its current name. Because the new charter did not come into effect until 1951, I refer to the group here as the Pan American Union.

his career by reporting for muckraking newspapers in Bogotá and Buenos Aires. He was reluctant to come anywhere near censorship. In addition, Vogt was a capable, industrious worker whose conservation reports had been useful to Mexico, Venezuela, Costa Rica, and other member nations; through his recommendation, Chile itself had established a national park in Tierra del Fuego. Trying to strike a balance, Lleras refused to fire Vogt but did ask for the rhetoric to be toned down.

Fairfield Osborn, 1960

Two months later, in August, *Road to Survival* appeared. Vogt had known that his focus on population control would be controversial. A few weeks before publication, he joked to a friend about its likely reception: "A lot people will feel that I should ring a bell and say, 'Unclean.'" To his surprise, the book became what his publisher called "the most dramatic and widely-discussed book of the year." The repercussions were international; years later, the noted French demographer Alfred Sauvy would recall that in Europe *Road to Survival* "set off a stir quite comparable to that raised in the beginning of the nineteenth century by Malthus's *Essay on the Principle of Population*."

At a stroke, Vogt and Osborn had put into public view concerns that had been building up for years in parts of the scientific community. Spurred by the war's environmental destruction and consumption of natural resources, ecologists around the world were warning about pollution, deforestation, erosion, and soil degradation. "Man's command over nature has grown more rapidly than his mastery of himself," the Yale botanist Edmund Sinnott claimed a month after *Road*'s publication. "Man, not nature, is the problem today."

Sinnott was president of the American Association for the Advancement of Science (AAAS), then the world's biggest and most influential scientific body. To scientists like him, Vogt and Osborn were simply affirming, loudly and publicly, what they already believed, and their books were all the more welcome for it. "Anyone with any technical knowledge understands that the dangers described in these books are real enough," the prominent Yale ecologist G. Evelyn Hutchinson argued.

Hutchinson praised Vogt and Osborn at a special AAAS symposium led by Sinnott and attended by thousands of scientists. The theme and tone of the session were conveyed by its title: "What Hope for Man?" The audience in the great conference hall, according to *The New Republic*: "Forty Thousand Frightened People." More than forty thousand, actually: Osborn's speech was broadcast on a popular nationwide radio program, guaranteeing millions of listeners in those days before television and the Internet.

It wasn't easy for a diplomatic organization like the Pan American Union to have an employee be the focus of an international uproar. Yet it wasn't easy to dump him, either. At the same time that the clamor was mounting, Vogt was organizing a large international symposium for the union. Held in Denver in September 1948, the Inter-American Conference on Conservation of Renewable Natural Resources was addressed by President Truman and attended by many of the hemisphere's conservation officials, including Vogt's former boss at the guano company. During the thirteen days of the meeting, the 1,500 conferees traveled hundreds of miles to see U.S. conservation projects in the Rocky Mountains and Great Plains. Tucked into a suit as gray as his hair, his baritone rumbling beneath his lightly tinted protective glasses, his limp and cane making him instantly identifiable, Vogt presided over the gathering like a kindly but careworn deity. Delegates repeatedly rose to thank him for his work. Not one of the sections mentioned birth control or overpopulation—a nod to Vogt's superiors at the Pan American Union. Giddy with success, Vogt said the conference "may well" come to be regarded "as one of the most important meetings of our decade, and even our century."

In today's world of cheap air travel and massive global symposia, Vogt's enthusiasm seems overblown. But after the Second World War a dozen or so international congresses did, in fact, lay out much of what has been the world order ever since. One such meeting in June 1945 chartered the United Nations. Five months later a second set up UNESCO and chose Huxley as its head. These were preceded by an international meeting in Bretton Woods, New Hampshire, that created the International Monetary Fund and the World Bank, followed by the signing of the General Agreement on Tariffs and Trade at a conference in Geneva. Hashing out conservation required almost as many sessions as the others combined, partly because the issues were poorly understood—and partly because Julian Huxley was picking a fight.

Huxley had managed to slip protecting nature into UNESCO's purview, even though it didn't fit into the organization's stated focus on education, science, and culture. But there was the vexed question of what, exactly, "protecting nature" meant. Huxley believed that the Truman administration's view—protecting nature meant using it wisely for human benefit—would end up putting a shiny green gloss on the same old piggish way of doing things. Huxley instead wanted to preserve the world's most beautiful places by fencing them inside parks and reserves—homelands for marvelous creatures like lions, tigers, elephants, and migratory birds. In these zones, industrial development would not be managed, it would be forbidden. And they would not be small, token efforts, but vast expanses—entire landscapes, permanently removed from human exploitation. An adroit politician, Huxley knew that these ideas would meet resistance.

It's tempting but not quite accurate to call Huxley's reaction to Truman a harbinger of the divide between Wizards and Prophets. At the time neither side had formulated its arguments. Vogt had written his book, but its message had not yet become common currency—Huxley, for instance, was trying to protect charismatic animals, rather than seeing himself, Prophet-style, as concerned with planetary ecological limits. Borlaug was still laboring in obscurity. Warren Weaver had not begun to think about hacking photosynthesis, still less to write his Wizardly manifesto about "usable energy." J. I. Rodale had promised that "the Revolution has begun," Albert Howard was thundering about the Law of Return, and Guy Callendar was warning about carbon dioxide, but few were paying attention. Truman was two years away from introducing the word "underdevelopment" to the world. Even further off—eight years—was the first commercial nuclear power plant, Calder Hall at Windscale, England.

Instead it might be better to posit that Huxley had unknowingly brought into the international realm a dispute that had been simmering for decades in the United States. The dispute was between John Muir and Gifford Pinchot, two of the more consequential figures in conservation history. Born in 1838, Muir was a Civil War draft dodger and college dropout who became a seer, a bearded, ragged mountain man who seemed to live on air and sleep on the stones of his beloved western peaks. When Muir was growing up, "wilderness" meant to most people wastelands full of dangerous creatures: places to be subdued. Muir came to view those unpeopled areas as spiritual homelands: places to be cherished and saved. "In God's wildness," he said, "lies the hope of the world—the great fresh unblighted, unredeemed wilderness." True meaning, he thought, could not be found in the world's increasingly crowded, noisy, and mechanized cities. Only in untouched nature could the spirit be redeemed. "Wildness is a necessity," he said. "Mountain parks and reservations are useful not only as fountains of timber and irrigating rivers, but as fountains of life." Muir's relentless advocacy led to the creation of the world's first national park, Yellowstone, in 1872, preserved largely to protect its geysers, hot springs, and other geological oddities, and the world's first wilderness park, Yosemite, in 1890. Soon after, wilderness parks were established across the globe, many of them in European colonies; Tsitsikamma, in South Africa, as an example, was set aside a few months after Yosemite.

Pinchot was the first professional forester in the United States. He admired and respected Muir but on the whole regarded the other man's mystic effusions as hooey. Instead of individual spiritual enlightenment, Pinchot sought the common material good—"the greatest good, for the greatest number, for the longest run." Born in 1865 to a wealthy family, he was a shrewd self-promoter, clever with other people's ideas, who cast himself as an avatar of Science (in fact, he had attended a year of forestry school in France, leaving before his professors thought he was ready). An inauthentic scientist but a visionary as authentic as Muir, he proclaimed that the world's prosperity depended on sustaining its resources, especially renewable resources like timber, soil, and freshwater. He wanted to protect them not by leaving great swathes of terrain free from human influence but by managing forests and fields with an elite cadre of scientific mandarins. "The first principle of conservation is development," he said. Development had to be conceived in the long term: "the welfare of this generation and afterwards the welfare of the generations to follow." He said, "The human race controls the earth it lives upon."

Although Muir and Pinchot agreed on many means, they disagreed about most ends, and became increasingly estranged. President Theodore Roosevelt, the first leader of any nation to place conservation at the heart of his agenda, had gone camping with Muir and thoroughly enjoyed his stories. But he chose to work with Pinchot, appointing him the nation's chief forester. In this position, Pinchot said, his goal was "perpetuation by wise use."

As it turned out, both Muir's rapture over wild beauty and Pinchot's thoughts of stewardship had a dark side: most of these "untouched" American landscapes in fact were inhabited by indigenous peoples. Yellowstone and Yosemite were turned into parks by expelling people who had been there for centuries. As the journalist Mark Dowie has documented, similar dispossessions in the name of Nature have taken place ever since. All too often, the results have been dreadful, both morally, because they involve tearing people from their homes, and practically, because these areas were molded in the shapes of their first inhabitants. The peoples of the U.S. West, for example, burned undergrowth frequently, to discourage insects and encourage the tender new growth that attracted animals. Eliminating fire in the name of forest protection has created a buildup of flammable material that in turn has led to devastating wildfires. Similarly, peoples in the Amazon forest reshaped their ecosystems by creating small clearings, which they filled with useful plants and fertilized with waste and charcoal. The results are some of the richest, most diverse areas in the forest—areas that now risk degradation without indigenous management.

"The Necessary Intellectual Scaffolding"

Following Pinchot's suggestion, Roosevelt convened a conference on U.S. natural resources in 1908 that led, as I discussed earlier, to an oil panic. Believing that resources for the hungry human enterprise needed to be managed worldwide, Pinchot urged Roosevelt to convene the first-ever global conference on conservation. Roosevelt agreed, but when he left office his successor, William Howard Taft, canceled the scheme. Pinchot lost favor with Taft in an ill-considered political fight and resigned. He bided his time until Roosevelt's distant cousin, Franklin Delano Roosevelt, became president. Badgered by Pinchot, the second Roosevelt agreed, like the first, to hold a global resources conference, though only after the war. He died while the war was still raging. The indefatigable Pinchot, then eighty-one and suffering from terminal leukemia, turned to Roosevelt's successor, Harry Truman. In September 1946, one month before Pinchot's death, Truman called for a conference.

The United Nations asked UNESCO to assist in preparing for the global summit, but Huxley was not inclined to play along with this Pinchot-inspired endeavor. In an amazing display of political legerdemain, Huxley managed in December 1947 to wrangle approval from the reluctant member nations of the United Nations to hold a second, competing global conference on resources. Paid for by the United Nations, like the first, but controlled exclusively by UNESCO, it would take place at exactly the same time and same place as the other conference. But it would trumpet a more Muir-like vision of the future, counteracting (or so Huxley hoped) the first conference. Approval in hand, the elated Huxley flew to Washington, D.C. That was when he met with Vogt and U.S. officials.

In the boardroom of the National Academy of Sciences, Huxley described his plans. Not only did he want UNESCO to stage a parallel conference, he wanted to create a parallel environmental bureaucracy, independent of UNESCO but financially supported by it. As the new organization's director later put it, it would "weld the tiny embryonic nucleus of European and American naturalists already converted to conservation into a powerful, constantly expanding, worldwide network of 'conservationists' . . . from all walks of life: politicians, economists, civil servants, pioneer ecologists, field workers, lawyers, directors of NGOs, and so on." Tens of thousands of people in Europe, North America, and their former colonies belonged to environmental groups, but these groups were separate from each other and had narrow agendas. If they were able to work together, share information, and expand their purview, Huxley believed, they could gather information on a global level and push back as a unified force against destruction wherever it occurred. UNESCO had been given the task of assembling the pure science necessary to understand ecosystems and landscapes. This new, parallel organization would undertake the applied science of conservation to safeguard natural systems against human depredation. Preliminary planning meetings had already occurred.

Huxley was helping to create a new kind of establishment: a self-appointed network of powerful individuals on a mission. Representing bands of concerned citizens, the environmental establishment would wield some of the functions of government but have little government oversight, subverting what Huxley saw as political leaders' destructive addiction to growth while benefiting from the clout acquired by that same addiction. Today the network's members include the Sierra Club, the Nature Conservancy, Greenpeace, the Rainforest Alliance, the World Resources Institute, 350.org, Conservation International, and the Worldwide Fund for Nature (known as the World Wildlife Fund in North America), and hundreds of other groups, local, national, and international, as well as countless environmental foundations, environmental journalism groups, environmental scientists, and environmental government agencies. Despite frequent quarrels and funding crises, these diverse entities have worked together with remarkable effectiveness. (Similar networks have sprung up in the fields of health and development.)

Over time, this loosely defined environmental establishment—British critics sometimes call it the "green blob"—would go on to lead campaigns against pollution, awaken the world to threats of extinction, acquire and set aside huge tracts of land, and play a prominent role in the sterilization of millions of women, under varying degrees of compulsion. These efforts have been both celebrated as profoundly democratic, in that they represent the views of voluntary groups, and attacked as profoundly undemocratic, in that citizens with opposing views have had little ability to check their actions. Either way, it has been a massive and unprecedented experiment in independent governance. Huxley's work was a powerful impetus in its creation.

Two days after the Denver conference finished, the meeting to establish this second organization began. Held in the former royal chateau of Fontainebleau, southeast of Paris, it brought together representatives of 23 governments, 126 nature groups, and 8 international organizations. The attendees—almost all of them white and male—met in the chateau's ornate Salle des Colonnes, a long room lined by black marble columns topped by gilded capitals. Outside the great windows was a view of the private forest where royal personages had once harried pheasants. As delegates took their chairs on the parquet floor, Huxley laid out his priorities in his opening address. In U.S. terms, he acknowledged Pinchot, but he embraced Muir. Yes, Huxley said, the natural world was a resource for humanity, but it had a greater value entirely apart from its potential for use. And he extolled the extraordinary parade of the world's creatures, which "are something in their own rights, are alien from us, give us new ideas of possibilities of life, can never be replaced if lost, nor substituted by products of human endeavor." At the end of the meeting, most of the attendees signed the constitution for what was called the International Union for the Protection of Nature (IUPN).*

Vogt was at Fontainebleau as an observer for the Pan American Union. Getting to France had not been easy; Vogt's immediate superior at the union, unwilling to give him an unsupervised public forum, had refused to let him attend. While Vogt was in Denver, Huxley reached out to union secretary general Alberto Lleras. Just nine days before Fontainebleau began, Lleras reversed the decision. Vogt hastily flew to France, arriving in time to lead the meeting's second session. No text exists of his remarks, but a secretary summarized his conclusions, startlingly apocalyptic to some attendees but familiar to readers of *Road to Survival*. In every continent, he said, resources were

being plundered, by the increasing number of consumers all trying to compensate for their previous economic failures by intensifying their methods of exploitation. . . . Unless he adopted a rational concept of land use capabilities, recognizing that certain lands could produce certain things with ease which could not be extracted from others without risk of exhausting their resources, man must end by destroying himself on a sadly plundered planet.

Education, Vogt said, was key. People had to learn that if all aspired to a Western standard of living, the pressure on ecosystems would be unbearable. They had to learn about carrying capacity and limits. To Vogt, IUPN's most important task "would be to promote knowledge of

---

* The IUPN became the International Union for the Conservation of Nature and Natural Resources (IUCN) in 1956. Today it has a membership of 1,200 governmental and private organizations and coordinates the work of 10,000 or more pro-bono scientists. Although IUCN has global reach, it is not well known in the United States, because the U.S. Fish and Wildlife Service performs many of the functions undertaken by IUCN elsewhere.

human ecology." He closed to applause. Vogt, the lead British representative said, "had gone to the root of the matter." Huxley agreed. Humanity, he said, "was decidedly a scourge, of which he himself and Nature were the first victims."

Between sessions, the delegates haggled in the corridors about who would lead the new organization. The Swiss, who had done much of the preparatory work, wanted a Swiss president. Britain demurred; the leading Swiss candidate was vocally sympathetic to the Soviet Union. Equally concerned were the Dutch, who believed their nation's long tradition of managing nature gave them the edge—hadn't they kept back the sea for centuries? Huxley wanted someone from the United States, which was paying most of the bills for the U.N., UNESCO, and the other new international institutions. He believed, probably correctly, that having a U.S. citizen in charge would help ensure IUPN funding. To his mind, the best candidate was Vogt: a proven administrator, as shown by his Pan American Union work, a patriot who had spied on Nazis in Latin America. He was, unusually, an American who was already an international civil servant.

Most important, Vogt was sympathetic to Huxley's ideas. A few weeks before coming to UNESCO, Huxley had written a sixty-two-page philosophical manifesto for the organization. The kind of high-toned rumination rarely associated with international bureaucrats, it posited that UNESCO had a single mission: "to help the emergence of a single world culture." Evolutionary biology, Huxley wrote, had given our species "the necessary intellectual scaffolding" for this great work. Guided by scientific authorities, humanity would control its own biological and social evolution, replacing random natural selection with purposeful human selection toward a peaceful, interconnected future. Two steps would be necessary to accomplish this "evolutionary progress," he said: "world political unity" (to create species-wide rules), and global population control (to control human development). UNESCO, he said, should lay the foundation for both. In this unified, self-controlled civilization, nature would be protected as a matter of course.

Unsurprisingly, Huxley's manifesto was criticized. U.N. member nations wanted no talk of world government; conservatives rejected his implicit advocacy of birth control; leftists loathed the notion of genetically molding humanity, which smacked of Nazi policies. Huxley officially backed off, though he remained convinced that society should be organized in light of biological principles—and that failing to do so could lead to environmental ruin.*

Vogt, for his part, was stirred by Huxley's plans for the IUPN. In some ways the new organization was a global version of what Vogt had tried to create with the Audubon Society a decade before. But rather than being a citizens' association, it was an elite corps of government-funded experts that would harness the efforts of volunteer groups. And it was focused on planning—something that interested Vogt, who had seen during the war the feats of mobilization that states could accomplish. To Huxley he suggested that the IUPN pick a Swiss director, but give him the position only until the two simultaneous United Nations conferences. Then Vogt could stand for office—backed, he hoped, by the Pan American Union, which might pay his salary. Huxley agreed to the short term of office, but chose a Belgian: Jean-Paul Harroy, secretary general of the Brussels Institut pour la Recherche Scientifique en Afrique Centrale and the former director of Belgium's parks in the Congo. The new organization would be based in Brussels, in the same building as Harroy's institute.

As befit his newly raised status, Vogt returned to the United States in a first-class cabin on the RMS *Queen Mary,* a luxury transatlantic liner, arriving on October 14 in New York. Four days later he was a keynote speaker at yet another symposium, this one in the grand ballroom of the Waldorf Astoria hotel, an art deco landmark in Manhattan. With Vogt that day on the dais were, among others, Fairfield Osborn; Bernard Baruch, a long-term presidential adviser; the best-selling, Pulitzer-winning historian Bernard DeVoto; and the best-selling, Pulitzer-winning novelist and organic farmer Louis Bromfield. The discussion moderator was the chair of the House Agricultural Committee. Addressing the group was New York governor Thomas E. Dewey, the Republican candidate for president; the election was two weeks away. Some of the addresses, including the one by Vogt, were broadcast nationwide. When Vogt decried "the obliviousness everywhere in the United States," he had the satisfaction of knowing that his words were being heard everywhere in the United States.

---

* Huxley's belief that society should be run by scientific experts was similar to the ideas of M. King Hubbert and Technocracy. But Huxley was influenced less by Technocracy than by the putatively scientific Five-Year Plans of Stalin's Soviet Union. In effect, Huxley wanted to introduce Stalinist planning but without the brutality of Stalinism itself.

Then, suddenly, he was back in his gray office in Washington, D.C. He bowed to his superiors' pressure and spoke no more in public. While the capital city preoccupied itself with the quadrennial extravaganza of the election, Vogt spent his days on the drudgework of preparing the Denver conference proceedings for publication. Truman won the election by a convincing margin, astonishing political handicappers. In the subsequent weeks of celebration in Washington, Vogt worked, almost covertly, with Huxley and others on the UNESCO/IUPN symposium (the second, competing meeting). Presumably, he also reacquainted himself with his wife; Vogt had traveled solo to Denver, Fontainebleau, and New York City. He was quietly buoyed by the reaction to his book and speeches and the possibility that he might lever himself into the leadership of the IUPN while still keeping his salary from the Pan American Union. There would be a force to educate the world, and Vogt would help direct its efforts. Then, on January 20, 1949, he heard Harry S. Truman's inaugural address.

Point Four

The speech was a call to arms in the battle against Communism. This struggle, Truman said, would have four major fronts, which he detailed in four "points." The first point was "to give unfaltering support to the United Nations." The second was to continue supporting Europe's recovery from the war. The third was to forge with Europe "a joint agreement designed to strengthen the security of the North Atlantic area"—an effort that would lead to the formation of the North Atlantic Treaty Organization (NATO), an alliance against the Soviet Union.

All of these were already part of U.S. policy. Point Four was something new, and Truman spent more time on it than on the others. "More than half the people of the world are living in conditions approaching misery," he said.

> Their food is inadequate. They are victims of disease. Their economic life is primitive and stagnant. Their poverty is a handicap and a threat both to them and to more prosperous areas. For the first time in history, humanity possesses the knowledge and skill to relieve the suffering of these people. . . . I believe that we should

make available to peace-loving peoples the benefits of our store of
technical knowledge in order to help them realize their aspirations
for a better life. . . . The old imperialism—exploitation for foreign
profit—has no place in our plans. What we envisage is a program
of development based on the concepts of democratic fair-dealing.

Technology-driven economic growth was the way to a better world, Truman said. "Greater production is the key to prosperity and peace. And
the key to greater production is a wider and more vigorous application
of modern scientific and technical knowledge."

Truman's Point Four, the historian Thomas Jundt has written, "was
the opening salvo in the U.S. postwar mission to modernize former
colonies through intensive economic and technological development."
Researchers, private groups, and federal officials would join hands to
reshape the new nations into affluent Western-style democracies. "The
goal was humanitarian—to improve the standard of living" in poor
places. But, Jundt notes, "it was also strategic." By helping former colonies, Truman hoped to prevent them from falling into the Soviet orbit.

Point Four's promise of science-driven development was electrifying. India, Pakistan, Egypt, Ghana, Brazil, Mexico—all embraced rapid
economic growth as a national goal. Like the agricultural scientists in
Mexico who wanted the newest hybrid maize, they wanted all the trap-
pings of modernity: dams, highways, steel mills, power plants, cement
factories, pulp and paper mills, universities crowded with STEM students, cities crowded with modernist cement-and-glass slabs. As an
ultimate accolade for Point Four, the Soviet Union began promising its
allies exactly the same kind of aid to help them reach exactly the same
kind of affluence.

To official Washington, Point Four was a complete surprise. Truman
had not consulted a single member of his staff, including his secretary
of state, before making this unprecedented commitment to promote the
welfare of other nations. The result, Vogt chortled, "was like lifting a
rock slab off an ants' nest. Bewildered bureaucrats, caught with their
ideas down, bumped into one another as they asked, 'What does the
president mean? What is new about this plan? Who is going to carry it
out? What will it cost?'"

Vogt's tone was sarcastic but his point was wholly correct: the very fact that Point Four was unprecedented meant that nobody had any idea how to do it. Nobody knew whether aid was simply a matter of passing on scientific expertise or if it also involved transferring the civil norms—Western ideas about private property, limited government, and the rule of law—that had accompanied the rise of that expertise. Nobody was sure whether "development" should focus on agriculture, so former colonies could feed themselves, or industry, so that they could trade and grow wealthy. On top of that, Point Four had no budget, no personnel, no legislative authority—nothing but Truman's conviction that economic growth had to happen fast for poor countries to prosper and fight Communism. The first post-inaugural meeting of State Department senior staff began with the words: "Well, gentlemen, what do you suppose the President meant?"

To Truman, the way to accomplish Point Four was simple: put scientists and policymakers together in a remake of the Manhattan Project that had built the atomic bomb. (Warren Weaver, who had just written his Wizardly manifesto about "usable energy," would have agreed.) And the opening step in this process was the United Nations conference on resources—"the first assault in a global campaign against want," as *The New York Times* put it in an article describing the multinational "advance guard" of scientists preparing for the meeting. "Day by day these technical experts send an ever-increasing number of charts, graphs, maps, and blueprints to the headquarters of the United Nations, forging the scientific weapons with which the United Nations may carry out the President's 'bold new program' of assistance to underdeveloped countries."

Scientific weapons! Vogt heard all this with horror, as did Huxley, Osborn, and the rest of the IUPN. Another dissenter, perhaps surprisingly, was Gifford Pinchot's widow, Cornelia Pinchot. In a forceful note to Truman in May 1949, she told him that the conference plan had next to nothing about actual conservation; it "ignores the fundamental purpose for which it was called." Osborn, too, wrote privately to Truman about his concerns. And he worked with Vogt and Huxley on UNESCO's competing conference, scheduled, like the first, for August.

Never one to avoid a megaphone, Vogt publicly lambasted Point Four in *The Saturday Evening Post,* the magazine that had excerpted his book four years before. The title—"Let's Examine Our Santa Claus Complex"—indicates his tone. Truman wanted to help people who, though poor, were proudly independent, Vogt said. But "American ostentatiousness and discourtesy" abroad would make aid targets "resent our do-gooding to the point of violence." Point Four would saddle poor nations with huge debts. Worst of all, it would lead to "destructive exploitation" of imperiled landscapes. "If Point Four results in speeding up soil erosion, raiding forests and land fertility, destroying watersheds, forcing down water tables, filling reservoirs [with dams] . . . and wiping out wildlife and other natural beauties, we shall be known not as beneficent collaborators, but as technological Vandals."

On a personal level, the article was unwise. The reaction at the Pan American Union was as immediate as it was negative. Latin American members resented his description of their citizens as "dominated by ancient superstitions and beliefs." The Washington officials who funded his Conservation Department didn't like his sneers at U.S. "bad manners and superciliousness." While the article was still on newsstands, union secretary general Lleras told Vogt that his rote disclaimer that it did not reflect the views of the Pan American Union was insufficient. Still even-toned despite the provocation, Lleras told Vogt that "there is an incompatibility, prone to all types of conflicts, between the active career of a published writer and the post of a chief of a section of the Pan American Union."

Wizards and Prophets

Four weeks later, the initial Pinchot-inspired, Truman-proposed U.N. conference opened in Lake Success, on Long Island, in a former gyroscope factory that had become the temporary headquarters of the United Nations. It was barely five miles from Vogt's birthplace in Garden City. It is interesting to speculate what he made of a meeting there devoted to spreading the industrial prosperity that had overwhelmed the rural landscape of his youth. Officially entitled the United Nations Scientific Conference on the Conservation and Utilization of Resources, the gathering was known by its acronym: UNSCCUR ("un-sker"). Point Four had given it a jolt of urgency. More than 1,100 official contributors, participants, and observers from fifty-two countries representing scores of learned societies, government agencies, and private groups

Garden City, the village where Vogt was born, was near the epicenter of the wave of suburbanization that transformed the Long Island landscape from the fields and forests of Vogt's childhood (top, the village in about 1905) to a sea of middle-class real estate (bottom, Garden City in the 1930s). The shockingly rapid change, which occurred all over the world in the twentieth century, was a powerful impetus to the creation of the environmental movement.

attended, orbited by a Kuiper belt of several thousand U.N. staff, journalists, minor diplomats, minions and toadies of all sorts, and security personnel. Not one delegate came from the Soviet Union or its satellites.

U.S. Interior Secretary Julius A. Krug gave UNSCCUR's opening address. "It's high time that we start a new era in conservation," he said, "an era consecrated to the development and wise use of what is available to the people of the world." Krug had read Vogt and Osborn, and scoffed. "There is not the slightest question in my mind," he said, "that scientists and engineers can find and develop food, fuels, and material to meet the demands of the world's increasing population with a greatly improved standard of living. I do not side with those who 'view with alarm' the increasing world population and the decreasing reserves of some things which now appear to be essential to our way of living." *Those who view with alarm*—that was Vogt and Osborn, sitting in the audience. Minutes later Antoine Goldet, head of the U.N. Department of Economic Affairs, promised that "the disciples of Malthus will be put to confusion."

The conference proceedings sprawl over eight volumes. With so many people at Lake Success discussing so many subjects, diversity of opinion was unavoidable. Osborn gave a talk, for example, insisting that humanity must forge a "new and enlightened" relationship with the environment. (*The* environment, one recalls, was a new concept, introduced by Vogt and Osborn; one sees it coming into focus in these talks.) And there were other expressions of concern. Nonetheless, as the historian Jundt has pointed out, the Borlaugian credo predominated: science and technology, properly applied, will allow us to produce our way out of our environmental worries. UNSCCUR discussions focused on mining oil from oil shale, spraying pesticides on a wide scale, extracting lumber from tropical forests more efficiently, manufacturing artificial fertilizers more cheaply, manipulating plant breeding, expanding nuclear power, and a host of other techno-fixes.

Typical of the gathering was the geologist A. I. Levorsen, dean of Stanford's School of Mineral Science, extolling the happy news that the world would never run out of fossil fuels. Undiscovered petroleum reserves, he said, "are on the order of 500 times the current annual world production." Shortfalls could only occur if societies failed to look for oil, and that could only occur if they failed to organize themselves in the "free enterprise—profit-incentive—system." This was the session that outraged M. King Hubbert and led to a public argument and, eventually, his Vogtian depiction of a peak of production. But few attendees paid attention to Hubbert's objections, or to Osborn's, or to anyone else's. The closing address summarized the meeting as conveying "the ways in which mankind can secure a larger return now and tomorrow, from the earth's resources." UNSCCUR was the public debut of Wizardly thought.

Vogt sat in the first conference's audience, but he was on the dais at the second. It was a smaller affair than the first: 173 delegates from thirty-two nations and a variety of agencies and private groups. Among those drawn to Lake Success were old friends like Robert Cushman Murphy, Ernst Mayr, Frank Darling, Aldo Leopold's son Starker, Clarence Cottam (Vogt's partner in the anti-mosquito-control debate), and most of the IUPN leadership—the cream of Western conservation. Julian Huxley was not present; he had been pushed out of UNESCO, partly because his drive to establish IUPN had ruffled U.N. sensibilities. Vogt led three of the eleven sessions, more than anyone else, and was one of the four members of the conference's general committee. The conference was formally known as the International Technical Conference on the Protection of Nature. It naturally had its own acronym: ITCPN.

The two conferences—UNSCCUR and ITCPN—took place in the same converted factory that housed the United Nations. Delegates to one rushed past delegates to the other in the hallways, but they were in different worlds. UNSCCUR was nature *utilization*; ITCPN was nature *protection*. UNSCCUR looked at the forces of industrialization and said, *Yes, but be smart about it*. ITCPN looked at the same forces and said, *No, there is a better way*. UNSCCUR was the hard path; ITCPN, the soft. UNSCCUR had money and official attention and the force of institutionalization. ITCPN had fervor and the energy of rebellion and a totalizing ideology. If UNSCCUR was the Wizard's manifesto, ITCPN laid out the Prophet's case. Each side was sincerely idealistic, privately dismissive of the other, and blind to its own faults. They were two groups of men with dark suits and dark ties and dark briefcases imagining the shape of the future in rooms cloudy with tobacco smoke.

"The environmentalism put forth at the ITCPN was an intellectual and moral broadside against the liberal international order that the United States was seeking to set in motion." (I am again quoting Jundt, the historian, whose work I am following here.) Osborn's address at UNSCCUR had been cautious. At ITCPN he also gave a speech—he was "practically a double agent," Jundt wrote—and there he pulled out all the stops. Darwin, Osborn said, "proved that man was an integral part of nature itself, and not a separate and independent being." People are not special! But the wonders of airplanes, radar, and the atomic bomb had "tricked" us into believing "that we are 'masters of the universe.'" Humanity's future could be bright, Osborn said, but only if people didn't think themselves "exempt from natural laws." The sole route to securing a future for our species, democratic society, and the planet itself was to understand our place within nature's limits and to base civilization on this knowledge.

Speakers decried environmental problems that would become the subject of crusades in decades to come: oil spills; predatory whaling; herbicide overuse; river-destroying dam projects; the introduction of exotic species; vanishing elephants, tigers, and other great creatures; and mindless consumption and materialism. A dozen papers questioned the use of DDT and other pesticides. It was like opening a window into the 1970s and 1980s and 1990s, a concatenation of headlines from the future. Many of the delegates represented small organizations that had never before encountered the others. Just as Vogt and Huxley had hoped, the conference was creating a network of the like-minded. Groups concerned with bald eagles met groups concerned with protecting shorelines met groups concerned with air pollution. Among the few commonalities was that everyone seemed to have read *Road to Survival*.

Typical was Ollie E. Fink of Friends of the Land. Formed in 1940 and based in Ohio, the organization's ten thousand members initially focused on degraded agricultural properties in the Middle West. But they were tiptoeing past that to something broader, and this movement was reflected in Fink's speech, which went beyond soil conservation to human affairs in general. "Conservation is a way of life . . . but it is not the traditional American way," Fink said (ellipsis in original). Instead, the nation and the world needed "a new culture," a new way of thought, "an ecological conscience." The boldness had limits, though. One attendee had asked, "Wouldn't it be useful to discuss openly the problem of the almost inevitable antagonism between protective measures and the interests of the economy?" The answer: No. Not yet, anyway.

Lake Success Conference No. 2 was as fruitful in its own way as Lake Success Conference No. 1. But Vogt was not able to lever its prestige into the director's chair at IUPN; the organization wouldn't take on someone who had annoyed the U.S. State Department, a major funding source. In October Pan American Union secretary general Lleras summoned Vogt to his office to tell him that he had used up his goodwill. Vogt agreed to resign. He doesn't seem to have borne any animosity toward Lleras, whom he had put into a difficult position. Vogt gave a final speech, defiantly titled "The United States Is Not the World." His last day at the union was November 15, 1949.

As in his dismissal from Audubon, he had angered powerful people, and they had thrown him onto the curb. He had reinvented himself then, going to the guano islands, but now he was forty-seven years old. He had a reputation, but it was not the kind that was welcomed by potential employers. He had no clear path forward. But the urgency hadn't gone away. On every continent factories were springing up, trees were coming down, and open land was being transformed into farms and suburbs. Vogt was standing on a beach, and the human tide was coming in.

"Starting a CONFLAGRATION!"

How to save the future? Hugh Moore knew the answer. Moore was a compact figure with arrow-straight hair and the preternatural energy of a self-made man. He was born in 1887, the youngest of six children on a Kansas farm. His father died when Moore was twelve. Moore left Kansas, became a reporter in New York City, then managed to get himself into Harvard. During his freshman year, his sister's husband, Lawrence Luellen, came to town. Everyone in the country, Luellen said, drank out of public dippers, which were rarely washed and never sterilized. In those days of rampant tuberculosis, America needed non-infectious drinking vessels. Luellen had a brainstorm: *Paper cups!* Hearing this, Moore had a second: *What a gold mine!*

Moore folded himself some test cups, origami-style. They looked so good he quit Harvard. The two men optimistically rented a room in the Waldorf Astoria. They used the hotel's gold-embossed stationery to write potential investors. One venture capitalist was so horrified by

Moore's portrait of the nation's water dippers, hanging on public drinking fountains like so many bacteriological time bombs, that he contributed a whopping $200,000. The Public Cup Vendor Company was born in 1909. Luellen soon quit to do other things. Meanwhile, Moore launched campaigns to abolish the common drinking cup. State after state banned them. Moore's cups—Dixie Cups, he later called them—made him rich.

Moore served in the First World War; his horrific experiences led him to become an ardent pacifist. When international tensions mounted in the 1930s, he sought to convene an international conference to head off another war. He also invented the paper ice cream cup and, later, the paper plate. The Second World War happened despite his efforts. Moore formed the Committee to Defend America by Aiding the Allies. He had the leisure; people were throwing away 25 million Dixie Cups a day. His ideology evolved, not always coherently. Fierce pacifism became fierce anti-Communism and fierce support of world government.

In 1948 he read *Road to Survival*. The book fell on him like a blow. Ever after, Moore credited Vogt with "really waking me up" to the truth: overpopulation was "the basic cause of wars" and "the spread of tyranny and communism." Vogt said that courageous people had to find a way out. Moore decided to be one of those people. He would devote his life and fortune to population control. He gave money to the Planned Parenthood Foundation of America. He formed the Population Action Committee in 1953 to—as he shouted at meetings—"come up with a plan for starting a CONFLAGRATION!" Seizing a typewriter, he wrote an impassioned pamphlet, "The Population Bomb." Its cover depicted a cartoon Earth, its continents jammed with cartoon people, a lighted fuse emerging from the North Pole. He sent more than a million copies to politicians, journalists, schoolteachers, and businesspeople. Moore promised that he was "not primarily interested in the sociological or humanitarian aspects of birth control. We *are* interested in the use which Communists make of hungry people in their drive to conquer the Earth." All this was after he stepped into William Vogt's life.

Moore wrote to Vogt as soon as he finished reading *Road* in November 1948. He kept in touch after Vogt left the Pan American Union and embarked on a tour of Scandinavia. While he was at Audubon and in Peru, Vogt had applied three times for a Guggenheim fellowship to write

Marjorie Vogt on a birdwatching excursion in Scandinavia in 1950

a popular book on ecology. After being turned down all three times, he had applied a fourth time in 1943, now promising to write a book about the guano birds. This idea finally convinced the Guggenheim Foundation to give him a fellowship, but he had never taken the money, because in the meantime he had been hired by the Pan American Union. Vogt had also received a Fulbright grant. Now he accepted the money from both, but not to study birds in Peru. Instead in May 1950 he traveled with his wife, Marjorie, to Denmark, Norway, and Sweden. The Scandinavian countries had legalized abortion and birth control and vigorously sterilized "deficient" people. Vogt wanted to see the results.

For nine months Bill interviewed officials, collected research papers, and tramped with his canes through the woods; at night, Marjorie typed up her husband's verbal notes as he lay, exhausted, on the couch or bed. All the while, he was being peppered with letters from Moore. Quick with languages, Marjorie spoke French and picked up some Swedish on the road; she often translated for Vogt. Ten weeks into the trip, the couple attended an ornithology conference, and she was fluent enough in Bird to join the give-and-take. Vogt's Fulbright was associated with the University of Oslo, which gave him an office and a comfortable apartment. Nonetheless, he worried obsessively about money—which may explain why he quit research instantly when Moore persuaded the birth-control pioneer Margaret Sanger to hire him as national director of the Planned Parenthood Federation of America.

Sanger would dominate Vogt's life for a decade. Born into a struggling working-class family in 1879, she was the sixth of eleven children (there were also seven miscarriages). Sanger was convinced that her mother's multiple pregnancies—a legacy, in part, of her Catholic faith—contributed to her early death. As a young woman Sanger was a nurse on Manhattan's Lower East Side, which exposed her to the realities of poor women's lives: lack of contraception, dangerous abortions, nonexistent prenatal care, risky childbirth. Becoming a campaigner for birth control—a term she invented—she published (illegal) birth-control pamphlets, delivered (illegal) birth-control speeches, and, in 1916, opened an (illegal) birth-control clinic, the first in the United States. Sanger ran afoul of the law and went to jail, but she kept publishing pamphlets, delivering speeches, and opening clinics. Her goal was for women to be able to run their own lives. Her opponents ranged from the Roman Catholic Church to the Indian Communist Party. In a mix of calculation and enthusiasm Sanger allied with anyone who offered to help her, supporting at one time or another anarchists, socialists, labor activists, race purifiers, conservationists, and Wall Street bluebloods. At various times she espoused racist sentiments that seem appalling today. Historians disagree on whether she truly embraced these ideas or was merely mouthing them to ingratiate herself with powerful people who would serve her larger cause.

By 1937 her movement had managed to have contraception legalized on the federal level, but forty states had birth-control bans of various sorts that had to be fought piecemeal (they endured until 1972). Despite the slow accumulation of political victories, Sanger was dissatisfied. The available birth-control methods were cumbersome, expensive, and often ineffective. She wanted a cheap, easy-to-use, oral contraceptive—a "birth-control pill," as she sometimes called it. Medical scientists were reluctant to take on the task, and many of her male allies didn't believe that women were competent enough to use a pill if one existed. The increasingly frustrated Sanger suffered a heart attack in 1949 and was confined to bed for six months, standard treatment at the time. Three more cardiac incidents rapidly followed, leaving her with agonizing angina. To alleviate her pain, Sanger's doctor son Stuart prescribed Demerol, an opioid to which she became addicted for the rest of her life. Despite poor health, Sanger remained determined to develop a contraceptive, and to do it under the aegis of the Planned Parenthood Federation of America (PPFA), the organization that had grown out of her birth-control clinics.

Road to Survival had impressed Sanger, who alluded to it in her speeches. The book had led Moore to become one of Planned Parenthood's chief benefactors. Largely at his suggestion, she hired Vogt as national director of PPFA in May 1951. It may have been a bigger step than he realized. Vogt had been an environmental advocate who thought that nature could be safeguarded through population control. Now he was a population-control advocate who believed that reducing birth rates would protect ecosystems along the way. The means had become the end.

PPFA had deteriorated during Sanger's convalescence, becoming so disorganized that its affiliates were in open revolt; Vogt, ever energetic, seemed like the man to bring it back. In the year after he arrived he hired researchers and staff, traveled thousands of miles to visit Planned Parenthood clinics, increased coordination between branch offices and headquarters, and, with money from Moore, moved its main office to a more suitable space. He set a goal of doubling the group's activity in ten years. Its officials and members were enthusiastic about his performance, with PPFA president Eleanor Pillsbury saying, "There is no one in the United States better qualified." Sanger and Vogt, though, were growing at odds.

While Sanger was recuperating from her first heart attack, she had been contacted by a longtime acquaintance, Katharine McCormick. A wealthy feminist who was the first woman to receive a science degree from MIT, McCormick had previously spent her philanthropic efforts on a futile effort to cure her husband's schizophrenia. When he died, she reached out to Sanger, asking where her millions could do the most good for birth control. Galvanized by the opportunity, Sanger spent the summer of 1952 in McCormick's house, poring over scientific literature. Both women focused on the work of Gregory Pincus, a medical researcher who had left academia to form his own for-profit laboratory. Funded by the drug company G. D. Searle, Pincus had spent five years trying to develop a synthetic version of the steroid cortisone. Another company had beaten him to the punch, and Searle had dropped him. Pincus knew that steroids could suppress ovulation, and thought that this could lead to a human contraceptive. In June 1952 the two women rode in McCormick's chauffeured limousine to Pincus's modest, one-story laboratory in Worcester, Massachusetts. McCormick was so pleased that she gave him a $10,000 check on the spot.

Vogt objected. Pincus, he said, was a discredited researcher who had lost out in both academia and industry. Whatever he did in that small, private laboratory wasn't real science; actual research required teams of M.D.s in clinical settings. Even if Pincus somehow came up with a contraceptive, neither he nor PPFA had the money or facilities to test it properly. In addition, Vogt didn't believe that an oral contraceptive could be produced cheaply enough to be used in poor countries. Other organizations, including the new Population Council, were sponsoring laboratory research. Vogt proposed that Planned Parenthood focus instead on education and clinics and raising public awareness, though he also agreed, in February 1954, to appoint a new research head to collaborate with Pincus. Hoping to please Sanger, Vogt chose John Rock, a retired Harvard Medical School professor of obstetrics and gynecology.

It didn't work. Sanger told friends that Vogt had "lost his mind": Rock was a devout Roman Catholic. Vogt told her, accurately, that Rock had been a birth-control advocate since the 1930s, but Sanger remained suspicious. To calm the waters, McCormick told Sanger, inaccurately, that Rock was no longer Catholic. Then she visited Vogt in New York. The meeting did not go well. Vogt dismissed Pincus as a hack and McCormick herself as a dabbling society lady. She knew that she had more scientific training than Vogt and was infuriated that he hadn't even bothered to visit Pincus in Worcester. It can't have helped that Vogt, who had been given an honorary doctorate by Bard in 1953, was asking people to call him "Dr. Vogt." McCormick stomped from the meeting. Nobody in PPFA, she wrote to Sanger, "is really concerned over achieving an oral contraceptive. It is to me vague and puzzling—really mystifying."

Later that year McCormick gave Pincus $50,000 to build an animal-testing facility. When Vogt complained about the cost, McCormick paid another visit to New York. Infuriating to her, Vogt never bothered to explain his reasons for objecting to Pincus's work.* Instead, he took the occasion to make a sales pitch: Would she fund another expansion of the New York offices? It is possible that he was beginning to feel defensive. In a time dominated by the Cold War, Vogt's attacks on capitalism were making him a pariah. The Conservation Foundation, which he had helped to found, had just ejected him from its advisory council and removed *Road to Survival* from its recommended reading list. Whatever the cause of Vogt's condescension, McCormick responded by taking PPFA out of the loop as Pincus and Rock tested birth-control pills in Puerto Rico and Hawaii. Contraceptive research being illegal in much of the United States, the work was conducted in secret. Pincus announced his first successes at a Planned Parenthood meeting in Japan in 1955. Sanger, though exhausted by the voyage to Tokyo, was beaming. Vogt wasn't even in the room.

Because he had the loyalty of Planned Parenthood employees, Vogt was able to hang on for another few years. But he was torn between the organization's simultaneous demands that he keep the U.S. affiliates running while establishing a global program of population control. The breaking point occurred when Hugh Moore, impatient with Vogt's caution, helped organize the World Population Emergency Campaign to launch a crash program to reduce human numbers by sending out cadres of trained fieldworkers equipped with subsidized contraceptives and specially printed brochures. To entice Sanger, the campaign transformed the PPFA's annual meeting into a "World Tribute to Margaret Sanger." Julian Huxley was the master of ceremonies. The luncheon, held in the Waldorf Astoria, simultaneously marked the end of Planned Parenthood's mission to empower individual women and Vogt's career in the organization. New blood was needed, Vogt was told. He was fired in September 1961.

There was, perhaps, some consolation. Two years before, Vogt had married for the third time. Six years younger than Vogt, Johanna von Goeckingk had been born in Brooklyn. Her German father died when she was young, and her Slovakian mother went to work as a seamstress

---

* It might have been useful if he had. As Vogt had feared, PPFA and Pincus didn't have enough resources to test the pill properly, and women were initially prescribed dangerously high dosages. Moreover, the drug—as Vogt predicted—was too costly for women in poor places.

Johanna von Goeckingk, 1929

in Holyoke, Massachusetts. A good student, Johanna went to Radcliffe College, Harvard's all-female sister school, where she was editor of the school yearbook and president of the campus Christian club. After graduation she moved to Manhattan, where she was a department-store manager. During the war she found a job with the State Department and then, after the war ended, with the new United Nations. She left to work for the Planned Parenthood Foundation of America. Presumably that is where she met Vogt. One can imagine a scenario—an affair, discovery, divorce—but Vogt's papers, picked through after his death by a loyal assistant, are silent about what transpired. All that is known is that Marjorie moved to California and that Bill and Johanna were married on December 26, 1959. She was fifty-one; he was fifty-seven.

Every indication is that the marriage was happy; Vogt at last may have found the right match for his obsessive, obdurate temperament. They even had a house in the country. That was lucky for him, for his life was entering a difficult period. After he was shown the door by Planned Parenthood, the Conservation Foundation gave him a temporary job as a researcher. In some ways, the position was a mismatch: Vogt was an activist by nature, an alarm-ringer who sought to disturb; the Conservation Foundation was a scholarly clearinghouse for environmental information. In other ways, though, the new job was perfect: it offered a chance to return to his old interest in Latin America.

Eighteen months after leaving Planned Parenthood, Vogt went to Mexico and El Salvador. He hadn't been to either place in sixteen years. In seven weeks of travel, he did see "small and encouraging signs of conservation progress." But mostly nothing had changed: "Ecologically triggered disaster, which may well find a forerunner in political explosions, lies not many years or decades ahead, for much of Latin America."

Under the aegis of the Conservation Foundation, Vogt visited Central America three times in an effort to create a network of concerned scientists there—a small, Hispanophone version of IUPN. Associated with a U.S.-sponsored institute in Panama, the initial meeting brought together eighteen Central American and Mexican officials and scientists. Vogt was the only outsider, but he still managed to dominate the room. Growing population, he said, was leading to environmental ruin.

The urgency was there, but the polio was catching up with him. He couldn't travel as readily as he had in the past. Seeing Vogt's difficulties, the Conservation Foundation's president, Samuel Ordway, gave him a permanent job in 1964 as its executive secretary. The conditions were that Vogt permanently sever all ties to Planned Parenthood and obtain Ordway's approval for anything he wrote for the public. Ordway didn't believe capitalism was incompatible with conservation; he vetoed Vogt's request to print his Latin American report in *Reader's Digest*. Vogt's papers are full of unpublished manuscripts from this time.

He managed to slip the leash long enough to write an article attacking foreign aid—"We Help Build the Population Bomb"—in *The New York Times Magazine* in 1965. But the article provoked little reaction. A year later he testified to the U.S. Senate against a foreign-aid bill. "There are many parts of the world where, lacking effective population control, we might better spend nothing on foreign aid since there we are literally exacerbating misery and destruction of the human habitat." Some of the senators feigned interest, he thought, but nothing happened. He was still shouting from the stage, but the audience wasn't there.

Why wouldn't they *listen*? The ecology was so clear, the implications so unavoidable. It was bewildering to Vogt.

The Senate was his last public appearance. He still maintained his easy baritone, but elsewhere the years were having their way. His once-bushy hair, gone from gray to white, was thin and straw-like; his collar sagged around his neck. It was easier for him to sit than stand. He had the bitterness of someone who didn't expect to be taken seriously. He struggled from his apartment on the Upper West Side of Manhattan to the Midtown offices of the Conservation Foundation. He didn't make it to work every day. At home, his wife was suffering from cancer.

Johanna died in January 1967. Vogt was devastated. Presumably his emotional state contributed to the stroke he suffered soon after. The canes were replaced with a wheelchair. He couldn't go to the office, which couldn't accommodate wheelchairs. He retired, age sixty-five, like one was supposed to do, though he had never intended it.

Hoping to rekindle the flame, Vogt wrote a sequel to Road *just as he was being pushed out of Planned Parenthood.* People! Challenge to Survival *was widely reviewed, but in the marketplace it disappeared without a trace.*

He didn't know what to do with his time. He had always been a fluent writer. Now the words came slowly, his shaking hands stabbing at his manual typewriter, the letters half hit and jagged on the paper. He couldn't go birding, except maybe for looking at the ragged Vs from his window as the geese went up and down the Hudson River in the spring and fall. Western civilization, he told acquaintances, was possessed by *dementia economica:* the insane substitution of "limited symbols—such as dollars, pesos, colones, lempiras, and quetzales—for such reality as topsoil, fertility, soil metabolism, available water, protein, and complex interdependencies within the ecosystem—including man." The symbols said that humanity was doing well, when actually "environmental deterioration is being accelerated."

In years past *The New York Times* had published two or three of his letters almost every year. They stopped taking his letters. In May 1968, he wrote dourly to the *Baltimore Evening Sun* about Robert Kennedy, a Roman Catholic, who was then running for president. There is something wrong with a man with ten children, Vogt said. An Associated Press reporter, remembering Vogt's name, telephoned him. Vogt didn't back down. Maybe he was pleased to have someone seeking his opinion. He said he wouldn't vote for Kennedy. "The last thing this country needs is more people," he told the reporter. "And the next to last, in my opinion, is a president of the United States who sets such a bad example." The comments were swiftly disavowed by Planned Parenthood. A month later, Kennedy was assassinated in Los Angeles. A month after that, on July 11, 1968, Vogt killed himself in his apartment. He left a note whose contents were never released.

There were no younger heirs; Vogt's estate went in its entirety to his mother, still living on Long Island. The obituaries were few and short. There was no memorial service. Who would give the eulogy? Vogt must have suspected this would happen. Maybe he thought it was the prophet's fate to die without honor.

The Swerve

Vogt spent his last days in the belief that all of his efforts had been futile. Humankind was making an idiot's march to destruction, resolutely deaf to his entreaties to change course. But none of this was true, at least in the way he seems to have imagined. Within years of his death his ideas had become part of the mental furniture of most educated Westerners. But in a different way it was all true: he had failed comprehensively. He had spent decades in the belief that a blind alley was an exit, setting back humanity's course, and his own.

It was not true that he had failed, because in May 1968, just days after Vogt's attack on Bobby Kennedy, the Sierra Club published *The Population Bomb*, by Paul Ehrlich, the Stanford biologist. When Ehrlich entered the University of Pennsylvania he had befriended some upperclassmen who were impressed by his refusal to wear the freshman beanie, then a demeaning tradition. Not wanting to join a fraternity in his sophomore year—another custom—Ehrlich had rented a house with his friends. They passed around books of interest, including *Road to Survival*. It helped push Ehrlich into ecology and population studies. When Ehrlich taught at Stanford, he talked about his ideas on population and the environment, which were mainly Vogt's ideas. Students mentioned Ehrlich to their parents. He was invited to speak to alumni groups, which in turn put him in front of bigger groups. After hearing him on the radio, the head of the Sierra Club asked Ehrlich to write a quick book, hoping— "naively," Ehrlich told me—to influence the 1968 presidential election. The Ehrlichs produced a manuscript in three weeks, basing it on his lecture notes. The publisher told them that joint bylines didn't sell; Paul's name alone would be on the book.

Published in May 1968, *The Population Bomb* attracted little initial notice. No major newspaper reviewed it for five months. *The New York Times* gave it a one-paragraph notice almost a year after its release. In February 1970, twenty months after publication, Ehrlich was invited onto *The Tonight Show,* a late-night talk show, then enormously popular. The invitation was a fluke; Johnny Carson, the comedian-host, was leery of serious guests like university professors because he feared they would be pompous, dull, and opaque. Ehrlich proved to be affable, witty, and blunt. Thousands of letters poured in after his appearance, astonishing the network. *The Population Bomb* shot up the best-seller list. Carson invited Ehrlich back in April, a week before the first Earth Day. For more than an hour he spoke about population and ecology to an audience of tens of millions. It was the moment that Vogt had dreamed of for decades.

Suddenly Vogt's ideas were everywhere. As Ehrlich put it, "No effort to expand the carrying capacity of the Earth can keep pace with unbridled population growth." Others—many others—echoed his words. If humankind continued to exceed its limits, the biophysicist John Platt warned in 1969, "we are in the gravest danger of destroying our society, our world, and ourselves in any number of different ways well before the end of this century." On the first Earth Day, in 1970, eighty-two-year-old Hugh Moore distributed hundreds of thousands of handbills about population and free audio tapes featuring Ehrlich. For the occasion he unveiled a new slogan: "People pollute." Its implication was clear: more people = more pollution. The ecologist Garrett Hardin summed up by proposing an Eleventh Commandment: "Thou shalt not transgress the carrying capacity."

These ideas were burnished with a sheen of digital precision in *The Limits to Growth* (1972), the Hubbert-inspired book by an MIT-based research team that had used computer models to predict that rising population and consumption would lead to catastrophe. Or, to use the team's italicized jargon: *"The basic behavior mode of the world system is exponential growth of population and capital, followed by collapse."* Population growth, the team emphasized, "must stop *soon.*" An amazingly influential tract, *The Limits to Growth* eventually sold 12 million copies in thirty-seven languages and stimulated passionate argument around the world. My experience, I suspect, was typical; *Limits* was assigned reading in my college ecology, economics, and political science classes.

Carried on the wave of population alarm, organizations from the International Planned Parenthood Federation and the Population Council to the World Bank, the United Nations Fund for Population Activities, and the Hugh Moore–backed Association for Voluntary Sterilization established a program to reduce fertility in poor places. It was the public-private network envisioned by Julian Huxley at Fontainebleau, joined in a common effort to save the world from the menace of overpopulation.

As a rule, anti-population campaigns were proposed by natural scientists: biologists, in the main, but also physicists and engineers. Opposition came from social scientists—anthropologists, sociologists, economists, and demographers. Anthropologists observed that, once again, rich people in one place were trying to reorder the lives of poor people in another place with little knowledge of their cultures. Sociologists pointed out that the intellectual justification for spending billions on family-planning programs was shaky—it tacitly depended on the notion that couples in the Third World were somehow too stupid to know that having lots of babies was a bad idea or, if they did know, had no idea how to avoid having them. Economists attacked the programs as intrusive, poorly planned interventions with perverse incentives. Demographers noted that the United States had gone from having, in Malthus's time, the highest fertility rate ever seen to having, in the twentieth century, one of the world's lowest. And this happened before the pill, before effective IUDs, before safe abortions, in an era when birth control was illegal.

The results of the campaigns were ghastly. Millions of women were sterilized, often coercively, sometimes illegally, frequently in unsafe conditions, in Mexico, Bolivia, Peru, Indonesia, Bangladesh, and, especially, India. In the 1970s and 1980s the Indian government, then led by Indira Gandhi and her son Sanjay, embraced policies that in many

states required sterilization for men and women to obtain water, electricity, ration cards, medical care, and pay raises. Teachers could expel
students from school if their parents weren't sterilized. More than 8 million men and women were sterilized in 1975 alone. ("At long last," World
Bank head Robert McNamara remarked, "India is moving to effectively
address its population problem.") All the while, the same programs
were pushing birth control with equal vigor. In Egypt, Tunisia, Pakistan, South Korea, and Taiwan, health workers' salaries were, in a system
that invited abuse, dictated by the number of IUDs they inserted into
women. In the Philippines, condoms and birth-control pills were literally pitched out of helicopters hovering over remote villages.

Inspired by *The Limits to Growth* approach, Song Jian, a Chinese specialist in ballistic-missile control, formed a research team in 1978 to create, in effect, a Chinese version of the *Limits* model. Not one member of
his team had experience in demography. This didn't stop the group from
using China's defense-industry computers to churn out population projections. Swooping curves on graphs showed China's population reaching 4 billion in 2080, an impossible burden. Borrowing methods from
Dutch computer scientists, Song's team calculated the desired trajectory for population as if it were aiming a missile. Out came the answer,
as inexorable as fan-folded paper unfurling from a dot-matrix printer:
the sole route that would allow China to stave off disaster would be for
all Chinese couples to have but a single child, beginning immediately.
Sneering at dissenting social scientists as useless hand-wavers, Song
pushed for the one-child program that was adopted by the government
in 1980. Tens of millions—possibly 100 million—of coerced abortions
occurred, often in poor conditions that led to infection, sterility, and
even death. Millions more women were forced to insert IUDs or be sterilized. (The anti-population campaigns have largely been abandoned.)

Vogt was not responsible for these cruelties—he died before they
began, and in any case was no longer in a position of influence. But his
intellectual guilt is heavy. If his great success was stoking concern about
the connection between population and environmental degradation,
his great failure was believing that the connection was simple, clear,
and pivotal. In *The Population Bomb*, Ehrlich, too, described the link
as straightforward. He devoted the first section of his book, "The Problem," to explaining the issues. After describing population growth, food

shortfalls, and environmental problems, the section ends with a summation: "The causal chain of degradation
is easily traced to its source. Too many
cars, too many factories, too much
detergent, too much pesticide, multiplying contrails, inadequate sewage
plants, too little water, too much carbon dioxide—all can be traced easily to *too many people*" (emphasis in
original). And then ends the chapter.
The white space on the page testified
to Ehrlich's view that the relationship
between population and environmental degradation was too obvious to be
worth belaboring.

Some parts of the case, to be sure,
are obvious. A world with zero people in it clearly would have no worry
about human impacts. Similarly, it
is easy to credit that a population of
20 billion would have a huge effect
on natural systems. But it's not at all evident what happens between
zero and twenty, or between zero and ten, where humanity has spent its
entire existence. The reason is that people are not fungible—the impact
of one person living one kind of life is completely different from that of
another person living another kind of life.

Looking back, Ehrlich told me, he wished he had "given more emphasis to consumption, and not just population"—an issue he rectified
in later writings. Consider the opening scene of *The Population Bomb*.
The first sentences describe a ride in an "ancient taxi" taken by Ehrlich
and his family through "a crowded slum area" in Delhi.

> The streets seemed alive with people. People eating, people wash-
> ing, people sleeping. People visiting, arguing, and screaming. Peo-
> ple thrust their hands through the taxi window, begging. People
> defecating and urinating. People clinging to buses. People herding

A Sierra Club-Ballantine Book 95¢

**POPULATION CONTROL OR
RACE TO OBLIVION?**

THE POPULATION BOMB

WHILE YOU ARE READING THESE WORDS
FOUR PEOPLE WILL HAVE DIED FROM
STARVATION, MOST OF THEM CHILDREN.

DR. PAUL R. EHRLICH

Foreword by David Brower
Executive Director, Sierra Club

So rapidly was The Population Bomb *rushed into print that nobody noticed that the fuse-lit bomb on the cover was captioned "THE POPULATION BOMB KEEPS TICKING."*

animals. People, people, people, people. . . . [S]ince that night, I've known the *feel* of overpopulation.

The Ehrlichs took the cab ride in 1966. How many people lived in Delhi that year? A bit more than 2.8 million, according to United Nations figures. By comparison, the 1966 population of Paris was . . . about 8 million. One can spend considerable time searching through libraries for expressions of alarm about the way the Champs-Élysées "seemed alive with people" without success. Instead, Paris in 1966 was an emblem of elegance and sophistication. What was different in India? Why was Delhi overcrowded, if Paris wasn't?

Look at the numbers from the United Nations. In 1970 Delhi's population was 3.5 million—an increase of 25 percent in the few years after Ehrlich's taxi ride. Five years after that, in 1975, Delhi had 4.4 million people—another 25 percent gain. Amazingly, Delhi's population was *eleven times* bigger in 2005 than it had been in 1950.

What was the cause of the city's extraordinary growth? I asked Sunita Narain, head of the Center for Science and Environment, a think tank in Delhi. "Not births," she said instantly. Instead, she said, the overwhelming majority of the new people in Delhi then were migrants drawn from other parts of India by the hope of employment. Delhi had jobs because the government of India was trying to shift people away from small farms into industry, which it saw as the vehicle for national prosperity. Many of the new factories were located in and around Delhi, and people were moving there to improve their lives. Because there were more migrants than jobs, Delhi had become jam-packed and often unpleasant, exactly as Ehrlich had described. But the crowding that gave him, ever after, "the *feel* of overpopulation" had little to do with birth rates and natural resources and density of numbers and much to do with laws and institutions and government plans. "If you want to understand Delhi's growth," Narain said, "you should study economics and sociology, not ecology and population biology."*

---

* For much of the last half of the twentieth century, Delhi was the second-fastest-growing city in the world. The fastest-growing was Tokyo. Tokyo was and is extremely crowded, but it is also clean, safe, and prosperous—a legacy, in part, of effective urban governance. The vicissitudes of Japanese history and culture have much more to do with whether Tokyo is "overpopulated" than the rate at which Japanese babies are coming into existence.

Few places better illustrate the complex relationship between population and environment than the lower Hudson River Valley, the area where Vogt got polio, went to college, and learned to love birds. To the west rise the Catskill Mountains, blue at sunset and blanketed by trees. Interstate 87 makes a black ribbon between the water and the mountains. In years past I spent some time driving back and forth on that road, and down long miles of its length the forest stretched out so far and so dark and so empty that I imagined I was looking at the America of 150 years past, before there were millions of people like me around. How wonderful, I thought, that so close to Manhattan is a huge piece of real estate that we never trashed.

I was wrong. If I had traveled through the Hudson Valley in the closing decades of the nineteenth century, I would have passed through an utterly different landscape. I would have been surrounded by hardscrabble farms and pastures ringed by stone walls. It might have looked picturesque—guidebook writers of the day thought so. But I wouldn't have seen many trees, because almost every scrap of land that wasn't vertical had been clear-cut or burned.

The forest was stripped to make way for agriculture and to supply New York's army of charcoal burners (who needed lumber to make charcoal), tanners (who extracted tannin from bark), and salt makers (who used wood fuel to boil down seawater). Loggers played a role too: Albany, the northernmost deepwater port on the Hudson, was the biggest timber town in the nation and possibly the world. When the first Europeans came to New York, the rolling uplands were almost entirely covered by open forest; by the end of the nineteenth century less than a quarter of the state was wooded, and most of what was left had been picked through, or was inaccessible, or was being kept by farmers as private fuel reserves. During the epoch that I, swooping along the tarmac in my minivan, was nostalgically picturing as a paradise, newspaper editorials were warning that deforestation would drive the valley toward ecological disaster.

Since then the collapse of small farming on the East Coast has allowed millions of acres to return to nature. When New York State surveyed itself in 1875, the six counties that make up the lower Hudson Valley—Columbia, Dutchess, Greene, Orange, Putnam, and Ulster—contained 573,003 acres of timberland, covering about 21 percent of their total area. A hundred years later trees covered almost 1.8 million acres, more than three times as much.

Back in 1875 these six counties had a collective population of 345,679. The U.S. Census says the figure for 2012 was 1.06 million. In other words, the number of people living there tripled in the same period that the local ecosystem climbed out of its sickbed and threw away its crutches. This wasn't just some odd thing that happened in New York. As a whole, U.S. forests are bigger and healthier than they were in 1900, when the country had fewer than 100 million people. Many New England states have as many trees as they had in the days of Paul Revere. Nor was this growth restricted to North America: Europe's forest resources increased by about 40 percent from 1970 to 2015, a time in which its population grew from 462 million to 743 million.

"People pollute," as Hugh Moore said. But more people ≠ more pollution. Eco-critics claim with some justification that the Hudson Valley recuperated because farmers abandoned it in order to wipe out the native grasslands of the Great Plains. But they can't explain away all the other good news. Seals and dolphins return to the Thames. White-tailed deer, almost extinct in 1900, plague New England gardens. Air quality in formerly polluted Japan improves remarkably. Wild turkeys have a greater range than they did when they were first seen by European colonists. If all this occurred during the population boom, why the belief, voiced by Vogt, Osborn, Leopold, and so many who followed them, that overpopulation will lead to an eco-catastrophe?

Years ago I had the chance to ask this question of Dennis Meadows, coleader of the research team that produced *The Limits to Growth*. "You can look at Lake Erie or Detroit and see it's gotten better," he said. "But to leap from that to the conclusion that there has been overall improvement is to look at one person getting rich and say that everybody is better off." Then a professor emeritus at the University of New Hampshire, Meadows and his co-workers had updated *Limits* several times, each update as pessimistic or more so than the original book. "When a rich country becomes concerned about environmental problems," Meadows told me, "then it can typically develop effective responses." Lead additives in gasoline became a subject of U.S. worries, he pointed out. Washington forced petroleum companies to pay to phase out leaded gas and car companies to pay to change their engines and drivers to pay more at the pump. Money was flung at the problem, and lead levels diminished.

I dutifully quoted Meadows to this effect in an article. Only a year after it was published did it occur to me that this argument might amount to saying that economic growth in fact allows societies to buy their way out of environmental problems—a Borlaugian stance. Curious, I again telephoned Meadows, who kindly took my call. Was affluence the answer, as Wizards said? No, he told me. He gave an example: the lush, beautiful evergreen forests in the mountains of Japan. Japan maintains them, he said, by importing all of its lumber from Southeast Asia, Australia, Brazil, and the U.S. West. Like the New Yorkers shifting agriculture to the Midwest, Japan was using its wealth to shift the ecological burden of deforestation. I asked if Japan couldn't simply ban wood imports and build its homes with plastic and foam and the weird new high-tech materials one sees on Tokyo streets. The shift might be costly, I suggested, but Japan is prosperous. Couldn't they buy their way out of this problem, too? No, Meadows said. When I asked why, he became exasperated with my stupidity. "Look, the basic facts are *obvious*," he said. "You can't keep growing forever on a finite planet—there are limits." But the exact relations among economic growth, environmental destruction, and planetary limits no longer seemed so obvious to me.

Compare a family in Delhi (2015 per-capita income: $3,180) to a family in, say, Copenhagen (2015 per-capita income: $47,750). Given their relative incomes, it seems safe to posit that the citizens of Copenhagen consume much more than the denizens of Delhi. But if people in Delhi burn coal to cook their food and heat their homes they may do more damage to the local and global environment than people in Copenhagen, who can get most of their power from the country's plentiful wind-power stations (wind supplies almost half of Denmark's electricity). In the fall of 2016 Delhi's smog was so bad that schools were closed all over the city for days. The pollution was due to coal in the city and small farmers burning crop residues in the adjacent states of Punjab and Haryana. Meanwhile the Copenhagen government has announced plans to cut the city's carbon dioxide emissions to zero by 2025.

The emissions difference does not necessarily mean that Delhi families cause more environmental damage than Copenhagen families. Danes eat so much meat and drive so many automobiles that in 2014 the Worldwide Fund for Nature claimed that Denmark had the fourth-biggest "ecological footprint" in the world. The main cause was the huge amounts of animal feed grown to support the Danish pork industry. To feed its meat habit, Denmark used more farmland per capita than anywhere else.

Which is more eco-friendly, Delhi or Copenhagen? The answer depends more on the weights assigned to air pollution and land use than the absolute level of economic growth or consumption. I can spend a million dollars paving over a magnificent redwood forest and that will appear in the statistics of gross domestic product as a million dollars of economic activity. But I can also spend that money buying front-row opera seats for poor schoolchildren and that, too, will appear in the gross domestic product as a million dollars of economic activity. The two activities contribute identically to the statistics, but their environmental impact is strikingly different. Presumably I could buy many million dollars' worth of opera tickets before coming close to the environmental impact of paving the forest—that is, I could increase GDP and have the net environmental impact shrink. *How many people?* is an important question, but it is less important than *What are those people doing?*

None of this is to deny that environmental problems are real. Overfishing, deforestation, soil degradation, contaminated groundwater, declining populations of mammals and birds, and, most alarming, the possibility of very rapid climate change—all of these are important. But the contribution of population growth to them is indirect, and the relationship to economic growth is equivocal. Focusing on them as a root cause, as Vogt did, is a distraction. It was a waste of two decades, and doubly unfortunate because the fight over population sometimes shrouded the more important part of Vogt's message, the part about limits. He denounced social scientists as fools, but he should have listened to them. And that, alas, applies to Borlaug, too.

NINE

The Wizard

Multiples

The world is various but the view from the laboratory bench is ever similar. Scientists speak different tongues and live in different places and worship different gods, but at any one time the array of available problems and techniques is limited and apparent to all. In consequence, discovery is a crowded business. "The pattern of independent multiple discoveries in science is in principle the dominant pattern," wrote the sociologist Robert K. Merton, in a passage well known to historians of science. Look hard at individual cases, Merton argued, and you will find that "all scientific discoveries are in principle multiples." Men and women emerge from the obscurity of the laboratory waving a flag in triumph and discover somebody ahead and somebody behind, each with the same flag. Almost always, even the most original scientists have doppelgängers—intellectual near twins who would have accomplished the same work, if circumstances had been slightly different. However different in temperament and circumstance, they were scrabbling along the same path.

Exceptions may exist: general relativity, developed in near solitude by Albert Einstein, could be one of them, though some writers have argued that the mathematician Hermann Minkowski would have found relativity first if he had not died young. More typical was the experience of the physicist William Thomson, who would become the British physicist Lord Kelvin (an important scientist despite his wildly erroneous predictions about coal supply, which I described earlier). As an eighteen-year-old freshman, proud of his precocious originality, he wrote a paper for the *Cambridge Mathematical Journal*—only to discover that its conclusions had been, according to Kelvin's son and biographer, "anticipated by M. Chasles, the eminent French geometrician." Later Kelvin learned to his chagrin that exactly the same mathematical ideas also had been "stated and proved by [Carl Friedrich] Gauss," the great German mathematician, and, later still, "that these theorems had been discovered and fully published more than ten years previously by [the British mathematician George] Green."

Newton and Leibniz, each separately working to develop the calculus, then fighting covertly over priority; Benjamin Franklin, marveling that the freethinking philosopher Claude Helvétius had so often had precisely the same ideas, "even though we had been born and brought up in the opposite parts of the world"; Charles Darwin and Alfred Russel Wallace, one on an English country estate, one in a malarial bog in the Malay Archipelago, formulating independently the theory of natural selection; Richard Feynman, Julian Schwinger, Sin-Itiro Tomonaga, and (probably) Ernst Stueckelberg, all without knowledge of the others, all managing to tame the unruly infinities of quantum electrodynamics—the list of multiples is distinguished and long, and it contains Norman Borlaug and M. S. Swaminathan.

Imperfect Double

Mankombu Sambasivan Swaminathan—"Swami," Borlaug called him—became Borlaug's friend and partner, but one can easily imagine an alternate history in which Swaminathan would have played the leading role. He led the introduction to India of the Green Revolution "package": the combination of agricultural chemicals (fertilizer and pesticides), carefully managed watering (usually by irrigation), and high-yielding seeds that could respond to both. Pioneered in Mexico in the 1950s, the package would have its greatest impact in Asia, and especially South Asia, where restructured wheat would meet similarly restructured rice. Hundreds of millions of lives would be changed. The predictions of imminent disaster made by Vogt and his disciples would fail to come true, though only at the price of social convulsion. Swaminathan himself would remain little known, especially in the West, although his work had direct impact on more people than Borlaug's work. Later he sometimes criticized the Green Revolution he had helped to establish. He was a Wizard who became a Prophet, or almost. ✓

He was born in 1925 in what was then the Madras Presidency, an administrative subdivision of British India that encompassed most of what is now southeast India. Mankombu, Swaminathan's first name, was his family's ancestral village; Sambasivan was his father's name; Swaminathan was his personal name. He grew up around his father's medical clinic and, during school holidays, his grandfather's rice farm. The family had owned the farm for generations—the land had been given to a long-ago ancestor by the local rajah in recognition of his knowledge of Hindu scriptures.

Swaminathan's father trained as a surgeon and set up his practice four hundred miles away from Mankombu, in a village, also in the Madras Presidency, which then had no qualified doctor. Cholera, plague, malaria, and other infectious ailments were rampant. Swaminathan's father set out to combat filariasis, a mosquito-transmitted disease that leads to grotesque swelling of the limbs. (One of its forms is known as elephantiasis.) He mobilized schoolchildren to identify mosquito breeding grounds, then asked their parents to fill in stagnant pools and move garbage dumps and disinfect drains and sewers. New cases of the disease ceased to occur—a lesson, Swaminathan said later, in the power of collective action.

It was the time of the Indian independence movement. Swaminathan's parents, ardent nationalists, hosted the movement's leaders, Mahatma Gandhi and Jawaharlal Nehru, several times in their travels around India. To show their support for the cause of freeing India from British rule, the family wore homespun clothing and boycotted British goods. Following Gandhi's precepts, Swaminathan's father treated everyone who came to the clinic, no matter what their position in society, no matter whether they were able to pay. The British didn't jail him for his anti-colonial opinions—he was the only trained doctor for two hundred miles.

Pancreatitis struck him suddenly when Swaminathan was eleven. There was no other doctor in the area to treat him. He died on the train to Madras, the nearest city with a hospital. He was not given a Hindu funeral because he had been excommunicated for treating "untouchables," people from the lower orders. Swaminathan and his siblings were taken in by his father's brothers. At the age of fifteen he graduated from a Catholic high school and went to college, intending to become a surgeon like his father.

The Second World War disrupted Swaminathan's plan to study medicine, as did the Bengal famine, among the century's worst disasters. Bengal was in the northeastern section of British India. The famine began in 1942, when Japan seized Burma (now Myanmar), India's eastern neighbor and another British colony. Burma had exported rice to Bengal; the conquest put an end to that. By chance, Bengal rice paddies were hit by a fungal disease at the same time. For the colonial government the resultant harvest shortfall was inconveniently timed. Kolkata (then called Calcutta), the Bengal capital, was a supply center for British forces. London had imported a million workers from elsewhere in India to the city to produce uniforms, shoes, containers, ammunition, and other military necessities. To feed this army of laborers, the colonial government requisitioned grain from the countryside. Nobody in power wanted to hear that there wasn't enough grain to do this. Even as famine mounted, waves of denial rippled through the bureaucracy. After months of insisting that the crisis was due to hoarding—Indian farmers' purported "tendency to withhold foodgrains from the market" to get better prices—Secretary of State for India Leo Amery reversed course and in November 1943 begged Prime Minister Winston Churchill to send food to Bengal. Churchill ignored the request with a quip about "Indians breeding like rabbits and being paid a million a day by us to do nothing about the war." No food shipments occurred. About 3 million people starved to death.

Taking his pre-medical courses in southern India, Swaminathan was not directly affected by the famine 1,500 miles away. But the images of the emaciated and dying in Bengal had a powerful impact. He decided to study agriculture, rather than stay in the more prestigious discipline of medicine. Initially his family was dismayed, but Swaminathan held firm. The struggle against British rule was reaching its height, he believed, and the famine had shown that the new nation would need food even more than it needed medicine. India had thousands of medical students but only 160 students seeking advanced degrees in agriculture. He agreed to finish his degree in zoology but afterward enrolled immediately in an agricultural college. Graduating at the top of his class, he won almost every award the school could give, then signed up for post-graduate research at the Indian Agricultural Research Institute in New Delhi.

India became independent on August 15, 1947, two months after his graduation. In a horrific convulsion, the nation immediately split, with two large, separate chunks of what had been northeastern and north-western India hiving off to form Pakistan. India was largely Hindu; Pakistan was largely Muslim. Spasms of religious violence led to hundreds of thousands of deaths and perhaps 15 million refugees. (After a brutal war in 1971, the eastern half of Pakistan became today's nation of Bangladesh.) Traveling to Delhi a month after partition, twenty-two-year-old Swaminathan was shaken to see the aftermath—refugee mobs in the train stations, bodies in the streets.

The sprawling green campus of the Indian Agricultural Research Institute was a refuge. Its striking red stone buildings, gathered around a big library with a clock tower, housed the nation's most advanced botanical laboratories. There Swaminathan's advisers directed him to work on the Solanaceae, a botanical family that includes tomatoes, potatoes, eggplant, tobacco, and both chili and bell peppers. At the same time, he concentrated on learning Dutch—a prerequisite for a UNESCO fellowship to study at Wageningen University, the Dutch national agricultural school. Swaminathan won the scholarship and took ship for Europe in December 1949.

The Netherlands was just recovering from the war; famine in the *Hongerwinter* ("hunger winter") of 1944–45 was a fresh memory. The university laboratories lacked heat and sometimes electricity. Swaminathan was asked to switch from eggplants, his focus in Delhi, to potatoes, a Dutch staple. Dutch potato fields were aswarm with parasitic nematodes. Wageningen was trying to fight them by breeding domesticated potatoes to wild, nematode-resistant relatives. But the wild and domesticated species had different numbers of chromosomes, which usually makes successful breeding impossible. Swaminathan figured out a workaround. The discovery was valuable enough to bring him to Cambridge University, in England, where he finished his Ph.D.

It was a propitious time to be a young geneticist at Cambridge. Swaminathan arrived in the fall of 1951, a few months after Francis Crick, a physicist turned molecular biologist at Cambridge, had begun

M. S. Swaminathan in his laboratory in the mid-1950s

to collaborate with James D. Watson, a U.S. post-doctoral fellow eleven years his junior. Soon they would publish the structure of DNA. Swaminathan's doctoral research on the genetics of Solanaceae fit neatly into the suddenly created field of molecular biology. Cambridge offered him a position after he finished his dissertation, but he took another offer, as a post-doctoral fellow at the University of Wisconsin.

Again, Swaminathan flourished. Quick, affable, crisply logical, able to juggle multiple lines of inquiry at once, he had a knack for splitting difficult problems into manageable chunks that were each susceptible to the knives of scientific inquiry. As scientists say, Swaminathan had *Fingerspitzengefühl*—"knowledge of the fingertips"—an intuitive flair for making balky lab equipment and recalcitrant plants do his bidding. His experiments just *worked*. Articles issued from Swaminathan's typewriter in an orderly flow and appeared in major journals. Believing that he was destined for a brilliant career, Wisconsin offered him a position as a professor. He turned down the offer, returning to India in 1954 to help his new country.

Although Swaminathan didn't know it, he was moving into rough terrain. For obvious reasons, Jawaharlal Nehru, the independence-movement leader who became India's first prime minister, wanted his nation to become strong, prosperous, democratic, egalitarian, and free of foreign domination. In the mode of the times he saw industrialization—the hard path, in energy terms—as the route to this goal. Steel, chemicals, coal, electricity, highways, machine tools—that was what India needed! Huge, busy factories and power plants! India would have had all of these already, Nehru believed, if it had not been stifled by British rule.

Where would the money come from to build up Indian heavy industry? Borrowing large sums from abroad was out of the question—India didn't have the stores of foreign exchange that would be necessary to repay the loans. Funds for industrialization therefore had to come from India itself. At the time, almost all of the nation's capital came from agriculture, because about nine out of every ten Indians were small farmers. Creating an industrial economy thus implied skimming off the profits from rural peasants and using that money to build steel mills, power plants, and highways in urban areas. Like Gandhi, Nehru was sympathetic to the plight of poor farmers. Nonetheless, his ideas about development unavoidably led him to drain the agricultural countryside to benefit the industrial cities.

Nehru promised that his new nation would forge its own, uniquely Indian path to development, but he was as much a devotee of science and technology as Truman and Warren Weaver. With the physicist Homi J. Bhabha, Nehru laid out a Wizardly manifesto that echoed every aspect of Truman's Point Four that had dismayed Vogt and Huxley. "The key to national prosperity, apart from the spirit of the people, lies, in the modern age, in the effective combination of three factors, technology, raw materials and capital, of which the first is perhaps the most important," Nehru's statement began. Industrialization can occur "only through the scientific approach and method and the use of scientific knowledge"—the physics, chemistry, and engineering that had created electric grids and radio networks and high-speed railroads. Called a "scientific policy resolution," the declaration was approved by the Indian parliament, made its way into the Indian constitution, and laid the groundwork for the nation's large network of technical schools.

But there was a lacuna in Nehru's concept of science: he saw it exclusively in terms of laboratory science, not field science; physics and molecular biology, not ecology, botany, or agronomy. He understood that India's farmers were poor in part because they were unproductive—they harvested much less grain per acre than farmers elsewhere in the world.

But unlike Borlaug, Nehru and his ministers believed that the poor harvests were due not to lack of technology—artificial fertilizer, irrigated water, and high-yield seeds—but to social factors like inefficient management, misallocation of land, lack of education, rigid application of the caste system, and financial speculation (large property owners were supposedly hoarding their wheat and rice until they could get better prices). This was not crazy: more than one out of five families in rural India owned no land at all, and about two out of five owned less than 2.5 acres, not enough land to feed themselves. Meanwhile, a tiny proportion of absentee landowners controlled huge swathes of terrain. The solution to rural poverty, Nehru therefore believed, was less new technology than new policies: give land from big landowners to ordinary farmers, free the latter from the burdens of caste, and then gather the liberated smallholders into more-efficient, technician-advised cooperatives. This set of ideas had the side benefit of fitting nicely into Nehru's industrial policy: enacting them would cost next to nothing, reserving more money for building factories.

As much as Nehru, Swaminathan wanted India to become a modern, secular nation in which everyone, regardless of origin, had a chance to prosper. At the same time, though, Swaminathan and his colleagues were focused on agricultural research, which implicitly endorsed the idea of investing in the farm sector—an idea supported by the conservative elites who opposed Nehru's plans for land and caste reform. In response, the Nehru government carefully examined every rupee allocated to farm research, to see whether it should instead be spent on promoting industry. Paradoxically, the research money was scrutinized even more when drought reduced Indian grain harvests in 1956 and 1957, and India began to import wheat from the United States through a special program of what amounted to subsidies. Although the food imports were effectively free, they were a painful demonstration of India's weakness. Because Nehru believed that India's freedom would ultimately derive from industrial might, the immediate effect of the agricultural shortfall was for the government to redouble its efforts to build up industry.

After his return, Swaminathan took a position in the Central Rice Research Institute in Cuttack, in the east Indian state of Odisha. Created by Britain just before independence, it was seen as a kind of tacit apology for the colonial government's failures in the Bengal famine. In its newly constructed laboratories Swaminathan was asked to investigate the possibility of crossing sticky, short-grained japonica rice, the kind of rice eaten in Japan, with fluffy, long-grained indica rice, the type favored in India. Partly because of intensive breeding programs in Japan, japonica was higher-yielding than indica—its spikelets burgeoned with short grain. The idea was to create rice that produced like japonica but looked and tasted like indica. As a result, Swaminathan began thinking systematically about what distinguished high-yielding and low-yielding varieties. After a few months he won a better position at his old base, Delhi's Indian Agricultural Research Institute. In his new job at IARI he continued thinking about high-yielding grain, though he turned his attention to wheat.

Wheat has been grown in India for at least four thousand years—archaeologists have found traces of it in Stone Age communities in the Indus Valley. Indian farmers had been working with wheat ever since, gradually creating varieties that were suited to the nation's climate and its culinary preference for the kind of amber-colored, hard-grain wheat that makes the thin, light-colored, puffy flatbread called *chapati* or *roti*. Like their Mexican counterparts, Indian farmers grew a scatter of different varieties in their fields, planting them sparsely to deter rust. At IARI, Swaminathan quickly discovered that these varieties responded to fertilizer so heartily that they lodged—the same problem faced by Borlaug a few years before.

Like Borlaug, Swaminathan decided that the answer was to breed wheat with shorter, stronger straw. Like Borlaug, he searched for short-straw varieties in India's botanical collections. Although these were much smaller than the massive storehouses in the United States, he found a few short varieties and planted them in test beds at IARI. He discovered, as had Borlaug, that these varieties' short straw led to small spikelets with little grain. Without access to huge numbers of different varieties, he was unlikely to find the genes he wanted. He kept on cross-breeding. But he also decided to try and make new genes himself.

In 1955 Swaminathan and his IARI collaborators began taking wheat grain to a small particle accelerator at the Tata Institute, a think tank in Mumbai. The accelerator sprayed out a beam of neutrons, which slammed into a target and produced gamma rays: ultra-high-energy photons. The researchers blasted wheat kernels with gamma rays for several hours, hoping that the gamma rays would tear into the DNA in the seeds and induce favorable mutations. In twenty-first-century terms, this was like trying to perform surgery with a chainsaw—a hopelessly crude procedure. In mid-twentieth-century terms, it was the most advanced method available. Most of the resultant plants failed to germinate or died quickly—the gamma rays had smashed up their DNA like wrecking balls. A few showed interesting characteristics, but none were short-strawed, at least that first year. Or the second. Or the third. Because Swaminathan couldn't get enough financial backing to irradiate and grow seeds in large numbers, the odds of success were especially poor. Like Borlaug in Mexico, he was throwing darts at moving targets. Unlike Borlaug, he could throw only a few darts at a time. Still, he saw no other way to proceed.

The frustrated Swaminathan showed his experimental plots at IARI in 1958 to a visiting Japanese wheat geneticist, Hitoshi Kihara. An Olympic skier and a pioneer in genetics, Kihara was the first to describe the structure of wheat's genome, along the way establishing the modern definition of the term "genome." He was an august figure who used retirement as an excuse to stop teaching, research nonstop, and add to his encyclopedic knowledge of plants. When Swaminathan described his difficulties, Kihara promptly informed him of the existence of some unusually short Japanese wheat varieties. Because Japan was still struggling to recover from the war, Kihara told Swaminathan that it would be easier for him to obtain samples of the dwarfs from a U.S. breeder, Orville Vogel.

Swaminathan wrote to Vogel, who replied that he was happy to send samples, but that he was working with winter wheat, which wouldn't grow well in hot India. Vogel also told him about a man in Mexico to whom he had sent samples. This man, working in the back end of nowhere, was engaged in a crazily massive effort to crossbreed the Japanese short wheat with local varieties to get the kind of spring wheat Swaminathan would want. The man was named Norman Borlaug. Swaminathan dispatched a letter to Mexico City.

"Each Digging His Little Gopher Hole"

It would not be accurate to say that the Cuban missile crisis was responsible for ending famine in India, but it would also be incorrect to say that the two events were unrelated to each other. On October 16, 1962, U.S. president John F. Kennedy learned that the Soviet Union had installed ballistic missiles in Cuba, setting off a confrontation that brought Washington and Moscow near nuclear war. Four days later, China invaded India. The two nations had quarreled for years over their Himalayan border. After Nehru placed Indian troops in the disputed territory, China unexpectedly attacked. Indian forces were heavily outnumbered.

Nehru implored Kennedy for immediate military assistance: hundreds of fighters, bombers, and radar planes and the thousands of airmen and logistical personnel needed to operate them. At stake, he told the president, was "the survival of India"—no, "the survival of free and independent governments in the whole of this subcontinent." His language was so nakedly desperate that the foreign ministry initially refused to send the letter to the White House.

Incredibly, Kennedy didn't respond. The president was "totally occupied with Cuba," complained John Kenneth Galbraith, the U.S. ambassador to India. "For a week, I have had a considerable war on my hands without a single telegram, letter, telephone call or other communication of guidance" from Washington. The requested planes and troops did not appear; the U.S. government, paralyzed by the missile crisis, did not even threaten to intervene. Nehru watched helplessly as China methodically tore apart India's border outposts. Once Beijing established secure control of the frontier, it declared a ceasefire on its terms a month after fighting began. It withdrew from some of the seized territory, but a large chunk of it remains in Chinese hands.

To Nehru, the debacle was shattering. Advisers saw his hunched posture and lurching gait and feared that he had suffered a stroke. Not only did the war represent a calamitous loss of face, it also ended Nehru's long-held hope of allying with China to counteract the influence of the United States and the Soviet Union, the two Cold War superpowers. So politically toxic was the loss that the official report on the war was kept secret for decades; at Delhi's request, Nehru's imploring letters to Kennedy were not made public until 2010. Nehru lost political support from others and confidence in himself; his health never recovered. (He died eighteen months after the war.)

Among the few beneficiaries of the defeat in India was M. S. Swaminathan. Swaminathan's initial letter to Borlaug in 1958 had gone astray, presumably because the Mexican Agricultural Program was in the process of shutting down. Believing that with the development of short-straw wheat the project had achieved its goals, the Rockefeller Foundation was slowly passing responsibility for it to the Mexican government, which later would help create CIMMYT. Borlaug had been let go. He had found a job with the United Fruit Company, which wanted to develop disease-resistant bananas in Honduras. Before he could move his family south, the U.N. Food and Agricultural Organization asked him to join a research group surveying wheat and barley research in North Africa, the Middle East, and South Asia. The team left in February 1960.

The results were dismaying. Everywhere Borlaug found senior scientists sunk in lassitude, using their education to create comfortable sinecures for themselves. At the same time he met energetic young people who wanted to help their countries, but who lacked what Borlaug regarded as proper training. One of the worst cases was India, where status-obsessed researchers were locking themselves in the laboratory to study minor crops—"each digging his little gopher hole of security in his own discipline." Borlaug reserved special disdain for Indian wheat breeders, who were focusing on "beauty of grain . . . rather than total yield." And he was appalled at the government's refusal to prioritize agriculture. One of the researchers he met was Swaminathan—they spoke for about an hour. Neither man seems to have been particularly impressed with the other. One can imagine Swaminathan paying little attention to the blunt, poorly educated foreigner who was trying to tell Indians what to do; Borlaug, for his part, may have seen the other man as the sort of smooth careerist he had spent years battling in Mexico.

When he returned to North America, Borlaug wrote a report detailing his findings and recommending that Rockefeller establish a training program in Mexico to inculcate what would later be called Green Revolution methods. The recent, resounding success of dwarf wheat ensured that his ideas were rapidly followed. Researchers from Afghanistan, Egypt, Libya, Iran, Iraq, Pakistan, Syria, and Turkey flew to Sonora. Arriving in early 1961, they received training in plant genetics, soil science, plant pathology, and other subjects from Mexican scientists who had been trained by Borlaug's team. Borlaug was hired to supervise; he never went to Central America to work on bananas.

India initially refused to participate in the training program—foreign "experts" had nothing to contribute. But food shortfalls were growing more common. As the country grew more dependent on U.S. aid, Indian planners reluctantly agreed in 1961 to test Green Revolution–style pesticides, fertilizer, irrigation, and technical advice in seven agricultural districts. Swaminathan had now seen a few examples of Borlaug's dwarf varieties; at his insistence, they were included in the trials. When the initial results were successful, Swaminathan began the lengthy process of officially inviting Borlaug to India.

The two men met again in March 1963. Recognizing that grain developed in Mexico might not be suited to Indian conditions, Borlaug wanted to visit the Indian wheat belt to see how people farmed there. Swaminathan took Borlaug and several of his Mexican students on a five-week harvest-time tour through northern India, visiting farms, talking to farmers, and looking at ag-research facilities. It was a version of the trip through Mexico taken by Rockefeller scientists before the Mexican Agricultural Program. This time the two men hit it off. Borlaug discovered that the man whom he had taken for a time-server "had one of the most brilliantly swift agricultural minds I had ever experienced." Swaminathan learned that what he had taken for brusque condescension was a directness and simplicity that he thought almost "child-like."

As he had been in the Bajío, Borlaug was heart-struck by the lives of poor smallholders: men reaping grain with sickles, threshing it by hand, and storing it in gunny sacks that let in disease and insects; women cooking chapati over fires fueled by cow dung outside mud-walled huts. Children with the dull gaze and discolored hair of chronic malnutrition. No schools, no electricity, no fertilizer, no running water, no access to credit. Impoverished soils: farmers often harvested less than half a ton of grain per acre, barely enough to survive even in good years. (Sonora was at the same latitude as North India and had a similar climate, but its farmers were taking in three tons per acre.)

Indian farmers could never overcome these obstacles on their own, Borlaug thought. Yet local politicians and scientists seemed uninterested in farmers' woes. As Borlaug later recalled, he was repeatedly told that poverty was the smallholders' lot; they were "traditional" and didn't want change. Knowing how much he had hated harvesting corn by hand, he found it impossible to believe that Indian smallholders wouldn't welcome improvements in their lives.

To Borlaug, the resistance seemed to be due, at least in part, to a bureaucratic fear of confronting high officials. The "first step" in the Green Revolution, he stated, was "the vigorous introduction" of "heavy rates of chemical fertilizers." (Nitrogen! Get that rubisco going!) Yet India was dragging its feet. The nation had few fertilizer plants, which meant that most fertilizer had to be imported, which meant in turn that it had to be paid for with foreign exchange. According to the Indiana University historian Nick Cullather, Nehru's development plan reserved four-fifths of India's foreign-exchange reserves for heavy equipment for industry. The remainder was for importing "essential raw materials." Fertilizer, an agricultural raw material, was given a lower priority than industrial raw materials like jute, the plant fiber in burlap or gunny cloth, because burlap and gunny cloth could be manufactured in India and sold abroad, recouping the foreign exchange used to buy jute. Meanwhile, fertilizer was a foreign-exchange loss—the food it produced was not exported, but used to feed Indians. As a result, so little fertilizer was imported that aid officials liked to say that India was not overpopulated—it was underfertilized.

In the West, businesspeople would see the fertilizer shortfall as an opportunity, and build their own fertilizer plants. But this didn't happen in India, partly because the nation had little investment capital, partly because fertilizer plants required large amounts of scarce electricity to operate, and partly because such plants were officially discouraged. In Nehru's government, planners assigned every industrial facility a numerical ranking. "Fertilizer factories ranked in the double digits," Cullather wrote, "well behind the high-priority steel mills and dam projects." Nobody wanted to tell Nehru and his ministers that lifting up Indian farmers necessarily would involve taking money away from heavy industry.

The inevitable confrontation occurred at the end of the trip, when Borlaug and Swaminathan came to IARI. Before a skeptical group of researchers and administrators, Borlaug talked about the need for scientists and the government to support farmers—above all, to help them get nitrogen into the fields. Then Swaminathan asked Borlaug the key question: Do you think your wheat could do for India what it has done in Mexico? The implication was: Are you certain enough of the value of this technology for us to fight the battle that would be required to import it? Borlaug hesitated. He had not tested his new varieties in India

Borlaug and Swaminathan in an Indian field in the mid-1960s

and still knew little about Indian conditions. I don't know, he admitted, chagrined.

The next day he flew to Pakistan. Two recent graduates from his training program now worked in a research institute outside Faisalabad, the nation's third-biggest city. Borlaug had previously sent them samples of his wheat varieties for testing. He hoped to quietly look at the results with his former students, which might give him some answers for Swaminathan's question. Instead he discovered that Pakistan's ministry of agriculture had arranged a special day to honor him. A crowd of dignitaries, civil servants, and newspaper reporters met Borlaug and the minister at the gates of the institute.

From Borlaug's point of view, it was an ambush. The head of the research institute, like some of his Indian counterparts, detested the thought of following orders from foreign experts. With the journalists in tow, he led Borlaug, the agriculture minister, and a group of researchers and officials on a tour of the facility. Rows of Pakistani native wheats and the Mexican varieties had been planted side by side. The Pakistani wheat grew straight and tall; the Mexican wheat was short and spindly.

"The weeds were nearly as high as the wheat," Borlaug later remembered. "The whole nursery was miserable." The director said, in Borlaug's recollection, "You see, the Mexican wheats don't fit here. Look how good the tall Pakistan wheats are." And he proceeded to tell Borlaug that his work would be useless in Pakistan—a dramatic presentation for the camera-wielding reporters and his superiors. Borlaug, "more and more irritated," said that the seed beds had not been properly prepared, the plants had been inadequately fertilized, and the plots had not been weeded. "This is how we grow wheat in Pakistan," the director said.

The argument continued into the evening, with Borlaug insisting that the trial had not been fair. With this kind of wheat, he said, Mexico had tripled its output in a few years. The director pointed to the evidence in the fields: the new varieties didn't work in Pakistan. Borlaug's former students watched but said nothing. As he later recalled,

> We were to leave the next morning on the plane at 10:00. As we walked toward the guest house, these two ex-students of mine, they came up and said, "We have something we want to show you tomorrow before your flight." I said, "OK, what time?" They said, "At daylight." [Early the next morning, there was a] tap on the window. I went out. It was just getting daylight. We walked to the most remote corner of the experiment station. And there were four beautiful plots, about the width of this room and maybe twice as long, of the best four new dwarf Mexican varieties that were commercial in Mexico. They said, "There they are. You see how they fit!" And I said, "Why didn't you plant the nursery like that?" They said, "They wouldn't let us."

Fuming, Borlaug left for Mexico. On the plane he wrote a memo to higher-ups at Rockefeller—an angry, barely literate screed, he said later. But his thrust was clear: the Mexican seeds could grow in South Asia. They just needed to be given a chance. The naysaying was directly harming poor farmers—literally snatching bread from their mouths. He attributed the foot-dragging to academic politics, bureaucratic laziness, and class-ridden careerism. And doubtless he was correct in some instances. It never seems to have occurred to him that any of the resistance might be attributable to something worth considering.

Rush Order

In November 1963, Swaminathan received the next shipment of Borlaug's wheat: 220 pounds each of four commercially released varieties and samples of another 600 breeding lines that were promising but not yet commercially available. IARI researchers divided the wheat among five-acre plots in four different experimental stations. The results were remarkable. Indian farmers typically reaped less than half a ton per acre. The four Mexican varieties yielded a per-acre average of about a ton and a half, and some plots came in at almost two tons.

Researchers, excited, tipped off the press. In March 1964 the extraordinary yields were trumpeted in India's largest newspapers, *The Times of India, The Statesman,* and *The Sunday Standard.* Capitalizing on the attention, Swaminathan asked the government to buy twenty tons each of the two best Mexican varieties—Sonora 63 and Sonora 64—to test in a thousand acres of demonstration plots around the country. In the usual course of events, the request would have been rejected by Nehru's industry-favoring ministers as a waste of foreign exchange. But things had changed. Nehru died in May 1964, his credibility as damaged by the war as his health. His successor, Lal Bahadur Shastri, quickly put the nation's steel minister, Chidambaram Subramaniam, in charge of the ministry of food and agriculture. Both Shastri and Subramaniam were independence activists who had been imprisoned by the British; both men had argued against Nehru's industry-first policies. And both men had seen the Mexican wheat at IARI. Now they had a chance to do something about it. They approved Swaminathan's request.

As the grain came in, Subramaniam proposed quintupling expenditures on agriculture, setting off a furious, months-long debate within the new government. While the government wrestled over policy, Swaminathan released the results of the newest tests of Borlaug's grain. Again they were positive. Now Swaminathan asked for two hundred tons of Sonora 64, five times more than before. Finance officials balked, but Swaminathan was able to circumvent them. His father-in-law, S. Bhoothalingam, was a high-ranking official in the ministry of finance. Bhoothalingam promised that the money wouldn't be lost in the bureaucracy or whittled down as a compromise. A request was sent to Rockefeller on July 2, 1965.

Swaminathan was lucky to get it through. The Shastri administration was ensnarled in a host of conflicts, inside and outside India. Nehru had been prime minister for seventeen years and had built the entire government in his image. Shastri and his ministers had different ideas, but trying to implement them set off a struggle with Nehru's bureaucrats. On another front, Lyndon Johnson was pressuring India to let U.S. companies build fertilizer factories in India, threatening to withhold food aid if Shastri did not go along. Relations with Pakistan were worsening; the two countries were skirmishing across their border. A drought had begun to settle into the northeastern state of Bihar, raising the prospect of hunger and food shortages.

For Borlaug, the sudden request for 200 tons of Sonora 64 was startling. More startling still was a second order, from Pakistan, for 250 tons—Borlaug's exhortations and the results of Swaminathan's tests in India had changed minds. Mexico had never exported any wheat across an ocean before. On top of that, the requests, received in July, were late. The grain had to be planted in November, which meant that it had to arrive in mid-October if the seeds were to be transported to the planting areas by that time. Bulky and heavy, the grain would have to travel by ship to South Asia, which would take at least two months. To make the planting deadline, the ship had to leave Mexico by mid-August. Borlaug therefore had a little more than a month to get the wheat on a ship. Sonora's seaport was the nearby city of Guaymas, about seventy miles from Borlaug's experimental site. A quick visit sufficed to establish that no ships were available there on short notice that could go to India. Instead Borlaug was forced to book passage for his grain from Los Angeles on "the last freighter that would get it in time to Pakistan and India for planting in the middle of October." It planned to depart on August 12, 1965.

Wellhausen, his supervisor, was on a long home leave. Borlaug was supposed to manage the Rockefeller office in Mexico City while his Sonora replacement, Ignacio Narváez, arranged to send the grain on the ship. To Borlaug's consternation he spent all of his time wrangling with government officials about paperwork. The contracts for the grain were between the governments of Pakistan and India and Mexico's state-owned Productore Nacional de Semilla (National Seed Producer), known as Pronase. The Pakistan agreement included provisions for unexpected accidents damaging the seed. The usual English legal term for these unexpected accidents was an "act of God." According to Vietmeyer, Borlaug's biographer, Pronase objected on religious grounds: "God didn't do bad acts." The company insisted that the text be changed to *acta de naturaleza*—act of nature. Pakistan insisted that "act of God" was demanded by legal precedent. Borlaug had the two parties sign two contracts, one in English with "act of God," one in Spanish with *acta de naturaleza*. Late in July the Indian government weighed in, asking Borlaug to change its order to one hundred tons of Sonora 64 and one hundred tons of another variety, Lerma Rojo 64, which had better rust resistance. Borlaug refused—Narváez and the grain, in twenty thousand sacks, were already on trucks roaring toward the U.S. border.

> And then everything went wrong. I thought I had the border fixed so that these 35 big trucks—and we didn't have much money for this operation—[would go through. Instead they] were held up at the Mexican border for two days. And I had to pay them to hold the freighter. Then we were held up for another day on the American side. Pure bureaucracy. Finally, my Mexican colleague that was supervising this, he called and he said, "They're on their way to the Los Angeles port." But they didn't get very far. It was the day of the Watts riot.

Ignited by accusations of police brutality, the riots in the African-American neighborhood of Watts turned a square mile of Los Angeles into a combat zone, complete with burning buildings, sniper fire, and thousands of armed police. The Mexican truckers wanted to turn back when they saw smoke rolling over the city and the National Guard with its rifles and tanks. The governor of California placed a curfew over an area of almost fifty square miles. Signs warned that troublemakers would be shot. Over the phone, Borlaug begged the owner of the shipping line to make the ship and its crew wait for his trucks. He told Narváez to order the truckers to drive through the fiery streets to the docks. Shouting and swearing, he called the L.A. police and demanded that his convoy be escorted through the blockades.

All the while he was fielding angry calls from Pronase. The check for the Indian grain had cleared. The Pakistani check, though, had bounced.

"Check" is a misnomer. The payment was a money order for $95,000, roughly equivalent to $700,000 today. Alas, the money order had misspelled Pronase's proper name, Productore Nacional de Semilla, and the bank in Mexico City wouldn't accept it. Pronase was insisting that it wouldn't allow the ship to be loaded until it received a new money order. Even as the trucks were rumbling toward the docks, Borlaug was frantically telephoning Rawalpindi, then Pakistan's capital. By that point it was Sunday, August 15. Reaching government officials on the weekend was no easier then than it is now. A flurry of cablegrams finally produced a request from a Pakistani official for more time.

At the docks, the truck drivers were surrounded by National Guardsmen. Narváez called Borlaug to ask if the grain should be loaded, given that it hadn't actually been paid for. By now Borlaug had learned a new legal term: *demurrage*, a charge payable to the owner of a ship if it is not loaded by an agreed-upon time. Rockefeller would owe demurrage to the shipping line if the grain was not loaded. But if Borlaug gave the go-ahead and Pakistan didn't replace the check, Rockefeller would be on the hook for $95,000. Leaving the sacks of wheat in the parking lot was not an option: grain used for seed, unlike grain used for flour, must be kept cool so that it will grow properly in the field. Frustrated and tired, Borlaug barked at Narváez in a way that would later make him feel ashamed. *Get the grain on the ship!* he snapped. *Send it on its damn way!*

When Narváez called to confirm that the ship was being loaded, Borlaug went to bed. Constantly on the telephone, he had stayed awake for seventy-two hours. When he awoke, he turned on the radio. War had broken out between Pakistan and India. His Pakistani contact lived in the city of Lahore, ten miles from the Indian border. Borlaug cabled him in shock and was astonished to receive a quick reply:

DON'T WORRY ABOUT THE MONEY STOP WE'VE DEPOSITED IT STOP AND IF YOU THINK YOU'VE GOT PROBLEMS YOU SHOULD SEE MINE STOP ARTILLERY SHELLS FALLING IN MY BACKYARD

The fighting was over Kashmir, a majority-Muslim region claimed by both India and Pakistan. A war in 1947–48 had ended up by partitioning Kashmir with a ceasefire line that neither side recognized as an official border. The new round of fighting had begun earlier in

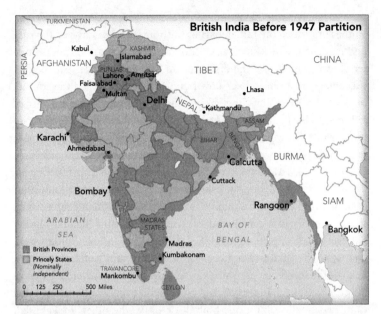

British India Before 1947 Partition

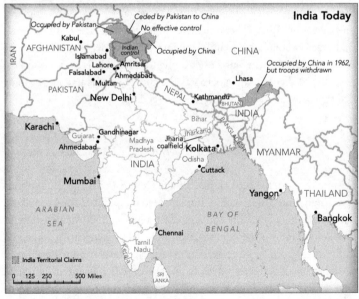

India Today

August, when about thirty thousand Pakistani troops dressed in civilian clothes infiltrated the Indian part of Kashmir. India discovered the ruse and launched its own incursion into Pakistani Kashmir. Prime Minister Shastri, mindful of the disastrous loss to China three years before, was determined not to back down. Indian forces escalated the conflict on September 6 by invading Pakistan itself, targeting the border city of Lahore. Pakistan launched a retaliatory assault aimed at the Indian city of Amritsar, a few miles from Lahore on the Indian side of the border. Large tank battles occurred on the ground; scores of planes shot at each other in the sky. More than six thousand soldiers died in the fighting.

Borlaug liked to proclaim his lack of political sophistication. But even he could see that his plan to unload India's portion of the grain in the Indian city of Mumbai and Pakistani's portion of the grain in the Pakistani port of Karachi was going to run into trouble. India would confiscate the Pakistani grain as war materiel. Again he telephoned the head of the shipping line. The company would have to divert the ship to Singapore, he said. In this neutral ground, the grain could be unloaded onto two smaller ships, one bound for India, the other for Pakistan. Even as he negotiated with the shipping company, he was receiving irate messages from Rockefeller executives in Manhattan. They had just realized that Borlaug had potentially incurred a large debt on behalf of the Pakistani government. Further pleading telephone calls came from Pronase, the seed company. Where was the check? Unable to cope, Borlaug fled the office and went fishing for several days.

Fumigation

Despite the war, the seed arrived in late October, more or less on time. Amazingly, both Pakistan and India planted it rapidly. In the first nation, about half the grain went into acre-sized plots at 2,500 farms as a public demonstration; the rest went onto 30 large government farms to multiply the seed for the next round of tests. The second nation, India, sowed it on demonstration plots at government research centers and government-selected "progressive" farms that were said to be more open to innovation. The planting was a month later than was optimal, but Borlaug was reasonably confident that the delay would not impede progress. In addition, Pakistan's second, corrected money order had arrived.

Narváez flew to Pakistan days after the seed arrived and reported that the new varieties had been planted properly and were being given sufficient water and fertilizer. Despite the ongoing war, he had obtained the necessary government approvals to travel through the militarized zones and stay in Lahore to provide advice during the growing season. Two weeks after his initial survey he and a Pakistani colleague looked at plantings in the north. Instead of healthy young shoots they found scattered, thin sprouts dotting fields that were mostly barren. The farmers had done everything according to instructions, but something was wrong. Narváez immediately sent a cable to Borlaug, who flew to Pakistan.

The two men decided to survey the whole country again, Borlaug looking at test plots in the south, Narváez in the north, meeting in the central city of Multan. (In those days Pakistan was not in the grip of sectarian violence; foreigners like Borlaug could travel in rural areas without taking elaborate precautions.) To Borlaug's dismay, Narváez's initial report proved accurate. The seeds had been planted about two weeks before, enough time for them to germinate and stick their heads above the surface. Half or more were effectively dead. Borlaug pelted the farmers with questions. As Narváez had reported, they had been following instructions to the letter. Something was wrong with the seed itself. The problems were sure to be replicated in India, too.

The two men met in Multan, in the middle of the country. Matching their despairing mood, the only hotel with available rooms was memorably unpleasant: cold, bleak, dirty, alive with vermin. They talked in their room until early in the morning, passing a bottle of whisky back and forth. Pakistan had invested a big fraction of its foreign-currency reserves in the seed, which now was failing. The huge effort to get the seed to South Asia had been a waste of time. Borlaug had let everyone down—officials in Pakistan and India who had taken a risk on him despite the opposition of their own scientists, farmers in Mexico who had believed in him enough to provide the grain at low cost, foundation executives in Manhattan who had backed his ideas despite his unwillingness to follow rules, villagers in India and Pakistan whose hopes he had raised. The two men couldn't understand what had gone awry. They finished the bottle just before dawn.

The next morning Borlaug had to write a batch of painful telegrams explaining the germination problems. All anyone could do at the moment, he said, was to double the planting rates, fertilize the seeds like mad, and hope that something could be salvaged. Communication between India and Pakistan was shut down by the war, so he had to send an embarrassing cable to Wellhausen in Mexico and ask him to pass the word to Rockefeller's India office, which would in turn send a note to Swaminathan. Later he would conclude that Pronase had wanted to ensure that its wheat wasn't eaten by rats or attacked by mold on what was for the company a voyage of unprecedented length. In its inexperience, it had fumigated the seeds aggressively with methyl bromide—so aggressively that it had damaged them. But that understanding wouldn't come for months. In the meantime, baffled and disheartened, Borlaug spent December 1965 anxiously shuttling between the two countries, the four-hundred-mile trip between Rawalpindi and Delhi extended by the now-necessary three-thousand-mile detour to Dubai—all direct flights had been canceled. The heavy fertilization seemed to be working, but that was no guarantee that the grain would come.

Meanwhile, President Johnson's foreign-aid program, the biggest element of which was food aid to India, was attracting opposition in Congress. Rather than displaying proper gratitude, from Johnson's point of view, India had spoken out against U.S. intervention in Vietnam—a personal affront. And it had got itself into a shooting war with Pakistan, another U.S. ally. Worst of all, in the president's eyes, the Nehru policy of prioritizing industry over agriculture still seemed to hold sway. He issued an ultimatum: if the Shastri government did not throw all available resources into agriculture, he would cut off food aid. To keep Delhi on a tight leash, aid would be issued on a month-by-month basis. The Indian government was furious. Kowtowing to foreigners was what Nehru, Shastri, and the rest of the Congress party leadership had spent decades fighting. And using food aid for political leverage—playing with the lives of millions of people—struck the inheritors of the Gandhian tradition as deeply immoral.

India and Pakistan slowly halted the fighting and signed another ceasefire at a conference in Tashkent in January 1966. India ended up in possession of 720 square miles of Pakistani territory; Pakistan won 220 square miles of India. Both sides claimed victory. Shastri represented India at the ceasefire conference. The morning after signing the agreement he suddenly died, giving rise to a generation of conspiracy theorists (most claim that the CIA assassinated him). Into his office stepped a formidable presence: Indira Gandhi, Nehru's daughter. (No relation to Mahatma Gandhi, she had married a man with that politically potent surname.)

Gandhi was immediately caught up in the tense back-and-forth between India and the United States—the former fiercely determined to maintain the independence it had struggled for, the latter seeking to draw India into the Cold War and viewing its industry-first policies as a humanitarian disaster. Eventually she caved in to Johnson's requirements, though with lasting bitterness. She agreed to spend more on agriculture. She kept Subramaniam as minister of agriculture. And she listened to his advice.

Despite the fumigation, some of the Mexican seed grew. What remained, though damaged, still outperformed traditional varieties grown with traditional methods. In Pakistan, Narváez had planted adjacent crops of Mexican wheat and local wheat on the grounds of the presidential palace. When the president took his afternoon walk, he could see the difference himself. For her part, Gandhi toured the experimental farms with Swaminathan. To Swaminathan, it was obvious that she knew little about agriculture. It was also obvious, he told me, that she was a quick study who was not afraid to make up her own mind. Journalists were writing that the foreign seed was an attack on Indian culture. A leading member of the powerful Planning Commission demanded that India terminate its relationship with Mexico.

Instead Gandhi decided to gamble. Three Indian officials went to Mexico in the summer of 1966 to buy eighteen thousand tons of Lerma Rojo 64—more than forty times the previous purchase. They refused to contract with Pronase, instead making a deal with a farmers' cooperative. Because Pronase was the only entity allowed to export grain, Borlaug arranged a special permit with the Mexican government, setting off a bureaucratic turf battle. The grain went out from the Sonora port of Guaymas in October, filling the holds of two big ships. Much of India's precious foreign-exchange reserves was consumed by the purchase. But Gandhi was feeling pressure to accelerate the wheat program, and not only from the agitated president in Washington. The monsoon rains had failed that summer in the northeast. Worst affected was the poor, populous state of Bihar, west of Kolkata, burned into Indian memories as the site of the horrific famine of 1943.

A small-scale academic contretemps has arisen over whether what

Loading and packing eighteen thousand tons of grain in 1966 was an enormous effort for Mexican distributors, who had never done anything like it before.

happened in Bihar in 1966 was an actual famine or a temporary short-fall exaggerated for political reasons. Between July and October of that year almost no rain fell in the state—except for a week in August, when heavy downpours caused catastrophic floods. Paradoxically, the rainfall didn't help Bihar's farmers. Instead it washed out their fields, carrying away the plants.

Local and parliamentary elections were scheduled across India for the following February. Fearing that voters would believe that it had let the situation get out of control, the Bihar government initially refused to admit the gravity of the situation. Some of Borlaug's travels took him to the edge of Bihar. Years later he would recall the trucks that plied the streets, picking up the nightly dead. Homeless children thronged the entrance of his hotels, begging for bread, thin hands plucking at his clothes. Physicians reported frightening increases in malnutrition and the diseases of dietary deficiency. Its hand forced by disaster, the Bihar government finally asked Delhi for help in the fall of 1966. But the

Gandhi administration, too, dismissed the alarms, partly because the resurgence of widespread hunger was an embarrassment to the nation, partly because Bihar was (accurately) viewed as horrifically corrupt—its politicians might exaggerate a famine just to pocket the relief money. Adding to suspicions, Bihar's leaders had not supported Indira Gandhi in the party caucuses after Lal Shastri's death.

As evidence of suffering mounted, Gandhi, too, was forced to reverse course. In November she gave a widely publicized speech calling on Indians to mobilize against the tragedy. "Countless millions of our people," she said, have "had the bread taken out of their mouths by an abnormal failure of the rains. . . . There is hunger and distress in millions of homes." Still, Gandhi didn't use the loaded word "famine," fearing the reaction it would trigger. Johnson demanded that she say it, because Congress would be more likely to dole out foreign-aid money for an official famine. People are starving in India, he proclaimed. He was furious when the Indian ambassador, mindful of the political situation in Delhi, refused to back his claim. Despite its unwillingness to utter the F-word, the government set up the biggest relief effort in modern Indian history, borrowing huge sums from the International Monetary Fund to import 20 million tons of grain, most of it from the United States. The operation was successful in that it prevented widespread death. It was unsuccessful in preventing sickness and misery. Notwithstanding Gandhi's efforts, unhappy voters punished her Congress party at the polls. The opposition won in Bihar. After much hemming and hawing, it declared in April 1967 that two-thirds of Bihar was afflicted by famine or scarcity. More than 34 million souls were affected.

At the time of the election, Borlaug and Swaminathan had been traveling across northern India, inspecting the test plots. This time, properly treated in Mexico, the wheat was germinating well. Giddy with success, the two men met with Subramaniam, the minister of agriculture. "A revolution is starting," Borlaug said, in his later recollection. "You must take action. Farmers are demanding more support." The country had to provide them with fertilizer, water, and financial support—the same demands that Johnson had been making. Subramaniam responded with fury. The election had cost Gandhi's Congress party multiple seats in the Lok Sabha, the lower house of India's parliament. One of the lost seats belonged to Subramaniam, who had been so preoccupied with arrang-

For Borlaug, the most meaningful part of winning the Nobel Peace Prize was probably the reaction from his long-ago neighbors in Iowa. Soon after the announcement, he was celebrated in Cresco, which put up a statue to its native son.

ing food aid that he had not gone home to campaign. Caught up in their work, Borlaug and Swaminathan had not realized what was happening. Subramaniam directed the two men to Deputy Prime Minister Ashok Mehta, director of the Planning Commission, who was still in office—and still the main opponent of building more fertilizer plants.

The meeting with Mehta was stormy. Borlaug told him that tens of thousands of farmers had seen what the new wheat could do. They want to produce more, he said, and they will demand the necessary fertilizer. India did not have the capital to build a lot of fertilizer plants, the deputy prime minister said. The Haber-Bosch process consumes enormous amounts of energy, and India did not have enough electricity. Borlaug said, in effect, feeding the citizenry is the first duty of any government. Fertilizer feeds people. If you don't have the money, let foreign companies build the factories. Since the days of Nehru, Mehta explained, government policy has been to refuse to permit foreign multinational corporations to control such vital sectors of the economy. Borlaug got angry. Unless the government somehow provided nitrogen to the fields, he shouted, the farmers would riot, vote out Gandhi and her Congress party, and install a new government that would respond to their needs.

They wouldn't care what was in the five-year plans, he said. They just wanted to have enough to eat.

Later Subramaniam came back into the government as Indira Gandhi's finance minister. "He always liked to say that the single finest hour in Borlaug's life was the time he spent shouting at Ashok Mehta," Swaminathan told me. "It may be true." To the end of his days, Subramaniam believed that the meeting convinced Mehta to throw the Indian government's support fully to agriculture. The government provided fertilizer and water. Yields soared in response—Indian harvests were 50 percent higher than they had been the year before. The country had never produced so much food. The next year yields were higher again. The year after that, Norman Borlaug won the Nobel Peace Prize. With the prize money, he bought his parents' house in Iowa—the first home he had ever owned. He gave it to his sisters.

"A Revolution That Failed"

In September 1970 a nineteen-year-old college student named James Boyce traveled on a fellowship to India. He first went to the central state of Madhya Pradesh, where the Peace Corps was helping villages test the new, high-yielding strains of wheat. As I mentioned earlier, state governments had asked "progressive" farmers to try out the wheat. "Progressive" meant that they supposedly weren't wedded to the hidebound ways of the past. In return for their willingness to take risks, the government provided them with seeds, fertilizer, and, most important, brand-new wells for irrigation.

Boyce is now an economist at the Political Economy Research Institute, at the University of Massachusetts. "There were about twenty-four of these villages with 'progressive' farmers," he told me not long ago. "And the remarkable thing was that in all twenty-four these supposedly 'progressive' farmers just happened to be the richest, most politically connected people in the village—the biggest landowners, the sons of politicians, you name it. People *hated* them." The project was supposed to target small farms, but these people had evaded the property ceiling by nominally breaking up their estates and registering the parcels under the names of relatives—living, dead, and fictitious. Several were members of the state legislature. None actually worked on the land. Boyce laughed, thinking about it. "And *that* was the introduction of the Green Revolution to this part of India!" The testing project was supposed to be aided by the Peace Corps volunteers, but most of them quit because it was being used to give comparatively wealthy people an additional leg up. The new varieties made a big difference in the fields—they increased the disparity between rich and poor.

Later Boyce and his wife, the population researcher Betsy Hartmann, lived in a village in northwest Bangladesh and wrote a book about it. The World Bank, Sweden, and the Bangladeshi government had funded three thousand irrigation wells in the region. All were supposed to be used by cooperative groups of small landholders. Instead, as they recount in *A Quiet Violence* (1983), every single one of them went to local rich people. The wells were a new technology, the deep tube well. One was acquired by the wealthiest landowner in Boyce and Hartmann's village. Because farmland in Bangladesh is typically broken into tiny parcels, even big landowners farmed many small, scattered plots. No matter where the landowner put the tube well, it would be able to irrigate some parcels that he did not own. These parcels, if irrigated, would be able to grow the new strains of wheat and rice, even in the dry winter. In other words, they would become valuable enough to steal, whether through fraud, violence, or bribing local officials. Fearing that their land was at risk, villagers smashed the tube well before it could be used.

Such stories occurred all over India, Pakistan, Bangladesh, the Philippines, Thailand, and Malaysia, according to witnesses like Boyce and Hartmann. Latin America saw them, too. The increase in yields made farmland more valuable, which made it worth seizing. Rich landowners, seeing an opportunity for advancement, evicted sharecroppers and renters and went into business for themselves, monopolizing local access to seeds and fertilizer. Increased harvests made prices fall. Big estateholders could more than make up for the decline with volume; smallholders were immiserated. All of this had begun in Mexico. In the name of increasing yields, Borlaug had shifted the program from the poor in the Bajío to a few prosperous landowners in Sonora. Those landowning farmers had worked hard, but they had reaped almost all the gains. The other big winners were corporate middlemen: grain processors like Archer Daniels Midland and Cargill and agrichemical suppliers like Monsanto and DuPont.

When I first visited Mumbai, in the mid-1980s, my friends and I spent a few days walking around the city. It was not hard to find evidence of extreme poverty. At one point a couple of kids in a schoolyard invited us to visit their class. Because we had nothing else to do, we agreed. In the classroom, the students goggled at us. It was like show and tell, and we were the show. The school was a charity effort, with most of the students from destitute families. They wore cast-off chinos and T-shirts from the charity. Afterward the teacher offered us tea. I asked where the students came from. Most were kicked out of their villages by the Green Revolution, he said, matter-of-factly. The city is full of these people.

Hard by the social costs were the environmental costs. The intensive fertilization mandated by the Green Revolution has heavily contributed to nitrogen problems on land and water. Pesticides have wreaked havoc on agricultural ecosystems and sometimes poisoned sources of drinking water. Poorly constructed and managed irrigation systems have drained aquifers. Soils have become waterlogged or, worse, loaded with salts when irrigation water evaporated. Possibly most worrisome, the energy costs of agriculture, mainly from making fertilizer, have soared. Industrial-style Borlaugian agriculture is a significant contributor to air pollution and climate change.

Criticisms like these began to appear soon after Borlaug won the Nobel. Between 1972 and 1979 the United Nations Research Institute for Social Development published fifteen analyses of the Green Revolution. Every single one was sharply negative. To Biplab Dasgusta, a prominent economist and Marxist politician in India, the major consequences of the Green Revolution included an "increase in the number and proportion of homeless households" and "growing concentration of land and assets in fewer hands." The Green Revolution, observed Per Olav Reinton of Oslo's International Peace Research Institute (PRIO), "demonstrates more clearly than most aid programs how good intentions produce misery." Perhaps the most influential and unequivocal study was *The Political Economy of Agrarian Change* (1974) by the Oxford economist Keith Griffin. His summation: "The story of the green revolution is a story of a revolution that failed." Between 1970 and 1989, more than three hundred academic studies of the Green Revolution appeared. Four out of five were negative.

Over time the tone became more vituperative. The journalist-activist Susan George was content to state in 1972 that the Green Revolution had "brought nothing but misery to the poor." Four years later the Dutch economist Ernest Feder described it as a "scheme for the self-liquidation of Third World peasantry." By 1991 the anti-globalization activist Vandana Shiva was charging that Borlaug's legacy to the world was nothing but "diseased soils, pest-infested crops, water-logged deserts, and indebted and discontented farmers." Two years before Borlaug's death in 2009, the celebrated left-wing journalist Alexander Cockburn accused him of mass murder. "His 'green revolution' wheat strains led to the death of peasants by the million." When Borlaug spoke at conferences, students sometimes booed.

Borlaug seldom replied directly, though the attacks stung. In private, he told friends that most of the criticism was sheer elitism. Somehow rich environmentalists in the West thought the world was better off if people in poor areas didn't improve their lives. He had nothing against organic this or that but it was unrealistic to promote it as a solution to hunger in the world of 10 billion. And it was immoral to stand in the way of feeding hungry people.

The most important point, in his view, was that the new methods and new crops had done what they were supposed to: increase yields. Economists have estimated that the global average productivity gains from Green Revolution crops are about 1 percent per year for wheat, 0.8 percent per year for rice, and 0.7 percent per year for maize. The numbers sound small, but over time the impacts grow large, compound-interest style. Between 1960 and 2000, wheat harvests in developing countries tripled. Rice harvests doubled. Maize harvests more than doubled. The extra food, Borlaug said, was why the population could increase while the proportion of hungry people went down.

Some of the environmental problems were real, he conceded. But they were due to bad policy and administration rather than anything inherent to Green Revolution technology. Farmers who had never used fertilizers and pesticides needed to be trained in their deployment; they needed to be educated about the need to calibrate dosages. Borlaug said this again and again and again, sometimes in numbing detail, but it never seemed to satisfy the critics.

To some extent, I suspect, the attacks confounded him. He had seen for himself the fruits of the Green Revolution. So had others. In 2008 and 2009 the journalist Joel Bourne toured through Punjab, in the

In Borlaug's view, many Green Revolution critics were blind to how bad the past had been. The India he saw in the early 1960s was little different from that captured in the 1940s by the photographer Margaret Bourke-White—a land of impoverished, drought-blasted farms like this rust-stricken wheat field in northern India.

north, talking to the farmers who had lived through the introduction of the new varieties. "There was nothing like it in their lifetimes," Bourne told me. "The first season they grew so much wheat they had no place to store it all. They closed the schools early and filled them full of grain." In 2016 the reporter Harish Damodaran traveled to the village of Jaunti, in the outskirts of Delhi, where Swaminathan had selected ordinary farmers—people with little land—to try the new seeds in 1964. The survivors were in their eighties, but they remembered that year well. Their harvests tripled, they said. "This was a miracle," one man told Damodaran. "It totally changed our lives." For an article, I traveled in the western state of Maharashtra in the mid-1980s. Farmer after farmer told me how the new seeds had increased their harvests. These were not wealthy people. One man told me with quiet pride that he and all of his brothers now had bicycles. In these places, whether by chance or design, the package had been distributed more equally.

Borlaug worked almost until his death in 2009. An eccentric Japanese rich person funded his efforts to develop high-yielding varieties for Africa, which had been little touched by the Green Revolution. He spent his ninetieth birthday at a reception in his honor given by the U.S. State Department. He had just flown back from Uganda, where he had been fighting a scary new strain of stem rust.

When I last spoke to Borlaug, a few years before he passed away, I asked him about the past criticisms. Critics, he said, never wanted to answer the counterfactual question: Where would the world be today if we had the same growth in population and affluence but none of the yield increases of the Green Revolution? Overuse of fertilizer, waterlogging soils, loading up land with toxic salts from badly run irrigation schemes—these were real issues, he said. But wouldn't you rather have these for problems than the kind of hunger we had in 1968?

He asked me if I had ever been to a place where most of the people weren't getting enough to eat. "Not just poor, but actually hungry all the time," he said. I told him that I hadn't been to such a place. "That's the point," he said. "When I was getting started, you couldn't avoid them."

The Workshop and the World

In February 1964 Borlaug went to a meeting in Pakistan of agricultural experts and government officials to explain why he believed the nation should adopt his new, high-yielding wheat varieties. The trip was difficult, and he arrived at the meeting having not slept for two days. The first speaker was the vice chancellor of the University of Sindh. He mocked Borlaug's announced hope of doubling the nation's wheat output. Especially ridiculous, he said, was the idea of doing it with Mexican wheat. Mexican varieties were too delicate for Pakistan. They were too short and needed too much water and fertilizer. Most of all, he said, the grain was the wrong consistency. It was even the wrong color. "Pakistanis," the vice chancellor shouted, "never ate red wheat!"

Borlaug's varieties, developed with Mexican bread in mind, were hard red wheat with high levels of protein and gluten and bitter undertones in the taste. (The "red" in red wheat refers to the color of the husk—the bran—around the kernel, rather than the kernel itself.) Combined with yeast, high-gluten flour produces bread with an "open crumb"—big, irregular holes of the sort seen in sourdough loaves. By contrast, the traditional South Asian *chapati* or *roti*—unleavened Indian or Pakistani flatbread—is made from soft white wheat, which has an amber husk, relatively little gluten and protein, and fewer bitter notes. On the dinner table, the results are light, fluffy, and almost sweet, with a closed crumb: tiny, fine holes in the bread.

The chapati is as much a part of daily life in much of South Asia as the baguette is part of daily life in France. It was such an emblem of Indian identity that it was used as a symbol of rebellion against the British in the nineteenth century. To South Asians, the whiteness of a chapati "suggested purity, luxury, even modernity" (I am quoting Cullather, the Indiana University historian). Because most Indian and Pakistani families ground their wheat at home between two circular stones, bran was mixed in with the flour. With amber-colored wheat, the bran doesn't change the color of the flour. The dark red bran in Mexican wheat produces dark flour, which to Indians has an aura of dirt and poverty. In addition, the texture of the bread and the feel of it in the mouth were wrong. Even the smell while it cooked was wrong.

For a Westerner, Iowa-born and -raised, to insist that Indians and Pakistanis make chapati with this strange Mexican wheat was as if a foreigner were demanding that French people make baguettes from pumpernickel. French people would regard the demand as a cultural affront. Similarly, the vice chancellor believed, South Asians would—and indeed should—reject this alien wheat.

Borlaug dismissed this kind of complaint as nit-picking. In none of his writing have I encountered any suggestion that this kind of "minutia" should be given more than "minor consideration." Borlaug seems to have viewed himself like a doctor faced with a patient's arterial bleeding—and a patient who refuses treatment because he objects to the doctor's nationality or the color of the bandages. That doctor would ignore the complaints and slap on the bandages. As for the farmers, he did not believe that they would refuse to plant more-productive, disease-resistant grain because it had a different color or smell. Nor would hungry people refuse to eat it. The widespread adoption of Mexican wheat varieties in India and Pakistan, in his view, testified to the correctness of this idea.

Women in northern India grinding grain between two circular stones in 1902

Only when I spoke to Swaminathan did I learn of the sequel to the meeting in Pakistan. After the first tests of Mexican wheat, Swaminathan and his associates realized that it would not fit well into South Asian culture for the reasons identified by the Pakistani vice chancellor. Without telling anyone at Rockefeller, they began irradiating the Sonora wheat at the particle accelerator in Mumbai in November 1963, three months before the meeting in Pakistan. Nothing happened the first year. The second year, Swaminathan got lucky. The color of wheat bran, we now know, is mainly controlled by four genes that are in turn switched on or off by a single gene known as R. By chance, the gamma rays passing through the seeds disabled some aspect of this mechanism; the bran color in the next, mutated generation was amber. Miraculously, its yield seemed to be unaffected.

A deft politician, Swaminathan called his new variety Sharbati Sonora—Sharbati is a celebrated traditional wheat variety from Madhya Pradesh. He introduced it with fanfare in 1967, emphasizing that Sharbati Sonora had been created by Indian scientists for Indian families in an Indian atomic-research facility. The variety turned out to be vulnerable to rust. Nonetheless, Swaminathan had removed the foreign red tint from the Green Revolution. By crossbreeding Sharbati Sonora with other local varieties, Swaminathan was later able to develop rust-resistant cultivars that seemed wholly Indian. This was the wheat that transformed Indian agriculture, the methodological cousin of the rice that became entangled with the social conflicts witnessed by Boyce and Hartmann.

As early as 1968, Swaminathan warned about smallholders' penchants for overusing fertilizer, pesticides, and irrigation water. As evidence of the ecological downsides of the Green Revolution accumulated, Swaminathan asked for increases in farmer training. By 1996 he was calling for a new transformation: changing the Green Revolution to an "evergreen revolution." The goal was to combine high technology with "traditional ecological prudence." Genetic engineering would create new varieties that needed less water and fertilizer and could tolerate salty soils. Farmers would install electronic monitors in their fields to monitor crop growth and ensure proper management of chemicals and water. Computers would combine the readings with weather data and crop-simulation models to produce individualized recipes for farmers to maximize their yields.

The critics weren't appeased—they had grown suspicious of the whole enterprise. They no longer gave much credence to the notion that the discoveries of laboratory white-coats would benefit ordinary people. Somebody like Swaminathan might view his work as a course correction in the long-run effort to improve agriculture for the common good, but the critics saw it as more of the same—a mad effort to make up for a disaster by doubling down on the methods that had caused it.

Agriculture was but one portion of a larger issue. As the philosopher Robert Crease puts it, the scientific enterprise studies phenomena—atoms, clouds, organisms, planets—by transferring them from the world we live in, with all of its confusion and sentiment, into a special workshop, a place where they can be reduced to abstract, measurable quantities and manipulated in a controlled manner. This method of working is incredibly powerful. It discovered the laws of electricity and created antibiotics and built the atom bomb and invented X-rays and generated techniques to harvest and store energy from the sun and wind. But it is also risky, as another philosopher, Edmund Husserl, observed in the 1930s. (His work didn't appear in English until decades later, under the forbidding title *The Crisis of European Sciences and Transcendental Phenomenology.*) The air in the scientific workshop is so clean and bracing and the results of researchers sequestering themselves inside so satisfying that they lose their bearings. They don't want to leave the workshop. They prefer to live in its world of abstraction, separate as angels from the messiness of life. Or, worse, the findings of the workshop seem so luminous and clear, so like beacons of truth, they forget that the workshop is a special place within the world and begin to think that it is above the rest of life and should control it. And here, Husserl said, lies peril, because the people outside the workshop will come to detest and disbelieve the people within its privileged walls.

Like Vogt, Borlaug was caught up in the clash between the workshop and the world. Farmers in India and Mexico thought of their wheat in terms of how they experienced it—as a plant that was (or wasn't) easy to grow and harvest, as grain that made flour with certain qualities, as the source of bread that, eaten daily, conveyed a statement about their lives, as a set of smells and tastes and colors, as a storehouse of memory and identity. Borlaug had taken this wheat into the workshop of science. There, in effect, he reduced wheat to a series of numbers: plant height, degree of rust resistance, spikelet number, flowering date, and so on. He measured these numbers in an effort to maximize another number: the weight of harvestable wheat. All of this was totally normal scientific procedure. And it *worked*—he created varieties of wheat that were resistant to multiple types of rust and yielded two or three times more grain.

But what was left out was the color of the bran, the texture of the grain, the pleasure of having several different types of flour, or, more important still, the relationship of the farmers to their land, and to each other, and the structure of power in a community or a nation. And then there was the omnipresence of greed. Borlaug was like a physicist who figures out how something should work on an idealized frictionless plane and then is startled when it doesn't function in the same way in the real world of hills and valleys.

Scientists say that because the workshop has found that *A* is true the world should do *B*. But people in the world notice that when scientists go into the workshop they strip their objects of study of everything but a few measurable quantities, and then the people object that what was stripped away was worthy, even essential. They see the scientists unable to understand the resistance to following course *B* as condescending aliens who don't share their values—and, all too often, they are correct. Why would you listen to people who have no idea what you consider important? Writing in the 1930s, Husserl believed that the ensuing rejection of expertise led to an embrace of the irrational and, ultimately, to the Nazis. Surely it has played a role in the rejection of scientific claims about genetically modified organisms, nuclear power, soil depletion, and climate change.

Vogt was saying that science, or at least ecology, dictated reaching into the most intimate aspects of people's lives. Borlaug was saying much the same thing. Both were doing it in the name of the future. And they were bewildered and hurt when the world of the present pushed back. It haunted their later days. They had laid out a path based on science as they understood it—a logical consequence of scientific rules. But people weren't heeding it. They weren't grateful. They behaved as if they thought they were exempt from the rules.

ONE FUTURE

TEN

The Edge of the Petri Dish

Special People

On June 30, 1860, Samuel Wilberforce, D.D., thirty-sixth Bishop of Oxford, attended the thirtieth annual meeting of the British Association for the Advancement of Science, held at Oxford University. Countless students have been taught that during the meeting Wilberforce attacked evolution, setting off an impromptu debate that became a "tipping point" in the history of thought. I was one of those students. The debate, my biology professor explained, was the opening salvo of the war between Science and Religion—and Religion lost. My textbook backed him up. Wilberforce's anti-evolution assault, it said, was swept aside by researchers' "careful and scientific defense." In a flourish unusual in an undergraduate text, it boasted that the pro-evolution arguments "neatly lifted the Bishop's scalp." That day the forces of empirical knowledge had beaten back the armies of religious ignorance.

None of this is accurate. There was no real debate that day in Oxford. Nor was there a clear victor, still less a scalping. Not that many people were paying attention; the debate was not mentioned by a single London newspaper. Still, it was important, though the quarrel was less between scientific theory and religious faith than between two conceptions of humankind's place in the cosmos. And far from being an enduring victory for rationality over faith, the debate inaugurated a conflict that continues to the present day, and is as much about the past as it is about the future.

In 1860 science was not assumed to be inaccessible to ordinary people; the attendees at the Advancement of Science meeting included many ordinary, middle-class Britons as well as Oxford students and faculty. The crowd packed a hall at the university's new museum, standing in aisles and doorways. In that jammed, sweltering space, the subject on attendees' minds was *On the Origin of Species*, by Charles Darwin. Published just seven months before, the book had created a public uproar, dividing educated Britons into pro- and anti-evolution camps.

Wilberforce's friends believed him to be an obvious candidate to lead the charge against Darwin. Ambitious, witty, and politically connected, the fifty-four-year-old cleric had a reputation for such smooth and convincing oratory that his detractors mocked him as "Soapy Sam." His allies were sure that an eloquent public condemnation from him would deal a severe blow to Darwinism.

As the bishop may have known, another audience member was poised for a counterstrike: Thomas Henry Huxley, almost twenty years Wilberforce's junior but already known as much for his vehement defense of Darwin, a friend, as for his contributions to comparative anatomy. A poor boy who had never been able to finish his university degree, Huxley had risen to a full professorship through ambition, brilliance, and dogged work. Prickly and quick to take offense, he enjoyed a good, vicious fight with plenty of character assassination.

In the version of events retailed in my college class, the bishop spoke for half an hour, his theatrical, booming voice filling the hall with Darwinism's supposed flaws. Much of the audience was delighted; every barb drew cheers and approving laughter. Goaded, perhaps, by the crowd, the bishop closed his peroration with the kind of snide jab that he would never have used in the pulpit. Turning to Huxley with a saponaceous smile, he asked grandly whether it was "through his grandfather or his grandmother that he claimed his descent from a monkey."

Huxley (my professor said) was delighted by this sally. He whispered to a neighbor, "The Lord hath delivered him into mine hands"—a line, fittingly enough, from Scripture. Then he stood. No, Huxley said, he would not be ashamed to have an ape for an ancestor, but he *would* be ashamed "to have sprung from one who prostituted the gifts of culture and eloquence to the service of prejudice and of falsehood."

Like the rest of my class, I chuckled when I heard this exchange. But I couldn't fathom why it counted as a victory for science. I understood that the bishop had overstepped by intimating that one of Huxley's grandparents had been sexually involved with an animal. Huxley had used this gaffe as an opening to strike back with an even more direct personal swipe. But neither man had said anything substantive about evolution. How could this be an advance for rational thinking?

Both Darwin and Wilberforce had lives marked by sorrow, Darwin losing two infants and his adored ten-year-old daughter to disease, Wilberforce losing not only an infant and an adult son but also his wife, who had just given birth to the couple's sixth child. But grief pushed them in opposite directions. Darwin was unable to reconcile the fact of his daughter's painful death with the idea of a just cosmos ruled by a benevolent deity. Even before her death he had been moving away from the faith of his youth; it was gone forever by the day of the funeral. In *On the Origin of Species*, the word "God" appears only once, incidentally.

For his part, the bishop tried to peer beyond the "paroxysms of convulsive anguish" caused by his wife's death to see healing in God, "the smiter of my soul." Ultimately, he came to believe that the loss of his wife—the "utter darkening of my life, which can never be dispelled"—was in fact "a call to a different mode of life . . . a more severe, separate, self-mortifying course." He prayed that her death would "kill in me all my ambitious desires and earthly purposes, my love of money and power and places." Purified by grief, he would become a warrior for Christianity.

No transcript exists of Wilberforce's remarks. But he had just written an eighteen-thousand-word takedown of *Origin*, soon to be published in a prominent literary journal. Most historians believe that in Oxford the bishop simply laid out the criticisms in his review. If so, Huxley faced a challenge, because Wilberforce's review was far from ignorant. Indeed, Darwin later admitted that it was "uncommonly clever; he picks out with skill all the most conjectural parts, & brings forward well all difficulties." Wilberforce's facility with scientific argument was unsurprising: the bishop had won a first-class mathematics degree at Oxford and, like Darwin, was a member of the Royal Society, Britain's premier scientific body.

Part of the bishop's cleverness lay in his decision to attack Darwin mainly on scientific grounds, rather than invoking Christian dogma. From the beginning of his review, Wilberforce targeted *Origin*'s greatest weakness: the paucity of direct evidence for the evolution of one species from another. If the history of life were filled with these "transmutations," the bishop reasoned, it must also be filled with in-between beasts, halfway evolved between old and new. Fossils of these in-between creatures should be everywhere. "Yet never have the longing observations of Mr. Darwin and the transmutationists found one such instance to establish their theory." If evolution was real, where were these intermediate entities?*

Only after spending more than fifteen thousand words critiquing the evidence for the *existence* of evolution did Wilberforce turn to his main concern: natural selection, the proposed *mechanism* for evolution. At bottom, the concept of natural selection is so simple that Huxley later claimed that his reaction to learning it was "How extremely stupid not to have thought of that!" Darwin contended that some offspring are, by chance, different from their parents; that some of these random differences—somewhat stronger muscles, say—will be beneficial (others will not be); and that individuals with the favorable variations will have a better chance of reproducing and passing on these favorable variations. In this way, Darwin argued, natural selection ensures that randomly appearing, advantageous features spread through populations. The process ensures that all species continually evolve through time, eventually giving rise, as the changes accumulate, to new species.

In the closing paragraphs of *Origin*, Darwin summed up his thoughts with an image: a hillside of untidy foliage that he often walked by. He asked readers to picture this "tangled bank," as he called it, alive "with many plants of many kinds, with birds singing on the bushes, with various insects flitting about, and with worms crawling through the damp earth." Although the inhabitants of the hillside—that is, Earth's living creatures—are "so different from each other, and dependent upon each other in so complex a manner, [they] have all been produced by laws acting around us." The important words here are "all" and "produced." Living things may look dissimilar to the casual eye, but they are identical on a deeper level—all of them. Each and every one was produced by natural selection, and natural selection will determine its future.

---

* Darwin argued that the fossil record was then too incomplete to show transitions, and that later discoveries would fill in gaps. Almost all scientists believe he was correct. Since then, paleontologists (dinosaur specialists) have uncovered many "missing link" species. One example: *Kulindadromeus zabaikalicus*, discovered in 2014. A small, bipedal dinosaur, it has both bird-like feathers *and* dinosaur-like scales, exemplifying how dinosaurs evolved into birds.

Each and every one—that would include human beings, wouldn't it? Here Darwin ducked. Throughout *Origin*, he sedulously avoided discussing whether his ideas also applied to people—if, that is, *Homo sapiens* were just another weed on his tangled bank.

Darwin's reticence didn't fool the bishop. If natural selection directs the course of life and people are part of life, Wilberforce wrote in his review, the clear implication is that "the principle of natural selection [applies] to MAN himself." (Note the sudden burst of capital letters; the bishop was thundering in disapproval.) Human beings, too, must have evolved through natural selection from some previous, not-quite-human species. And this notion, Wilberforce said, is "absolutely incompatible" with a true understanding of the "moral and spiritual condition of man."

Human beings, the bishop believed, had been created by God and endowed with a unique spark. But if, as Darwin suggested, people were created by unthinking natural forces, they could not possibly have any high standing. Should *Homo sapiens* share its nature with all other creatures—should our species be, as Wilberforce facetiously suggested, just a bunch of over-achieving "mushrooms"—we would be, by definition, nothing special.

As the University College London geographer Simon Lewis has pointed out, Darwin's argument—that *humankind* owed its existence to the same processes that gave rise to flatworms and amoebae—was a second, biological Copernican Revolution. The original Copernican Revolution is usually said to have begun in the early sixteenth century, when the Polish-German polymath Nicolaus Copernicus, drawing on data from Arab and Persian geometers and his own examination of the sky, proposed that Earth orbits around the sun, and not the sun around Earth. Because Earth moved, it could not be, as had been thought, the focal point of the cosmos. The Copernican Revolution, the science historian Dick Teresi has noted, was neither a revolution, in the sense that it occurred over a long time, nor particularly Copernican, in that it was also the product of thinkers other than Copernicus. Still, it had great impact on our conception of ourselves and our place in the cosmos. Earth, our home, was no longer the pivot of existence. It was just a place, one among many, without particular distinction.

Unlike the first Copernican Revolution, the second happened rapidly and was largely the product of a single mind. But it, too, nudged our species out of the spotlight. "We are not even at the heart of life on Earth," as Lewis put it. Because humankind owed its existence to the same processes that produced every other organism, Darwin implied, *Homo sapiens* was a species like any other species. This new Copernican Revolution was what had attracted Wilberforce's ire.

To the bishop, there was a fundamental line between human beings and all other creatures. Naturally, he described that difference in Christian terms: people had souls, animals did not; people were endowed by God with the capacity for change and redemption, animals were not. But Wilberforce's view can be put in broader, more general terms, which do not depend on religious belief: *Homo sapiens* has an inner flame of creativity and intelligence that allows it to burn down barriers that would trap any other species. Or, more succinctly: human beings are not wholly controlled by the natural processes that control all other creatures. We are *not* simply another species.

Wilberforce's remark about Huxley's ape ancestors was thus more than a snarky gibe. Consciously or not, the bishop was effectively asking whether Huxley was prepared to affirm that he and all other people were prisoners of biology. Blinded by contempt, Huxley seems not to have realized that his adversary was posing, however rudely, an important question. (The "great question," the great conservationist George Perkins Marsh called it a few years later: "whether man is *of* nature or *above* her.") Not grasping the underpinnings of the dispute, Huxley didn't even try to engage them. Darwin later shuddered at the "awful battles which have raged about 'species' at Oxford," but there was no actual debate. At least not in the sense of a genuine attempt to hash out diverging beliefs.

Both Huxley and Wilberforce thought they had come off well. Three days after the encounter, the bishop bragged to a friend, "I think I thoroughly beat him." Certainly his supporters in the audience, "cheering lustily," thought so. Equally pleased, Huxley later boasted that he "was the most popular man in Oxford for full four and twenty hours afterwards." In the years to come, Huxley and Wilberforce ran into each other from time to time. The meetings were always cordial. Both viewing themselves as the winner, they could be magnanimous in victory.

Over the decades Huxley came to be seen as triumphant. In his school in the 1960s, the writer Christopher Hitchens was taught that "Huxley cleaned Wilberforce's clock, ate his lunch, used him as a mop for the floor, and all that." In my college a few years later, I learned much the same thing. The Wilberforce-Huxley dustup was presented as a morality play ending in a straightforward victory for rational thought. Only much later did I realize that from today's perspective the implications of our species' lack of specialness were different from what my teacher and textbook had presented.

In Wilberforce's day, those who hoped for a better future cheered on Huxley, because science and technology seemed to promise a better life. But now that science and technology have allowed the human enterprise to risk its own survival, the partisans of hope have stepped back from some of Huxley's implications. Wizards and Prophets each have a separate blueprint for the future. But both assume that Wilberforce, not Huxley, was correct—that human beings are special creatures who can escape the fate of other successful species. If the Oxford debate was a morality play, the vices and virtues have slipped offstage and switched masks.

Slight Amendment of the Foregoing, Whooping-Crane Edition

Lynn Margulis was a Huxleyite. When she told me that human beings, like Gause's protozoa, would wipe themselves out, she was affirming her belief in Darwin's view: biological laws apply to every creature. After talking with her I sometimes told people about these ideas. Few accepted them, and even those who agreed did not fully endorse Margulis's perspective. They told me that the human race was doomed because people are greedy and stupid, not because, as Margulis thought, overreaches and crashes are the natural way, as much a part of the wonders of life as coral reefs and tropical forests. But I also never met anyone who had a convincing argument that she was wrong.

A year after her death, I bumped into Daniel B. Botkin, an ecologist who had recently retired from the University of California at Santa Barbara. Botkin has worked in many areas but is perhaps best known for *Discordant Harmonies* (1990), a classic study debunking the long-held belief that ecosystems will exist in a timeless balance unless people disturb them. He had known Margulis well and respected her. "But she's wrong on this one," he said.

Not all species would multiply themselves out of existence if given the chance, he said. Among the exceptions is the whooping crane, *Grus americana*. The subject of one of the longest conservation efforts in North America, the whooper, as it is called, is a sister species to the Eurasian crane, *Grus grus*; geneticists believe the two species split off from a common ancestor 1 to 3 million years ago. Despite their physical similarity, the birds behave differently. Hundreds of thousands of Eurasian cranes exist, despite human hunting. The bird aggressively expands its territory when possible, sometimes infuriating farmers by taking over their fields. Whoopers, by contrast, are shy creatures of the marsh, rarely seen in groups bigger than two; as far as is known, the entire species has never numbered more than 1,500 individuals. "Explosive growth is evidently not part of its evolutionary strategy," Botkin said.

There are other examples—not many, but they exist. Another example, from Botkin: the Tiburon mariposa lily (*Calochortus tiburonensis*). Native to northern California, it lives only on soils made from serpentine, a relatively rare kind of stone that produces soils filled with chromium and nickel, which are toxic to most plants. Serpentine soils occur usually in isolated patches with relatively defined borders—natural petri dishes, one might say. The lily reproduces slowly enough that it never overwhelms its environment. It never hits the edge of the petri dish.

Was there any known case, I asked Botkin, of a species *changing* its evolutionary strategy? A creature that went from rapid, Gause-style expansion to quiet adjustment to its environment? Of a protozoan transforming itself, so to speak, into a whooping crane? Or of a plant that somehow makes its own serpentine soil? Isn't this what Borlaug and Vogt were each advocating in their different way? That people had a special something—call it a soul, as Wilberforce did—which allowed them to do this?

"That's the question, isn't it?" Botkin said.

Exemption

One possible answer to the question is provided by Robinson Crusoe, hero of Daniel Defoe's famous novel. Shipwrecked alone on an uninhabited island off Venezuela in 1659, Crusoe is an impressive example of fictional human resilience and drive. During his twenty-seven-year exile he learns to catch fish, hunt small mammals, tame goats, prune citrus trees, and create "plantations" of barley and rice from seeds salvaged from the wreck. (Defoe didn't know that citrus and goats were not native to the Caribbean and thus probably wouldn't have been on the island.) Rescue comes in the form of a shipful of mutineers, who plan to maroon their captain on the supposedly empty island. Crusoe helps the captain recapture his ship and offers the defeated mutineers a choice: permanent exile on the island or trial in England. All choose the island. Crusoe has harnessed so much of its productive power to human use that even a gaggle of inept seamen can survive there in comfort.

Robinson Crusoe's first three chapters recount how its hero ended up on his ill-fated voyage. The youngest son of an English merchant, Crusoe has a restless spirit that leads him to become an independent slave trader. On a voyage to Africa his ship is captured by a "Turkish rover" captained by a Moor from Morocco. "As his proper Prize," Crusoe becomes the captain's house slave. After two years of servitude, Crusoe steals his master's fishing boat and escapes. He bumbles in the boat down the West African coast without food or water and is rescued by a Portuguese slave ship bound for Brazil. There the enterprising Crusoe establishes a small tobacco plantation. But he is short of labor, and decides with some other plantation owners to obtain that labor by taking a ship to Africa and buying some slaves. The ship wrecks on the return voyage. Except for Crusoe, all hands perish, slaves included. He ends up alone on his island.

What is striking to a modern reader is that Defoe saw nothing remarkable about expecting readers to sympathize with a man in the slave trade. Crusoe has no qualms about slaving even after having been, most unhappily, a slave himself. Here, character echoes author: Defoe extolled slavery as "a most Profitable, Useful, and absolutely necessary Branch of our Commerce." Backing words with deeds, he owned shares in the Royal African Company, created in 1660 to buy men and women in Africa and transport them in chains to the Americas. When the company was attacked in Parliament, he offered to write the equivalent of editorials in its favor. It paid him the rough equivalent of $50,000 for his public-relations services.

Defoe was a person of his time. Three centuries ago, when he was writing *Robinson Crusoe*, societies from one end of the world to another depended on slave labor, as had been the case since at least the Code of Hammurabi, in ancient Babylon. Customs differed from one place to another, but slavery was sanctioned and practiced everywhere from Mauritania to Manchuria. Unfree workers existed by the million in the Ottoman Empire, Mughal India, and Ming China. In classical Athens, two-thirds of the inhabitants were slaves; imperial Rome, the historian James C. Scott has written, "turned much of the Mediterranean basin into a massive slave emporium." Slaves were less common in early modern Europe, but Portugal, Spain, France, England, and the Netherlands happily exploited huge numbers of them in their American colonies. In the last half of the eighteenth century alone, almost 4 million people were taken from Africa in chains. In colonies throughout the Americas at that time, in places ranging from Brazil to Barbados, from South Carolina to Suriname, slaves were so fundamental to the economy that they outnumbered masters, sometimes by ten to one.

Then in the nineteenth century, slavery almost stopped entirely. The implausibility of this change is stunning. In 1860, slaves were the single most valuable economic asset in the United States, collectively worth more than $3 billion, an eye-popping sum at a time when the U.S. gross national product was less than $5 billion. (The slaves would be worth as much as $10 trillion in today's money.) Rather than investing in factories like northern entrepreneurs, southern businessmen had sunk their capital into slaves. Rightly so, financially speaking—slaves had a higher return on investment than any other commodity available to them. Enchained men and women had made the region politically powerful, and gave social status to an entire class of poor whites. Slavery was the foundation of the social order. It was, thundered South Carolina senator John C. Calhoun, "instead of an evil, a good—a positive good." (Calhoun was no fringe character; a former U.S. secretary of war and vice president, he would become secretary of state.) Yet despite the institution's great economic value, part of the United States set out to destroy it, wrecking much of the national economy and killing half a million citizens along the way.

Incredibly, the turn against slavery was as universal as slavery itself. Great Britain, leader of the global slave trade, banned its market in human beings in 1807 after a tireless campaign by abolitionists. Two laws enacted in 1833 and 1838 freed all British slaves. Denmark, Sweden, the Netherlands, France, Spain, and Portugal soon outlawed their slave trades, too, and after that slavery itself. Like stars winking out at the approach of dawn, cultures across the globe removed themselves from the previously universal exchange of human cargo. Slavery still exists; the International Labor Organization estimates that almost 25 million people are still forced to work as captives. But in no society anywhere is slavery a legally protected institution—part of the social fabric—as it was throughout the world two centuries ago.

Historians provide many reasons for this extraordinary transition, high among them the fierce opposition of slaves themselves. But another important cause is that abolitionists convinced people around the world that slavery was a moral disaster. An institution fundamental to human society for millennia was made over by ideas and a call to action, loudly repeated.

In the last few centuries, such profound changes have occurred repeatedly. Another, possibly even bigger example: since the beginning of our species, almost every known society has been based on the subjugation of women by men. Tales of past matriarchal societies abound, but there is little archaeological evidence for their veracity. In the long run, women's lack of liberty has been as central to the human enterprise as gravitation to the celestial order. The degree of suppression varied from time to time and place to place, but women never had an equal voice. Union and Confederacy clashed over slavery, but they were in accord on the status of women: in neither state could women attend school, have a bank account, or, in many places, own non-personal property. Equally confining in different ways were female lives in Europe, Asia, and Africa. Nowadays women are the majority of U.S. college students, the majority of the U.S. workforce, and the majority of U.S. voters. Again, historians assign multiple causes to this shift, rapid in time, confounding in scope. But a central element was the power of ideas—the voices and actions of suffragists, who through decades of ridicule and harassment pressed their case. In recent years something similar may have occurred with gay rights: first a few lonely advocates, censured and mocked; then victories in the social and legal sphere; finally, perhaps, a slow movement to equality.

Every whit as profound is the decline in violence. Ten thousand years ago, at the dawn of agriculture, societies mustered labor for the fields and controlled harvest surpluses by organizing themselves into states and empires. These promptly revealed an astonishing appetite for war. Their penchant for violence was unaffected by increasing prosperity or higher technological, cultural, and social accomplishments. When classical Athens was at its zenith in the fourth and fifth centuries B.C., it was ever at war: against Sparta (First and Second Peloponnesian Wars, Corinthian War); against Persia (Greco-Persian Wars, Wars of the Delian League); against Aegina (Aeginetan War); against Macedon (Olynthian War); against Samios (Samian War); against Chios, Rhodes, and Cos (Social War). Greece was nothing special—look at the ghastly histories of China, sub-Saharan Africa, or Mesoamerica. Look at early modern Europe, where war followed upon war so fast that historians bundle them into catch-all titles like the Hundred Years' War or the even more destructive Thirty Years' War. The brutality of these conflicts is difficult to grasp; to cite an example from the Israeli political scientist Azar Gat, Germany lost between a fifth and a third of its population in the Thirty Years' War—"higher than the German casualties in the First and Second World Wars *combined*." The statistic is sobering: Germany lost a greater percentage of its people to violence in the seventeenth century than in the twentieth, despite the intervening advances in the technology of slaughter, despite being governed for more than a decade by maniacs who systematically murdered millions of their fellow citizens.

As many as one out of every ten people met a violent death in the first millennium A.D., the archaeologist Ian Morris has estimated. Ever since, violence has declined—gradually, then suddenly. In the decades after the Second World War, rates of violent death plunged to the lowest levels ever seen. Today, humans are far less likely to be slain by other members of their species than a hundred years ago, or a thousand—an extraordinary transformation that has occurred, almost unheralded, in the lifetime of many of the people reading this book. Given the mayhem documented in every day's headlines, the horrors in the Middle East and the ghastly strife in northeast Africa, the idea that violence is diminishing may seem absurd. Nonetheless, every independent effort to collect global statistics on violence suggests that we seem to be winning, at least for now, what the political scientist Joshua Goldstein calls "the war on war."

Multiple causes for this turnaround have been suggested. But Goldstein, a leading scholar in this field, argues that the most important is the emergence of multinational institutions like the United Nations, which owe their origins to peace activists from the last century. These organizations have by no means stopped all fighting. But over time, Goldstein says, they have snuffed out, almost invisibly, conflicts that in previous eras would have led to horrific brutality.

Past successes do not guarantee future progress. Violence has ticked upward in the last decade, and may get worse. One can readily imagine some ghastly political or religious insurgency that reinstates slavery; many insurrectionary forces go out of their way to brutalize women. Global poverty has fallen dramatically in recent decades, but could rebound. Lunatics with nuclear weapons may yet strike—a possibility that will never go away. There is no permanent victory condition for being human, as the writer Bruce Sterling has remarked.

Given this record, though, even Lynn Margulis might pause. No European in 1800 could have imagined that in 2000 Europe would have no legal slavery, women would be able to vote, and same-sex couples would be able to marry. No one could have guessed that a continent that had been tearing itself apart for centuries would be largely free of armed conflict, even amid terrible economic times. No one could have guessed that Europe would have vanquished famine.

Preventing *Homo sapiens* from destroying itself à la Gause would require a still greater transformation, to Margulis's way of thinking, because we would be pushing against Nature itself. Success would be unprecedented, biologically speaking. It would be a reverse Copernican Revolution, showing that humankind is exempt from natural processes that govern all other species. But might we be able to do exactly that? Might Margulis have got this one wrong? Might we indeed be special?

Consider, again, Robinson Crusoe. He was a slaver—but also, in the end, he had a special spark. Confronted with a threat to his survival, he changed his way of life, root and branch, to meet it. Working alone, he transformed the island, enriching its landscape. And then, to his surprise, he realized that he "might be more happy in this Solitary Condition, than I should have been in a Liberty of Society, and in all the Pleasures of the World."

Living alone on a large, biologically rich island, Crusoe was able to take as many of its resources as he wanted—he was, so to speak, barely past the first inflection on Gause's curve. Margulis's presumption is that if he and the mutineers had stayed, they would eventually have hit the second inflection point and wiped themselves out. (I am making the unrealistic assumption that they would not have left the island.) Wizards and Prophets both believe that Margulis is wrong—that Crusoe and the others would have gained enough knowledge to save themselves. They would have either used this knowledge to create technology to soar beyond natural constraints (as Wizards hope) or changed their survival strategy from expanding their presence to living in a steady-state accommodation with what the island offered (as Prophets wish).

Of course, Crusoe was a fictional character (though Alexander Selkirk, the castaway whose story apparently inspired Defoe, was not). And the challenge facing the next generation is vastly larger than Crusoe's challenge. But is it so unlikely that our species, a congeries of changelings, would be able to do exactly as Crusoe did—transform our lives to meet new challenges—before we round that fateful curve of the second inflection point and nature does it for us? I can imagine Margulis's response: You're imagining our species as some sort of big-brained, hyper-rational, cost-benefit-calculating computer! A better analogy is the bacteria at our feet! Still, Margulis would be the first to agree that removing the shackles from women and slaves has begun to unleash the suppressed talents of two-thirds of the human race. Drastically reducing violence has prevented the waste of countless lives and a staggering amount of resources. Is it really impossible to believe that we wouldn't use those talents and those resources to draw back before the abyss?

Our record of success is not that long. In any case, past successes are no guarantee of the future. But it is terrible to suppose that we could get so many other things right and get this one wrong. To have the imagination to see our potential end, but not have the cultural resources to avoid it. To send humankind to the moon but fail to pay attention to Earth. To have the potential but to be unable to use it—to be, in the end, no different from the protozoa in the petri dish. It would be evidence that Lynn Margulis's most dismissive beliefs had been right after all. For all our speed and voraciousness, our changeable sparkle and flash, we would be, at last count, not an especially interesting species.

Appendix A: Why Believe? (Part One)

Years ago I sat in on an introductory course given by Lynn Margulis. At a certain point she referred to evolution as the cornerstone of modern biology. A student raised her hand and said she didn't believe in evolution. "I don't care whether *you* believe it," Margulis replied. "I just want you to understand why *scientists* believe it." After that, she said, the student could decide whatever she wanted.

Not long after, I spoke to John Maynard Smith, a distinguished British evolutionary biologist. In our conversation I recounted Margulis's injunction to her student. Maynard Smith laughed heartily. The reason for his amusement, he explained, was that Margulis was a prominent critic of mainstream evolutionary theory. She didn't dispute the existence of evolution by natural selection. But Margulis thought that natural selection was just part of the picture—in the long run, symbiosis and chance were more important sources of evolutionary innovation. Maynard Smith thought Margulis was off base. "But I'll grant her one thing," he said, as I remember it. "She's a remarkably good skeptic. Even if she's wrong, she's *fruitfully* wrong."

In the main text, I tried to follow Margulis's example in my discussion of climate change—lay out why the great majority of scientists believe that it is occurring and caused by human activity. This belief is the culmination of a century and a half of investigation into atmospheric chemistry and physics. But I also argued that this belief, by itself, doesn't compel specific action—it's hard to know what duties we owe to faraway descendants. People have to decide what to think on their own.

Now I would like to follow Margulis again, by arguing for the role of skepticism. By skepticism I don't mean the ad hominem claim that climate science is a "hoax." About four thousand scientists from eighty countries and countless representatives from those countries were involved in the last report of the Intergovernmental Panel on Climate Change. To imagine that these thousands of researchers and government officials were all part of a devilish plot—and that they all kept silent about it, as did their predecessors in previous IPCC reports—is foolish. It is particularly hard to figure out why oil states like Iran and Saudi Arabia, which participate in IPCC reports, would enlist in a fraudulent endeavor to rid the world of fossil fuels.

At the same time, though, the "hoax" rhetoric reflects something real. It is a way, however inexact or distorted, of expressing the fear that the risks of rising carbon-dioxide levels are being systematically exaggerated by environmentalists. The activists, in this view, are using climate change as a tool to win social changes that they cannot get in other ways. In 2007 the activist-journalist Naomi Klein published *The Shock Doctrine,* which argued that right-wing elites were using—or even manufacturing—economic crises as a pretext to force societies to adopt corporations' preferred policies (slashing social-welfare programs, reducing taxes, cutting regulations, and so on). The policies were presented as solutions to the crisis but in fact were designed to enrich the already-wealthy people who proposed them. Some global-warming opponents see climate change in the same way—as an overstated or even fictitious "crisis" that left-wingers like Klein use as an excuse to force other people to do what they want (reducing consumption, overturning capitalism). When activists retort (accurately) that the great majority of researchers agree with them—well, that proves the researchers, too, are in on the plot.

To avoid this, decouple for a moment the science (rising carbon dioxide levels leads to a warmer planet) from the proposed remedies (getting rid of fossil fuels).

If the basic physical understanding of atmospheric carbon dioxide were proven to be incorrect, it would be a remarkable event in the history of science. As a rule, entire disciplines don't get big things wrong if they are part of the arena of study. True, physicists wrongly believed for several centuries that outer space was filled with a mysterious substance called the ether, but the longevity of that error was due mainly to researchers' inability to test for the ether's existence. In the nineteenth century appropriate testing methods were developed, and the belief was quickly exploded. Climate change has been studied off and on since Tyndall, systematically since about 1960, and intensively since about 1990. Given that long effort, it would be highly unusual for the general consensus—pouring lots of carbon dioxide into the air increases average global temperatures—to be wrong.*

It would be especially unusual given that the models have made many successful quantitative predictions. An early example was the 1967 prediction by Syukuro Manabe and Richard T. Wetherald—two researchers at the National Oceanic and Atmospheric Administration, in Washington, D.C.—that the lower atmosphere and the stratosphere would act in opposite ways: warming in the former would be accompanied by cooling in the latter. Because the stratosphere is hard to observe, confirmation did not occur until 2011. But it did occur. Twelve years after that initial prediction, Manabe and two other scientists predicted something else: land areas would heat up faster than ocean areas, with the slowest warming being around Antarctica. That, too, has turned out to be the case. There are many other examples.

Yet despite these successes, we still do not understand either the rate at which climate change is occurring or its precise effects. (I discussed this when I looked at climate sensitivity.) Here there is room for skepticism. It may be that the atmosphere won't respond to carbon dioxide nearly as fast as the doomsters fear. Or that the impact will be distributed in ways that we don't yet understand. As I wrote this book, I asked half a dozen climate scientists to identify what they see as the biggest uncertainties—the reasons that their fears could be wrong. Here are some of their answers.

The first stems directly from the fact that no computer today can handle calculations that cover the entire surface of the earth and its atmosphere. In consequence, researchers simplify their models by treating the atmosphere and surface as an array of cubes, each perhaps fifteen or twenty miles on a side (different models have different sizes). The cubes are treated as if they were uniform, but in the real world, of

---

* A possible counterexample is the theory that dietary fat causes obesity. This idea held sway from about 1970 to about 2000, but has since been strongly challenged. If the dietary-fat hypothesis truly proves to be incorrect, it will be an outlier in scientific history—and it will have been believed for much less time than the carbon-dioxide theory.

course, a cube of air many miles on a side can hold many different types of clouds. Or the supposedly uniform ground in a cube could be partly covered by a lake and partly by a mountain. To handle such variables, scientists write equations intended to approximate what is going on. Inevitably, small errors build up. To account for them, the models must be "tuned." This means manually tweaking the parameters (the change in temperature as altitude changes, say, or the way heat moves in the ocean). The tweaked models are then compared to twentieth-century weather records to see how well they reproduce them. Unfortunately, the models are also supposed to *explain* those same weather records. The unavoidable risk is that scientists will fool themselves, inadvertently using the tweaks to mask the inaccuracies of a model. It also means that other scientists using the model won't know whether a specific prediction arises directly from the underlying science (good) or is largely due to tuning (bad). None of this is shady in any way—all large physical models must be tuned. But doing it correctly is a constant worry.

All the models have difficulties with clouds, especially clouds over the oceans (because the oceans cover most of Earth's surface, this means most clouds). As the ocean grows warmer, the air above it becomes more humid, which encourages clouds to form. The clouds, constantly changing and turbulent, interact in a highly complex way with the constantly changing winds and currents in the atmosphere. Turbulence, convection, airflow—all are notoriously intractable physics questions. Indeed, the reason that the military has built so many costly wind tunnels is that aeronautical engineers have wanted to watch what happened when a plane was exposed to wind turbulence because they couldn't predict the consequences well with mathematics. This problem has only gotten worse—the need now is to account for entire global systems of clouds. Low-altitude clouds tend to reflect sunlight, cooling the air around them. High-altitude clouds tend to trap infrared radiation, heating the air around them. Will warmer oceans end up driving clouds higher into the atmosphere? Will they create more low-altitude thunderstorms? Will air and water currents in a warmer planet push cloud formations north or south of where they would have been? Which effects will dominate? More than a few scientists think this will long be a source of uncertainty.

In these descriptions of climate models, one word has not appeared: biology. Climate models are constructed by physical scientists. Yet one lesson that biology has taught is that living creatures profoundly shape the world. Consider the role of methane hydrates in the Arctic. The land sheds organic molecules into the water like a ditchdigger taking a shower. Sewage plants, fertilizer-rich farms, dandruffy swimmers—all make their contribution. Plankton and other minute sea beings flourish where the drift is heaviest, at the continental margins. When these creatures die, their bodies drizzle slowly to the seafloor. Microorganisms feed upon the remains. In a process familiar to anyone who has seen bubbles coming to the surface of a pond, the microbes emit methane gas as they eat and grow. (Methane, one recalls, is a potent agent of climate change.) Under the high pressure of these cold depths, water and methane react to each other: water molecules link into crystalline lattices—"methane hydrates," in the jargon—that trap methane molecules. Such vast amounts of methane are stored in methane hydrates that researchers fear that their release into the air could set off catastrophic changes in climate.

In 2017 scientists from the Woods Hole Oceanographic Institute tried to measure the methane seeping up from Norway's Arctic coast, where the oceans are warmer than they used to be. They discovered that the warming that was pushing up the methane was also pumping nutrients from the seafloor to the surface. The nutrients were feeding huge blooms of phytoplankton. To the scientists' astonishment, the plankton took in so much carbon dioxide via photosynthesis that they more than canceled out the effects of the methane. This is a small example of a general problem: our continuing ignorance about the impacts of life.

The issue is not restricted to microorganisms—no one has an accurate figure for how much carbon dioxide is absorbed in forests around the world. All in all, raising carbon-dioxide levels tends to increase plant growth—a negative feedback, because it tends to reduce those same carbon-dioxide levels. But increasing the area covered by plants ("leaf-area index," in the jargon) also makes some bare areas darker—a positive feedback, because dark areas tend to reflect less sunlight into space. The effects are complicated and not uniform. In northern areas, thicker forests seem to change global wind patterns. In southern areas, higher leaf-area indexes affect groundwater levels, and in turn are affected by them. In 2017 an international research team estimated in *Nature Climate Change* that "the greening of the earth" had "mitigated" surface warming by about one-eighth in the previous thirty years. No one knows if this will continue.

To repeat, none of this means that climate change is not occurring, or that the science underpinning it is incorrect. It means instead that the basic physical mechanisms elucidated by Callendar and his successors are modulated, in ways that we do not yet understand, by other factors (clouds, plant growth), which could reduce—or, just as likely, amplify—their impacts. Because of these uncertainties, climate systems are full of "natural variation"—changes with causes that scientists can't yet identify. In Chapter 7, I described my friend Rob's model, published in 2016, for how rising carbon dioxide could cause the Antarctic to melt quickly. Temperatures in the West Antarctic—roughly speaking, the portion of Antarctica that fronts onto the southern Pacific Ocean—have risen in recent years. It was easy to couple the rising temperatures and Rob's theoretical calculation and produce a picture: the Antarctic is responding to our carbon dioxide, and that could flood coastal cities from Miami to Mumbai. Quite naturally, the result was newspaper headlines around the world.

But a year later two other research teams looked at the historical data and concluded that the rising temperatures in the West Antarctic were within the range of natural variation (with the possible exception of the temperatures in the Antarctic Peninsula, the tongue of land that extends almost to the tip of South America). These two teams did not contradict my friend's work. What they did was look at different data, and come up with a different picture: the Antarctic is warming, but not in a way that is different from the past.

Everyone involved believed that human-caused warming is occurring—the question is how fast, how long, and what it will do. Rob was saying: the speed and effects look like *this*. The two other teams were saying: no, we think they look like *that*.

Here, to my mind, is the great value of skepticism. No matter whether Wizards or Prophets prevail, whether the future is full of thousands of nuclear plants or millions of networked solar installations, reducing the risks of climate change will be pricey and politically difficult. If the costs of fighting climate change are offset by the benefits of (say) avoiding rapid sea-level rise, they will be well worth paying. But if large sums are spent on a chimera, the effects will be negative—spending too much can be as bad as spending too little.

People sometimes try to wish this problem away by saying that some preferred measure could be easily paid for by gutting this or that pointless or corrupt government program. But much of the supposed "abuse" actually reflects different beliefs about what government should be doing—one person's "wasteful spending" is another person's "vital government function." For people of one disposition, spending too much on reducing carbon-dioxide levels could involve the peril of leaving the nation open to hostile incursions or failing to improve its physical infrastructure. For people of another disposition, the risk involves cheating the poor today of education, sanitation, and social investments in the name of a purported benefit to their descendants. This worry can be put more bluntly. If, as most economists believe, people tomorrow will be more affluent than people today, the hazard is that we end up valuing tomorrow's rich more than today's poor.

The great value of fruitful skeptics—the reason that, as Maynard Smith suggested, they should be celebrated even if they prove to be wrong—is that they force advocates to think through these issues. And, of course, sometimes the skeptics are right. Maynard Smith didn't think that Lynn Margulis's skeptical ideas about natural selection's relative import to evolution were correct. But he had been equally dismissive about Margulis's skeptical ideas about the lack of importance of plants and animals in the evolutionary tree and—well, that was spot on.

Appendix B: Why Believe? (Part Two)

The similarities between disbelief in the scientific evidence for the risk of human-caused climate change and the scientific evidence for the safety of genetically modified food plants and animals are inexact but striking. In both cases skeptics argue that large bodies of scientific evidence are untrustworthy because the scientists themselves have bad motives—they are, variously, paid shills of greedy corporations or witting tools of anti-human agitators. And in both cases the skeptics often claim that the other side is acting in bad faith—that it has drummed up a phony crisis (a heating world, a risk of famine) to which it can present its preferred ideas (anti-industrialism, corporate capitalism) as the only possible solution.

There is another parallel. In the previous appendix, I argued that recognizing a problem doesn't automatically imply accepting either its seriousness or any particular solution to that problem. In useful discussions, this leads to seeking to manage risks: the risk that the problem is as serious as activists think, the risk of not achieving other goals if too much time, money, and attention is devoted to solving a problem that turns out not to be very serious.

Listed below are excerpts of statements from nine different scientific groups, representing thousands of researchers, about the safety risks of genetically modified organisms (all are available online, if one wants to read the reports in their entirety). Taken together, these groups are for genetically modified food what the IPCC is for climate change—conscious attempts by large numbers of scientists to state what is generally accepted in their field. Given the statements below, there is little reason to suppose—from what we know today—that ingesting GMOs poses any unusual risk to human health.

Does that mean that everyone should accept them? Not necessarily. But it would be useful if the discussion moved from the safety of GMOs, almost a nonissue, to the actual object of contention: whether the current version of industrial agriculture can, with the addition of new technologies, provide for the world of 10 billion in a long-lasting way—or if the perils involved (ecological, economic, spiritual) are large enough to require it to be radically revamped.

The National Academies of Sciences (2016)

"The research that has been conducted in studies with animals and on chemical composition of GE foods reveals no differences that would implicate a higher risk to human health from eating GE foods than from eating their non-GE counterparts. . . . The committee could not find persuasive evidence of adverse health effects directly attributable to consumption of GE foods."

> *(Committee on Genetically Engineered Crops: Past Experience and Future Prospects. National Academies of Sciences, Engineering, and Medicine. 2016. Genetically Engineered Crops: Experiences and Prospects. Washington, DC: National Academies Press.)*

World Health Organization (2014)

"GM foods currently available on the international market have passed safety assessments and are not likely to present risks for human health. In addition, no effects on human health have been shown as a result of the consumption of such foods by the general population in the countries where they have been approved."

> *(Frequently Asked Questions on Genetically Modified Foods, World Health Organization, May 2014 ["prepared by WHO in response to questions and concerns from WHO Member State Governments"].)*

American Association for the Advancement of Science (2012)

"The science is quite clear: crop improvement by the modern molecular techniques of biotechnology is safe."

> *(Statement by the AAAS Board of Directors on Labeling of Genetically Modified Foods, 20 Oct 2012.)*

American Medical Association (2012)

"Bioengineered foods have been consumed for close to 20 years, and during that time, no overt consequences on human health have been reported and/or substantiated in the peer-reviewed literature. However, a small potential for adverse events exists, due mainly to horizontal gene transfer, allergenicity, and toxicity. . . . [But] thorough pre-market safety assessment and the FDA's requirement that any material difference between bioengineered foods and their traditional counterparts be disclosed in labeling, are effective in ensuring the safety of bioengineered food."

> *(Report 2 of the Council on Science and Public Health [A-12]; Labeling of Bioengineered Foods [Resolutions 508 and 509-A-11], 2012.)*

European Union (2010)

"The main conclusion to be drawn from the efforts of more than 130 research projects, covering a period of more than 25 years of research, and involving more than 500 independent research groups, is that biotechnology, and in particular, GMOs, are no more risky than e.g., conventional plant breeding technologies."

> *(Directorate-General for Research and Innovation. European Commission. 2010. A Decade of EU-Funded GMO Research [2001-2010]). Brussels: European Union.)*

American Society for Cell Biology (2009)

"Far from presenting a threat to the public health, GM crops in many cases improve it."

> *(American Society for Cell Biology. 2009. ASCB Statement in Support of Research on Genetically Modified Organisms. Press release, 30 Jan.)*

Researchers from the Department of Cellular and Molecular Medicine, St. George's University of London (2008)

"Foods derived from GM crops have been consumed by hundreds of millions of people across the world for more than 15 years, with no reported ill effects (or legal cases related to human health), despite many of the consumers coming from that most litigious of countries, the

USA. There is little documented evidence that GM crops are potentially toxic. . . . The presence of foreign DNA sequences in food per se poses no intrinsic risk to human health."

> *(S. Key, et al. 2008. "Genetically Modified Plants and Human Health." Journal of the Royal Society of Medicine 101:290–98.)*

Union of German Academies of Sciences and Humanities (2006)
"Because of the rigour with which they must be tested and the controls to which they are subject, it is extremely unlikely that GMO products approved for market in the European Union and other countries present a greater health risk than the corresponding products from conventional sources."

> *(InterAcademy Panel Initiative on Genetically Modified Organisms. Union of the German Academies of Science and Humanities. 2006. "Are There Health Hazards for the Consumer from Eating Genetically Modified Food?" Statement, International Workshop, Berlin, 2006.)*

French Academy of Science (2002)
"All [health] criticisms against GMOs can be largely rejected on strictly scientific criteria."

> *(R. Douce, ed. 2002. Les Plantes Génétiquement Modifiées. Paris: Tec & Doc Lavoisier [Rapports de l'Académie des Sciences sur la Science et la Technologie 13].)*

Acknowledgments

I began this book by mentioning my daughter. In the interests of fairness and family comity, let me mention that I have two other children whose futures are of equal concern to me. This big book about many things is for Newell, Emilia, and Schuyler.

Countless people helped me along the way: Rollie Natvig of the Sons of Saude guided me through Scandinavian Iowa. Kent Mathewson drove with me to Pithole. Helen Burggraf hosted me in London. Josh Ge took me to carbon-capture plants in Inner Mongolia and water systems in Shanghai and Liuzhou and (in a section that, alas, I cut from the manuscript) rubber farms in Laos and Thailand. Shankhmala Sen helped me get through the coalfields in India, logistically and linguistically; Aqeel Khan arranged for me to meet Shankhmala. Lance Thurner and Matt Ridley shared their unpublished research. Rick Bayless let me visit his garden. Schuyler Mann worked to check the notes and bibliography. Bruce Lundeby helped me with Norwegian archives. Westher Hess and Norm Benson provided me with photographs and insight about Walter Lowdermilk, Westher's father. (Benson's forthcoming biography of the fascinating Lowdermilk is high on my to-read list.) Mark Johnson of the Borlaug Foundation took me through the homestead on a frigid winter morning. Madhura Swaminathan invited me to speak to her father in Chennai—my thanks to her and the M. S. Swaminathan Research Foundation for their hospitality, their willingness to let me poke through their papers, and a host of other things. Transcripts of my interview with Prof. Swaminathan will be available on the foundation website.

An assortment of genealogists ferreted out facts in North Carolina (Anne H. Lee), Iowa (Jan Wearda), Ohio (Dottie Nortz), Long

Island (Ottman Research Services), and online (Cynthia and Glenn Clark). Please let me doff my cap to Roger Joslyn, Certified Genealogist, who guided me through the wilds of New York civic bureaucracy with aplomb. Leland Goodman, ace of aces, did the illustrations; Nick Springer, once again, the maps. I am lucky to know them.

At various times I discussed *The Wizard and the Prophet*, in part or whole, with Kevin Kelly, Neal Stephenson, Bill McKibben, Joyce Chaplin, Tyler Cowen, Larry Smith, Peter Kareiva, Cassie Phillips, Wen Stephenson, Ellen Ruppel Shell, Bob Pollin, Dava Sobel, Dick Teresi, Gary Taubes, Lydia Long, Tyler Priest, Daniel Botkin, and, most of all, Ray Mann (partner in thought as in life). Some of these people disagreed with me, vehemently but usefully; others may be surprised to see their names here. In one such conversation Stewart Brand suggested the first version of the title; in another, Cullen Murphy suggested the final version, the second time he has done this for me.

The more I write, the more I am grateful for first readers: the brave, kind souls willing to read part or all of an unfinished manuscript. My thanks to Daron Acemoglu, Joel Bourne, James Boyce, Stewart Brand, Ana Caicedo, Bob Crease, Rob DeConto, Ruth DeFries, Erle Ellis, Dan Farmer, Betsy Hartmann, Susanna Hecht, Jeffrey Kegler, Maggie Koerth-Baker, Jane Langdale, Mike Lynch, Ted Melillo, Narayanan Menon, Oliver Morton, Ramez Naam, Sunita Narain, Ray Pierrehumbert, Michael Pollan, Matt Ridley, Ludmila Tyler, and Carl Zimmer. All found infelicities, misapprehensions, logical fallacies, and plain old howlers. Oliver Morton noted, ever so gently, that by mistakenly typing "million" instead of "billion," I was erring by three orders of magnitude. I appreciate his—their—forbearance. Remaining goofs are, of course, my own.

This book would not have been possible without the facilities of the archives (sometimes online) at the Rockefeller Foundation (where Michele Beckerman and Lee Hiltzik helped me find images and documents), Guggenheim Foundation (my heartfelt thanks to Andre Bernard), the Denver Public Library (Coi Drummond-Gehrig found marvelous pictures of Juana, Marjorie, and Bill Vogt), Princeton University, Smith College, the Library of Congress, the University of Iowa, the University of Wisconsin, and the University of Minnesota. I am grateful for their aid.

I am lucky in my editors. First, those who commissioned articles that fed into this manuscript: Corby Kummer and Scott Stossel at *The Atlantic;* Andrew Blechman at *Orion;* Susan Murcko at *Wired;* Barbara Paulsen and Jamie Shreeve at *National Geographic;* Cullen Murphy at *Vanity Fair;* Maria Streshinsky at *Pacific Standard;* Tim Appenzeller, Colin Norman, and Elizabeth Culotta at *Science;* and Luke Mitchell at *The New York Times.* Following them were my editor and his team at Knopf—the superbly generous Jon Segal, the patient Kevin Bourke, and the remarkable Susanna Sturgis. This is my sixth book with Jon. It is my ninth with my agent, Rick Balkin. To say that I treasure their care, attention, and friendship is an understatement.

Special thanks come last:

To Andrew Blechman (for lending his editorial skills at a crucial stage, which helped me arrange and think through this material). To Susanna Hecht (for a zillion discussions and reading suggestions and hospitality in Geneva and Los Angeles). To Michael Pollan (for letting me steal his organizational scheme, which in turn he stole from his wife, Judith Belzer). To Lynn Margulis, who jolted my life in the space of a few conversations. How I wish she were here to tell me how wrong I am about everything!

As I was fumbling away at this project, Mark Plummer, my oldest friend, passed from this world. Twenty-plus years of conversation with Mark—twenty-plus years of his incisive criticism—informed every line of this book. *The Wizard and the Prophet* is the first substantial effort of mine in many years that he did not read, a fact that makes me melancholy in ways that I find hard to express. How I miss those phone calls that began, "What I hear you saying here is . . ."

Abbreviations

(for Sources and Notes)

ALP = Aldo Leopold Papers, University of Wisconsin*

AM = *Atlantic Monthly*

AOA = Norman Borlaug oral history interview, 12 May 2008, American Academy of Achievement, Washington, DC*

BCAG = *Boletín de la Compañia Administradora del Guano*

BDE = *Brooklyn Daily Eagle**

BestR = Vogt, W. 1961–62(?). Best Remembered (unpub. ms.) Ser. 2, Box 4, FF2, VDPL

CCD = Correspondence of Charles Darwin*

CIMBPC = Norman Borlaug Publications Collection, International Center for Wheat and Maize Improvement, Texcoco, México (portions*)

GEC = *Global Environmental Change*

GFA = Vogt files, Guggenheim Foundation Archives, New York, NY

HS = *Hempstead (N.Y.) Sentinel**

HOHI = M. King Hubbert oral history interview by Ronald Doel, 4 Jan–6 Feb 1989, Niels Bohr Library and Archives, American Institute of Physics*

IEA = International Energy Agency

LHNB = Norman Borlaug oral history interview by Paul Underwood, 2007(?), www.livinghistory.net*

LPMJS = *London, Edinburgh, and Dublin Philosophical Magazine and Journal of Science*

NBUM = Norman E. Borlaug papers, University Archives, University of Minnesota, Twin Cities*

NYT = *New York Times*

OGJ = *Oil & Gas Journal*

PNAS = *Proceedings of the National Academy of Sciences* (many articles*)

PPFA1/2 = Planned Parenthood Federation of America Records, 1918–1974/1928–2009, Sophia Smith Collection, Smith College, Northampton, MA

PTRS = *Philosophical Transactions of the Royal Society* (sometimes A or B)

QJRMS = *Quarterly Journal of the Royal Meteorological Society*

RFA = Rockefeller Foundation Archives, Tarrytown, NY

RFOI = Norman Borlaug oral history interview by William C. Cobb, 12 June 1967, RG 13, Oral Histories, Box 15, Folder 7, RFA (also at TAMU/C*)

Some Notes = Vogt, W. W. 1950s(?). Some Notes on WV for Mr. Best to Use as He Chooses, Series 2, Box 5, FF21, VDPL

TAMU/C = Norman Borlaug papers, Texas A&M, CIMMYT records*
VDPL = William Vogt Papers, Denver Public Library Conservation Archives
VFN = Field notes, William Vogt, Ser. 3, Box 7, VDPL
VIET = Vietmeyer, N. 2009–10. *Borlaug.* 3 vols. Lorton, VA: Bracing Books.
VvV = *Frances Bell Vogt vs. William Walter Vogt*, Nassau County, L.I., Index no. 3959, Microfilm roll 117, civil cases 3937–3975
WP = *Washington Post*

* = available gratis online at time of writing

In addition, I abbreviate some publishers' names:

CUP = NY: Cambridge University Press
HUP = Cambridge, MA: Harvard University Press
MIT = Cambridge, MA: MIT Press
OUP = NY: Oxford University Press
UCP = Berkeley, CA: University of California Press
YUP = New Haven, CT: Yale University Press

Notes

Prologue

4 Import of science: Brand (2010:16) makes this case elegantly.

4 Hunger decline, longevity rise: United Nations Food and Agricultural Organization 2009:11 (~24% in 1969–70); idem. 2017:2–13 (11.0% in 2016). Life expectancy from World Bank (http://data.worldbank.org/indicator/SP.DYN.LE00.IN). See also R.D. Edwards 2011; Riley 2005; and World Health Organization 2014 (www.who.int/gho/mortality_burden_disease/life_tables/en/).

4 Global growth will continue: Every economic study I have seen projects overall growth, though many worry about distribution. See, e.g., Johansson et al. 2012 (global GDP "could grow at around 3% per year over the next 50 years," 8).

7 "Ecocide": The term is due to Arthur W. Galston (in Knoll and McFadden 1970:47, 71–72).

7 Differences in perceptions and beliefs: Luten 1986 (1980):323–24. My thanks to Antoinette WinklerPrins for drawing my attention to this work.

9 "and durable": Mumford 1964:2. Technofuturist F. M. Esfandiary called the two sides "up-wingers" and "down-wingers" (Esfandiary 1973).

10 Environmentalism: For most of the twentieth century the term referred to a school of psychology that emphasized individuals' environments rather than their genetic inheritance (Worster 1997 [1994]:350). The modern use of the term to indicate a belief in protecting the natural world came in about 1966 (Jundt 2014a: 250n).

10 "one billion": "Borlaug's Revolution," *Wall Street Journal*, 17 July 2007. The claim seems to have originated in Easterbrook 1997.

11 Four elements: Though associated with Plato, the earliest known discussion was by the philosopher Empedocles in about 460 B.C.

11 Predictions of disaster: Devereux 2000:Table 1 (actual famine toll); Meadows et al. 1972 ("one hundred years," 23); Ehrlich 1969 (pesticides, life expectancy; "42 years by 1980," 26), 1968 ("is over," "hundreds of millions," 1). The "utter breakdown" claim is from a 1970 interview with CBS News; my thanks to Kyra Darnton of Retro Report for sending it to me. Four years later Ehrlich promised "that a great increase in the death rate due to starvation will occur well before the end of the century, quite possibly before 1980" (Ehrlich and Ehrlich 1974:25).

Chapter One: State of the Species

18 Vogt's visit: Vogt travel diary, 18 Apr 1946, Ser. 3, Box 6, FF27, VDPL. See also idem., 22 Dec 1943, Ser. 3, Box 6, FF26, VDPL. Chapingo is a small village on the edge of the bigger town of Texcoco.

21 "or more pressing": Letter, L. S. Rowe to W. Weaver, 2 Aug 1946, Ser. 1, Box 1, FF38, VDPL; undated drafts are in the same folder.

21 Margulis's life: Sagan 2012.

22 Scale of microworld: Whitman et al. 1998 is the classic source of the 90% estimate, based on Luckey 1972:1292; 1970:Tab. 1. Later modifications include McMahon and Parnell 2014; Serna-Chavez et al. 2013; Van der Heijden et al. 2008. In conversation, Margulis gave me the ten-times-as-many figure, but I have updated as per Sender et al. 2016.

22 Amazing microworld: See the excellent Yong 2016; Zimmer 2011.

23 Gause: Gall 2011; Israel and Gasca 2002:211–15; Gall and Konashev 2001; Brazhnikova 1987; Vorontsov and Gall 1986; Kingsland 1986:244–46; Gause 1930 (first article).

23 Pearl and debate over his work: de Gans 2002; Kingsland 1995 (1985): chaps. 3–4 (dozen articles, 3 books, 75–76); Pearl and Reed 1920; Pearl 1925, 1927 ("characteristic course," 533). Pearl's work was anticipated by Pierre-François Verhulst (1838) and built on Alfred Lotka (e.g., Lotka 1907, 1925 [esp. chap. 7]).

24 *Struggle for Existence*: Gause 1934.

25 Time-lapse video: Author's visit. See also Mazur 2010 (2009):266.

25 Natural selection: See Epilogue. For Margulis's slightly unusual definition, see Mazur 2010 [2009]:265–67 [dachshunds, human reproductive potential]); Margulis and Sagan 2003 (2002):9–10.

26 Zebra mussels: Author's visits, Hudson Valley; Strayer et al. 2014, 2011 ("1%," 1066); Carlsson et al. 2011. Zebra mussels hit the Great Lakes in 1986–87 (Carlton 2008); their populations exploded and then collapsed (Karatayev et al. 2014).

27 Same laws as the rest: Margulis and Dobb 1990:49.

27 Lice: Toups et al. 2011; Kittler et al. 2003; Travis 2003. Toups et al.'s date is somewhat earlier than that of Kittler et al., though they view the two teams' work as "largely consistent" (30).

29 Becoming human: Pettit 2012:59–72 (burials); Henshilwood et al. 2011 (ochre); Henshilwood and d'Errico 2011 (ostrich eggs); Bouzouggar et al. 2007 (beads); Yellen et al. 1996 (harpoons).

30 Human uniformity compared to bacteria: 1000 Genomes Project Consortium 2015; Li et al. 2008. Li and Sadler 1991 is a classic study. For *E. coli* single-base diversity, see Jaureguy et al. 1981; Caugant et al. 1981.

31 Comparisons to mammals, apes: Prado-Martinez et al. 2013; Leffler et al. 2012 (lynxes and wolverines); Salisbury et al. 2003; Kaessman et al. 2001 (primates). My thanks to Carl Zimmer for generously correcting errors in an earlier version of this section and suggesting references.

31 Breeding population: Henn et al. 2012; Fagundes et al. 2007.

31 Blue eyes: Eiberg et al. 2008; Frost 2006.

32 Fire ants: Author's interview, Wilson; Ascunce et al. 2011; Mlot et al. 2011; Yang et al. 2010; King and Tschinkel 2008; Wilson 1995 (1994):71.

32 Super-colonies: Goodisman et al. 2007; Holway et al. 2002:195–97; Tsutsui and Suarez 2002; Suarez et al. 1999. Super-colonies are rare in the ant's native range.

32 *L. humile* super-colony: Van Wilgenburg et al. 2010; Sunamura et al. 2009; Kabashima et al. 2007. See also Moffett 2012; Pedersen et al. 2006; Suarez et al. 1999.

32 Early human history: Richter et al. 2017 (possible early human emergence); Kuhlwilm et al. 2016 (humans 100,000 years ago in Siberia); Liu et al. 2015 (humans 80,000–120,000 years ago in Asia).

33 Population 10,000 years ago: Haub 1995 (a standard estimate). See also "Historical Estimates of World Population," available at www.census.gov.

33 105,000-year-old sorghum: Mercader 2009.

33 Trajectory of agriculture: Heun et al. 1997; Lev-Yadun et al. 2000; Tanno and Willcox 2006; Willcox 2007. See also Scott 2017: Chap. 1; Mann 2011b; Burger et al. 2008. For fertilizer, see chapter 4.

35 Human share of Earth's production: Smil 2016:48 (25 percent), 2013:183-97; Vitousek et al. 1997 ("terrestrial productivity," 495); Vitousek et al. 1986 ("39 to 50%," 372).

35 Anthropocene: Crutzen and Stoermer 2000. See also Biello 2016; Steffen et al. 2011; Crutzen 2002.

Chapter Two: The Prophet

40 Guano background: Gregory Cushman's work (2014, 2006) helped form this chapter. I have also benefitted from discussions with Ted Melillo, Daniel Botkin, and Susanna Hecht.

40 Uric acid in guano: Cushman 2014:23–27; Ñúñez and Petersen 2002:71–84, 170–72. Guano deposits in Peru were first noted by von Humboldt, who sent samples to chemists in Paris (von Humboldt and Berghaus 1863:228–47; Fourcroy and Vauquelin 1806). A fine Humboldt biography is Wolf 2015.

40 Early fertilizers: Pomeranz 2000:583–84 (bean cakes); Wines 1986; Roberts and Barrett 1984 (poudrette); Braudel 1981 (1979): 116–17, 155–58 (nightsoil).

40 Peruvian chemist: Cochet 1841. Andean peoples had used guano as fertilizer for centuries, but their conquerors didn't learn about guano until its value was rediscovered by European scientists. An indigenous term for bird excrement, *wanu*, is the origin of the word "guano" (Whitaker 1960; Murphy 1936:1:286–95).

41 Guano industry, war: Cushman 2014 (2013): chap. 2; Melillo 2012; Mann 2011a:212–20; Inarejos Muñoz 2010 (war); Vizcarra 2009:370 (revising Hunt); Hollett 2008; Miller 2007:147–55; Miller and Greenhill 2006; Hunt 1973:70 (guano income).

41 Guanay cormorants: King 2013: chap. 10; Murphy 1954, 1936, 2:899–909 ("into guano," 901); 1925:71–125 ("given point," rafts, 74–75); Hutchinson 1950:18 (35 lbs.). Vogt's measure was 34.8 lbs./bird/yr (Letter, W. Vogt to A. Leopold, 11 Jun 1941, ALP). "Gregarious beyond imagining" draws on language quoted by King (200).

41 Formation of Compañía: Cushman 2014: 148–52, 168–90 (seeking Murphy, 190); Duffy 1994:70 (sustainable-management programs); Coker 1908a, b, c (U.S. scientist).

41 Vogt's arrival: Travel journal, Ser. 3, Box 5, FF36, VDPL.

43 "Environmental thought": Cushman 2014:190.

43 Vogt's idyllic childhood: Vogt, W. W. 1943? Background Information about Dr. William Voght [sic]. Ser. 2, Box 5, FF21, VDPL; Some Notes, 1–5 ("hot dog joints," 2); BestR, chap. 1 ("Nebraska," 1). My thanks to Roger Joslyn for helping me find the marriage certificate and much else about the Vogt family.

44 Vogt's father: Application for Headstone, Pension C2326861, 11 March 1944 (discharge from navy); Hempstead, Floral Park, Nassau, NY, 1940 U.S. Census, entry

for Frances B. Brown (leaving school); "Garden City," HS, 7 Nov 1901 (attendees at wedding); Certificate and Record of Marriage No. 21813, William Walter Vogt and Frances Bell Doughty, Hempstead, Nassau, NY, 31 Oct 1901, Registered No. 4216; "Engagement Announced," BDE, 7 Dec 1900; "Three Jolly Rovers," NYT, 5 Aug 1900 (arrest); Magisterial District No. 1, Indian Hill Precinct, Jefferson, KY, 1900 U.S. census, entry for William F. Vogt; Enumeration District 151, Louisville, Jefferson, KY, 1880 U.S. Census, entry for Fred. Wm. Voght [sic]. Vogt Sr.'s siblings: "John H. L. Vogt" (obituary), Vista Press (San Diego, CA), 25 Apr 1970; "Former Soloist with Orchestra Here Dies," Louisville Courier-Journal, 4 Apr 1952; Certificate of Death, Commonwealth of Kentucky, Registration District 755 (Jefferson County), File 116527724. Thanks to Anne H. Lee for Kentucky genealogical research.

44 Vogt's birth: Certificate and Record of Birth, William Walter Vogt, Mineola, N. Hempstead, Nassau, State of New York Bureau of Vital Statistics, No. 18569 (Registered No. 4192). At birth, he was Frederick William Vogt; he became William Walter Vogt in 1904.

45 Saga of Vogt Sr., Schenck: Letter, Vogt to Robert Cushman Murphy, 6 Feb 1964, Ser. 1, Box 2, FF4, VDPL (World's Fair grounds); "Garden City," HS, 10 Aug 1911 (postmistress, vacation); Action for Absolute Divorce, Affidavit of Clara Doughty, 2 May 1907, VvV; Parker 1906 (Doughty family); "Wanderers Heard From," BDE, 9 Oct 1902; "Drugs Under the Hammer," BDE, 7 Jun 1901; "Wife and Babe Deserted by Vogt for Married Woman," Duluth Evening Herald, 5 Jun 1902; "Vogt's Stock to Be Sold," BDE, 4 Jun 1902; "Sheriff in Possession of Voght's [sic] Drug Store," BDE, 30 May 1902; Enumeration District 0713, North Hempstead, Nassau, NY, 1900 U.S. Census, entry for Geo. W. Schenck (Schenck family). Died as baby: Some Notes, 1.

46 Vogt and Schenck divorces: Testimony of Mary J. Schenck, 21 March 1908, VvV; "Garden City," HS, 26 March 1908; "Schoolgirl Wife Sues." NYT, 22 March 1908; "Decree for Mrs. Vogt," BDE, 22 March 1908. See also: legal notices, Frances Bell Vogt, Plaintiff, against William Walter Vogt, defendant, Sea Cliff (NY) News and Glen Cove News, 8 Jun 1907, 22 Jun 1907. The divorce was not finalized until 15 Jan 1909 (Order, VvV). New York divorce law: DiFonzo and Stern 2007, esp. 567–69; O'Neill 1969:140–45. Schenck divorce: "Schenck Wants Divorce," Philadelphia Inquirer, 19 Jul 1908; "Co-respondent Sued Now," BDE, 20 May 1908; "Schenck Divorce Case Up," BDE, 19 Jul 1908; "East Williston," HS, 23 Jul 1908; "Final Decree Granted," BDE, 11 Dec 1908.

46 Vogt family finances: Rasky 1949:6 (Fannie); Bureau of the Census, Official Register, Persons in the Civil, Military, and Naval Service of the United States, and List of Vessels. Vol. 2: The Postal Service (Washington, DC: Government Printing Office, 1909), 323 (Clara).

46 "of my life": BestR, chap. 1, 1–2.

46 Brooklyn: Some Notes, 2 ("27 cents"); Brooklyn Assembly District 9, Kings County, N.Y., 1920 U.S. Census, Enumeration District 478, Page 9B, entry for Lewis Brown.

47 Vogt and books: Some Notes, 2 ("my generation"); Nabokov 1991:213 ("glutton for books"); Burroughs 1903 (scientists' dismay); Seton 1898:18 ("among wolves").

47 Scouting: Some Notes, 2–5 ("ever since," 2); Wadland 1978:419–45 (Seton and Scouting).

48 Vogt and polio: Some Notes, 3 ("until morning"); "1 Death, 1 New Case in Epidemic Here," BDE, 26 Sept 1916; "94,000 Absences as Schools Open," NYT, 26 Sept 1916; "Lowest Friday Epidemic Record," BDE, Sep 1 1916. According to Peterson, Vogt's mother, "scanning the hospital bulletin in the morning, read that her son had succumbed during the night" (1989:1254).

48 Vogt recuperates: Some Notes, 4 ("worth it"), 6 ("rather badly," courage); BestR, chap. 1, 3 ("in my life"); Peterson 1989:1254; McGrory 1948; letter, R.V. Mattingly to Secretary, Guggenheim Foundation, 3 Apr 1943, GFA (weak legs, spine, lungs).

48 Vogt's college career: McCormick 2005:24; Some Notes, 6 (scholarship); letter, W. Vogt to A. Leopold, 29 Jul 1939, ALP; Anon., ed., 1921, *The Manual Anvil*, NY: Manual Training High School (high school literary club, 151).

49 Vogt's jobs: Some Notes, 8–10 ("Executive Secretary"); McCormick 2005:25; Delacorte 1929; "The Funnies," *The Writer*, Jan 1929; "Literary Market Tips," *The Author & Journalist*, Dec 1928. My thanks to Will Murray and John Locke for these sources.

49 Mary Allraum, marriage: Marriage License and Affidavit for License to Marry No. 16999, County of New York, 7 Jul 1928; Certificate and Record of Marriage No. 18002, State of New York, 7 Jul 1928; Florida Passenger Lists, 1898–1964, National Archives, Washington, DC, Roll Number 7, Immigration record, S.S. *Gov. Cobb*, 3 Oct 1923; "Women Students of the University of California Will Present Annual Partheneia Faculty Glade Masque and Pageant in April," *San Francisco Chronicle*, 26 Feb 1922 (p. 1); "Campus Pageant of Youths Tempting," *San Francisco Chronicle*, 12 March 1922 (p. 1); "L.A. Girl Picked for Stellar Role in Partheneia at U. of C.," *Oakland Tribune*, 20 March 1922; list of junior college sophomores, *The Southern School* (UCLA) yearbook 1921, 158; Index to Marriage Licenses and Certificates, Alameda County, Vol. 10, 1902–1904, California State Archives, Sacramento, CA, Dec. 13, 1903 (Allraum marriage).

50 Ornithology embraces amateurs: Weidensaul 2007:127–85 passim; McCormick 2005:25; Barrow 1998:178–79, 193–94; Ainley 1979; Vogt 1961 ("odd-ball").

51 Vogt's duties: Anonymous, 1943?, Background Information About Dr. William Voght [*sic*]. Series 2, Box 5, FF 19, VDPL; Biographical Note, Vogt Papers website, idem; McCormick 2005:31n51; Some Notes, 8 (college birding).

51 Peterson's life: Carlson 2007; Graham and Buchheister 1990:130–35; Devlin and Naismith 1977.

52 Publication of *Field Guide*: Carlson 2007:46–70; Weidensaul 2007:200–10 (copies sold, 209); McCormick 2005:26–28; Peterson 1989; Vogt 1961 ("would not sell").

52 Juana Broadway debut: Pollock 1929; "The Channel Road," *Playbill*, 21 Oct 1929.

52 Broadway decline: Broadway fell from 233 productions in the 1929–30 season to 187 in the 1930–31 season. In 1938–39, it hosted 96 (Matelski 1991:148).

52 Jones Beach sanctuary: Vogt 1938a, b; 1933. For background, see Caro 1974:145–56, 182–225 (developing Long Island parks), 233, 310 (sanctuary).

52 Vogt projects: Vogt 1938b; Cushman 2013:191 (Mayr's advice); McCormick 2005:31–32 (Mayr's advice); Anonymous. 1940:80 (prize); W. Vogt, n.d., "A Preliminary List of the Birds of Jones Beach, Long Island, N.Y.," Ser. 2, Box 4, FF19, VDPL.

53 Dovekie research: Murphy and Vogt 1933 ("herd psychosis," 348). Vogt had previously published two brief, un-peer-reviewed reports in *The Auk* (48:593, 606).

53 Long Island ducks: Different authorities provide different counts. See, e.g., the websites of the Coastal Research and Education Society of Long Island (11 species), the New York Department of Environmental Conservation (10), and the National Wildlife Reserve complex (15).

53 Duck decline: Greenfield 1934 ("dangerously low").

53 Long Island development: C. W. Leavitt, Garden City (map), 2 Apr 1914, available at Garden City Historical Society; Some Notes, 2 ("Robert Moses"). Vogt had worried about the decline of marshes as early as 1931 (Vogt 1931).

54 Oyster Bay closes sanctuary, Vogt loses job: Confoundingly, Oyster Bay is the name

for both a township that runs from north to south across the middle of Long Island and an upscale hamlet within the township, on the North Shore; the township, not the hamlet, owned the property. Embarrassed by the news coverage, the township turned about and presented the land to the U.S. Biological Survey. The survey had no experience with parks and no money for them. Oyster Bay was griping again by fall ("L.I. Bird Sanctuary's Needs," BDE, 27 Oct 1935; "Bird Sanctuary Leased to U.S.," BDE, 3 Jul 1935; "See Sanctuary Action Mistake," BDE, 5 Jun 1935; "Deplores Fate of Bird Haven," BDE, 31 May 1935; "Bird Reserve Dispute Looms," BDE, 30 May 1935; "Bird Refuge Is Facing Abandonment Saturday," NYT, 21 May 1935; "Deplore Bird Sanctuary End," BDE, 20 May 1935; Pilat 1935 ["Vogt has done a great job"]).

54 Baker becomes NAAS director, hires Vogt: Carlson 2007:74–75; Graham and Buchheister 1990:117–19, 128–29, 140–41 (buys *Bird-Lore*); Anon. 1938. "Bird is 'Boid' in the South, but in a Genteel Way." BDE, 8 Apr (Juana teaching).

54 New contributors like Leopold: Leopold 1938. Leopold's first *Bird-Lore* essay became a major chapter in his famed *Sand County Almanac* (Leopold 1949). Murphy wrote a monthly column from 1937 to 1940.

54 Vogt's Audubon activities: Carlson 2007:75–77; Cushman 2006:238; Duffy 1989; Vogt et al. 1939; Moffett 1937.

54 U.S. malaria problem: Webb 2009:146–50 (scope, Map 5.3); Cottam et al. 1938:93 (5 million cases). It was partly due to increased damming (Patterson 2009:127; Shah 2010:185–89).

55 Mosquito control: A thorough overview is Patterson 2009: esp. chap. 6 (effects of Depression, 120–29); Cottam et al. 1938; [Vogt] 1935. Reiley 1936 and Peterson 1936 give examples of the transition to federal money. See also Webb 2009:153–54.

55 Long Island: Butchard 1936 (Suffolk); Froeb 1936:128 (Nassau); Cottam et al. 1938 (artificial ponds, 95).

55 "Thirst on the Land": Vogt 1937a ("rackets," 15; "erysipelas," 7).

56 Ecosystem services: People have known for millennia that nature was useful, but Marsh (1864) was the first to discuss its uses systematically, noting, for instance, that deforestation led to soil and water degradation (254–64) and that birds and insects helped with pest control and fertilization (87–103). The concept then lay largely dormant until Vogt picked it up. It did not become popular until the 1970s (Gómez-Baggethun et al. 2010). The term was coined by Ehrlich and Ehrlich (1981:102).

56 Loss of services: Vogt claimed in addition that the benefits were overstated. Many drainage projects occurred in areas with no malaria. Even those in actual malaria zones were usually ineffective, because the ditches weren't maintained and thus quickly turned into stagnant pools. As a result, malaria rates were rising, despite all the ditches. This was true but misleading; Vogt had not understood that draining programs needed years to exert their full impact.

56 Mosquito control debate: Cottam et al. 1938: 81 ("misdirected"); 94 ("*control*," emphasis in original). Patterson (2009:138–43) has a fine account. For Cottam, see Bolen 1975.

56 Audubon and early conservation movement (footnote): Holdgate 2013 (1999):17–21, 2001; [Vogt] 1936 (proposing movement).

57 *Bird-Lore* becomes depressing: Cushman 2006:238–39; Peterson 1973:49 ("destruction"); Graham and Buchheister 1990:143–44. Examples of Vogt's editorials are [Vogt] 1935, 1937b ("dinner table").

57 Vogt alienates contributors: See, e.g., Letter, T. S. Roberts to W. Vogt, 30 Oct 1936,

Bell Museum of Natural History Records, University of Minnesota Libraries, University Archives, uarc00876-box37-fdr322; W. Vogt to T. S. Roberts, 3 Nov 1936, ibid., uarc00876-box25-fdr234; letter, T. S. Roberts to W. Vogt, 11 Nov 1936, ibid., uarc00876-box37-fdr322.

57 Vogt's failed coup: letter, Margaret Nice to Vogt, 9 Dec 1937, Ser. 1, Box 2, FF5, VDPL ("exhaustion"); Devlin and Naismith 1977:71–72; Fox 1981:197–8; Peterson 1989:1255; 1973:50 ("petrel that he was"). Graham and Buchheister (1990:117–18, 142–44) provide a somewhat different version.

58 Vogt at North Chincha: Author's visit; VFN; BestR, chap. 4 ("measure it," lack of hat, 2); Cushman 2013:191–95; Rasky 1949 ("of science", 7); J. A. Vogt 1941:23–24 (loss of tools); Letter, W. Vogt to A. Leopold, 11 Feb 1940, ALP (tools); Murphy 1925b:103–4 (shoveling roof).

58 Vogt's enjoyment: BestR, chap 5 (coffee, 8; "scallop bed," 9); J. A. Vogt 1940:268 ("Doctor Pajaro"). To his pleasure, the coffee was *carocolillo*, an Andean variant that rather than having two seeds—coffee beans—in each fruit has a single, oddly wrinkled bean, with an especially concentrated taste. Don Guano: Duffy 1989:1257.

59 Vogt's fascination at profusion of life: VFN, e.g., 17 March 1939 [FF38]; 13 Feb 1939, 6 Feb 1939 ("entirely appreciate"), 4 Feb 1939, 31 Jan 1939 (doesn't mind smell) [FF36]; Vogt 1942:310; BestR, chap. 4 ("on the island," 5–6).

59 Research questions: Rasty 1949:7 ("11,000,000 guano birds"); Vogt 1939; BestR, chap. 4, 7 ("increment of excrement").

62 Juana graduates, comes to Peru: Diary, J. A. Vogt, 3 Jul 1939 ("pick their teeth,"), 4 Jul 1939, 13 Jul 1939, Ser. 4, Box 8, FF12, VDPL; J. A. Vogt 1940 ("pure good luck," 265; "family feuds," 267; "such a spot," 273); "Audience of 20,000 Attends Annual Outdoor Ceremony at Columbia University," NYT, 7 Jun 1939; Columbia University in the City of New York, Catalogue Number for the Sessions of 1939–1940 (NY: Columbia University, 1940).

62 Discovery of El Niño, Murphy's visit: Fagan 2009:31–44; Cushman 2004, 2003; Hisard 1992; Hutchinson 1950:49–58; Vogt 1942a ("Peruvian coast"); Murphy 1926 ("guano birds," 32), 1925a ("Current," 433); Lavalle y García 1917. The three original articles were Eguiguren Escudero 1894; Carrillo 1893; and Carranza 1892. The realization that El Niño was an oscillating system of Pacific-spanning currents did not come till decades later.

63 Vogt's El Niño: VFN, FF40, 43, 46, 49, 51; Vogt 1960:124–25 ("China and India"); 1942a (77°F, 509; "no indication," 511; bird counts, 510); 1942b (surface temperature, Fig. 10); 1942c:9–10, 86–88 (birds go north and south); letter, W. Vogt to A. Leopold, 29 Jul 1939, ALP ("*all* gone"); Murphy 1936 1:96 (around 60°F). See also Cushman 2006:240–55; McCormick 2005:71–79; Hutchinson 1950:54. At the time, the three islands probably still held about 10 million birds (Jordán and Fuentes 1966:Fig. 1).

63 Birds come and go: VFN.

64 Vogt's explanation for Guanay movements: Cushman 2014:195; Vogt 1942a ("wholesale destruction," 521); 1942b:88–89; 1942c:9–12; letter, W. Vogt to A. Leopold, 31 Dec 1941, ALP. Murphy had come to similar conclusions (Murphy 1925a:433). Later Vogt handed plankton analysis to Mary Sears, a pioneering Wellesley College biologist working at Pisco (Vogt 1942c:4–5; letter, W. Vogt to A. Leopold, 15 Dec 1940, ALP).

64 Leopold: Meine 2010 (1988) is still the standard biography; also useful is Flader 1994. Cornell actually established a forestry school in 1898, ahead of Yale, but it closed.

65 Clements: Botkin 2012:134–37; Worster 1997 (1994):209–20; Clements 1916: esp. 104–7 (climax); 1905 (superorganism). Similar ideas were expressed at about the same time by Henry C. Cowles, another plant biologist. For a classic attack on the "balance of nature," see Botkin 1992 (1990).

66 Biotic potential, environmental resistance: Cushman 2014 (2013):194–97; Vogt 1942b:25. Vogt took the terms from entomologist Royal Chapman (Chapman 1926:143–62).

66 Clements, Elton, and ancient ideas of order: Botkin 2016:35–55, 2012:106–14; Simberloff 2014; Egerton 1973; Odum 1953; Elton 1930 ("is remarkable," 17); Clements and Shelford 1939. As Botkin (2012:135) and Worster (1994:378–87) note, in the 1980s the superorganism was extended to claim that biosphere functions as a kind of entity called Gaia. Elton's *Animal Ecology* (1927) provided some of Vogt's basic concepts, including the ecological niche and the pyramid of numbers (see Vogt 1948b:86–95). See also Worster 1997 (1994):388–420.

67 Leopold balances Clements and Elton: Callicott 2002; Meine 2010 (1988):410; Leopold 1949:214–17; 1939 (quotes, 728); A. Leopold, 1941?, "Of Mice and Men: Some Notes on Ecology and Politics," ALP, Writings: Unpub. Mss, Ms. 110, 1186–92. Leopold's early Clementsianism: e.g., Leopold 1924, 1979 (written in 1923).

67 Leopold and Vogt: Cushman 2006:345–50; Meine 2010 (1988) (colleagues' uncertainty about his type of ecology, 394–95; discussions with Vogt, 477–80; "for years," 495); Letters, Leopold to J. Darling, 31 Oct 1944 ("my thought"), ALP; Leopold to E.B. Fred, 27 Jan 1943, GFA.

69 Vogt's recommendations: Cushman 2014:195 ("by Nature"); Vogt 1942b:83–85, 88–89, 118–29 ("from their nests," 84), 1942c ("birds themselves," 11); Letter, W. Vogt to A. Leopold, 9 May 1941, ALP ("the throat").

70 Vogt's decisions: Cushman 2006:345 (early Ph.D. thoughts); see the many Vogt-Leopold letters from 1939–42 in ALP and, for another job possibility, Barrow 2009:197–98.

70 Vogt decides to go to Wisconsin: Letters, W. Vogt to A. Leopold, 31 Dec 1941, 4 Feb 1942, 28 Apr 1942; A. Leopold to W. Vogt, 16 March 1942; Vogt, W. 1942. Application for University Fellowship, 30 Jan, ALP.

70 Vogt hunts Nazis: Letters, W. Vogt to A. Leopold, 26 March 1942 ("placed Nazis"), 16 May 1942; A. Leopold to E. B. Fred, 22 May 1942; W. Vogt to A. Leopold, 8 Aug 1942, ALP; Recommendation, Major T.L. Crystal, n.d., 1943 Guggenheim application, GFA; "Bird Watchers Back," *Dunkirk (NY) Evening Observer*, 13 April 1942.

71 Vogt hired by Union: Bowman et al. 2010:241–61; Union Internationale pour la Protection de la Nature 1949:61; "Erosion Is a World Problem," *El Nacional* (Caracas), 27 Sep 1947, FF8, Box 6, VDPL ("precisely"); [Vogt] 1946:2. The treaty is the Convention on Nature Protection and Wildlife Preservation in the Western Hemisphere (161 United Nations Treaty Series 193). Vogt also spent a few months as associate director of the Division of Science and Education of the Office of the Coordinator in Inter-American Affairs.

71 Vogt sees deforestation, erosion: Williams 2006:371–77; Leopold 1999:76 ("The destruction of soil is the most fundamental kind of economic loss which the human race can suffer"); Vogt 1945; Letter, W. Vogt to J. Vogt, 27 Apr 1945, VDPL; Zon and Sparhawk 1923:2:558–666.

72 Vogt in Mexico: McCormick 2005:102–12; [Vogt] 1946:3–6; Vogt 1945a ("hundred years," 358); Vogt 1944 (conservation guide); W. Vogt, 1945?, "Man and the Land in Latin America," unpub. ms. Ser. 2, Box 3, FF29, pp. 5–6, VDPL. See also W. Vogt, 1964, "A History of Land-Use in Mexico," Typescript, Ser. 3, Box 7, FF27, VDPL.

72 "lies ahead": W. Vogt, 1944, Confidential Memorandum, Ser. 2, Box 4, FF29, VDPL.

72 Vogt in South, Central America: Vogt 1948a ("skin disease," "bad situation," 109); [Vogt] 1946 ("ground for optimism," 7; "any possible means," 14); Vogt 1946b ("growing worse," 28).

72 Crisis in El Salvador: Vogt 1946c ("cultivable land," 1; "at once," 3); Vogt 1945b:110 (train simile). See also Durham 1979, esp. chap. 2.

73 Mexico fertility rate, 1940s: Mendoza García and Tapia Colocia 2010, Chart 2.

73 Growth as social goal: Collins 2000:1–32 ("scarcity economics," 6; "more production," 22); Vogt 1948b ("must balance," 110–11). As Collins notes, Leon Keyserling, effective leader of the Council of Economic Advisers, believed that emphasizing growth was "the one really innovating factor" in policy since the New Deal (21). Smith on growth: e.g., Smith 1776:1:85. See also Robertson 2012a:346–56; 2005:26–34. The Employment Act is Public Law 79-304 ("purchasing power," Sec. 2).

74 "Hunger at the Peace Table": Vogt 1945b (all quotes); letter, W. Vogt to F. Osborn, 19 May 1945, Box 3, FF16, VDPL (senators meeting). Leopold called it "the best job of explaining land ecology so far" (Letter, A. Leopold to W. Vogt, 21 May 1945, ALP).

74 Impact of bomb, war: Hartmann 2017; Jundt 2014a:13–17; Allitt 2014:25–23; Robertson 2012a:36–38; Worster 1997 (1994):343–47.

75 "Be told otherwise": Leopold 1993 (1953):165.

75 Malthus's life: Bashford and Chaplin 2016, chap. 2; Mayhew 2014:49–74; Heilbroner 1995 (1953):75–85; James 2006 (1979):5–69 (a classic biography); Chase 1977, esp. 6–12, 74–84 (a sharply negative take); [Malthus] 1798. Six editions appeared in Malthus's lifetime, the last four with only minor changes from the second. I have adapted several sentences from Mann 2011a:179.

76 Franklin: Franklin 1755. Malthus initially hadn't read Franklin, but took the figures from English political theorist Richard Price's quotes; later Malthus gave proper attribution (Bashford and Chaplin 2016:43–47, 70–72, 118). See also Zirkle 1957.

76 Malthus's argument: [Malthus] 1798 (farm increase, 22; U.S., 20–21, 185–86; preventive and positive checks, 61–72; "of the world," 139–40).

77 Inevitability of misery: Malthus 1826:2:29 ("laws of nature"); Malthus 1872:412 ("somewhere else"). This quotation is from the posthumous seventh edition of the *Essay*, but the idea was present from the first (Malthus 1798:15).

77 Malthus's predecessors: The earliest generally noted is Giovanni Botero (2017 [1589], esp. Book 7). Others include Buffon, Franklin, Graunt, Herder, and Smellie. Hong Lianje had a modern anticipation of Malthus in 1793 (Mann 2011a:177–80).

77 *Essay* at bad time: Mayhew 2014:63–65.

77 Anti-Malthus invective: Mayhew 2014:86–88 (Southey); Coburn and Christenson 1958–2002:3 ("wretch"), 5:1024 ("*defy* them!"); Shelley 1920 (1820):51 ("tyrant"); Marx 1906–1909:1:556 ("plagiarism").

78 Malthus inspires Darwin and Wallace: Osprovat 1995 (1981):60–86; Browne 1995:385–90, 542 ("individual against individual"); Bowler 1976. Chapter 3 of *The Origin of Species* is called "The Struggle for Existence."

79 Impact of Malthus, Darwin: Bashford and Chaplin 2016, chaps. 6, 7; Hartmann 2017, chap. 3; Bashford 2014, part 1; Mayhew 2014; Robertson 2012a; Connelly 2008 (esp. chaps. 1–2); Chase 1977. As for Darwin, Timothy Snyder has noted, interpretations of his ideas "influenced all major forms of politics" (Snyder 2015:2).

79 Stoddard: Cox 2015:36–38; Gossett 1997:388–99; Stoddard 1920 ("*finally perish,*" 303–4 [italics in original]). Modern readers are most likely to encounter *Rising Tide*

as the subject of an approving speech by the wealthy brute Tom Buchanan at the beginning of *The Great Gatsby*.

79 Population books: Grant 1916; Marchant 1917; More 1917 (1916) (arguing that uncontrolled breeding was the chief obstacle to feminism); East 1923; Ross 1927; Thompson 1929; Dennery 1931. Although the population movement was international, it was dominated by Anglophone writers (Connelly 2008:10–11). See also Josey 1923.

80 Ross fired from Stanford: Mohr 1970; "Warning Against Coolie 'Natives' and Japanese," *San Francisco Call*, 8 May 1900 ("them to land"). See also Connelly 2008:42.

81 Hitler's biological ideas (footnote): Snyder 2015, esp. 1–10 ("as biology," 2); Weinberg 2006 (1928), esp. 7–36 ("land . . . remains," 21; "land area," 17). Hitler was not one to cite his sources, but the use of Darwin and Malthus in his "second book" is striking and obvious.

82 Left, right, and conservation: Purdy 2015; Allitt 2014:72 ("snow job"); Nixon 2011:250–55; Spiro 2009; Fox 1981:345–51; Chase 1977; Grant 1916:12 ("lower races"). Madison Grant wrote the introduction to *Rising Tide of Color*. One of my sentences is a reworked version of a Purdy sentence. As late as 1994, the polemical literary theorist Edward Said scoffed at environmentalism as an "indulgence of spoiled tree-huggers" (Nixon 2011:332n). Exceptions existed, notable among them Murray Bookchin's *Our Synthetic Environment* (1962). See also Dowie 2009.

82 Vogt's scornful attitudes: Vogt 1948 ("spawning," 77; "copulation," 228; "populations," 47; "codfish," 227; "abroad," "despoilers," 164; "parasites," 202; "Free Enterprise," 15; "one blood," 130). Robertson describes Vogt as "paternalistic but not racist." He observes that Vogt sneered at Latin America's scientific expertise, though at the same time insisted that the low level was due not to "lack of intelligence or ability" (2012a:54) but colonialism and elite corruption. As Robertson notes, Vogt wrote in 1952 that "industrial development should be withheld" from poor countries as a form of birth control (ibid., 157), but the argument was not directly based on race—though that would have been little consolation to the people involved, who were overwhelmingly non-white. Powell believes that Vogt and Leopold simply drew weaker connections than their predecessors "between environmental health and the racial vigor of white Americans" (2016:202).

82 People as biological units: [Vogt] 1946:48. Like Stoddard (1920 ["must die," 174]), Vogt stressed human equivalence to other species: "Studying the amoeba, one can forecast much of the behavior of such complex creatures as [famed economist] Rex Tugwell or Albert Einstein" (Vogt 1948:17).

82 Marjorie Wallace: Washington Births, 1891–1929. Washington State Department of Health Birth Index: Reel 6. Washington State Archives, Olympia, WA; Enumeration District 46, Sheet 5A, Entry for Marjorie E. Wallace, 1920 U.S. Census, Washington, King County, Union Precinct; Enumeration District 41–35, Sheet No. 3A, Entry for Marjorie Wallace, 1930 U.S. Census, California, San Mateo County, Precinct 14; "S.M. High to Graduate 63 in December," *The Times* (San Mateo), 14 Nov 1932; University of California (Berkeley), 1938, *The Seventy-Fifth Commencement*, Berkeley: University of California, 51; Office of the City Clerk, City of New York. Certificate of Marriage Registration No. 3559. George Devereux and Marjorie Elizabeth Wallace, 27 March 1939; idem., Certificate of Marriage Registration M0222434, 30 March 1939; Devereux 1941 (thesis); Letter, W. Vogt to E. Vollman, 23 Dec 1946, Ser. 1, Box 2, FF21, VDPL (contribution to Vogt's work).

83 Devereux: Laplantine 2014; Murray 2009; Gaillard 2004 (1997):191.

83 Vogt divorce and remarriage: Application for Marriage License, Washoe County,

NV, No. 209221. William Vogt, 4 Apr; Second Dist. Court, Washoe County, NV, Decree of Divorce No. 99170, *Marjorie Devereux vs. George Devereux*, 4 Apr 1946, Washoe County (Nevada); "Decrees Granted," *Nevada State Journal*, 2 Sep 1945; telegram, W. Vogt to J. A. Vogt, 25 May 1945, Ser. 1, Box 2, FF23, VDPL; letters, J. A. Vogt to W. Vogt, 28 Mar 1946, Ser. 1, Box 2, FF21, VDPL; W. Vogt to H.A. Moe, 12 Jul 1945, 1 Apr 1946, GFA; J. Vogt to H.A. Moe, 12 Sep 1945 GFA. After the divorce, Juana worked as a diplomatic attaché in Mexico City and Paris and a public affairs officer for the U.S. Information Agency in Seville. She retired in the 1960s and moved to Phoenix, where she died in 2003 at the age of 100 (U.S. Dept. of State. 1949. Foreign Service List, Publication 3388. Posts of Assignment, 45; idem. 1951:20; idem. 1953:79; U.S. Social Security Death Index, Juana A. Vogt, 21 July 2003). Vogt seems not to have told any of his friends about Marjorie's existence before the marriage (see, e.g., letter, A. Leopold to S. Leopold, 22 Aug 1945, ALP). Ingram and Ballard 1935 (Reno as divorce capital).

83 "dinner table": Letter, W. Vogt to WP, 25 Aug 1947, Ser. 1, Box 3, FF24, VDPL.

83 Vogt's movements: Letters, W. Vogt to E. Vollman, 5 Aug, 23 Dec 1946, Ser. 1, Box 2, FF21, VDPL; W. Vogt to J. Vogt, 27 Apr 1945, Ser. 1, Box 3, FF23, VDPL; W. Vogt to A. Leopold, 29 May, 5 Aug, 27 Sep 1946, ALP; A. Leopold to W. Vogt, 5 Jun 1946, ALP.

84 Sloane: Letter, W. Vogt to A. Leopold, 5 Aug 1946, ALP ("big advance"); Hutchens 1946; "On Their Own," *Newsweek*, 15 Jul 1946; letter, Sloane to G. Loveland, 19 Feb 1948, Box 2, FF11, William Sloane Papers, Princeton University Archives; letter, Sloane to H. Taylor, n.d. [Jan 1948?]. Box 3, FF1, idem.

84 Osborn's life: Cushman 2014:272–74 (Conservation Foundation); Robertson 2005:35–44; Regal 2002 (Osborn Sr.); "Fairfield Osborn, 82, Dies," *Berkshire (Mass.) Eagle*, 17 Sep 1969 (sparrow hawk); "Conservation Unit Set Up to Warn U.S.," NYT, 6 Apr 1948.

85 Osborn's ideas for book: Osborn 1948:vii ("conflict with nature"); letter, F. Osborn to W. Albrecht, 15 Aug 1947, ALP ("processes of nature").

85 Vogt and Osborn credit each other: Osborn 1948:204 ("to the problem"); letters, W. Vogt to F. Osborn, 31 Mar 1948, Ser. 2, Box 3, FF16 ("thought of that"); Osborn to Vogt, 3 Apr and 22 May 1948, Box 2, FF8; telegram, Osborn to Vogt, 12 Feb 1948, idem, all VDPL. Osborn's description of Latin America (164–75) is drawn from Vogt.

85 Reaction to *Road to Survival, Our Plundered Planet*: Cushman 2014:262–63; Robertson 2012a:56–57, 2005:22 ("present century," "against the sun"); Desrochers and Hoffbauer 2009:52-55; McCormick 2005:125–27; Linnér 2003:36–38; "Ten Books in Prize Race," NYT, 8 March 1949; Lord 1948 ("lack of it"); E.A.L. 1948 ("hopeful"); North 1948 ("best written"); Memorandum, William Sloane Associates, 18 November 1948, Ser. 2, Box 2, FF17, VDPL (list of schools).

86 Personal congratulations: Letter, R. T. Peterson to Vogt, 30 July 1948, Box 2, FF12, VDPL; letter, G. Murphy to Vogt, n.d. [1948?], Box 2, FF4, VDPL ("new Bible"); Leopold to Vogt, 25 Jan 1946, Box 2, FF1, VDPL ("excellent"); Leopold's praise is on the dust jacket.

86 Criticisms of Vogt, Osborn: Flanner 1949:84 ("crime wave"); Hanson 1949 ("modern problems"); "Eat Hearty," *Time*, 8 Nov 1948 ("unprovable"). The Soviet Union officially denounced Vogt in multiple fora (see, e.g., *Boletín de Información de la Embajada de la U.R.S.S.*, Mexico City, 28 May 1949). Vogt's supporters suspected that the unsigned *Time* article was written by Charles Kellogg, the pro-industry U.S. secretary of agriculture (letter, D. Wade to J. Hickey, 2 Dec 1948, ALP; letter, J. Hickey to W. Vogt, 23 Nov 1948, ALP).

86 *Concerned report on the global condition:* I paraphrase Warde and Sörlin 2015:38. See also Mahrane et al. 2012:129–30.

87 Vogt and Osborn bring population-environment nexus to public: I owe this point to Gregory Cushman (2006:290), whom I paraphrase here. See also Robertson 2005:23–26; Chase 1977:406. Most of the arguments in *Road* were anticipated in an unpublished Leopold article, "In the Long Run: Some Notes on Ecology and Politics" (Leopold 1991 [1941]). Vogt probably never saw it (Powell 2016:172–74).

87 Influence of *Road:* Of Vogt, Gregory Cushman has written: "No single figure was more influential in framing Malthusian overpopulation of humans as an ecological problem—an idea that became one of the pillars of modern environmental thought" (2014:190). To Matthew Connelly (2008:130), Vogt "helped set an agenda that would persist for thirty years." John H. Perkins (1997:136) says that *Road* was "probably of more influence" than *Plundered Planet*; Mahrane et al. highlight Vogt's anti-capitalism (2012:130n), seeing it as a precursor to the 1960s. Thomas Robertson (2005:23) gives equal credit to Osborn: "Vogt and Osborn played as big a role as anyone—including Carson—in spurring the shift from conservation to environmentalism." Vogt's book helped push Carson from natural history to activism (Lear 2009 [1997]:182–83). Ehrlich, Moore, and Vogt: see chapter 7. "educated Americans": Chase 1977:381.

87 Vogt's rhetoric: Vogt 1948b ("wiped out," 17; "shot," 117; "shambles," 114; "responsibility," 133).

88 "The Land Ethic": Leopold 1949:201–26 ("tends otherwise," 224–25).

89 Environment determines character: The view that environment causes character is today called *environmental determinism*. Europeans long believed that, in the words of the sixteenth-century alchemist Richard Eden, "all the inhabitants of the worlde are fourmed and disposed of suche complexion and strength of body, that euery one of them are proportionate to the Climate assigned vnto them" (Chaplin 1995:66). These ideas continued into the twentieth century; the geographer Ellen Churchill Semple's widely used textbook, *Influences of the Geographic Environment*, told students that hot climates tend "to relax the mental and moral fiber, induces indolence, [which makes] not only the natives averse to steady work, but start the energetic European immigrant down the same easy descent" (Semple 1911:627). Overviews include Hulme 2011; Fleming 1998:11–32; and Glacken 1976 (1967).

89 Inventing *the* environment: Warde and Sörlin 2015:39–43 ("idea of *the* environment," 39); Robin et al. 2013: 157–59, 191–93; Robertson 2012b; Worster 1997 (1994):191–93 (power of naming), 350; Glacken 1976 (1967): esp. chap. 2 (Hippocrates quotes, 87); Vogt 1948b ("world scale," x). Osborn, too, covered the globe, but portrayed environmental problems more as a collection of local issues. "transform it": Freire 2014:88.

89 *Carrying capacity:* Robertson 2012b:339–46; Mallet 2012; Sayre 2008; Kingsland 1995 (1985), chaps. 3–4. The concept originated in Chapman 1928. Leopold's prime example of carrying capacity, a study of deer, has been heavily criticized (Caughley 1970).

89 Vogt and carrying capacity: Sayre 2008:130–32 ("The neo-Malthusian use of carrying capacity appears to have its origins in the book *Road to Survival*," 130); Vogt 1948:16–45 (quotes, 16); letter, W. Vogt to A. Leopold, 16 Apr 1947, ALP. Osborn, by contrast, a straight-up Malthusian, warned that overbreeding would overwhelm food and water supplies. To ecologist/activist Garrett Hardin carrying capacity is "central to all discussions of population and environment" (1993:204).

90 Malthus's lack of data on farm production: Author's conversation, Chaplin. Chap-

lin covers some of the implications in Chaplin 2006 and Bashford and Chaplin 2015.

90 Franklin: Franklin 1755 (similarity of people and plants, 9), 1725.

91 "Nature bats last": Ehrlich 1969:28.

91 Odum's textbook and carrying capacity: Mallet 2012:631–33 ("undergraduates," 632–33); Odum 1953 ("can occur", 122).

92 *Planetary boundaries:* Rockström et al. 2009:32 (all quotes); Rockström 2009. A fine popular treatment is Lynas 2011. "be obeyed": Bacon 1870 (1620):8:68.

92 "ecology or conservation": Vogt 1950.

93 Leopold's death, plans to hire Vogt: Lin 2014:107-11; Meine 2010 (1988):479, 519–20; Letter, S. Leopold to B. Leopold, 24 Apr 1948, ALP.

93 Publication of *Sand County Almanac:* Meine 2013, 2010 (1988):523–27.

Chapter Three: The Wizard

95 Borlaug's first days in Mexico: VIET2:27, 56; Hesser 2010; Borlaug 1988 ("dreadful mistake," 24); Bickel 1974:118–19.

97 Borlaugs emigrate: Author's visit, Saude cemetery; Hesser 2010:4–5, 217–19 (Ole and Solveig marriage); VIET1:35, 70; Clodfelter 2006 (1998):35–65 (Dakota war); N. B. Larkin, 1981, Genealogical chart, Borlaug Family Genealogical Material, uarc01014-box01-fdr02, NBUM; Bickel 1974:34–36; RFOI:120; [H.M. Tjernagel]. 1930. Obituary of Ole Borlaug. The Assistant Pastor (Jerico and Saude Lutheran churches), Feb; Flandreau 1900:135–92 (Dakota war); Sogn og Fjordane fylke, Leikanger, Ministerialbok nr. A 6 (1810–1838), Fødte og døpte 1821, p. 128, available at www.arkkiverket.no (Ole's birth). Thanks to Bruce Lundy for help with Norwegian archives.

98 Saude: Author's visit, interviews; Borlaug interview with Matt Ridley, 27 Dec 04; VIET1:64–65; Hildahl 2001 (church services); Bickel 1974:28–31; S. Swenumson, 1921, Childhood Memories as Written by Rev. Stener Swenumson. Borlaug Family Genealogical Material, uarc01014-box01-fdr02, NBUM; Fairbarn 1919, 1:243–47, 322, 361, 403–4, 434, 437–38, 449–50; United States Bureau of Commerce 1912–14, 2:588, 2:620. My thanks to Matt Ridley for providing me with his interview notes.

98 Borlaug's family: Hesser 2010:7–11; A. S. Borlaug (2006?), Memoir of Ole and Solveig Borlaug, unpub. ms.; Bickel 1974:25 ("Norm boy"); RFOI:119–23; LHNB; Enumeration District 134, New Oregon, Howard, Illinois, 1920 U.S. Census, entries for Nels and Thomas Borlaug, Annie Natvig; Anon. 1915a (plat map of New Oregon), 1915b (plat map of Utica); "A Double Wedding Yesterday," *Cresco Plain Dealer*, 15 Aug 1913; Enumeration District 128, New Oregon, Howard, Illinois, 1910 U.S. Census, entries for Nels, Thomas, John Borlaug, Annie Natvig. My thanks to Rollie Natvig for sending me a copy of the Borlaugs' wedding picture, Anna Sylvia Borlaug's memoir, and obituaries.

98 Borlaug's birth: Record of Births, No. 3, Howard County, Iowa, filed 10 Apr 1915. Thanks to Jan Wearda for obtaining this record for me.

99 Henry's home: Author's visit, Saude. I thank Mark Johnson of the Borlaug Foundation for showing me the home. The foursquare style: Gowans 1986:84–93.

99 Isolation of Saude: VIET1:26–28 ("to the world"); *Cresco Plain Dealer*, 9 Jul 1915 (Nels as subscriber).

99 Saude school: Author's visit, Saude; VIET1:36–40 ("corn grows!" 37; "nearly died," 40); AOA; Bickel 1974:21–25; RFOI 21–23; U.S. Department of Commerce 1921–

23, 3:324 (racial statistics). The Borlaug Foundation has moved the school, New Oregon Township No. 8, from its original location to the Borlaug homestead.

102 Work, hoeing thistles, maize harvest: Borlaug interview with Matt Ridley, 27 Dec 04; VIET1: 59–60, 68–69, 74–75, 95 ("horror"). Hoeing Canadian thistles could not be avoided; Iowa banned them as a noxious weed in 1868.

102 Vietmeyer biography (footnote): Some original manuscripts available at TAMU/C (e.g., N. Vietmeyer, 2002, "Hunger Fighter," typescript, Dallas Home Records, Box 6, FF1–3; N. Vietmeyer, 1998, "Hunger Fighter," Dallas Home Records, Box 10, FF2; [N. Vietmeyer?], 1996, Working Outline, Professional Memoirs, Norman E. Borlaug, typescript, Texas A&M Office Records [1], Box 10, FF33; Memorandum of Understanding, 12 Aug 1996, idem).

103 Borlaug and education: Author's interview, Borlaug, 1998 (perspiration vs. inspiration); Hesser 2010:8; VIET1:35–37; RHOI:122.

103 Sina pushes high school: VIET1:79–84; Northwestern University. 1949. Ninety-first Annual Commencement (program), 56 (Sina higher ed); W. Libbey, ed., *The Tack* (Cresco, IA: Cresco High School, 1926), 14 (Sina graduation).

103 Fordson tractor: VIET1:94–98 ("and night," "from servitude"); Wik 1973 (1972):82–102 ("free man," 101).

104 Cresco: Author's visit; vintage photographs in author's possession. Population figures from Iowa State Data Center (iowadatacenter.org).

104 Borlaug's athletic career: Detailed in M. Todd, ed. 1932. *The Spartan* (Cresco, IA: Cresco High School), 13, 34, 42, 44, 48; G. Baker, ed. 1931. *The Spartan* (Cresco, IA: Cresco High School), 12 (Bartelma hired), 79; W. Hoopman, ed. 1930. *The Spartan* (Cresco, IA: Cresco High School), 49. See also Hesser 2010:8; VIET1:35–37, 62 (radio), 86–88; Anonymous 1984:16; Chapman 1981; RFOI:122; LHNB ("my objective", radio) .

105 Bartelma: Author's visit, Cresco; Borlaug interview with Matt Ridley, 27 Dec 04; LHNB; RFOI:126; Anonymous 1984:2, 15–23; Bickel 1974:44–45 ("to compete"). Teachers College is now the University of Northern Iowa.

106 Champlin: G. Hess, ed., *The Spartan* (Cresco, IA: Cresco High School, 1933), 10, 23; Baker, G., ed. 1931. idem, 80; K. Baker, ed., *The Tack* (Cresco, IA: Cresco High School, 1928), 15, 38, 52–54, 58–60, 84–85, 89–90. Champlin would work as a marketer at General Mills; he is credited with coining the name Cheerios.

106 Borlaug's departure: LHNB ("State Teachers"); Vietmeyer, N. 1983. Mr. Wheat. unpub. ms., arc01014-box01-fdr11, NBUM; Bickel 1974:49–50; RFOI:125 ("from Friday"); "Many Students Leave Homes in Cresco for Colleges, Universities," *Mason City Globe Gazette*, 27 Sept 1933.

106 Admissions test: University Calendar, Bulletin of the University of Minnesota 36:3–4 (10 Oct 1933). Because the test was on 24 September, it seems likely that the riot occurred on the weekend of 16–17 September, during the Chicago milk strike.

107 Midwest dairy unrest: White 2015, chap. 2; Block 2009, esp. 143–45; Lorence 1988 (Wisconsin fights); Skocpol and Finegold 1982 (New Deal programs); Perkins 1965; Hoglund 1961 (prices, 24–25); Dileva 1954 (Iowa role); Jesness et al. 1936 (milk prices in Minneapolis, Tables 4, 5, 10); Murphy et al. 1935, esp. 12–15; Byers 1934. Also useful: Czaplicki 2007 (rise of pasteurization); United States Department of Agriculture 1933.

107 Chicago milk strike: "Appeal to State Police for Guard in Milk Strike," *Brainerd (MN) Daily Dispatch*, 15 Sep 1933; "Violence Flares in Illinois Milk Strike," *Edwardsville (IL) Intelligencer*, 16 Sep 1933; United Press International, "Milk Strike Ends in Chicago Area," *Moorhead (WI) Daily News*, 19 Sep 1933.

107 Minneapolis riot: VIET1:125–34; AOA; Norman Borlaug, interview with Mary Gray Davidson, Common Ground (Program 9732), 12 Aug 1997, commongroundradio.org ("triggered it"); LHNB; Bickel 1974:55–58.

108 Initial struggles at Minnesota: Borlaug interview with Matt Ridley, 27 Dec 04; VIET1:123–25, 137–38; LHNB ("liked the outdoors"); RFOI: 128–29; Bickel 1974:58–62.

109 Minnesota vs. Wisconsin program: Green 2006; Miller 2003; Miller and Lewis 1999; Chapman 1935:xiii, 10, 14, 69–73; University of Minnesota. 1934. Bulletin 46:112–14 (curriculum); Leopold 1933; Leopold 1991 (1941):181–92. Gifford Pinchot is credited with establishing the first U.S. forestry program, at Yale, in 1900. Cornell set up a program in 1898, but disbanded it in 1903.

109 Off-campus jobs: VIET1:138–46 ("was gone," 145); RFOI: 126–28; LHNB ("little better").

110 Margaret Gibson and family: VIET1:135–36, 144; L. D. Wilson, 2009, "Medford," *Encyclopedia of Oklahoma History and Culture*, available at www.okhistory.org; Bickel 1974:53–54; "Four All-American Gridders Playing with Red Jackets," *Post-Crescent* (Appleton, Wisc.), 24 Oct 1930; "George Gibson Is Elected Captain of Gopher Eleven," *Brainerd (MN) Daily Dispatch*, 8 Dec 1927; Enumeration District 68, Fayette Township, Seneca, NY, 1910 U.S. Census, entry for Thomas R. Gibson; Archives of Ontario, Registrations of Marriages, RG 80-5-0-317, MS-932 (1903), No. 20331; Enumeration District 90, Town of Romulus, Seneca, NY, 1900 U.S. Census, entry for Robert Gibson; Enumeration District 162, Romulus, Seneca, NY, 1880 U.S. Census, entry for Robert Gibson; Town of Romulus, Seneca, NY, 1870 U.S. Census, entry for Robert Gibson.

110 Marriage, loss of job: VIET1:200–02; Hesser 2010:23–25; Bickel 1974:82–83; Hennepin County (Minn.), Marriage License and Certificate, Norman E. Borlaug to Margaret G. Gibson, No. 200-277, recorded 6 Nov 1937.

111 "Dr. Stakman": RFOI: 131 (last sentence rearranged for clarity).

112 Stakman: Borlaug interview with Matt Ridley, 27 Dec 04 (attending lecture); Dworkin 2009:19–22; Perkins 1997:89–91; Christensen 1992; Stakman oral history interview with Pauline Madow, 29 May–6 June 1970, RG 13, Oral Histories, Boxes 9–11, RFA; Stakman 1937:117 ("essential branches").

112 Rust history: Kislev 1982; Carefoot and Sprott 1969 (1967):41–47; Theophrastus 1916: 2:201–3. Several rust species exist and rust can also strike oats, barley, and rye, but because stem rust and wheat are the most economically important I concentrate on them.

113 Stem-rust spores: Anikster et al. 2005:480 (size); Carefoot and Sprott 1969 (1967):39 ("the universe"); Stakman 1957:261 (50 trillion).

113 Stem-rust life-cycle: Leonard and Szabo 2005; Roelfs et al. 1992 (jam, 92); Petersen 1974.

114 1916 epidemic: Campbell and Long 2001:19; U.S. Senate 1922:11–12.

114 Barberry eradication: Dworkin 2009:19–22; Dubin and Brennan 2009; Campbell and Long 2001 ("pro-German," "alien," 26–27; "rustbuster" 29); Leonard 2001; Perkins 1997:89–92; Roelfs 1982; Large 1946 (1940):366–70; E. C. Stakman, 1935 "A Review of the Aims, Accomplishments and Objectives of the Barberry Eradication Program," Cereal Rust Laboratory Records, typescript, uarc00037-box15-fdr31, University of Minnesota Archives; Stakman and Fletcher 1930 (1927); Stakman 1923 (1919) ("outlaw," "it is," 3–4); Beeson 1923 ("menace," 2); "News of the Nursery Trade," *Florists' Review*, 27 Jun 1918.

114 Thatcher: Kolmer et al. 2011; Hayes et al. 1936.

115 Wallace visits Mexico, advocates for assistance: Olsson 2013:202–14; Cullather 2010:54–59; Culver and Hyde 2001, esp. 246–51; W. C. Cobb, 1956, "The Historical Backgrounds of the Mexican Agricultural Program," typescript, RG 1.2, Ser. 323, Box 9, Folder 62, RFA, esp. II-1–3, 11; Crabb 1947: chaps. 7, 10; Wallace 1941; Alexander 1940 (diets).

116 Foundation beginning: Chernow 2004 (1998): 550–83; Farley 2004; Fosdick 1988.

116 Foundation's leeriness, Wallace presses, General Education Board: Olsson 2013:64–71, 182–200; Harwood 2009:387–88; W. C. Cobb, 1956, "The Historical Backgrounds of the Mexican Agricultural Program," typescript, RG 1.2, Ser. 323, Box 9, Folder 62, RFA ("could be done"); General Education Board 1916 (1915). Additionally contributing to the foundation's leeriness was that the GEB had been attacked as a Rockefeller plot, and barred by Congress from working with the government. The GEB, founded in 1902, preceded the foundation.

117 Sauer and Mexican maize: Letter, Carl Sauer to Joseph Willits, 5(?) Feb 1941, RG 1.2, Ser. 323, Box 10, Folder 63, RFA (all quotes). I have put his argument in contemporary terms. For a fuller description of maize's cultural diversity, see Mann 2004 and references therein.

117 Stakman and Mexican rust: Dworkin 2009:22–24 (Stakman Mexico visits); Stakman et al. 1940 (Mexico as rust reservoir).

117 Maize, population figures: Instituto Nacional de Estadística, Geografía e Informática 2015, Cuadro 9.27 (maize); Mendoza García and Tapia Colocia 2010, Chart 1 (pop.); Cotter 1994:235–38 (maize imports); Wylie 1941, Table 1 (maize from U.S.). For 1920 maize figures, see the discussion in the 2000 version of the INEGI report. See also Myren 1969:439–40.

118 Bradfield, Mangelsdorf, and Stakman report: Survey Commission. 1941. Agricultural Conditions and Problems in Mexico: Report of the Survey Commission of The Rockefeller Foundation. RG 1.1, Ser. 323, Box 1, Folder 2, RFA ("judicious advice," 14); idem. 1941. Summary of Recommendations. RG 1.2, Ser. 323, Box 10, Folder 63; idem. 1941. Rockefeller Foundation's Survey of Agriculture in Mexico. RG 1.1, Ser. 323, Box 11, Folder 70 ("pitifully low").

118 Ecological consequences of land reform: González 2006; Dwyer 2002 (U.S. anger); Sonnenfeld 1992: esp. 31–32; Esteva 1983:266; Yates 1981:48 (2.5 million acres); Venezian and Gamble 1969:54–62 (50 million acres). Further pushing environmental degradation, many *ejidos* could not obtain credit to buy fertilizer, irrigation equipment, or better farm tools, because the new National Bank of Ejidal Credit was undercapitalized (Olsson 2013:332–33).

119 Harrar: McKelvey 1987.

120 Mexican Agricultural Program begins: Waterhouse 2013:18–29, 98–109; Olsson 2013:215–32; Harwood 2009:392–93; Perkins 1997:106–15; Fitzgerald 1986:459–64; Baum 1986:5–7; Hewitt de Alcántara 1978:33–37; Anon., 1978, "Chronology of the Development of CIMMYT," uarc01014-box33-fdr34, NBUM; W. C. Cobb, 1956, "The Historical Backgrounds of the Mexican Agricultural Program," RG 1.2, Ser. 323, Box 9, Folder 62, RFA, esp. II-3–11; F. B. Hanson, diary, 4 March, 10–11 Jul, 10–12 Aug 1942, RG 12, F-L, Box 194, Reel M, Han 3, Frame 585, RFA (Stakman and Harrar appointments); letter, R. Fosdick to J. A. Ferrell, et. al., 31 Oct 1941, RG 1.1, Series 323, Box 11, Folder 72, RFA; Ferrell, J. A. 1941. Memorandum, Vice President Wallace, RBF and JAF, Regarding Mexico, Its Problems and Remedies, 3 Feb. RG 1.1, Ser. 323, Box 1, Folder 2, RFA; Letter, H. A. Wallace to R. E. Fosdick, 13 May 1941, RG 1.1, Ser. 323, Box 12, Folder 79, RFA. Wallace's intervention was decisive; Rockefeller president Raymond Fosdick quoted his views almost word

for word in staff meetings (minutes from staff conference, 18 Feb 1941, RG 1.2, Ser. 323, Box 10, Folder 63, RFA). Stakman's War Emergency Committee work is detailed in Box 13, Folders 1, 3–8, 11, 20–22, Stakman papers, University of Minnesota, Twin Cities.

120 U.S.-Mexico relations: Dwyer 2002; Schuler 1998:155–98.

120 Maize as focus: Olsson 2013:239–40, 255–73, 279–81; Harwood 2009:391–92; Matchett 2006:360–62; Secretaria de Agricultura 1946:93; Survey Commission. 1941. Agricultural Conditions and Problems in Mexico: Report of the Survey Commission of The Rockefeller Foundation. RG 1.1, Ser. 323, Box 1, Folder 2, RFA (focus on maize); Summary of the Survey Commission's report, 4 Dec 1941, RG 1.1, Ser. 323, Box 11, Folder 70, RFA. The deemphasis on wheat, compared to later writings, is reflected in the foundation's annual reports (e.g., Rockefeller Foundation 1946:160–62; 1945:21–24, 167–69; 1944:170–71).

120 Multiple goals of Rockefeller program: Lance Thurner, pers. comm.; Waterhouse 2013:98–99; Singh et al. 1994:19–20 (Stakman focus on stem rust); E. J. Wellhausen oral history interview with William C. Cobb, 28 Jun–19 Oct 1966, RG 13, Oral Histories, Box 25, Folders 1–2, RFA; Advisory Committee for Agricultural Activities. 1951 (21 Jun). The World Food Problem, Agriculture, and the Rockefeller Foundation. Typescript, RG 3, Ser. 915, Box 3, Folder 23 ("deliver more").

120 Failure of maize program: Olsson 2013:299–310; Harwood 2009:398–400; Matchett 2006 ("parties involved," 365); Fitzgerald 1986:465–67; Aboites et al. 1999 (Mexican researchers' beliefs); Cotter 1994 (demand for science); Myren 1969; E. J. Wellhausen, oral history interview with William C. Cobb, 28 Jun–19 Oct 1966, RG 13, Oral Histories, Box 25, Folders 1–2, RFA (political utility of hybrid maize); Stakman, E. C. 1948. Report of Mexican Trip with Confidential Supplement Regarding Mexican Agricultural Program. RG 1.2, Series 323, Box 10, Folder 60, RFA (foundation frustration with Mexican officials).

121 Borlaug and Stakman: VIET1:216–33 (eye damage, 218; "fire and light," 233); Bickel 1974:86–89 ("my boy" 88); Borlaug 1941.

122 Borlaug's coursework, thesis: Borlaug 1945; Transcript File No. 103665 (Dest. 239359), University of Minnesota Registrar's office. I am grateful to Barb Yungers for sending me Borlaug's academic records.

124 DuPont job, arrival in Delaware: VIET1: 235–36; Bickel 1974:89–91.

125 Borlaug and DDT (footnote): Russell, pers. comm.; Russell 2001:86, 124–48; Kinkela 2011: chap. 1; Perkins 1978; Borlaug 1972; Borlaug. 1973? DDT and Common Sense. Typescript, TAMU/C 002/003[1]009; Knipling 1945. DDT was developed in the 1930s, after a Geigy researcher accidentally discovered its properties. (The researcher, Paul Hermann Müller, received a Nobel Prize in 1948.) Geigy tried to market DDT in the U.S., only to have its U.S. subsidiary decide in early 1941 that it couldn't compete against existing insecticides like pyrethrum. Pyrethrum was extracted from chrysanthemum blossoms grown in Asia. When Japanese conquests cut off the Asian chrysanthemum supply, the U.S. military, which feared the loss of troops to insect-borne diseases, directed Agriculture Department researchers to look for a substitute. In November 1942 Geigy sent over samples of DDT, which the government tested, obtaining favorable results. DuPont refused to make DDT unless Geigy gave up its patent rights; strong-armed, the Swiss company caved in; DuPont made a lot of money.

126 Borlaug takes on Mexico: VIET1:251–55, 2:23; Bickel 1974:91–92, 96–100; RFOI 138–39; Harrar travel diary, 17 Feb 1942, RG 1.2, Ser.464, Box 1, FF3, RFA; Diary, F. B. Hanson, 7 Apr 1942, RG 1.1, Ser. 205, Box 12, FF179, RFA.

126 Initial MAP staff: Rockefeller Foundation 1944:170–71; Harrar diary, 25 Feb 1944, RG 12.2, Ser. 1.1, Box 18, FF 45, RFA (Colwell); Letter, C. Sauer to J. Willits, 23 Aug 1943, RG 1.1, Ser. 323, Box 1, FF 6, RFA (Wellhauser).

126 Initial planting: VIET2: 27–32 ("we thought," 28); N. Borlaug, 1981, "The Phenomenal Contribution of the Japanese Norin Dwarfing Genes Toward Increasing the Genetic Yield Potential of Wheat," address, 30th Anniversary of the Founding of the Japanese Society of Breeding, Tokyo. Typescript, CIMBPC, B0051-R, 15.

127 Borlaug at sea: RFOI: 138–140 ("to DuPont," 140).

127 Second child: VIET2: 32–34 ("I can"), 72–73; Hesser 2010:39–40; Bickel 1974:111–14, 128–30, 157; RFOI: 141–42.

127 Borlaug takes over stem-rust project: Bickel 1974:118–19; RFOI: 150.

128 Wheat conditions in Bajío: Instituto Nacional de Estadística y Geografía 2015: Cuadro 9.37; Bickel 1974:121–22; Borlaug 1958:278–81, 1950:170–71; Rupert 1951; Borlaug et al. 1950; N. E. Borlaug, 1945, "Wheat Improvement in Mexico," typescript, CIMBPC, B5533-R. See also, Hewitt de Alcántara 1978: 37–40.

128 Farm survey: N. E. Borlaug, 1945, "Annual Survey of Wheat Growing Areas of Mexico for Determination of Severity of Damages Caused by Diseases," typescript, CIMBPC, B5528-R; idem., 1945, "Outline of the Diseases of Wheat," typescript, CIMBPC, B5530-R. Borlaug summarizes conditions in Borlaug 1950a:171–73.

129 Three men planting 8,600 varieties: VIET2:48–56; Bickel 1974:141–45 (clothing); Paarlberg 1970:5–6; Borlaug 1950a:177–87; RFOI: 155–57, 161–62 (agronomists' attitudes). Borlaug provided slightly different accounts of the origins and numbers of the seeds. I have mostly followed Borlaug 1950a, as this was written closest to events. In addition to the main two locations, Borlaug tried planting small amounts of wheat at seven other locations in Mexico.

130 Borlaug and Bajío poverty: VIET2: 49–53 (quotes, 51); Bickel 1974:110–11 (resistance to new ideas), 143–44 (metal tools); E. J. Wellhausen, oral-history interview with William C. Cobb, 28 Jun–19 Oct 1966, RG 13, Oral Histories, Box 25, Folders 1–2, RFA, 46–48.

132 Failure of second crop: VIET2: 54–57 ("to the ground," 56); N. Borlaug, 1981, "The Phenomenal Contribution of the Japanese Norin Dwarfing Genes Toward Increasing the Genetic Yield Potential of Wheat," address, 30th Anniversary of the Founding of the Japanese Society of Breeding, Tokyo. Typescript, CIMBPC, B0051-R, 14–15.

133 Bajío not enough: Author's interview, Borlaug; VIET2: 38–39 ("whole populace," 39 [emphasis mine]), 65; Borlaug 2007:288.

133 *Shuttle breeding:* Hesser 2010:48–51; AOA; LHNB; Borlaug 2007:288–89, 1950a (initial outline); Ortiz et al. 2007; Rajaram 1999; RFOI: 152–88 passim. The name was coined in the 1970s (Centro Internacional de Mejoramiento de Maíz y Trigo 1992:14).

133 Mexico wheat areas: Borlaug, N., et al.(?) 1955. FA003, Box 87, FF1755, Rockefeller Foundation Photograph Collection, RFA (map); Borlaug and Rupert 1949; N. E. Borlaug, 1945, "El Mejoramiento del Trigo en México," typescript, CIMBPC, B5529-R.

134 Sonora and first visit: Cerruti and Lorenzana 2009 (Table 3, acreage; Map 2, description of area); Borlaug 2007:288–89; Cotter 2003:125; Dabdoub 1980; Bickel 1974:120–27.

134 Breeding dogma: Kingsbury 2011:294; Dubin and Brennan 2009:11; Borlaug 2007:289; Perkins 1997:226.

134 Harrar-Borlaug argument: VIET2: 67–69; AOA; McKelvey 1987:30–31.

139 First season at Sonora: VIET2: 69–73; RFOI: 169–70 ("complete disaster"); Hesser 2010:46–48; Borlaug 2007:289–90; Borlaug 1950.

139 Passage across U.S. to Sonora: VIET2:74–78; RFOI: 163–64.

141 Conflict with Hayes: RFOI 188; AOA; VIET2:102–03; Borlaug 2007:289; Bickel 1974:180; Hayes and Garber 1921:111, 113, 281–86ff. (e.g., "The field selected for the comparative trials should be representative of the soil and climatic conditions under which the crop will be grown," 51).

142 Borlaug quits: VIET2:104–09, 112–26; RFOI: 166–68 ("walked out," "own organization!," "like children!"); AOA; Perkins 1997:228; Bickel 1974:180–84.

143 State of plant breeding: Perkins 1998, chap. 3. The Rockefeller University experiment that showed that DNA was the mechanism of heredity was published in 1944 but not widely believed until Watson and Crick showed how DNA could carry genetic information.

143 Four times as many genes: Brenchley et al. 2012 estimates wheat has ~95,000 genes. Humans are thought to have 20,000 or fewer (Ezkurdia et al. 2014). Both figures are tentative. Borlaug describes the difficulties in RFOI: 307–8.

144 Eye color: White and Rabago-Smith 2011.

144 Photoperiodicity mutation: Baranski 2015; Guo et al. 2010 (mutation); Kingsland 2009:299 (discovery, lack of attention); Borlaug 2007 ("serendipity," 289); Beales et al. 2007; Cho et al. 1993.

145 Beginnings of success: VIET2:170–72; Borlaug 1968, Table 1.

145 *Lodging:* Borlaug estimated that "at the time of harvest, 85 percent of all the wheat [was] flat on the ground in the Yaqui Valley" (RFOI: 214).

146 15B: VIET2:158–61; Dubin and Brennan 2009:5; Stakman 1957:264; Anonymous 1954:1–3 ("was susceptible," 2).

147 1950 testing, conference: Kolmer et al. 2011; Borlaug 1950b.

147 500 researchers: Rockefeller Foundation, *Annual Report*, 1959:30.

147 1951 results: Borlaug et al. 1952.

148 All but two: VIET2:189–90; Borlaug 1988:27; Rodríguez et al. 1957:127; Borlaug et al. 1953, 1952; Rupert 1951: Table 9; Borlaug, N.E., et al. 1953. Stem and Leaf Rust Reaction of Wheats in the 1951 International Wheat Nursery when Grown at Mexe, Hidalgo, Mexico in the Summer of 1952. Typescript, CIMBPC (B5564-R).

148 Resuscitation of maize program: Emails to author, Lance Thurner; Matchett 2002; Myren 1969.

148 Foundation to expand: W. Weaver, Memorandum, 11 Dec 1951, RG 3.1, Ser. 915, Box 3, FF20, RFA (Harrar promoted); W. Weaver, Memorandum, "Agriculture and the Rockefeller Foundation," 12 Jul 1951, idem; J. G. Harrar, Memorandum, "Agriculture and the Rockefeller Foundation," 1 Jun 1951, idem; Letter, W. Weaver to J. G. Harrar, 31 May 1951, idem; "Excerpt from Minutes of Meeting of the Advisory Committee on Agricultural Activities," 19 May 1951, idem.

148 Borlaug's situation, Argentina trip: VIET2:174–76; Baranski 2015:58; Perkins 1997:230; Borlaug 1988:27; Borlaug et al. 1953:10–11 (race 49). Race 49 is genetically close to Race 139; the two are often referred to interchangeably.

148 Borlaug and Bayles: RFOI: 198–200; Bickel 1974:197–99 ("with it").

149 Norin 10: Lumpkin 2015; VIET2:181–83, 195; Reitz and Salmon 1968.

149 First two Norin 10 trials: VIET2:202–03, 208–09, 224–33; RFOI: 200; Borlaug 1988:27–28. Norin 10 was winter wheat. By early spring, when it flowered, the rest of Borlaug's test varieties were producing grain. He had nothing to cross it with. Scouring his fields, Borlaug found a single plant, a variety sent by Rupert from Colombia, with a misfiring biological clock; it, too, had late flowers. He was able to

use it to pollinate the Japanese plants—only to see rust overwhelm them. By failing to take the variety's internal calendar into account, he had seemingly wasted all of Vogel's genetic material.

149 Rising yields in Sonora: Cerruti and Lorenzana (2009), Salinas-Zavala et al. (2006), and Hewitt de Alcántara (1978) collect figures from INEGI, FAOSTAT, and the Comisión Nacional del Agua.

152 Field day chaos: VIET2:234; RFOI 201–05, 214–16 ("away right there"); Bickel 1974:236–38.

152 Borlaug and mills: VIET3:33–34; Borlaug 1988:27–28 (grain problems); Baum 1986:7 (release of better varieties); Bickel 1974:239.

153 Package: Borlaug apparently first referred to the "package" in summer 1968 (Borlaug 1968:27); by 1969 it was common parlance at CIMMYT (Myren 1969:439). Borlaug said that "75 to 80 percent" of the package in Mexico was applicable in India and Pakistan, a figure he later applied to other nations (Borlaug 1968:13; Borlaug 1970).

154 "Green Revolution": Speech, W. S. Gaud, 8 March 1968, available at agbioworld.org.

154 Victory-lap speech: Borlaug 1968:Table 1 (Mexico), Table 2 (India), 33 ("fellow men").

Chapter Four: Earth: Food

159 Weaver, Rockefeller, and molecular biology: E. O'Sullivan 2015; Hutchins 2000; Kay 1993 (Weaver coins name, 4); Rees 1987 (Nobels, 504); Priore 1979; "Warren Weaver, 84, Is Dead After Fall," NYT, 25 Nov 1978; Weaver 1970, 1951; Memorandum, Warren Weaver, Translation, 15 Jun 1949, Weaver papers, Ser. 12.1, Box 53, FF476, RFA; Shannon and Weaver 1949.

160 "[Vogt's] strictures?": Memorandum, Chester Bernard to Warren Weaver, 31 Aug 1948, RG 3.2, Ser. 900, Box 57, FF310, RFA. The two men met that same day, presumably in part to discuss Road (Diary, Warren Weaver, 31 Aug 1948, RG 12, S–Z, Reel M, Wea 1, Frame 8, Box 502–03, RFA). See also Cullather 2010:64–66. Barnard became president on July 1.

160 First modern statement: Cullather 2010:66 (Weaver "articulated the post-Malthusian counterargument the foundation would use for the next thirty years").

160 Weaver's report: Memorandum, W. Weaver, Population and Food, 8 Jul 1953 (orig. 17 Jul 1949), RG 3, Ser. 915, Box 3, FF23, RFA (all quotes). Weaver used the "large" calorie: the energy to raise the temperature of 1 kilogram of water by 1°C (that is, to raise 2.24 pounds of water by 1.8°F). There is also a "small" calorie, used occasionally in chemistry and physics. I use the large calorie in this book.

161 80 billion: Weaver thought the estimate was low, because it didn't include non-solar energy sources and fossil fuels. But these were dwarfed by the energy from the sun, so he set them aside for the purposes of his analysis.

162 IngenHousz and photosynthesis: Magiels 2010; Morton 2009 (2007):319–43.

163 Story of N: I lifted this subtitle from Hugh S. Gorman's fine book (2013).

164 Humus theory: Jungk 2009; Manlay et al. 2006:4–6; Fussell 1972, chap. 5; Gyllenborg 1770 (1761), esp. 13–17, 21–28, 48–50; Aristotle 1910:467b–468a. Gyllenborg was Wallerius's student; different editions identify one or the other as author.

165 Attacks on humus theory, law of minimum: Jungk 2009; Sparks 2006:307–10; Brock 2002:32–35 (fake dissertation), 74 (taking credit for others' work), 107–24

(fake experiments), 146–49 (law of minimum), 160–66; Van der Ploeg et al. 1999 (Sprengel); Liebig 1840:64–85 (main objective of farming, 85), 1855 (23–25, law of minimum); Sprengel 1828 (93, law of minimum).

165 Nitrogen abundance: Galloway et al. 2003.

166 Liebig and N: Gorman 2013:58–63; Brock 2002:121–24, 148–79ff.; Smil 2001:8–16.

166 Organic machine: White 1995.

166 Liebig fertilizer fiasco (footnote): Brock 2002:120–28, 138–40. The fullest version of Liebig's scoffing at nitrogen fertilizer is in his third edition ("superfluous," 213). His switch to the nitrogen camp occurred in 1859 (Liebig 1859:264–66).

167 Chilean nitrates: Pérez-Fodich et al. 2014; Gorman 2013:66–69; Melillo 2012; Smil 2001:43–48 (half as explosives, 47).

167 Crookes: Morton 2009 (2007):178–82; Smil 2001:58–60; Crookes et al. 1900 ("of the world," 16; "a few years," 43; "general scarcity," 194–95).

170 Haber and Bosch: Smil 2001:61–107 ("and hydrogen," 72; "liquid ammonia," 81). Haber led what was, in effect, the first national weapons lab. Under his enthusiastic direction, it developed a cyanide gas, Zyklon B, that was used in Hitler's gas chambers. (Haber and his wife were born into Jewish families, but they converted to Protestantism.)

170 Haber-Bosch superlatives: Naam 2013:133 ("can grow"); Melillo 2012 (1930s); Smil 2011b ("world's population" [updating Smil 2001:157]); Von Laue 1934 ("from air"). See also Morton 2015:193–95.

172 Nitrogen downsides: Bristow et al. 2017 (Bengal); Morton 2015:194–201 (biggest problem, 197); Canfield et al. 2010; Galloway et al. 2002; Smil 2001:177–97. See also Guo et al. 2010.

173 "cultural movement": Conford 2001:20.

173 McCarrison and Hunza: Wrench 2009 (1938), esp. 28–46, 56–66 ("of cancer," 33); Vogt 2007:24–25; Fromartz 2006:12–16; Conford 2011:178, 2001:50–53; McCarrison 1921 ("extraordinarily long," 9).

174 New scientific genre: I take these examples from Taubes 2007:89–95. The studies were conducted by Albert Schweitzer, Aleš Hrdlička, A. J. Orenstein, and Samuel Hutton. This kind of work is hard to evaluate, because the subjects typically don't keep precise personal records and are often related. In addition, it is rarely possible to have control groups.

175 McCarrison, Viswanath, Suryanarayana: Viswanath 1953; Viswanath and Suryanarayana 1927; McCarrison and Viswanath 1926. The relative contributions are my interpretation, but Viswanath's annoyance shimmers from the pages of his articles. My thanks to Ellen Shell for helping me with this section.

176 McCarrison on soil: McCarrison 1944 (1936) ("constituted food," 17; "our needs," 12).

176 Albert Howard: Wrench 2009:153–58; Pollan 2007:145–51; Fromartz 2006:6–12; Conford 2011:95–98, 2001:53–59 ("soil fertility," 54–55); L. E. Howard 1953: esp. chap. 1; A. Howard 1945:15–22, 151. Howard's pre-India work focused on the hops plant, used to flavor beer. At the time researchers believed hops were best propagated artificially, by grafting. Howard obtained better results by pollinating with bees. This "amounted to a demand that Nature no longer be defied," he said. "It was for this reason highly successful" (ibid., 16). Even his critics recognized his centrality, e.g., Hopkins 1948:96, 181.

177 Indore process: Howard and Wad 1931: esp. chap. 4.

178 "is gold": Hugo 2000:1086.

178 Louise Howard: Oldfield 2004.

178 Rule or Law of Return: Manlay et al. 2006:10; Howard 1945 ("human wastes," 5; "Nature's farming," 41); Balfour 1943.

179 Howard's claims: L. E. Howard 1953:26 ("brutality"); A. Howard 1940 ("research organization," 160; "less and less," 189; "and mankind," 220). Louise Howard appears to be quoting from W. J. Locke's *At the Gate of Samaria*, a popular novel from 1894.

179 Aristocratic, conservative Christians: Conford 2011:327–34, 351–56; 2001:146–63, 190–209, 217; Moore-Colyer 2002 (a study of one organic leader, Rolf Gardiner). Conford lists 73 "leading figures" in the early organic movement (2001: Appendix A). Of these, 27 were deeply religious or spiritual, 18 were either hereditary aristocrats or rich landowners, and 16 were either fervently right-wing or fascist. To be sure, some members were socialists, others were ordinary farmers, and not all were inspired primarily by Howard. Northbourne, for instance, was mainly inspired by Rudolf Steiner and (like Howard) was not a member of the Soil Association (Paull 2014). My thanks to Philip Conford and Oliver Morton for helping me with this section.

179 Balfour: Gill 2010 (New Age Christianity, 171–82); Conford 2001:88–89; Balfour 1943 ("each other," 199).

180 North American Christian inspiration: Lowe 2016. One difference is that North American Christian advocates typically focused on preserving rural lifeways rather than agriculture per se.

180 Rodale: R. O'Sullivan 2015: chap. 1 (lived longer, 95); Cavett 2007 ("in my life!"); Fromartz 2006:18–21; Conford 2001:100–103; Rodale 1952, 1948 (Hunza).

180 Rodale builds empire: O'Sullivan 2015, esp. 18–20, 26–27, 58–59, 222–27 (subscriptions, 32, 88); Northbourne 1940 ("organic," 59, 103).

182 Industry pushes back: O'Sullivan 2015:56–58 ("powder keg," 18; "Every Bug," 57); Conford 2011:289–95, 325–26; 2001:38–43; Throckmorton 1951 ("misguided people," 21); Bowman 1950 ("is no!").

182 Criticisms, defenses of organic viewpoint: Pollan 2007:146–49 (I borrow his descriptor, "airy crumbs"); Hopkins 1948 ("extremist views," 115); Balfour 1943 ("animal's body," 18); Howard 1940 ("and women," 31; "own being," 45; "forest humus," 68; "of Liebig," 220).

183 FAO degradation study: United Nations Food and Agricultural Organization 2011a, Fig. 3.2 (25% is highly degraded, 8% is moderately degraded).

183 Organic vs. chemical war: O'Sullivan 2015 ("organiculturalist," 56); Picton 1949:127 ("the phalanx"); Rodale 1947 ("has begun"); Northbourne 1940 ("chemical," 81, 99, 101; "very hard," 91; "and laborious," 115).

184 Discovery, import of rubisco: Morton 2009 (2007):39–47 ("care about," x); Benson 2002; Portis and Salvucci 2002; Wildman 2002.

184 Eighty-three thousand enzymes: Placzek et al. 2016. BRENDA is at www.brenda-enzymes.org.

185 Rubisco's incompetence: Walker et al. 2016; Zhu et al. 2010; Mann 1999 (quotes).

186 Rubisco abundance: Raven 2013; Phillips and Milo 2009 (11 lbs.); Sage et al. 1987; Ellis 1979 (most abundant protein). It has been suggested that collagen is more abundant.

186 Evolution of photosynthesis: Cole 2016; McFadden 2014.

187 Margulis and symbiosis: Author's conversations, Margulis; Weber 2011; Sagan 1967. Bhattacharya et al. 2004 and Raven and Allen 2003 review multiple acts of symbiosis.

187 Gene migration (includes footnote): Raven and Allen 2003; Huang et al. 2003 (tobacco); De Las Rivas et al. 2002; Martin et al. 2002 (one-fifth), 1998; Sugiura

1995. The reference genome in Cyanobase (genome.microbedb.jp/cyanobase) has 3,725 genes.

187 Multiple rubisco varieties: Tabita et al. 2008.

188 IRRI founding: Bourne 2015:62-64; Cullather 2010:159–71; Chandler 1992: chap. 1 ("contributions," 3).

189 Political hopes for Asia: Cullather 2010:146–58 ("for food," 162). Hunger and income statistics for 1960 are uncertain. I draw on discussions in Dyson 2005 (esp. 55–56); Ahluwalia et al. 1979.

190 IR-8: Bourne 2015:66; Hettel 2008 ("sheer luck"); Chandler 1992:106–17; Jennings 1964.

190 Gibberellin (footnote): My thanks to Ludmila Tyler for drawing my attention to this point.

191 Spread, impact of IR-8: Bourne 2015:66-69; Cullather 2010:167–79 ("on hunger," 171; "Mao books," 176); Mukherji et al. 2009: Table 2 (irrigation use); Hazell 2009:7–14; Dawe 2008; Abdullah et al 2006:35; Alexandratos 2003:22; Dalrymple 1986:1068–72. Historic rice production and fertilizer use from FAOSTAT (faostat. fao.org).

191 Projections: Hunter et al. 2017 (rise of "25%–70% above current production levels may be sufficient to meet 2050 crop demand"); Fischer et al. 2014 ("world demand for staple crop products should grow by 60% from 2010 to 2050," 2); Foley 2014 ("population growth and richer diets will require us to roughly double the amount of crops we grow by 2050"); Garnet 2013 ("food production may need to rise by as much as 60–110% by 2050 overall," 32); Alexandratos and Bruinsma 2012 ("increase by 60 percent from 2005/2007–2050," 7); Tilman et al. 2011 ("a 100–110% increase in global crop demand from 2005 to 2050," 20260); Godfray et al. 2010 ("Recent studies suggest that the world will need 70 to 100% more food by 2050", 813); Royal Society 2009:1 ("even the most optimistic scenarios require increases in food production of at least 50%"), 6 (effects of meat consumption); World Bank 2008 ("cereal production will have to increase by nearly 50 percent and meat production by 85 percent from 2000 to 2030," 8). On the role of affluence, see Weinzettel et al. 2013.

192 Meat consumption: Author's interviews, email, Walter Falcon, Joel Bourne, Michael Pollan; Vranken et al. 2014; Rivers Cole and McCoskey 2013 (decrease with affluence); Smil 2013 (10 to 40%, 133). Meat production from FAOSTAT (faostat.fao.org).

193 Widely cited study: Ray et al. 2013, 2012. See also Grassini et al. 2013; Jeon et al. 2011:1; Dawe 2008; Hibberd et al. 2008:228.

193 Actual/potential yields: I simplify the formulation in Ittersum et al. 2013.

193 Food waste (footnote): Bellemare et al. 2017; Gustavsson et al. 2013.

194 Lack of arable land, inability to expand irrigation: Author's interviews, IFPRI, CIMMYT, IRRI; United Nations Food and Agricultural Organization 2013:10; Murchie et al. 2009:533; Mann 2007.

194 Early atmosphere: Kasting 2014; Lyons et al. 2014. See also Lane 2002, chap. 3.

194 Ordinary photosynthesis: Ordinary photosynthesis is known as C3, after another molecule, this one with three carbon atoms.

197 C4 project: Author's interviews, Jane Langdale, Paul Quick, Peter Westhoff, Thomas Brutnell, John Sheehy, Julian Hibberd. Project overviews include Wang et al. 2016; Furbank et al. 2015.

197 Audacity of project: Surridge 2002:576.

199 Shotgunning genes and CRISPR: Hall 2016; Specter 2015 (fine popular accounts of CRISPR); Vain et al. 1995 (shotgunning cereals); Klein et al. 1987 (invention

of technique). Before CRISPR, the rice project used a different method, infecting plants with Agrobacterium, a bacterium that inserts genes from a plasmid (a free-floating DNA-containing body, like a chloroplast, in the cell) into the DNA of a plant cell. The genes are switched on and the plant cell produce nutrients for the bacterium. By adding genes to the plasmid, geneticists can use this mechanism to insert new genetic information into plant cells. Overall, though, this method seems to have been less common than shotgunning.

200 Possibilities: Jez et al. 2016.

200 Other ways to improve photosynthesis (footnote): Taylor and Long 2017; Pignon et al. 2017; Krondijk et al. 2016. My thanks to Ruth DeFries for drawing my attention to this work.

201 Vandalism and test: Hall 1987; Maugh 1987a, b; "Genetic Tests to Proceed in Face of Protest," *San Bernardino County Sun*, 15 Apr 1987.

202 Asilomar conference and regulations: Berg et al. 1975; Berg and Singer 1995; Frederickson 1991:274–83, 293–98 ("should not," 282).

202 Lack of diversity: Vettel 2006:220–22 ("political motivation"); Frederickson 1991:293–98 (participant list). Reporters attended only after legal threats from the *Washington Post*.

202 Ice-minus controversy: Author's attendance at Rifkin lectures; Bratspies 2007:109–11; Thompson 1989 ("the Boys"); Hall 1987 ("human beings," 134); Joyce 1985; Complaint, *Foundation on Economic Trends v. Heckler*, 14 Sep 1983, in *Biotechnology Law Report* 2:194–203 [1983] ("mutant bacteria", ¶19); Lindow et al. 1982. Natural "ice-minus" *P. syringae* exist, but they spontaneously revert to the common form. The Berkeley researchers made the change irreversible by removing part of the gene that promotes ice nucleation.

203 Scientific studies, GMO bans: See Appendix B and the online database of GMO bans at the Borlaug-founded International Service for the Acquisition of Agri-Biotech Applications (www.isaaa.org).

204 Public fears: Pew Research Center 2015, chap. 3; Gaskell et al. 2006; Blizzard 2003; Hall 1987 ("about it").

206 Borlaug's claims for GMOs: See, e.g., Borlaug 2004.

206 Nichols farm: Author's interviews, visits. Nichols's farm is certified as sustainable, because the certification is required to sell at some Chicago markets.

209 Organic/conventional yield comparisons: Hossard et al. 2016 ("Maize yields are on average 24% higher for tested low-input [and conventional] relative to organic . . . Wheat yields are on average 43% higher for tested low-input relative [or conventional] to organic"); Kniss et al. 2016 (corrected version: "across all crops and all states, organic yield averaged 67% of conventional yield"); Ponisio et al. 2015 ("organic yields are . . . 19.2% [±3.7%] lower than conventional yields"); de Ponti et al. 2012 ("organic yields of individual crops are on average 80% of conventional yields"); Seufert et al. 2012 ("overall, organic yields are typically lower than conventional yields" [5–34% lower depending on "system and site characteristics"]); Badgley et al. 2007 ("the average yield ratio (organic : non-organic) of different food categories .. was slightly <1.0 [8.6%] for studies in the developed world and >1.0 for studies in the developing world"). See the critiques in Kirchmann et al. 2016; Kremen and Miles 2012; Connor 2008. My discussion is scribbled in the margins of Pollan 2007:176–84. Brecht's line: *Erst kommt das Fressen, dann kommt die Moral*. Engels anticipated these arguments, famously, in chap. 9 of his *Dialectics of Nature*.

210 Annuals vs. perennials: González-Paleo et al. 2016; Smaje 2015; Crews and DeHaan 2015; Cox et al. 2006. Perennial grasses typically have evolved better protective

measures against pests and diseases (pathogens that afflict annuals seldom infect their perennial relatives). But most also ripen asynchronously, making harvest difficult.

212 Domesticating wheatgrass: Zhang et al. 2016, Fig. 1; Lubofsky 2016; Scheinost et al. 2001; Wagoner and Schaeffer 1990; Lowdermilk and Chall 1969:232–33 (U.S. introduction).

212 Wheat-wheatgrass hybrids: Author's interviews, Jones, Curwen-McAdams; Curwen-McAdams and Jones 2017; Curwen-McAdams et al. 2016 (*T. aaseae*); Hayes et al. 2012; Larkin and Newell 2014; Wagoner and Schaeffer 1990; Tsitsin and Lubimova 1959.

213 Cassava: Author's interviews, emails: Botoni, Larwanou, Wenceslau Teixiera, Susanna Hecht. Production data from FAOSTAT; USDA (2016 crop production summary); Howeler ed. 2011.

213 Trees: See, e.g., Dey 1995 (acorns); Garrett et al 1991 (walnuts); Robinson and Lakso 1991 (apples). Because tree crops exist in many cultivars and are grown with many different cropping regimes, production numbers vary tremendously. Overall temperate-zone tree-crop production data is available at the USDA website.

214 Deprecating agriculture: Author's conversations, James Boyce, Vern Ruttan, Daron Acemoglu; Cullather 2010:146–48; Boulding 1963, 1944.

214 Farm employment: Dmitri et al. 2005:2–5.

Chapter Five: Water: Freshwater

216 California tomatoes: USDA Economic Research Service, 2010, *U.S. Tomato Statistics (92010)*, http://usda.mannlib.cornell.edu. See also overviews at www.ers.usda.gov.

217 California water projects: A classic if polemical history is Reisner 1993, esp. 9–10, 194–97, 334–78, 499–500; see also Prud'homme 2012:240–51.

219 Freshwater: Gleick and Palaniappan 2010:11155–56; Babkin 2003:13–16; Shiklomanov 2000, 1993 (water volumes, 13–14). About two-thirds of groundwater is saline (Gleick 1996).

219 Human appropriation of water: McNeill 2001:119–21; Shiklomanov 2000, Tables 2,4; Postel et al. 1996: Fig. 2.

220 Brazil and India: Figures from AQUASTAT (www.fao.org/nr/water/aquastat/main/index.stm).

220 Flow and stock: I draw here on distinctions described by, among others, Malm (2016:38–42), Gleick and Palaniappan (2010), and Wrigley (2010:235). My thanks to Mark Plummer, Jim Boyce, Daron Acemoglu, and Mike Lynch for helping me with this section. For salmon, other factors must also be taken into account, like the number of salmon that die in the ocean before spawning. Still, the principle remains: catching a fish one spring does not reduce the fish supply the next.

221 Lack of substitutability of water: My last line paraphrases a remark made to me by Oliver Hoedeman of Corporate Europe Observatory.

221 Ogallala: Peterson et al. 2016 (flow, Fig. 6); Reisner 1993:435–55.

222 Ruining aquifers: Hertzman 2017: Table 1; Sebben et al. 2015 (saltwater intrusion); Famiglietti 2014 (overview); European Environment Agency 2011: Chap. 8.

222 Water shortages: Mekonnen and Hoekstra 2016.; Comprehensive Assessment of Water Management in Agriculture 2007 (IWMI study); Shiklomanov and Balonishnikova 2003:359; Shiklomanov 2000. Along with IWMI, Shiklomanov, of the

Russian State Hydrological Unit, is probably the most widely cited source of global water-demand projections.

223 Water consumption: Figures from AQUASTAT (http://www.fao.org/nr/water/aquastat/water_use/index.stm).

223 Irrigation losses: Lankford 2012.

223 Up to 50 percent higher: Leflaive et al. 2012:216; Amarasinghe and Smakhtin 2014 (esp. Table 1).

223 Groundwater for irrigation: Siebert et al. 2010.

224 Lowdermilks to Promised Land: Mané 2011:65; R. Miller 2003:56–57; Lowdermilk and Chall 1969, 2:314–16; Lowdermilk 1940:83–91. Promised Land: Exodus 23:31, Genesis 15:18–21. The first paragraph of this section rewrites Mané's first paragraph.

226 Lowdermilk's life: Helms 1984; Lowdermilk and Chall 1969 (realization, 1:61–63; fleeing China, 1:100–108); Lowdermilk 1944:11–13 (dream of visiting Palestine).

227 Decline of Mesopotamia: Lowdermilk and Chall 1969, 2:328–32 ("salty desolation," 331); Lowdermilk 1948 ("radiant gold"), 1940:92–100 ("dirty place," 96; "of civilization?" 97), 1939; Deuteronomy 4:45–49 (location), 8:7–9 ("and hills"); Psalms 104:16 ("full of sap"); Song of Solomon 5:15 ("the cedars"). All quotes from Revised King James Edition.

227 Lowdermilk's theory: Rook 1996:98–103 ("'Allah'"); Lowdermilk 1944:53–65, 135–39 ("settled areas," 136–37); 1942:9–10, 1939.

227 Modern perspectives: Wilkinson and Rayne 2010; Hughes 1983; Wertime 1983.

228 Absorptive capacity of Palestine: Siegel 2015:20–22; Alatout 2008b:367–74; Anglo-American Committee of Inquiry 1946, 1:185 (immigration tallies); United Kingdom 1939 ("Arab population").

228 Engineers vs. "plant men": Lowdermilk and Chall 1969, 2:207, 218–19.

229 Lowdermilk's vision: Rook 1996:115–31, 139–42; Lowdermilk 1944 ("amazed," "twenty-four countries," 14; "our day," 19; industries and electrification, 68–75, 85–87; "thriving communities," "splendid opportunity," 121; "from Europe," 122; "a million," 124; Jordan plan, 121–28; "artificial lakes," 139–40).

230 Lowdermilk's influence: Siegel 2015:35–41; Alatout 2008b:379–82; Rook 1996: 142–52, 159–62; Lowdermilk and Chall 1969, 2:543–44; Anglo-American Committee of Inquiry 1946, 1:411–14 (British rejection).

230 National Water Carrier: Author's visit; Siegel 2015:39–40 (Panama Canal); Cohen 2008; Alatout 2008a. Other information from Mekorot website (www.mekorot.co.il).

232 Influences on Howard: Marx 1909, esp. 3:945 ("rift"); Kropotkin 1901 (1898); Morris 1914 (1881); Liebig 1859:176–79 ("be collected"). Howard may have read Hugo.

232 Howard: Clark 2003; Beevers 1988; Evans 1997 (1989):111–13 (aqueduct); Howard 1898 ("new civilization," 10; water plans, 153–67); Howard 1902.

234 Howard and Tel Aviv: Katz 1994.

234 Tel Aviv wastewater: Author's interview, Oded Fixler; Siegel 2015:78–85; United Nations Economic and Social Commission for Western Asia and Bundesanstalt für Geowissenschaften und Rohstoffe 2013; Loftus 2011; Aharoni et al. 2010.

236 Hard vs. soft path: Brooks et al. 2010 ("sustainable future," 337); Brooks et al. eds. 2009; Brooks and Holtz 2009; Brandes and Brooks 2007 ("and attitudes," 2); Gleick 2003 ("both paths," 1527); 2002; 2000 ("freshwater runoff," 128); 1998; Brooks 1993.

237 Israeli soft-path: Author's interviews, Noam Weisbrod, Ittai Gavrieli, Yoseph Yechieli. Siegel 2015:11–12, 46–50, chaps. 4–5 (drip irrigation, water reuse).

239 Red-Dead canal: Author's interviews, Oded Fixler, Nobil Zoubi, Munqeth Mehyar; Donnelly 2014. Government of Jordan 2014; World Bank 2013.

239 Israel hard-soft conflict: Author's visit, interviews, Siegel 2015: 116, Berck and Lipow 2012 (1995):140

240 Urban growth, water failures: United Nations Population Division, World Urbanization Prospects (https://esa.un.org/unpd/wup/); United Nations Human Settlements Programme 2016: Table E.2.

241 Fecklessness: Kunkel Water Efficiency Consulting 2017 (Pennsylvania); Milman and Glenza 2016 (30 cities); Water Integrity Network 2016 (overpumped aquifers, 39; KwaZulu, 64; Lixil Group 2016 (India, see web addendum "Findings"); Bundesverband der Energie- und Wasserwirtschaft 2015 (France); Siegel 2012:190–95 (Israel/Palestine).

242 Veolia in Pudong and Liuzhou: Mann 2007; see also Prud'homme 2012:269–70.

243 "pure economical": Boulding 1964.

247 Loeb's career: Siegel 2015:119-21; Cohen and Glater 2010; Hasson 2010.

249 Growth, potential of desalination: International Desalination Agency 2017:esp. 72–76; Goh et al. 2017; Delyannis and Belessiotis 2010; Delyannis 2005. Desalination plant numbers from International Desalination Agency website (idadesal.org).

249 Almonds and alfalfa: Holthaus 2015.

250 Carlsbad, California desalination, critiques: Author's visits and interviews, San Diego Water County Authority; International Desalination gency 2017: 12–13, 42; Cooley and Ajami 2014; Cooley et al. 2006.

Chapter Six: Fire: Energy

251 Birth and rise of Pithole: Author's visits, Pithole; Knickerbocker and Harper 2009:108–14 (population); Burgchardt 1989:78–82 (population); Darrah 1972, chap. 3 (bars and brothels, 34); Cone and Johns 1870:75–76; untitled description of Pithole, Boston Daily Advertiser, 24 Jul 1865; Viator 1865. Darrah is the classic history of Pithole. See also Crocus 1867. I am grateful to Kent Mathewson for accompanying me to Pithole and the Drake Well Museum.

251 Pennsylvania as first oil patch: Earlier oil wells had been dug in Galicia and Azerbaijan (then controlled by Russia), but they didn't lead to much for a while. Pennsylvania had the first modern wells—drilled by engines, as opposed to dug by hand, and encased in pipes to prevent flooding. And discoveries there led to the creation of today's fossil-fuel industry (Vassilou 2009:195–96).

252 Oil-Dorado and Petrolia: e.g., Cone and Johns 1870; "Fire in the Oil Regions," NYT, 15 Feb 1866.

252 Fire-fighting dredge, prostitute parade: Burgchardt 1989:80; "Crocus," 1867:36-37.

253 Decline and fall of Pithole: Darrah 1972:133–37 (oil wells stop producing), 178–82, 205–31 (281 people, 227; $4.37, 231); Philips 1886; Taylor 1884:14–18 passim.; "Deserted Villages," Boston Daily Advertiser, 21 Oct 1878; Cone and Johns 1870: 82–84; "Story of a Once Famous City," Wisconsin State Register, 26 Jun 1869; "Petroleum Matters," Daily Cleveland Herald, 5 Sep 1865 (end of first Pithole well).

254 Energy demand forecasts: BP 2015 (37% by 2035); IEA 2014b (37% by 2040); World Energy Council 2013 (61% by 2050); Larcher and Tarascon 2015 (100% by 2050).

254 Carboniferous coal formation: Nelsen et al. 2016; : Martin 2013:392–96; Floudas et al. 2012; DiMichele et al. 2007.

255 Early Chinese coal: Dodson et al. 2014.

255 Early history of fossil fuels: Yergin 2008 (1991):7–9; Williams 2006, chap. 7 (defor-estation); Richards 2005 (2003):194–95, 227–41; Freese 2004 (2003), chaps. 2–3.

255 Early British coal: Freese 2004 (2003):21–42 (Nottingham, 24); Gimpel 1983 (1976):80–84; Braudel 1981 (1979):367–72.

255 Increase in well-being post-1800: Clark 2007:1–16 ("remote ancestors," 1 [empha-sis added]) is a fine summary. Clark makes clear that prosperity was not equal and for all; Malm (2016) focuses on the human costs.

256 Versailles: Williams 2006:164.

256 Jefferson: Hailman 2006:219; Letter, Thomas Jefferson to Thomas Mann Randolph, 28 Nov 1796, in Oberg 2002:211.

256 Transformative power of fossil fuels: Gallagher 2006:192–95; Lebergott 1976:100, 1993, Tables II.14 and II.15 (U.S. running water, central heating). Rybczynski (1986) discusses how new heating technologies helped create the modern expec-tation of a comfortable home; "portable climate": Emerson 1860:74–75 (I have reversed the order of the sentences).

256 Average U.S. car horsepower: http://www.epa.gov/fueleconomy/fetrends/1975–2014/420r14023a.pdf (Table 3.3.1).

257 "Dregs": Carll 1890:24.

257 Shares of energy production: IEA 2015b:6.

258 Abundance is problem, not scarcity: I am echoing arguments in Labban 2008:2 and Radkau 2008 (2002):251. Radkau discusses, fascinatingly, pre-industrial worries about what one might call "peak wood" (201–14).

259 Carnegie's lake of oil: Nasaw 2006:76–78; Carnegie 1920:138–39 ("would cease").

259 Standard Oil pessimism: Yergin 2008 (1991):35–36; Chernow 2004 (1998):283–84 ("crazy?"). There was also a small but growing oil industry in Azerbaijan.

259 Pennsylvania peak: Harper and Cozart 1992: Fig. 4.

260 Beaumont: Yergin 2008 (1991):66–79; McLaurin 1902 (1896):459–63.

260 Perception of vulnerability: See, e.g., Shuman 1914; "Liquid Fuels," *Chemical Trade Journal and Chemical Engineer*, 8 Feb 1913 ("it is by no means certain that as the present fields become depleted new ones will be opened"); "Liquid Fuels for the Navy," *Chemical Trade Journal and Chemical Engineer*, 29 March 1913 ("there is no immediate prospect of more plentiful [oil] supplies being found"); Thurston 1901 ("a time must come, and that within a few generations at most, when some other energy other than combustion of fuel must be relied upon," 283). Clayton 2015: chap. 2 and DeNovo 1955 give further examples.

260 Roosevelt at governors' meeting: McGee 1909:3–12 ("the nation," 3; "imminent exhaustion," 6); Clayton 2015:39; Bergandi and Blandin 2012:113–15. The meeting, organized by pioneering forester Gifford Pinchot, was attended by 36 governors.

261 Repeated warnings: Clayton 2015:40–43; Olien and Olien 1993:42–44; Day 1909b (quotes, 460). Unofficially, survey officials were even more pessimistic. In 1919 chief geologist David White warned in a popular magazine that "the peak of pro-duction will soon be passed—possibly within three years" (White 1919:385).

261 British coal debate: Jonsson 2014:160–64, 2013, chap. 7; Madureira 2012:399–404; "The Coal Question," *Saturday Review* 21(1866):709–10; McCulloch 1854 (1837):596–600; Holland 1835:454–63; Great Britain House of Lords 1830; Bakewell 1828 (1813):178–81 (2,000 years, 181); Williams 1789:158–79 ("fortunate island," 172).

261 Jevons paradox: Madureira 2012:406–13; Heilbroner 1995 (1953):172–76 passim; Black 1972–81, 1:203 (Gladstone); Courtney 1897 (Mill, 789); Thomson 1881

("not slowly," 434); Jevons 1866 (paradox of increased consumption, 122–37; "of consumption," 242).

261 British coal peaks, world output rises: U.K. historical coal production: www.gov .uk/government/statistical-data-sets. World historical coal production: Smil 2008:219–21.

262 Churchill advocates oil: Churchill 2005 (1931):73–76. For the later, similar U.S. move to naval oil, see DeNovo 1955.

262 Government buys BP: Jack 1968; Statement, W. Churchill, in Great Britain House of Commons 1913:1465–89 ("we require," 1475).

262 British involvement in Iran: Yergin 2008 (1991):118–33; Zirinsky 1992.

262 Race to control Middle East supplies: Yergin 2008 (1991):160–89 is a good over-view. Also useful are Dahl 2001; Marzano 1996; Shwadran 1977:2; Cohen 1976; Mejcher 1972; DeNovo 1955.

263 U.S., Russia oil production: Ferrier 2000 (1982), 1:638.

263 U.S. oil production 1920–29: "U.S. Field Production of Crude Oil," Energy Infor-mation Agency; "U.S. Ending Stocks of Crude Oil," Energy Information Agency (both at www.eia.gov).

263 Russia and Venezuela: Maugeri 2006:30–32.

263 Newspaper search: The archive was at www.newspapers.com.

263 Newspaper warnings: "Nation Faces Oil Famine," *Los Angeles Times*, 23 Sep 1923 ("the nation"); "Oil Famine Within Two Years Is Scouted by Students of Indus-try," *Houston Post-Dispatch*, 12 Dec 1924 ("two years"); Stevenson 1925 ("twenty years"); R. Dutcher, "Prices of Oil Are Kept Down Only by Vast Overproduction," *Times-Herald* (Olean, NY), 13 Feb 1928 ("of oil").

264 "A Giant Lampshade, Reversed": The best general histories of solar power I have come across are Perlin 2013; Madrigal 2011; Kryza 2003. Also useful are Johnson 2015 (emphasizing the connection to fossil-fuel fears I focus on here); Perlin 2002; Hempel 1983.

264 Mouchot's early life: Pottier 2014; Quinnez 2007–2008; Bordot 1958; Mouchot 1869a:193 ("expectations").

264 French coal fears, Mouchot's solution: Jarrige 2010:86–88; Mouchot 1869a:214–15 ("do then?"), 230–31 ("mechanical applications"). See also Kryza 2003:151–53.

265 Early solar use: Perlin 2013:3–35, 57–78 (ancient China, 3–8; ancient Greece, 13–14; Vitruvius, 23; Pompeii, 32); de Saussure 1786, 4:36–48, 261–63 ("through glass").

265 Early use of burning mirrors: Perlin 2013:36–55. In a probably apocryphal inci-dent, the Greek mathematician Archimedes used parabolic mirrors to burn up an attacking Roman fleet (Kryza 2003:37–48).

266 Mouchot's first research: Pottier 2014; Simonin 1876:203; Ebelot 1869; Mouchot 1869a:193; 1869b; 1864. A number of Italian researchers had similar ideas before Mouchot, but apparently they didn't actually build any solar engines (Silvi 2010).

266 Experiments in Paris and Tours: Pottier 2014; Perlin 2013:88–91; Jarrige 2010:88–89; Quinnez 2007–2008:306–9; Simonin 1876:204–9 ("the sky," 204); Bontemps 1876:105–7; Mouchot 1875. Simonin, the journalist, was a longtime solar enthusi-ast (Jarrige 2010:87).

266 Comparison with coal: Anonymous 1870:310–11 ("of coal"). The engineer was Paul-Théodore Marlier (*Mémoires et Compte Rendu des Travaux des Société des Ingénieurs Civils de France* 21 (1873):54, 64). A flotilla of Mouchot solar engines big enough to run a factory would cover 100,000 square feet, a big area in an urban setting (Perlin 2013:91).

267 Algeria and exhibition: Pottier 2014; Perlin 2013:92–95 ("in the world"); Jarrige 2010:89–91; Quinnez 2007–2008:309–16.

267 Mouchot gives up: Jarrige 2010:92. His assistant, Abel Pifre, took over for a few years, eventually constructing a solar-powered printing press (Quinnez 2007–2008:316–18; Collins 2002; Pifre 1880; Crova 1880).

268 Mouchot's last years (footnote): Pottier 2014; Quinnez 2007–2008:319–20; Bordot 1958; "Louis [sic] Mouchot in Poverty," NYT, July 27, 1907.

268 Britain's steam-engine growth: Tunzelman 2003 (1986):74–78.

268 Ericsson's solar projects: Johnson 2015:22-31; Perlin 2013:99–108 ("mere toy," 104); Kryza 2003:106–23; Collins 2002; Hempel 1983:47–50; Church 1911 (1890), 2:260–301 ("the solar rays," 265; "and complex," 271); Anonymous 1889 ("should go on," 191); Ericsson 1888 ("perfected"); "The Coal Problem and Solar Engines," NYT, 10 Sept 1868.

268 Ericsson's vision: Ericsson 1870 (all quotes). Ericsson admitted that not all of the Earth got enough sunlight for his engines. But the suitable area included a band "from northwest Africa to Mongolia, 9,000 miles in length and nearly 1,000 miles wide" and an equivalent sunbelt in the Americas, enough to permanently change the human condition.

269 Lovins and soft path: Parisi 1977; Yulish 1977; Lovins 1976. Lovins did not indicate a source for the term "soft," but may have taken it from Robert Clarke (1972).

270 Winsor and Gas Light and Coke Company: Tomory 2012:121–238 ("of the world," 121; 30,000 lamps, 234); Mokyr ed. 2003 2:393–94 (Baltimore).

271 Pasadena and Cairo installations, Massachusetts textbook: Johnson 2015:41–57, 64–69; Perlin 2013:109–17, 129–42; Kryza 2003, chaps. 1, 3–5, 8 (a fascinating account); "Rev. Charles Henry Pope," Cambridge Tribune, 23 Feb 1918; "American Inventor Uses Egypt's Sun for Power," NYT, 2 Jul 1916; Pope 1903.

272 Father Himalaya and Pyrheliophoro: Tinoco 2012; Pereira 2005; Rodrigues 1999; Graham 1904; "Father Himalaya and the Possibilities of His Prize-Winning Pyrheliophor," NYT, 12 Mar 1904; "Pyrheliophor, Wonder of St. Louis Fair," NYT, 6 Nov 1904 ("necessary to lie").

273 Hubbert's life: The principal sources are Inman (2016) and HOHI.

274 Hubbert and Technocracy: Inman 2016:35–121 passim; Yergin 2012 (2011):236 (Great Engineer); [Hubbert] 2008 (1934) (principal text); Session 4, 17 Jan 1989, HOHI; Akin 1977 passim. The Technocracy Study Course was anonymous, but Inman (2016:344n) makes a convincing case for Hubbert's authorship.

274 Technocracy Study Course: [Hubbert] 2008 (1934) ("essentially different," 99; "135,000,000 people," 158; "North American Continent," 220).

275 Hubbert quits Technocracy: Inman 2016:120–21 ("Technocracy meeting").

276 Hubbert vs. Stanford geologist: Inman 2016:114–18; Levorsen 1950 (1.5 trillion, 99); Session 6, 23 Jan 1989, HOHI ("of oil"); Hamilton 1949 ("metaphysics"). The Stanford geologist was A. I. Levorsen (see chapter 8). Hubbert appended a sketch of his curve and predicted the peak in 50–75 years (Levorsen 1950:104).

276 State of geology: Oreskes 2000.

276 "minds of men": Pratt 1952:2236.

276 Hubbert publishes formal model: Hubbert 1949 (all quotes). The model was revised and expanded in Hubbert 1951.

277 Hubbert and petri dish: Hubbert made the comparison explicitly in Hubbert 1962:125–26. See also Hubbert 1938.

277 Hubbert's Gause-like curves: Hubbert 1951 ("matter and energy," 271; "to zero," 262). Hubbert's "curve-fitting" method is dissected in Lynch 2016:75–82; Sorrell and Speirs 2010.

277 Hubbert moves to Shell: Session 4, 17 Jan 1989; Session 5, 20 Jan 1989, HOHI. Inman (2016:73–98) recounts his life between Columbia and Shell.

277 Shell's unhappiness: Session 7, 27 Jan 1989, HOHI ("parts out?").

278 Hubbert predicts peak: Hubbert 1956. Later Hubbert predicted the peak "should occur in the late 1960's [or] early 1970's" (Hubbert 1962:73). See also Priest 2014:50–52.

278 Peak in 1970: U.S. Field Production of Oil (1859–present), Energy Information Agency, available at www.eia.gov.

278 Hubbert, McKelvey, Udall: Author's interviews, Priest; Inman 2016:183–86, 212–13, 267–70; Priest 2014 (fight with McKelvey, 53–63; "McKelvey out," 66; "ninety-eight-year history," 67); interview, David Room (Global Public Media) with Udall, 8 Feb 2006, transcript at www.mkinghubbert.com ("a Hubbert man").

278 Oil blockades: Overviews include Clayton 2015:106–16; Mitchell 2011, chap. 7; Yergin 2008 (1991):570–614; Bryce 2008:93–97; Adelman 1995:99–117 (see esp. Table 5.4); Grove 1974 ("Oil Age," 821). Previously, during the 1967 Arab-Israeli war, half a dozen Arab nations launched an oil embargo. It had next to no effect, because the U.S. then still produced much of its own oil.

279 Impact of price controls: Author's interview, Michael Lynch; Lynch 2016:33–36; Hamilton 2013:13–15; Bryce 2008:93–95; Adelman 1995:110–17 ("The shortages were created entirely at home, the result of price controls and allocations," 112). Bryce (2008:95) notes that U.S. oil companies had big stockpiles of oil that they did not want to refine into gasoline, because price controls ensured they would lose money on every gallon.

279 *Limits to Growth*: Meadows et al. 1972; Schoijet (1999:518–19) says William Behrens, one of the coauthors, used Hubbert's work as a starting point. See also Inman 2016:232–35; Sabin 2013:86 ("*Limits to Growth*").

279 Shortage rumors: Salmon: "Salmon Shortage," *The Times* (San Mateo, CA), 9 Nov 1973; Cheese: Associated Press 1974; Onions: Charlton 1973; Raisins: "New Breakfast Blow: Raisin Shortage Hits," *Milwaukee Journal*, 2 Feb 1977; Toilet paper: Lynch 2016:33; Malcolm 1974. The Japanese shortage was experienced by the author's wife and her parents, who lived in Japan at the time.

280 Carter speech: J. Carter, Speech to nation, 18 April 1977, in *Public Papers of the Presidents of the United States: Jimmy Carter*, Book I, January 20 to June 24, 1977, 655–61. Washington: U.S. Government Printing Office.

280 Hubbert limit in 1995: Grove 1974:821.

280 Carter boosts coal: Blum 1980 ("The policy of the Carter Administration is to burn three times more coal by the year 1995," 4); Carter 1977a, b. Carter called the energy crisis the "moral equivalent of war." Critics seized the acronym and referred to his program as MEOW.

280 Historical oil prices: http://inflationdata.com/Inflation/Inflation_Rate/Historical_Oil_Prices_Chart.asp.

281 Kern River output, reserves: Reserve figures from OGJ and the California Division of Oil, Gas, and Geothermal Resources (Marilyn Tennyson, USGS, pers. comm.). Also useful: Tennyson et al. 2012; Takahashi and Gautier 2007 (first drilling, 6-9); Adelman 1991:10–11; Roadifer 1986; "U.S. Fields with Reserves Exceeding 100 Million Bbl," OGJ, 27 Jan 1986. I am grateful to Sarah Yager at the *Atlantic Monthly* for helping me sort through these numbers and contacting Dr. Tennyson.

281 1998 blowout and aftermath: Waldner 2006; Singer 1999.

282 "Never": Adelman 2004:17.

283 Victorian Internet: Standage 2013 (1998) ("of information," xvii–xviii).

283 Smith and selenium: Perlin 2013:302–4; 2002:15–16; Smith 1891:310–11 ("in the

morning"), 1873a, b. One source says that the actual discovery was made by a telegraph clerk, who then told Smith (Anonymous 1883).

283 Adams and selenium: Adams and Day 1877; 1876 ("of light," 115).

283 Fritts's selenium panels: Perlin 2013:305–8 ("at that time," 307); Fritts 1885, 1883. Several other inventors received patents on "solar cells" at about this time, but none seem actually to have built them.

284 Einstein's papers: Pais 1982 (photoelectric effect, 380–86). Later that year a fifth paper introduced the famous equation, $E = mc^2$. The term "photon" was coined in 1926.

284 Chapin's test: Perlin 2002:26.

284 Development of transistor: Isaacson 2014:136–52; Riordan and Hoddeson 1998; Hoddeson 1981.

285 Fuller and Pearson invent silicon panels: Johnson 2015:137–51; Perlin 2013:310–25; 2002:25–36 ($1.43 million, 35). Fuller and Pearson built upon earlier work by Russell Ohl. The gold-leaf comparison uses 1,200 sq. ft. as the average size of a suburban home in the 1950s and $35 as the price of gold.

285 Oil shocks as solar catalyst: Johnson 2015:179–80; Jones and Bouamane 2012:16–18; Fialka 1974.

286 Sunlight, human energy use: Solar incidence and reflection taken from the foundational Sørensen (2011 [1979]:174); human use from IEA 2014b:48. The ratio of incident sunlight to human consumption depends on the estimate of the latter; some researchers (e.g., Pittock 2009 [2005]:177) say that the sun produces ca. 10,000 times human energy production. See also Smil 2008, chap. 2. Morton (2015:62–71) has an elegant popular discussion.

286 Counterculture and solar: Johnson 2015:185–90; Baldwin and Brand eds. 1978 ("even lovable," 5); Lovins 1976; Commoner 1976 ("be possessed," 153); Grove 1974:792–93 (Hubbert backs solar). Commoner's "concept of solar energy utilization as naturally conducive to democracy became a central tenet of 1970s energy politics" (Johnson 2015:203). Similarly, Earth Day organizer Denis Hayes claimed in *Rays of Hope* that peak oil would lead to a "post-petroleum age" of solar-powered liberty (Hayes 1977). These arguments were anticipated by Pope (1903:139, 154) and Huxley (1993 [1939]:148–65ff.).

286 The Pentagon, Big Oil, and photovoltaics: Johnson 2015:156–74; Nahm 2014:55–61; Lüdeke-Freund 2013 (BP); Jones and Bouamane 2012:14–16, 21–38, 51–53 (Exxon and Mobil, 23–24); Perlin 2002:41–46 (space), 61–69 (70%, 68). Nonpetroleum firms made PV for space and offshore platforms before Big Oil stepped in; the petroleum companies set up the first firms that manufactured terrestrial solar panels. Not all the financial muscle behind PVs was tied to oil: In 1999 the then-biggest U.S. solar firm, First Solar, was acquired by the investment arm of the Walton family, which owns Wal-Mart. In Europe and Japan PVs were the province of huge electronics firms like Siemens and Sharp.

286 Campbell and Laherrère: Campbell and Laherrère 1998 ("Before 2010," 79; "of it", 81; "nations depend," 83). They didn't use the term "peak oil," which was not coined until 2002.

287 Peak-oil predictions: Clayton 2015:155 (Pickens); Bush, G. W. 2008. Statement, World Economic Forum, 18 May, available at georgewbush-whitehouse.archives .gov ("oil is limited"); Simmons 2005:xvii ("irreversible decline"); Kunstler 2005: 26 ("unimaginable austerity").

287 Public fears of peak oil: Clayton 2015:159 (highest price, 78 percent); Kemp 2013 (83 percent); Swartz 2008 ("take care of that"); Sesno 2006 (three-quarters); "a given time": DeLillo 1989:66.

288 Modi bio: Price 2015 ("religious freedom," 207); J. Mann 2014. The visa decision was reversed in 2014.

289 Modi's solar program: Author's visits, interviews, Gujarat; Mann 2015; Moon, B-K. 2015. "Remarks at 10 MW Canal Top Solar Power Plant," 11 Jan (available at www.un.org/sg/en/content/sg/speeches); Modi 2011 (green autobiography).

289 Efficiencies of photovoltaics, coal: The National Renewable Energy Laboratory tracks efficiency improvements over time at www.nrel.gov/ncpv/; for coal, see IEA (Coal Industry Advisory Board) 2010:90 (Table II.7). Mann 2014 (costs of CCS); Prieto and Hall 2013 (low solar efficiency).

289 Cost of solar and coal plants: Bolinger and Seel 2015; Energy Information Agency 2013. The EIA (2013:6) estimated the capital cost of a single-unit advanced PC coal plant at $3.25/watt; Bolinger and Seel (2015:13), of Lawrence Berkeley Laboratory, put the median price of a utility-scale photovoltaic plant at $3.10/watt. These are snapshots of a moving, vaguely defined target.

290 India solar-power share: Installed/Derated Capacity of Gujarat, Gujarat Energy Transmission Corporation (31 Jul 2015), available at www.sldcguj.com (5%); Bhat 2015 (10%); Power Grid Corporation of India 2013 (35%).

290 India energy storage: Author's visit, interviews, Gujarat; Choudhury 2013; Muirhead 2014.

290 German energy-storage projects: Department of Energy Global Energy Storage Database, available at www.energystorageexchange.org.

291 Crescent Dunes: Author's visit, interviews; Crescent Dunes production and Nevada energy data from Electricity Data Browser and Nevada State Profile, both at www.eia.gov. Basin and Range critiques: www.basinandrangewatch.org. Beetles: "Endangered and Threatened Wildlife and Plants; 12-Month Finding on a Petition to List Six Sand Dune Beetles," 77 *Federal Register* 42238 (18 Jul 2012). Concentrated solar power and storage was pioneered in Spain, but these projects did not store enough power to last an entire night.

292 Prophets protest renewables: Author's interviews, Center for Biological Diversity; Montgomery 2013, Woody 2012 (Mojave); Clark 2013 (England); Griffin 2014 (Ireland); Ouellet 2016 (Canada); Hambler 2013 ("climate change"); Nienaber 2015, Hollerson 2010 (Germany).

Chapter Seven: Air: Climate Change

295 Great Oxidation Event: Author's conversations, Margulis; Schirrmeister et al. 2016; Lyons et al. 2014; Bekker et al. 2004. Some have argued that the evidence for mass death is weak, *contra* Margulis (Lane 2002, chap. 2); Margulis's version is Margulis and Sagan 1997 (1986):99–113 ("holocaust," 99). Technically, the name "Oxygenation Event" should be used. Cyanobacteria were initially vulnerable to increasing oxygen, but quickly evolved mechanisms to cope with it.

297 Fourier's life: Christianson 1999: Chap. 1 (box, 3); Fleming 1998:62–63 ("all of space," 63). Herivel 1975; Grattan-Guinness 1972: esp. 1–25, 475–90.

298 Fourier's climate papers: Fourier 1824, 1827 ("innumerable stars," 569; "polar regions," 570). Useful analyses include Pierrehumbert 2004; Fleming 1998:55–64. In the previous century scientists had proposed that heat was a substance, sometimes called phlogiston or caloric. In 1804 Benjamin Thompson, Count Rumford, disproved this notion. At about the same time William Herschel discovered that there was more to sunlight than the visible spectrum. Fourier built on these findings.

299 "Greenhouse effect" (footnote): Hay 2013:264; van der Veen 2000; Mudge 1997; Von Czerny 1881:76. French physicist Claude Pouillet compared the atmosphere to a greenhouse in 1838, but didn't actually use the word "greenhouse." Instead he talked of "diathermanous screens" of glass, attributing the comparison, incorrectly, to Fourier (Pouillet 1838).

299 Tyndall's life: Hulme 2009a; Weart 2008 (2003):3–5; Bowen 2005:81–87; Fleming 1998:65–74; Eve and Creassey 1945.

300 Discovery of the Ice Ages: Rudwick 2008: esp. chaps. 13, 34–36.

300 Tyndall's work: Tyndall 1861 ("81 per cent," 178; "absorption of 15," 276; "geologists reveal," 276–77).

301 Water vapor's big punch: I am indebted to Raymond Pierrehumbert and Rob DeConto for help in this section. See also Pierrehumbert 2011.

303 Foote (footnote): Sorenson 2011; Reed 1992:65–66; Foote 1856.

303 CO_2 measurements: Mudge 1997: Fig. 1; Crawford (1997:9) lists some prominent examples.

303 Careers of Högbom and Arrhenius: Christianson 1999:105–09; Fleming 1998:74–75; Crawford 1997; Hawkes 1940.

304 Högbom's CO_2 research: Crawford 1997:7–8; Berner 1995; Arrhenius 1896:269–73 ("very great," 271).

305 Arrhenius's year of calculation: Weart 2008:5–6; Christianson 1999:113–15; Crawford 1997 ("full year," 8); Arrhenius 1896 ("tedious calculations," 267). The U.S. scientist was Samuel P. Langley, who first described the CO_2 spectral lines (Langley 1888). Arrhenius's wife, Sophie Rudbeck, was a theosophist, an anti-smoking activist, and linguistic reformer who wrote only with a special phonetic alphabet; intent on her own work, she refused to be Arrhenius's assistant. Arrhenius, a man of his time, did not take this well.

306 Arrhenius's estimate: Arrhenius 1896 (he also published a Swedish version). I have slightly simplified his assessment. His values were not far from present estimates, mainly by luck.

306 "were granted": Crawford 1997:11.

306 Critiques of Arrhenius theory: Fleming 1998:111–12; Mudge 1997:14–15; F.W.V. and C.A. 1901; Ångström 1900 ("Arrhenius did," 731 [quotation from translation by Peter L. Ward]). Ångstrom was backed by Charles Greeley Abbot, director of the Smithsonian Astrophysical Observatory, who improved his measurements (Abbot and Fowle 1908, esp. 172–73).

307 Ice-age theories (and footnote): Weart 2008: 10–18, 44–48, 72–75, 126–28; Fleming 2007:68–69 ("on the climate"). For a brief biography of Simpson, see Gold 1965. In 1941 the prominent climatologist William Jackson Humphreys was equally scornful: "No possible increase in atmospheric carbon dioxide could materially affect either the amount of insolation reaching the surface or the amount of terrestrial radiation lost to space" (quoted in Fleming 1998:112).

307 Warmer winters: Fleming 1998:118-21. The trend was remarked upon in the popular press, but received little academic study. Among the few efforts was Kincer 1933.

307 Callendar's life: Hulme 2009b:48–53; Fleming 2007, esp. 1–32; Callendar 1939:16 ("such a change").

307 Arrhenius's approximations: Arrhenius 1896 ("as if," "is introduced," 241; "to assume," 252; "not differ," 256).

308 CO_2 and H_2O absorption: I skip some nuances for the sake of readability, hoping that doing so will not mislead the reader. For example, the image depicts the absorption of radiation in the upper atmosphere, where the gaps in the water-vapor spectrum are clearest. Closer to the surface, the spectral bands smear out,

something not understood until the early 1950s (Weart 1997:333–34). The best simple explanation I have encountered is Richter 2014 (2010): chap. 2.

308 Callendar's views, criticisms: Fleming 2007: chap. 5, 1998:114–18; Weart 1997: 324–32; Callendar 1938 ("deadly glaciers," 236; critical quotes, 237-40). Ultimately, Callendar wrote 38 papers on climate change (Fleming 2007:99–108).

311 Military boosts atmospheric science: Doel 2009:151–58; Weart 2008:54–56, 1997:332–43; von Neumann 1955 ("climatological warfare"); Kluckhohn 1947 ("dominate the globe"). For the Pentagon's special interest in polar climate change, see Sörlin 2016 (2013):40–47; Doel 2009:142–47.

312 Revelle and Suess: Fleming 1998:122–28; Weart 1997:339–47; Revelle and Suess 1957 ("in the past," 19). Similar work was performed at almost the same time by Harmon Craig and the team of James Arnold and Ernest Anderson, but Revelle and Suess, right or wrong, got most of the credit.

312 Keeling's life: Bowen 2005:110–24; Keeling 1998 ("distressed about," 27; "had been enough," 29; "backyard incinerators," 33).

312 Keeling measures CO_2: Hulme 2009b:54–56; Keeling 1998:32–46, 1978, 1960; Weart 1997:350–53. As a check, Keeling set up a second station in equally remote Antarctica.

313 CO_2 data: R. F. Keeling et al., "Scripps CO_2 Program" (available at scrippsco2.ucsd .edu).

314 "atmospheric carbon dioxide": Revelle and Suess 1957:26 (emphasis mine).

314 "alarming or steep": "A Warmer Earth Evident at Poles," NYT, 15 Feb 1959. See also Fleming 1998:118-21; Weart 1997:319–20.

314 DeConto and Pollard: DeConto and Pollard 2016. Previdi and Polvani (2016) later came to roughly similar conclusions.

315 IPCC and Antarctic ice: Previdi and Polvani 2016 (little previous evidence of melting); Church et al. 2013 (IPCC), esp. Fig. 13.27, Table 13.3.

316 Climate change as moral conundrum: Jamieson 2014 ("of morality," 156); Gardiner 2011.

316 17th-century Manhattan: Jamieson 2014:173; Sanderson 2009. Hans Jonas (1984) has argued that these paradoxes mean that we must construct an entirely new morality.

317 Weitzman: Weitzman 2007 ("paternalistic," "future utility," 707; "of the world", 712).

317 *Discount rate:* The discount rate is a composite of several parameters: the relative importance of future benefits; attitudes toward risk; uncertainty about the future; and the potential inequality between members of different generations. For the sake of simplicity, I focus on the first.

317 Chichilnisky: Chichilnisky 1996 ("an apartment," 235; "of the present," 240).

318 *Children of Men* scenario: Scheffler 2013, 2013:38–42 ("our own deaths," 75–76; "of altruism," 79). One of my sentence rephrases a sentence in the introduction by Niko Kolodny.

320 "even notice": Jamieson 2014: 111.

320 Feedback: Feedback loops were recognized by Arrhenius, but meaningful efforts to assay their impact did not occur until the 1950s and 1960s, e.g., Möller 1963.

321 Butterfly effect: Lorenz 1972.

322 Lorenz's "glitch": Gleick 1988:11–31; Lorenz 1963. See also Weart's webpage "Chaos and the Atmosphere" at www.aip.org. In the 1970s, mathematicians came across Lorenz's discovery, and it became a foundation stone for the new discipline of chaos theory.

322 Colorado conference, challenge of instability: Weart 2008 (2003):8–11, 58–61 ("by

definition," 10; "statistical average," 59); R. Revelle et al., "Atmospheric Carbon Dioxide," in United States President's Science Advisory Committee 1965:111–33 (Appendix Y4). In March 1963 the Conservation Foundation held a smaller conference, "Implications of Rising Carbon Dioxide Content of the Atmosphere."

323 Climate science conflicts with rest of academia: Allan et al. 2016; Jamieson 2014:25–28; Guillemot 2014; Hulme 2011; P. N. Edwards 2011.

324 Bryson: Peterson et al. 2008:1325–28; Weart 2008 (2003):63–79ff.; Wineke 2008; Hoopman 2007.

324 Rasool and Schneider: Rasool and Schneider 1971 ("an ice age", 138); Cohn 1971 ("five to ten"). See also Schneider 2009:17–21; 2001.

325 Warnings on global cooling: Mathews 1976 ("Climate?"); Ponte 1976; Will 1975 ("of the century"); Gwynne 1975 ("Cooling World"); Ehrlich and Ehrlich 1974:28; Colligan 1973 ("Ice Age"). See also Boidt 1970. Wrap-ups are in Peterson et al. 2008; Bray 1991. Gwynne later retracted his story (Gwynne 2014). Two weeks after quoting Rasool, the *Post* reported that a second MIT expert panel had dismissed all climate fears, warming and cooling alike (Sterling 1971; Study of Critical Environmental Problems 1970). Remarkably, Bryson's preface to Ponte (1976) says, "There is no agreement on whether the earth is cooling." See also Morton 2015:274–79.

325 CIA report: Central Intelligence Agency 1974:26–42. Annex II is labeled "Climate Theory" but devoted entirely to Bryson's ideas; it cites British meteorologist H. H. Lamb's mistaken claim that most scientists then favored cooling.

325 Cooling/warming papers: Peterson et al. 2008. At the time, according to Norwine (1977:9), "most" climatologists believed that impacts were "in the direction of surface *warming*, not cooling" (see also Norwine:13, 25–27). Still, a National Academies of Science panel hedged its bets (United States Committee for Global Atmospheric Research Program 1975:186–90).

325 Schneider redoes calculations: Schneider 2009:42–43; Kellogg and Schneider 1974 ("estimate," 1167). See also Schneider 1975: 2060.

326 Mount Agung: Peterson et al. 2008:1328–29; Hansen et al. 1978. Although some at the time were skeptical of Hansen et al.'s calculations, most regard them today as essentially correct (Self and Rampino 2012; Self et al. 1981).

326 Hansen testimony and its impact: Pielke 2011 (2010):1–3; Hulme 2009b:63–66; Weart 2008:149–50; Fleming 1998:134–35; Usher 1989; McKibben 1989; Hare 1988 ("act now," 282); Shabecoff 1988; Weisskopf 1988; United States Senate 1987–88: 2:39–80 (Hansen quotes; emphasis added, from watching video of the event). Numbers of climate-change articles from Web of Science, updating Goodall 2008: Fig. 1. Wirth's schemes: Interview, Tim Wirth, PBS Frontline, available at www.pbs.org/wgbh/pages/frontline/hotpolitics/interviews/.

328 Logical culmination: Environmental leaders agree. "In recent years, the environmental movement has morphed steadily into the climate-change movement" (McKibben 2007:42). See also Brand 2010 (2009):1.

328 Environmental consensus: Allitt 2014:67–79; Sabin 2013:44–52 ("beyond factions," 46); Sills 1975 ("radical left," 4); Soden ed. 1999: Table 5.5 (major bills).

329 New eco-threats: Oreskes and Conway 2010: chaps. 4–5 (ozone, acid rain); Robock and Toon 2010 (nuclear winter); Environmental Protection Agency 2004 (acid rain); Morrisette 1989 (ozone); Levenson 1989:214–18 (nuclear winter).

329 Dysfunctional dance: Allitt 2014 (initial environmental claims often exaggerate, 49–61; "were manageable," 12–13); Simpson 2014 (initial cost estimates often exaggerate). See also Sabin 2013; Harrington et al. 2000; Mann and Plummer 1998.

330 NRC estimate: Wagner and Weitzman 2015:50; Nierenberg et al. 2010:320–25; Schmidt and Rahmsdorf 2005; Charney et al. 1979:16. A second report in 1979 from the defense program JASON concluded that doubling CO_2 would lead to a 2–3°C increase (MacDonald et al. 1979).

330 Climate sensitivity: Freeman et al. 2015; Wagner and Weitzman 2015:12–14, 35–36, 48–56, 176n, 179–81n; Roe and Baker 2007 (showing that large uncertainty is "an inevitable consequence of a system where the net feedbacks are substantially positive," 631); Hulme 2009:46–48. To be fair, some of the uncertainty is due to our inability to predict human actions—how fast we will dump CO_2 into the air, for instance, and how much deforestatioin we will cause.

331 Bumpers and Domenici: United States Senate 1987–88:37 (Bumpers), 157–58 (Domenici).

331 Hebei: Author's visit, interviews.

332 China coal pollution: Vogmask.cn (mask); "Chinese City of Harbin Blanketed in Heavy Pollution," Agence France Presse, 21 Oct 2013; advertisement, *Beijing Times*, 24 Oct 2013 ("Taking Action!").

332 China coal use: IEA 2016; Best and Levina 2012:7; Wang and Watson 2010:3539.

333 Health costs of coal to China: Cohen et al. 2017; Chen et al. 2013 (life expectancy); Shang et al. 2013 (Fig. 5b, city mortality); Anderson 1999: Table 22 (cancer impact).

333 India pollution: Mann 2015; Lelieveld et al. 2015.

334 U.S. coal deaths: Caiazzo et al. 2013; Schneider and Banks 2010.

336 Coal and petroleum shares of emissions: IEA 2015a:xv (Figs. 6, II-2). See also Nordhaus 2013:158 (calculating a similar estimate with data from Oak Ridge Laboratory's Carbon Dioxide Information Analysis Center).

336 China and U.S. coal and petroleum use: *China Statistical Yearbook 2016*, available at www.stats.gov.cn/tjsj/ndsj/2016/indexeh.htm; Energy Information Agency Annual Energy Outlook 2017, available at www.eia.gov/outlooks/aeo/. About 3% of Chinese coal consumption is at the household level.

336 Vehicles and households: Vehicles: International Organization of Motor Vehicle Manufacturers, available at www.oica.net. About a fifth of them are in the United States. Households: Author's interviews, Population Reference Bureau.

336 Coal-plant numbers: Pers. comm., Paul Baruya, IEA Clean Coal Centre (existing plants); pers. comm, Antigoni Koufi, World Coal Association (planned coal plants); Global Coal Plant Tracker (endcoal.org/tracker/). The U.S. has 491 big coal power plants (Energy Information Agency, www.eia.gov), 130 steel plants (American Iron and Steel Institute, www.steel.org), and 107 cement factories (the Portland Cement Association, www.cement.org).

336 Not a new insight: See, e.g., Nordhaus 2013:160; Keith 2013:37.

337 Black carbon: Bond et al. 2013; Streets et al. 2013; Menon et al. 2010.

337 CCS: Overviews include Liang et al. 2016; Shakerian et al. 2015; MacDowell et al. 2010. This section is drawn from Mann 2014.

338 Parasitic costs, efficiency: Cebrucean et al. 2014:21; Wald 2013; Cormos 2012:444; IEA 2012:9 (average coal efficiency is "under 30% to 45%"); Carter 2011:5; MacDowell et al. 2010:1647; Haszeldine 2009:1648; Ansolabehere et al. 2007:ix (3 million tons/yr).

339 Indians without electricity: Mayer et al. 2015:9 (citing 2009–10 National Sample Survey); Kale 2014:178 (citing 2011 India census).

339 India coal mortality: Brauer et al. 2016 (1.3 million); Chowdhury and Dey 2016 (~5-800,000). Another study argues that the yearly fatality tally is ~80,000 people a year in Mumbai and Delhi alone (Maji et al. 2017).

339 Nuclear supporters: Brand 2010 (2009):85–89. Brand lists some Prophets who grudgingly endorse nuclear power, too.

340 Capacity factors, cost per kilowatt-hour, death rates: Hirschberg et al. 2016, Figs. 2, 10A; Brook and Bradshaw 2015: Table 1; Energy Information Administration 1990 (Tables 6.7.A, B in 2016 ed.); Energy Information Administration 1970 (Table 8.4, 2016 ed.). Burgherr and Hirschberg's mortality analysis (2014: Fig. 8A) shows new-model nukes with fewer deaths per gigawatt-year than any other power source; old-model nukes were second, behind wind-power installations (a few wind workers die by falling off the high towers). Fukushima: United Nations Scientific Committee on the Effects of Atomic Radiation 2016.

341 Land use: Brook and Bradshaw 2015: Table 1; Hernandez et al. 2014; McDonald et al. 2009 (Nature Conservancy study).

341 French nukes: http://data.worldbank.org/indicator/EN.ATM.CO2E.PC (per-capita emissions); http://www.world-nuclear.org/information-library/country-profiles/countries-a-f/france.aspx (exports, electricity prices); http://www.world-nuclear.org/information-library/facts-and-figures/nuclear-generation-by-country.aspx#. UkrawYakrOM (nuclear share of electricity); Brand 2010:111.

341 Wizards' critique of renewables: A fine example is Frank 2014.

342 CCS projects: Global CCS Institute (www.globalccsinstitute.com); Willberg 2017 (Saskatchewan); Joint Statement by G8 Energy Ministers, Aomori, Japan, 8 Jun 2008, http://www.g8.utoronto.ca/energy (quotes). Several CCS coal projects have failed, notably a Mississippi plant that in 2017 gave up CCS and switched to natural gas after spending $7.5 billion.

342 Mountaintop removal: Epstein et al. 2011; Palmer et al. 2010.

342 Coal-mine fires: Author's visit, Jharia; Stracher et al. 2011–15.

343 Georgia, South Carolina nukes: Plumer 2017. The Watts Bar Unit 2 plant in Tennessee, switched on in 2016, cost just $4.5 billion, but most of the plant was built in the 1970s and mothballed. Construction resumed in 2007.

343 Amount of high-level waste: International Atomic Energy Agency 2008: Table 5. I have extrapolated from these 2005 figures, using their estimate of 12,000 tons/year (14). The number of operating reactors worldwide has not changed dramatically in the intervening years. For the number of nukes, see the World Nuclear Association (www.world-nuclear.org). In all cases I have converted metric to imperial units.

344 Factor of a million: By far the most important components of high-level waste are strontium (^{90}Sr) and cesium (^{137}Cs), both with half-lives of about 30 years. A physicist's rule of thumb is that every 20 half-lives corresponds roughly to a million-fold reduction in radioactivity. ^{90}Sr and ^{137}Cs would hit that level after 600 years. I thank Alan Schwartz for reminding me of this rule of thumb.

344 Jacobson-Delucchi road maps: Jacobson et al. 2015a, b (list of projects, 2114–15); Jacobson and Delucchi 2011a, b, 2009. Criticisms are gathered in Clack et al. 2017.

346 Smil's critiques: Smil, pers. comm.; Smil 2011a; Smil 2008:380–88.

347 Sweden: Pierrehumbert 2016.

347 Mount Katrina: Author's visits, interviews (especially Dane Summerville, Army Corps of Engineers); Mann 2006.

349 Coastal city studies: Hallegatte et al. 2013 (population, supp. inf.); Joshi et al. 2015 ($2.9 trillion); Hinkel et al. 2014 (9.3%GDP); Jongman et al. 2012 (1 billion people). There are many other studies, almost all with results in this line.

349 Big cultural losses: Nordhaus 2013:108–13; Coletta et al. 2007.

349 Shanghai: Author's visit; Fuchs 2010:3-4; Xu et al. 2009.

349 Protecting Chicago: Adelmann 1998; Cain 1972.

350 Venice protection, population: Ross 2015; Magistro 2015; details at www.mosev-enezia.eu.

350 Asian coastal flood risks: Fuchs 2010 (second report, 3).

350 Pinatubo: Morton 2015, chap. 3 ("the Sahara," 85); Hansen et al. 1992; Newhall and Punongbayan 1997. This was not the Philippines eruption used earlier by Hansen and his colleagues to study global cooling, but another one.

351 Warming since 1880: GISS Surface Temperature Analysis, NASA Goddard Institute for Space Studies (data.giss.nasa.gov/gistemp/); Clark 1982:467, updated at ESS-DIVE (lbl.gov).

352 Droplet size: Morton, pers. comm.; Morton 2015:85 ("the Sahara"); Keith 2013:88–94.

353 Regional airlines: www.ryanair.com (Facts and Figures); www.alaskaair.com (Company Facts); U.S. 76 FR 31451 (Special Conditions: Boeing Model 747-8 Airplanes).

353 Cost and methods of geoengineering: Keith 2013:94–116 ("sea level rise," 100); McClellan et al. 2012, 2011.

353 Taking the edge off: Caldeira and Wood 2008; Wigley 2006.

353 Coining of "geoengineering": Marchetti 1977.

353 Technical fix: Strictly speaking, geoengineering *isn't* a technical fix, because it doesn't fix the climate, just veils the symptoms (Pielke 2011 (2010):234–35). I use the term anyway, because it is seen as a cheap technological approach to a complex problem.

354 Frauds: Goodell 2011 (2010):53–69; Fleming 2010, chap. 3 (Hatfield quotes, 90–91).

354 Dyrenforth: Fleming 2010:53–74; Hoffman 1896; letter, Dyrenforth to Sec. of Ag., 19 Feb 1892, in U.S. Senate, Executive Documents, 1st Sess., 52nd Cong., v.5, Doc. No. 45. See also Le Maout 1902.

354 Early climate geoengineering: Goodell 2011 (2010):75–87 ("suit us," 77); Fleming 2010:194–200, 212–40; Keith 2000:250–51; Weart 1997; R. Revelle, "Atmospheric Carbon Dioxide," in United States President's Science Advisory Committee 1965:111–33 (Appendix Y4); Teller 1960:280–81.

355 Qualms at geoengineering: Keith 2013: esp. chap. 5; Crutzen 2006 ("pious wish," 217); Wagner and Weitzman 2012 (chemotherapy); Kintisch 2010:13 (time has come). Morton (2015) is especially good on this subject.

355 Geoengineering side effects: Keith et al. 2016 (particles); Morton 2015: esp. 107–23; McCusker et al. 2014 (risks of stopping); Kravitz et al. 2014 (rainfall); Curry et al. 2014 (temperature extremes); Keith 2013:68–72; Pielke 2011 (2010):125–32; Robock et al. 2009; Robock 2008 (a brief, comprehensive negative brief).

356 Rogue geoengineering: Wagner and Weitzman 2015:38–39, 116–27; Keith 2013:111–13, 152–56 ("weapons states," 115); Victor 2008 ("on his own," 324). The *Forbes* billionaire list is published annually at www.forbes.com.

357 Planting eucalyptus or jatropha: Heimann 2014; Becker et al. 2013; Ornstein et al. 2009.

358 Political feasibility: Becker and Lawrence 2014 ("local populations," 32).

358 Sahel drought: Joint Institute for the Study of the Atmosphere and Ocean, 2005—"Sahel Precipitation Index (20–10N, 20W–10E), 1900–November 2016." Available at jisao.washington.edu/data_sets/sahel/; Hulme 2001; Mellor and Gavian 1987:235 (100,000 deaths). This is a conservative estimate. Winslow et al. (2004:5) estimated a death tally for the first famine wave alone of 200,000.

358 Burkina Faso: Author's visit, interviews, Chris Reij, Mathieu Ouédraogo, Aly Ouédraogo; Reij et al. 2005; Kabore and Reij 2004.

359 Sawadogo: Author's visits, interviews, Sawadogo, Ouédraogo, Reij; Fatondji et al. 2001.

360 Jatropha calculation: Becker et al. 2013 (carbon estimates, 241). Per-capita emissions from World Bank (data.worldbank.org).

361 Global greening: Zhu et al. 2016 and references therein.

361 Reforesting the Sahel: Author's visits, Niger, Burkina Faso, Mali; interviews, Reij, Ouédraogo, Larwanou, Edwige Botoni; Reij 2014; Mann 2008; Nicholson et al. 1998. In East Africa, Ethiopia has permanently reforested hundreds of square miles of formerly barren land (Reij, pers. comm.).

361 Rice sterility (footnote): Jagadish et al. 2015.

361 Carbon farm sustainability: Bowring et al. 2014; Becker et al. 2014, 2013; Ornstein et al. 2009.

362 Charcoal and climate change: Mao et al. 2012; Woolf et al. 2010 (1/8th); Mann 2008, 2006:344–49; Lehmann 2007; Lehmann et al. 2006; Okimori et al. 2003. Several sentences in this section are reworked from Mann 2006.

362 Too many or too few people: Based on Wagner and Weitzman 2015: chap. 5.

Chapter Eight: The Prophet

365 Washington meeting: Memorandum, Informal Summary of Minutes of Meeting Held at the Request of Dr. Julian Huxley in the Board Room of the National Academy of Sciences, Washington, D.C., at 10.00 a.m., December 23, 1947, Box 2, FF1, VDPL.

365 Huxley and his family: Clark 1968. A eugenicist but (eventually) an antiracist, Huxley wanted "to ensure that mental defectives shall not have children" (1993 [1931]:98) even as he insisted that race was "a pseudo-scientific rather than scientific term"—it had no biological reality (1939 [1931]:216). Under his leadership UNESCO committed itself to combating racism, though he continued to hope that humankind would waken to the need to purge itself of "unfit" stock.

366 Science-driven growth for all: Macekura 2015:17–30; Rist 2009 (1997): 69–79; Collins 2000:1–32; Public Law 79-304 ("purchasing power," Sec. 2).

366 Huxley's fears: Macekura 2015:32–35; Deese 2015:150–54; Bashford 2014:273–78.

366 Huxleys' failures to make themselves heard: Deese 2015:155–56; Toye and Toye 2010:326–28.

368 *Harper's* article: Vogt 1948a ("hundred years," 481; "million acres," 484; "exploiters," 486).

369 Furor over *Road*: Sauvy 1972 ("*Law of Population*," 968 [he reviewed *Road* in the same journal in 1949]); Memorandum, The Editorial Program, n.d. (1948), unsorted papers, William Sloane papers, Princeton University Archives ("the year"); letter, Vogt to G. Murphy, 29 Jun 1948, Box 2, FF4, VDPL ("'Unclean'").

369 Scientists' support for Vogt and Osborn: Bashford 2014:278–80; Robertson 2012:59; Nichols 1948 ("problem today"); Hutchinson 1948:396 ("real enough"). Ecological Society of America president Paul Sears called *Road* "the most convincing account of man's material plight that has yet appeared" (Sears 1948).

370 AAAS symposium: Jundt 2014a:17–26; Bliven 1948 ("Frightened People"); Department of State [1949]; [Associated Press?], "What Hope for Man?" *Fitchburg (MA) Sentinel*, 17 Sept 1948. At about the same time, the British Association for the Advancement of Sciences, the second-most influential scientific body, held an equally worried meeting on the same theme, addressed by U.N. Food and Agriculture head John Boyd Orr (Connelly 2008:131–33).

370 Inter-American Conference: United States Department of State [1949] ("our century," 1). Cushman (2006:348) notes the absence of birth control.

370 Foundation of UNESCO, choice of Huxley: Maurel 2010:16–28; Toye and Toye 2010:322–30.

371 UNESCO and "nature protection": Holdgate 2013:30–36; Mahrane et al 2012:130–33; UNESCO 1949:9–14; Coolidge 1948; Informal Summary of Minutes of Meeting Held at the Request of Dr. Julian Huxley in the Board Room of the National Academy of Sciences, Washington, D.C., at 10.00 a. m., December 23, 1947, Box 2, FF1, VDPL; Huxley 1946:45. To grab the issue for UNESCO, Huxley maneuvered past Boyd Orr at the U.N. Food and Agriculture Organization, who wanted conservation for his agency.

371 Muir-Pinchot clash: Bergandi and Blandin 2012:109–16; Miller 2001; Smith 1998; Shabecoff 1993:64–76; Fox 1981; Nash 1973 (1967):123–40, 162–81 (Yosemite as first wilderness park, 132).

372 Muir quotations: Gifford 1996:301 ("the wilderness"); Muir 1901:1 ("of life").

372 Pinchot: Miller 2001 (leaves before ready, 88; "longest run," 155; efforts to set up conference, 372–75, 441–42n); Pinchot 1909:72–73 (other quotes); 1905:2 ("wise use").

373 Dispossession at Yosemite and Yellowstone: Powell 2016:58–59, 76; Dowie 2009:4–11; Nabokov and Loendorf 2002, esp. 53–56, 87–92, 179–92, 227–36. For a general survey of indigenous environmental modification, see the sources cited in Mann 2005: Chaps. 8–9.

374 Steps to U.N. conference: Jundt 2014b:44–48; Mahrane et al. 2012:4–7; Robertson 2009:33–36; Linnér 2003:32–35; Miller 2001:359–64; McCormick 1991:25–27; Nixon 1957: 2:1153, 1154, 1163–66, 1170–72; United Nations 1950:vii (Truman letter), 1947:491–92, 1947:469, 491–92 (conference announcement).

374 Huxley's plans: Holdgate 2013:18–28 (preliminary meetings), 39 ("and so on"); Wöbse 2011: 338–40; Informal Summary of Minutes of Meeting Held at the Request of Dr. Julian Huxley in the Board Room of the National Academy of Sciences, Washington, D.C., at 10.00 a.m., December 23, 1947, Box 2, FF1, VDPL.

376 Founding IUPN: Holdgate 2013 (1999):29–38; Wöbse 2011:340–41; Mence 1981:1–9; [Bernard?] 1948; Coolidge 1948; "green blob": Paterson 2014.

376 Vogt at Fontainebleau: McCormick 2005:179–83; Union Internationale pour la Protection de la Nature 1950 ("human ecology," 28; "first victims," 31).

377 Huxley's manifesto: Huxley 1946 ("scaffolding," 8; "evolutionary progress," 12; "world political unity," 13). Huxley never directly mentioned birth control or abortion but his support for both is clear (45).

378 Huxley's continued conviction: See, e.g., J. Huxley, "What Are People For? Population Versus People," address to Planned Parenthood, 19 Nov 1959, PPFA1, Box 14, FF13.

378 Huxley inspired by five-year plans (footnote): Deese 2015:71–73.

378 New York forum: New York Herald Tribune Forum 1948:11–46 (Osborn-Vogt session); Associated Press, "Unity-for-Peace Plea Is Renewed by Dewey," 21 Oct 1948; Passenger Lists of Vessels Arriving at New York, 1820–1897, Microfilm Publication M237, Roll 7666, p. 75, U.S. Customs Service Records, RG 36, U.S. National Archives (ancestry.com).

379 Point Four speech: Text from Truman Library (www.trumanlibrary.org).

380 Point Four as surprise: Macekura 2015:26–32; Jundt 2014b: ("also strategic," 47); Cleveland 2002:117–18 ("'President meant?'"); Perkins 1997:144–51; U.S. Department of State 1976, 1:757–88 (Acheson not consulted, 758n); Vogt 1949:17 ("'it cost?'"); "Blueprints Drawn to Effect Point 4," NYT, 6 May 1949.

381 Point Four critiques: Robertson 2009:41–42 (Cornelia Pinchot); Vogt 1949 (other quotes).

382 Conflict with PAU: Letter, Lleras to Vogt, 21 July 1949, Box 2, FF1, VDPL.

382 UNSCCUR: United Nations 1950 ("of living," 7; "to confusion," 8; "enlightened," 15); Levorsen 1950 ("world production," 94); Hamilton 1949; Teltsch 1949; McGrory 1948 (Krug reads Vogt and Osborn). My discussion follows Jundt 2014b:48–52; one of my sentences is a rewritten version of one of his.

385 ITCPN: Jundt 2014b:58–67 ("in motion," 44; "double agent," 53); Holdgate 2013:41–43; Wöbse 2011:341–47; Beeman 1995 (Friends of the Land); Union Internationale pour la Protection de la Nature 1950 (Osborn quotes, 17–19; Fink quotes, 215–16), 1949:68–69, 84–85 ("the economy?"); "Talks on Nature Slated," NYT, 21 Aug 1949; "Deer in North America Starve, Wildlife Parley Is Told," Evening Star (Washington, D.C.), 9 March 1949.

387 Vogt resigns: Anonymous 1949; "Conservationist to Speak," Evening Star (Washington, D.C.), 23 Oct 1949; letter, Vogt to Lleras, 17 Oct 1949, Box 2, FF1, VDPL; letter, Vogt to H.A. Moe, 12 Mar 1950, GFA.

387 Moore: Bashford 2014:268–69 ("the Earth"); Critchfield 16–17, 30–33; Mosher 2008:36–40 ("CONFLAGRATION!," 37); Fowler 1972; "The History of Dixie and the Dixie Cup," James River promotional brochure (James River now owns the Dixie Cup company).

389 Guggenheim and Fulbright: Letters, Moe to Vogt, 21 Dec 1950, 28 Mar 1950, GFA; Vogt, applications for 1938, 1939, 1940, and 1943 Guggenheim fellowships, GFA; ; Memorandum, Fulbright Awards for the Academic Years 1950–51: American Citizens, Fulbright Archives (libraries.uark.edu/SpecialCollections/FulbrightDirectories/).

389 Scandinavian fertility laws: Connelly 2008:67, 103–4.

389 Vogt in Scandinavia: Journal, Marjorie Vogt, Box 6, FF28, VDPL.

390 Sanger: Good biographies include Baker 2011; Chesler 1992. See also Reed 1983 (1978): Part 2.

391 Vogt at PPFA: Minutes, Annual Membership Meeting, 7 May 1952, Box 14, FF12; W. Vogt, "Report of the National Director," 6 Mar 1952, Box 23, FF6; Minutes, Annual Membership Meeting, 23 Oct 1951, Box 14, FF11 ("better qualified"); Release, "World Population Authority Named Director of Planned Parenthood," 18 May 1951. Box 23, FF26; Minutes, Executive Committee Meeting, 15 May 1951, Box 23, FF5; letter, R. L. Dickinson to M. Sanger, 26 Nov 1948. Box 70, FF4, all at PPFA1; letter, L. Campbell to W. Vogt, 26 Feb 1949, Box 1, FF13, VDPL. For examples of Sanger's praise, see Sanger 1950, 1949.

393 McCormick, Vogt, and the pill: Baker 2011:290–94; Chesler 1992:407–12, 430–34; Lewis 1991:107 (Bard); Reed 1983 (1978):335–45; letter, M. Sanger to K. McCormick, 23 Feb 1954; letter, M. Sanger to M. Ingersoll, 18 Feb 1954 ("his mind"); letter, K. McCormick to Sanger, 17 Feb 1954 ("mystifying"), all at PPFA1.

393 Poor testing of pill (footnote): Liao and Dollin 2012; Leridon 2006.

393 Vogt fired: McCormick 2005:198–202; letter, Vogt to Moe, 13 Jun 1961, GFA.

393 Johanna: "Miss von Goeckingk Wed," NYT, 27 Dec 1959; 1929 Radcliffe Prism; "Prayer Service to Open at Radcliffe," Boston Herald, 29 Sep 1929; Enumeration District 7-155, Holyoke, MA, 1930 U.S. Census, entry for Marie von Goeckingk; Enumeration District 573, Ward 22, Kings County, NY, 1910 U.S. Census, entry for Leopold von Goeckingk. The marriage may have been in trouble for a while; the couple was taking separate vacations by 1955 (letter, Vogt to Moe, 16? Sep 1955, GFA). Summer house: letter, Vogt to Moe, 31 Jul 1962, GFA.

394 Vogt's work at Conservation Foundation: Lewis 1991:109–16 ("including man," 113); United States Senate 1966:717–27 ("human habitat," 725); [Vogt et al.] 1965; Vogt 1965; letter, Vogt to G. Heiner, 23 Oct 1964, Box 1, FF31, VDPL; letter, S.

Ordway to Vogt, 31 Jan 1964, Box 2, FF6, VDPL; Vogt 1963 ("progress," 13; "Latin America," 16). The unpublished essays are in Box 4, FF14–17, VDPL.

395 Denunciations of aid: Vogt 1965; Vogt 1966.

397 Vogt's last days, death: Duffy 1989; "William Vogt, Former Director of Planned Parenthood, Is Dead," NYT, 12 Jul 1968; "Bobby's Brood Gives Wrong Image for Victory," Associated Press, 9 May 1968; letter, Vogt to B. Commoner, 18 May 1967, Box 1, FF17, VDPL ("being accelerated"); Obituary notice, Vogt—Johanna von Goeckingk, NYT, 29 Jan 1967.

397 *Population Bomb:* Author's interview, Ehrlich; Sabin 2013:10–49; Cushman 2013:272 (Vogt's influence); Robertson 2012a:126–51; Ehrlich 2008, 1968; Tierney 1990; Goodell 1975:13–21; Webster 1969; Rosenfeld 1968. Ehrlich's *Tonight Show* appearances from Wikipedia and IMDb.com. The oft-repeated claim that Ehrlich was on 20 times or more appears to be incorrect.

398 Vogtian warnings: Hardin 1976 ("carrying capacity," 134); Ehrlich and Holdren 1969:1065 ("population growth"); Platt 1969:116 ("this century").

399 *Limits to Growth:* Author's interviews, D. Meadows; Meadows et al. 1972 ("*by collapse,*" 142; "stop *soon,*" 153); sales/translations from Club of Rome (the sponsors), clubofrome.org.

399 Public-private network in population control: Connelly (2008) provides a remarkable portrait of the institutions in action.

399 Critiques of population control: Connelly 2008; Hartmann 1995.

400 Runaway population control programs: Jiang et al. 2016 (abortions); Greenhalgh 2008, esp. chaps. 4, 6 (rise of one-child policy); Connelly 2008, esp. 289–326 ("population problem," 323); Song 1985 (influence of Western computer modelers, 2–3). An additional issue is that many Asian families, wanting a male child and only allowed one, aborted girls who would otherwise have been wanted. Estimates of the "missing" girls in China reach as high as 10 million (Ebenstein 2010: Table 2). I thank Betsy Hartmann for many discussions of these topics.

401 "*too many people*": Ehrlich 1968:66–67.

402 Ehrlich in Delhi: Ehrlich 1968:15–16 ("of overpopulation"), 84. Vogt made the same argument in his Senate testimony, using New York's then-polluted water and Los Angeles's then-polluted air as examples. "Diminish the population of either city sufficiently and the problems would largely vanish" (United States Senate 1966:720). Ehrlich's Delhi story was attacked as racist, a charge that pained him (author's interview).

402 Delhi, Paris, Tokyo population growth: Author's interview, Narain; United Nations Department of Economic and Social Affairs 2006: Table A.11.

403 Hudson Valley, Europe forests: Forest Europe 2015 (current Europe forests); U.S. Census Bureau, Annual Estimates of the Resident Population: April 1, 2010 to July 1, 2012 (available at www.census.gov); Canham 1999 (historic NY forests); Kauppi et al. 1992 (1970-1990 Europe forests); Considine and Frieswyk 1982: Table 87; Seaton 1877 (population).

404 Replenishment of nature: Deer, turkey (Sterba 2012:87–89, 104–05, 150–60); Thames (see annual survey from Zoological Society of London, sites.zsl.org/inthe thames); Japan (United Nations World Health Organization 2016:Annexes 1, 2).

405 Delhi/Denmark comparison: Wind-power data from energinet.dk; Delhi farm fires from worldview.earthdata.nasa.gov (1 Nov 2016, "fires and thermal anomalies" overlay); per-capita Delhi income (Rs212,219) from *Economic Survey of Delhi 2014–15* (delhi.gov.in); per-capita Copenhagen income (DKK 322,000) from state website (denmark.dk); Copenhagen climate plan from the city website, www.kk.dk; *WWF Living Planet Report* 2014 (wwf.panda.org).

Chapter Nine: The Wizard

407 "In principle multiples": Merton 1961:477.

408 Examples of multiples: Skousen ed. 2007, 2:173 ("of the world"); Browne 2002: 14–33 (Darwin, Wallace); Crease and Mann 1996:140–44 (Stueckelberg); Thompson 1910: 1:44–45 (quotes), 113.

408 Swaminathan: Biographies include Dil 2004; Iyer 2002; Gopalkrishnan 2002; Erdélyi 2002. Many of his writings are collected in Rao 2015.

410 Bengal famine: Ó Gráda 2015:38–91 ("the market," 49; "the war," 92). Refining a classic analysis by Amartya Sen, Ó Gráda concludes that the harvest shortfall "would have been manageable in peacetime . . . The famine was the product of the wartime priorities of the ruling colonial elite" (91).

410 160 agriculture students: Saha 2012:xxii.

411 Indian Agricultural Research Institute campus: Author's visit, IARI.

413 Nehru and science: Singh 2014; Government of India 1958; see also Nehru 1994 (1946), esp. 31–33. The decree is in Part IVA, 51A(h) of the Indian constitution.

414 Nehru's industrialization plans and agriculture: Cullather 2010:135–52ff, 198–200; Varshney 1998:25–47ff (land ownership, 29); Perkins 1997:161–75ff.

414 Subsidy program: The program, authorized by the Agricultural Trade Development and Assistance Act of 1954—or Public Law 480—was a compromise that allowed U.S. farm states to keep subsidy-driven wheat, maize, and rice production high while disposing of the excess in Asia (author's interviews, James Boyce; Cullather 2010:142–43).

415 Swaminathan at IARI: Swaminathan 2015:1–2; 2010a:2–3; Dil 2004: Appendix IX (list of papers).

415 Early wheat in Indus Valley: Fuller et al. 2007.

416 Initial fertilizer and radiation experiments: Author's interviews, Swaminathan, P. C. Kesavan; Chopra 2005; Pal et al. 1958; Swaminathan and Natarajan 1956.

416 Swaminathan, Kihara, and Vogel: Author's interviews, Swaminathan; Swaminathan 2015:2–3; 2010a:3; Crow 1994.

417 JFK and Sino-Indian War: Reidel 2015: chap. 4 ("of guidance," 119; "this subcontinent," 138).

417 Nehru's weakness: Cullather 2010:196–97; Brown 1999:160–64.

418 Foundation ends Mexico program: Rockefeller Foundation 1916–, *Annual Report*, 1959:30.

418 Borlaug, bananas, FAO: VIET3:19–25; Borlaug 1994:iv; Bickel 1974:225–28.

418 Borlaug India report: RFOI 206; Cullather 2010:192 (all quotes). In interviews, Swaminathan said that he hadn't formed any special impression of Borlaug.

418 Training program: VIET3:24–27, 35–37; Borlaug 1994:v–vi; Bickel 1974:233–36.

419 Seven districts: Cullather 2010:195. A group of Indian and U.S. academics argued in an influential Ford Foundation report in 1959 that hunger would be a bottleneck on Indian development. In the long run, they warned, importing low-priced U.S. wheat would be counterproductive. By effectively setting a ceiling on domestic prices, it would discourage domestic production (Government of India 1959).

419 Borlaug-Swaminathan trip: Author's interview, Swaminathan ("child-like"); Swaminathan 2010b:4–5; VIET3:67–76; Cullather 2010:198–99; Bickel 1974:244–46 ("ever experienced").

420 Fertilizer struggle: Saha 2013; Cullather 2010:198–201 ("raw materials," "dam projects," 199); N. E. Borlaug, "Indian Wheat Research Designed to Increase Wheat Production," typescript, CIMBPC, 11 Apr 1964 (B5634-R) ("chemical fertilizers,"

2). Increasing fertilizer imports would mean decreasing jute imports. And India's finance minister, a powerful figure in the government, as Cullather put it, "guarded the jute allotment like a mastiff" (200).

421 Confrontation in Pakistan: VIET3:76–81 (all quotes).

423 1964-5 tests: Swaminathan 1965; Swaminathan 2010b:4–5, 1965; Perkins 1997:236.

423 Shastri, Subramaniam, Bhoothalingam: Author's interviews, Swaminathan; Saha 2013, esp. 302–5; Bhoothalingam 1993: 108.

425 Sending grain to India and Pakistan: LHNB ("Watts riot"); VIET3:112–18; Bickel 1974:272–79; Paarlberg 1970:15.

428 India-Pakistan war: VIET3:119–20 ("MY BACKYARD," 119).

430 Methyl bromide disaster: VIET3:130–31.

432 Bihar famine: Rubin 2009:703-06; Dréze 1995:48-63, appendixes; Dyson and Maharatna 1992; Brass 1986; Berg 1971; Ramalingaswami et al. 1971; Gandhi 1966 ("of homes," 63).

434 Borlaug, Subramaniam, Mehta: Author's interviews, Swaminathan; VIET 2:167-169.

435 Parents' home: Author's interview, Mark Johnson (Borlaug Foundation); Bickel 1974:346-47.

435 Boyce's story: Author's interview, Boyce; Hartmann and Boyce 2013.

437 Criticisms of Green Revolution: Cockburn 2007 ("by the million"); Freebairn 1995 (4 out of 5); Shiva 1991 ("discontented farmers," 12); Pearse 1980; Griffin 1974 ("that failed," xi), 1972; Hewitt de Alcántara 1978; Dasgupta 1977 ("fewer hands," 372); Feder 1976 ("Third World peasantry," 532); Bickel 1974 (boos, 350–51); Reinton 1973 ("produce misery," 58); Byres 1972; George 1986 (1976) ("to the poor," 17); Cleaver 1972; Palmer 1972 (the Green Revolution has turned "parts of the Near East" into "genetic disaster areas," 95); Frankel 197. UNRISD books listed at unrisd.org.

438 Borlaug's response: Author's interviews, Borlaug; VIET3:107–08.

438 Productivity gains: United Nations Food and Agriculture Organization 2004; Evenson and Gollin 2003.

439 Success stories: Author's visits, Pune region; Bourne, pers. comm.; Bourne 2015:78-81; Damodaran 2016.

440 Pakistan meeting: VIET3:90–91.

442 Sharbati Sonora: Author's interviews, P. C. Kesavan, Swaminathan; Austin and Ram 1971; Varughese and Swaminathan 1967. Details of other types of adaptation in M.S. Swaminathan, "Can We Face a Widespread Drought Again without Food Imports," Address to Indian Society of Agricultural Statistics, 1972. Typescript, M.S. Swaminathan Foundation archives. The color change and other adaptations were overshadowed by Swaminathan's announcement that the mutated grain had, compared to ordinary grain, high levels both of protein and the essential amino acid lysine, and thus was more nutritious. The claim seems to have stemmed from erroneous laboratory tests. In any case, Swaminathan reported the increase overenthusiastically (e.g., Swaminathan 1969:73). After the lysine claim was disproved, he was charged with spreading false data. Borlaug emphatically disputed the charge and subsequent investigations found no basis for it (Saha 2013:309–10; Parthasarathi 2007: 235–40; Borlaug and Anderson 1975; Hanlon 1974). Wheat bran color: Metzger and Silbaugh 1970.

443 Evergreen revolution: Swaminathan 2010a, 2006 ("prudence," 2293), 2000, 1996 (esp. 232); M. S. Swaminathan, "The Age of Algeny, Genetic Destruction of Yield Barriers and Agricultural Transformation," Address to 55th Indian Sci-

ence Congress, 1968, typescript, M. S. Swaminathan Foundation archives (1968 worries).

444 Workshop and the world: Crease forthcoming; Husserl 1970.

Chapter Ten: The Edge of the Petri Dish

449 Wilberforce: Biographies include Meacham 1970; Wilberforce 1888; Ashwell and Wilberforce 1880–82. For his reputation, see the obituary tribute by Prime Minister William Gladstone (Ashwell and Wilberforce 1880–82, 3:450–51).

449 "the Bishop's scalp": Case and Stiers 1971:297; Case 1975:90.

449 Victory for science: See, e.g., Hitchens 2005 ("tipping point"); Brooke 2001 ("one of the great stories of the history of science," 127); Glick 1988 ("a key chapter in the mythology of English science," xvi); Lucas 1979. Smith (2013) describes the common portrayal of the debate as "the day when . . . science threw off the shackles of religious authority." Gauld found sixty-three accounts of the debate (1992a:151). Their "purpose," he said, was to celebrate "the triumph of Darwinism over uninformed religious prejudice" (1992b:406). This portrayal of the debate dates back at least to White 1896 (Huxley's sally "reverberated through England" [1:71] and "secured [Wilberforce] a fame more lasting than enviable" [2:342]). Other accounts of the debate include Hesketh 2009; Depew 2010:338–43; Browne 2006:95–97, 2002:153–70; Brooke 2001; Thomson 2000; Jensen 1991:68–86, 1988; Gilley 1981; Altholz 1980; Meacham 1970:212–17; Wilberforce 1888:247–48; Ashwell and Wilberforce 1880–82, 2:450–51; Anonymous 1860a:18–19, 1860b:64–65; letter, J. D. Hooker to C. Darwin, 2 Jul 1860, CCD 8:270; letter, C. Darwin to T. H. Huxley, 3 Jul 1860, CCD 8:277; letter, J. R. Green to W. B. Dawkins, 3 Jul 1860, in Leslie ed. 1901:43–46; letter, C. Darwin to T. H. Huxley, 5 Jul 1860, CCD 8:280; letter, T. H. Huxley to F. Dyster, 9 Sept 1860, Foskett 1953.

449 Not a single newspaper: Ellegard 1958:380; Jensen 1988:170–71.

450 *Origin* creates uproar: Browne 2002: chap. 3.

450 "into mine hands": 1 Samuel 23:7 ("And Saul said, God hath delivered him into mine hand" [King James Bible]) It seems worth noting that the first record of Huxley saying this is in his son's biography, published forty years after the event. An eyewitness wrote that Huxley was "white with anger" (Tuckwell 1900:52).

451 Story of Huxley-Wilberforce: The common version is based on accounts by Huxley's son Leonard (1901: 2:192–204 ["monkey," 197]) and Darwin's son Francis (1893: 1:251–53 ["falsehood," 252]).

451 Darwin and Wilberforce background: The standard Darwin biography is Browne 1995, 2002; see also Browne 2006, F. Darwin 1887. If anything, Wilberforce and his family were even more attached to science than Darwin and his family. Not only Bishop Wilberforce, but two of his three brothers took first-class mathematics degrees (Ashwell and Wilberforce 1880–82, 1:32). By contrast, Darwin found mathematics "repugnant" and avoided it as a student (F. Darwin 1887: 1:46). Their elections to the Royal Society appear in the lists of the Fellows published in the *Philosophical Transactions of the Royal Society of London*.

451 Darwin and Wilberforce reactions to loss: Hesketh 2009:43–46. Darwin: Keynes 2002 (2001); Browne 1995:498–504. Wilberforce: Ashwell and Wilberforce 1880–82:1:50, 177–92 (quotes from 180–81).

451 Wilberforce's essay: [Wilberforce] 1860. The review was anonymous, but Darwin, Huxley, and many others knew its authorship.

451 "all difficulties": Letter, C. Darwin to J. D. Hooker, 20 Jul 1860, CCD 8:293. Darwin went back and forth on the strength of the critique (Letter, C. Darwin to C. Lyell, 11 Aug 1860, CCD 8:319 ["the Bishop makes a very telling case against me by accumulating several instances, where I speak very doubtfully"]; letter, C. Darwin to T. H. Huxley, 20 Jul 1860, CCD 8:294; letter, C. Darwin to A. Gray, 22 Jul 1860, CCD 8:298; letter, C. Darwin to C. Lyell, 30 Jul 1860, CCD 8:306). By contrast, Huxley always sneered at the "foolish and unmannerly" review, which "eked out lack of reason by superfluity of railing" (Huxley 1887:183–84).

452 Incompleteness of fossil record: [Wilberforce] 1860 ("their theory," 239); Darwin 1859: chap. 9.

452 Natural selection: Huxley 1887 ("thought of that!" 197); Darwin 1859: chap. 4 ("their kind," 81, "surviving," 61); Darwin and Wallace 1858.

452 "tangled bank": Darwin 1872:429. Darwin may have been inspired by the Orchis Bank, which he often walked past on his morning stroll (Keynes 2002:251). The first edition of *Origin* (1859:489) called it an "*en*tangled bank" (emphasis mine).

453 Avoided mention: Two pages from the end, Darwin did write, obliquely, that in "the distant future . . . light will be thrown on the origin of man and his history" (Darwin 1859:488). His decision to avoid discussing humankind was part of his view "that direct arguments against christianity & theism produce hardly any effect on the public" (letter, C. Darwin to E. B. Aveling, 13 Oct 1880, in Feuer 1975:2). Darwin added that he wanted to avoid upsetting his family.

453 Wilberforce's objection to downgrading human status: [Wilberforce] 1860:256–64 ("condition of man," 257; "mushrooms," 231). See also Cohen 1985:598, 607n22; Meacham 1970:213–14.

453 Non-revolutionary Copernican revolution: Dick Teresi, pers. comm. According to Teresi, "the earth being special has long been misinterpreted." In the Christian conception of the day, the Earth was a fallen place. It was at the center of the cosmos, but not admirable. "It's special in the sense of 'isn't that *special*?'" Teresi explained. In the eighth century B.C., thinkers in northern India had put the sun at the center of the cosmos; 500 years later, so did the Greek astronomer Aristarchus of Samos. By A.D. 1000, the Maya had a heliocentric system. Nonetheless, Copernicus's rigorous methods were an advance.

453 Second Copernican Revolution: Lewis 2009. I thank Oliver Morton for drawing this to my attention and Simon Lewis for kindly allowing me to lift his idea.

454 "or *above* her": Marsh 1864:549 (emphasis added).

454 "at Oxford": Letter, Darwin to T. H. Huxley, 3 Jul 1860, CCD 8:277.

454 Reactions to debate: Jensen 1988:171–73; Lucas 1979:323–25; Altholz 1980:315 ("beat him"); letter, J. R. Green to W. B. Dawkins, 3 Jul 1860, in Leslie 1901:43–46 ("cheering lustily"); letter, T. H. Huxley to F. Dyster, 9 Sept 1860. In Foskett 1953 ("hours afterward"). Cohen (1985:597–98) points out that the debate convinced the ornithologist Henry Baker Tristam, the first scientist to use Darwin's natural selection in an article, to switch sides and oppose evolution.

454 "and all that": Hitchens 2005. Even the bishop's sympathetic biographer describes his performance as "inept" (Meacham 1970:215).

455 *Discordant Harmonies*: Botkin 1992 (1990). See also Botkin 2012.

456 Crane evolution, range, population: Meine and Archibald 1996:159–62 (Eurasian crane numbers), 175 (whooper numbers); Krajewski and King 1996:26 (evolution), Krajewski and Fetzner 1994 (evolution); Doughty 1988:4 (range), 15–18 (numbers).

456 Tiburon lily: Botkin 2016:171–72.

457 *Robinson Crusoe*: Defoe 1719 ("rover," 19; "Prize," 20).

457 Defoe and slavery: Richetti 2005:18 (shares); Keirn 1988 (editorials); Defoe 1715 ("our Commerce," 5). He was paid £12 10s 6d (~$50,000 today, according to Measuringworth.com).

458 Ubiquity of slavery: Scott 2017:esp. 155-82 (Greece, "emporium," 156); Mann 2011:Chap. 8 (early modern slavery); United States Bureau of Census 1909:139–40 (U.S. slave populations). Middle Passage figure from Trans-Atlantic Slave Trade Database (www.slavevoyages.org). A recent global history is the ongoing Cambridge World History of Slavery project.

458 Value of slaves: Williamson and Cain 2015; Ransom and Sutch 1990 (value, 39; profitability, 31). GDP figure from Gallman (1966: Table A-1), taking a rough midpoint between his values for 1849–58 and 1869–78. More recent estimates from the Angus Maddison project place the GDP at $2.24 billion in 1990 Geary-Khanis dollars (Bolt and van Zanden 2013). Extrapolating backwards from Balke and Gordon (1989: Table 10) puts the figure at about $7 billion. Inflation values from www .measuringworth.com. "a positive good": John C. Calhoun, "Speech on Slavery," U.S. Senate, Congressional Globe, 24th Congress, 2nd Sess (Feb. 6, 1837), 157–59.

459 Abolition of slavery: Many books tell this tale. Among the best are Drescher 2009 and Davis 2006; 24.9 million slaves: International Labor Organization 2017.

459 Paucity of matriarchal societies: Summaries of the common view are Harari 2015:152–59; Balter 2006:36–40, 107–14, 320–24; Christian 2005:256–57, 263–64.

459 Women's status in past: A beginning point for this complex subject is Smith 2008. For U.S. women, see Evans 1997 (1989). For European women, see Anderson and Zinsser 2000 (1988).

460 Decline in war and violent death: Morris 2014; Diamond 2012: chap. 4; Pinker 2011; Goldstein 2011; Gat 2006; Keeley 1996; Richardson 1960.

460 Classical Greece warfare: Van Wees 2004.

460 Germany casualties: Gat 2013 ("Wars *combined*," 152).

460 Levels of violence in early societies: Many well-known researchers, including Steven Pinker, Jared Diamond, and Ian Morris, maintain that organized violence extends past the invention of agriculture to the foraging bands of our oldest ancestors. Evidence for this assertion comes from archaeological reports of ancient settlements and anthropological studies of today's remaining bands of hunter-and-gatherers, all of which are replete with traces of war. Both types of evidence have been criticized. First, critics say, archaeologists as yet have found only one site with warfare— Jebel Sahaba, in northern Sudan—that is older than 10,000 years, which means that the physical evidence for warfare dates almost entirely from later, agricultural societies. Because the switch from foraging to farming changed society profoundly, one can't assume that levels of warfare remained the same. For their part, anthropologists have described some strikingly violent modern foragers, but in every case these groups were studied long after they had begun interacting with bigger, more technological societies. Here the contention is that the anthropologists' subjects were not unchanged from ancient days, but contemporary people with guns and steel blades; present-day warfare among them shouldn't be viewed as evidence about the past. To sidestep this dispute, I focus on the last 10,000 years, which nobody seems to think were peaceful. Recent pro-early-war arguments include Morris 2014:52–63, 333–38 (1 out of 10); Diamond 2012: chap. 4, 2006: esp. 294– 98; Pinker 2011; Tooby and Cosmides 2010; Gat 2006: Part 1; Fukuyama 1998:24– 27. All base their work on earlier studies, among them Bowles 2009; Otterbein 2004; LeBlanc and Register 2003; and, especially, Keeley 1996. Anti-early-war argu-

ments include Thorpe 2005; Layton 2005; and the essays in Fry 2013, esp. Ferguson 2013a, b. Anthropological criticisms are generally based on ideas from Wolf 1982.

460 Violent death rates since Second World War: Themnér and Wallensteen 2013; Lacina et al. 2006. These represent, respectively, the Uppsala Conflict Data Program and the Peace Research Institute Oslo Battle Deaths Dataset, the leading efforts to quantify war casualties globally. Naturally, their methodologies have been attacked (e.g., Gohdes and Price 2013), but the defenses have been, to my eye, robust (Lacina and Gleditsch 2012).

461 Peacekeeping operations' success: Goldstein 2011.

461 Violence uptick: Institute for Economics and Peace 2017.

461 Global poverty fall: According to a World Bank economics research group, 1.96 billion people lived in destitution (>$1.90/day, 2011 PPP) in 1990; in 2015, the figure was a projected 702 million, a drop of more than two-thirds (Cruz et al. 2015). "being human": Interview, B. Sterling, Slashdot.org, 23 Dec 2013.

461 "Pleasures of the World": Defoe 1719:132.

Appendix A: Why Believe? (Part One)

465 Successful predictions: Stouffer and Manabe 2017; Gillett et al. 2011; Stouffer et al. 1989; Manabe and Wetherald 1967. For a discussion of successful predictions, see Raymond Pierrehumbert's 2012 Tyndall lecture (available at www.youtube.com).

466 Tuning: Voosen 2016; Curry and Webster 2011.

466 Clouds: Author's interview, Pierrehumbert; Ceppi et al. 2017; Voosen 2012.

467 Methane hydrates: Pohlman et al. 2017.

467 Biological impacts: author's interviews, Daniel Botkin; Ahlström et al. 2017 ("Vegetation processes such as [tree] mortality and fires are poorly captured in most ESMs [Earth systems models] . . . the ESMs generally predict tropical forest extent and Amazonian biomass that are too low compared to observations"); Tröstl et al. 2016 (volatiles); Zeng et al. 2017 ("mitigated"); Zhu et al. 2016 (greening).

468 Antarctic models: Stenni et al forthcoming; Smith and Polvani 2016; DeConto and Pollard 2016. I thank Matt Ridley for drawing my attention to this work.

Works Cited

1000 Genomes Project Consortium. 2015. "A Global Reference for Human Genetic Variation." *Nature* 526:68–74.

Abbot, C. G., and F. E. Fowle. 1908. "Income and Outgo of Heat from the Earth, and the Dependence of Its Temperature Thereon." *Annals of the Astrophysical Observatory of the Smithsonian Institution* 2:159–76.*

Abdullah, A. B., et al. 2006. "Estimate of Rice Consumption in Asian Countries and the World Towards 2050." In Pandey, S., et al., eds. *Proceedings for Workshop and Conference on Rice in the World at Stake.* Los Banos, Philippines: IRRI, 2:28-43.

Aboites, G., et al. 1999. "El Negocio de la Producción de Semillas Mejoradas y su Rol en el Proceso de Privatización de la Agricultura Mexicana." *Espiral* 5:151–85.

Adams, W. G., and R. E. Day. 1877. "The Action of Light on Selenium." *PTRS* 167:313–49.

———. 1876. "The Action of Light on Selenium." *Proceedings of the Royal Society of London* 25:113–17.

Adelman, M. A. 1995. *Genie Out of the Bottle: World Oil Since 1970.* MIT.

———. 1991. "Oil Fallacies." *Foreign Policy* 82:3–16.

Adelmann, G. W. 1998. "Reworking the Landscape, Chicago Style." *Hastings Center Report* 28:S6-S11.

Aharoni, A., et al. 2010. "SWITCH Project Tel-Aviv Demo City, Mekorot's Case: Hybrid Natural and Membranal Processes to Upgrade Effluent Quality." *Reviews in Environmental Science and Biotechnology* 9:193–98.

Ahlström, A., et al. 2017. "Hydrologic Resilience and Amazon Productivity." *Nature Communications* 8:387.*

Ahluwalia, M. S., et al. 1979. "Growth and Poverty in Developing Countries." *Journal of Development Economics* 6:299-341.

Ainley, M. G. 1979. "The Contribution of the Amateur to North American Ornithology: A Historical Perspective." *The Living Bird* 18:161–77.

Alatout, S. 2008a. "Bringing Abundance into Environmental Politics: Constructing a Zionist Network of Water Abundance, Immigration, and Colonization." *Social Studies of Science* 39:363–94.

———. 2008b. "'States' of Scarcity: Water, Space, and Identity Politics in Israel, 1948–59." *Environment and Planning D: Society and Space* 26:959–82.

Alexander, J. 1940. "Henry A. Wallace: Cornfield Prophet." *Life,* 2 Sep.

Alexandratos, N., and J. Bruinsma. 2012. World Agriculture Towards 2030/2050: The 2012 Revision. ESA Working Paper No. 12-03. Rome: United Nations Food and Agricultural Organization.*

Allan, R., et al. 2016. "Toward Integrated Historical Climate Research: The Example of Atmospheric Circulation Reconstructions over the Earth." *WIREs Climate Change* 7:164–74.

Allitt, P. A. 2014. *Climate of Crisis: America in the Age of Environmentalism.* NY: Penguin Press.

Altholz, J. L. 1980. "The Huxley-Wilberforce Debate Revisited." *Journal of the History of Medicine and Allied Sciences* 35: 313–16.

Amarasinghe, U. A., and V. Smakhtin. 2014. *Global Water Demand Projections: Past, Present, and Future.* IWMI Research Report 156. Colombo: International Water Management Institute.*

Anderson, B. S., and J. P. Zinsser. 2000 (1988). *A History of Their Own: Women in Europe from Prehistory to the Present.* 2 vols., 2nd ed. OUP.

Anderson, R. N. 1999. *U.S. Decennial Life Tables for 1989–91: United States Life Tables Eliminating Certain Causes of Death*, vol 1, no. 4. Hyattsville, MD: National Center for Health Statistics.*

Anglo-American Committee of Inquiry. 1946. *A Survey of Palestine.* 3 vols. Jerusalem: Government Printer.*

Ångström, K. 1900. "Ueber die Bedeutung des Wasserdampfes und der Kohlensäure bei der Absorption der Erdatmosphäre." *Annalen der Physik* 308:720–32.*

Anikster, Y., et al. 2005. "Spore Dimensions of *Puccinia* Species of Cereal Hosts as Determined by Image Analysis." *Mycologia* 97:474–84.

Anonymous. 1984. *The History of Wrestling in Cresco.* Cresco, IA: Cresco High School.

———. 1954. *Race 15B: Stem Rust of Wheat.* Washington, DC: Agricultural Research Service, U.S. Department of Agriculture (ARS 22-10).*

———. 1949. "Vogt's Stand Costs Job." *Science News Letter* 56:424.

———. 1940. "Report of the Secretary of the Linnaean Society of New York for the Year 1938–1939." *Proceedings of the Linnaean Society of New York* 50/51:79–82.

———. 1915a. *Atlas of Howard County, Iowa.* Chicago: W. H. Lee.

———. 1915b. *Standard Historical Atlas of Chickasaw County, Iowa.* Chicago: Anderson Publishing Co.

———. 1889. "John Ericsson" (obituary). *Science* 13:189–91.*

———. 1883. "Photometry—No. IV." *Engineering* (London) 35:125.*

———. 1870. "Utilisation Industrielle de la Chaleur Solaire." *Le Génie Industriel* 39:309–12.*

———. 1860a. "Science: British Association." *Athenaeum Journal*, 7 July, pp. 18–32.*

———. 1860b. "Science: British Association." *Athenaeum Journal*, 14 July, pp. 59–69.*

Ansolabehere, S., et al. 2007. *The Future of Coal: Options for a Carbon-Constrained World.* MIT Interdisciplinary Study Report. MIT.*

Aristotle. 1910 (ca. 350 BC). *De Iuventute et Senectute, de Vita et Morte, de Respiratione*, trans. J. I. Beare and G. R. T. Ross. In *The Works of Aristotle*, vol. 3, ed. W. D. Ross. Oxford: Clarendon Press.

Arrhenius, S. 1896. "On the Influence of Carbonic Acid in the Air upon the Temperature of the Ground." *LPMJS* 51:237–76.

Ascunce, M. S., et al. 2011. "Global Invasion History of the Fire Ant *Solenopsis Invicta.*" *Science* 331:1066–68.

Ashwell, A. R., and R. Wilberforce. 1880–82. *Life of the Right Reverend Samuel Wilberforce, D.D.* 3 vols. London: John Murray.*

Associated Press. 1974. "Milk Producers Think Cheese Shortage Coming." *Terre Haute (IN) Tribune*, 5 Jan.

Austin, A., and A. Ram. 1971. *Studies on Chapati-Making Quality of Wheat* (ICAR Technical Bulletin 31). New Delhi: Indian Council of Agricultural Research.

Babkin, V. I. 2003. "The Earth and Its Physical Features." In *World Water Resources at the Beginning of the Twenty-First Century*, ed. I. A. Shliklomanov and J. C. Rodda, 1–18. CUP.

Bacon, F. 1870 (1620). *Novum Organum*. In *The Works of Francis Bacon*, 15 vols., ed. and trans. J. Spedding et al. New York: Hurd and Houghton, 1869–72.

Badgley, C., et al. 2007. "Organic Agriculture and the Global Food Supply." *Renewable Agriculture and Food Systems* 22:86–108.

Bakewell, R. 1828 (1813). *An Introduction to Geology*, 3rd ed. London: Longman, Rees, Orme, Brown, and Green.*

Baker, J. H. 2011. *Margaret Sanger: A Life of Passion*. NY: Hill and Wang.

Baldwin, J., and S. Brand, eds. 1978. *Soft-Tech*. NY: Penguin.

Balfour, E. B. 1943. *The Living Soil: Evidence of the Importance to Human Health of Soil Vitality, with Special Reference to Post-War Planning*. London: Faber and Faber.

Balke, N. S., and R. J. Gordon. 1989. "The Estimation of Prewar Gross National Product: Methodology and New Evidence." *Journal of Political Economy* 97:38–92.*

Balter, M. 2010. "Of Two Minds About Toba's Impact." *Science* 327:1187–88.

———. 2006. *The Goddess and the Bull: Çatalhöyük—an Archaeological Journey to the Dawn of Civilization*. Walnut Creek, CA: Left Coast Press.

Baranski, M. 2015. "The Wide Adaptation of Green Revolution Wheat." Ph.D. dissertation, Arizona State University.

Barrow, M. V., Jr. 2009. *Nature's Ghosts: Confronting Extinction from the Age of Jefferson to the Age of Ecology*. Chicago: University of Chicago Press.

———. 1998. *A Passion for the Birds: American Ornithology After Audubon*. Princeton, NJ: Princeton University Press.

Bashford, A. 2014. *Global Population: History, Geopolitics, and Life on Earth*. CUP.

Bashford, A., and Chaplin, J. 2016. *The New Worlds of Thomas Robert Malthus*. Princeton, NJ: Princeton University Press.

Baum, W. C. 1986. *Partners Against Hunger: The Consultative Group on International Agricultural Research*. Washington, DC: World Bank.

Beales, J., et al. 2007. "A Pseudoresponse Regulator Is Misexpressed in the Photoperiod Insensitive Ppd-D1a Mutant of Wheat (*Triticum aestivum* L.)." *Theoretical and Applied Genetics* 115:721–33.

Beaton, K. 1955. "Dr. Gesner's Kerosene: The Start of American Oil Refining." *Business History Review* 29:28–53.

Becker, K., and P. Lawrence. 2014. "Carbon Farming: The Best and Safest Way Forward?" *Carbon Management* 5:31–33.

Becker, K., et al. 2013. "Carbon Farming in Hot, Dry Coastal Areas: An Option for Climate Change Mitigation." *Earth System Dynamics* 4:237–51.

Beeman, R. 1995. "Friends of the Land and the Rise of Environmentalism, 1940-1954." *Journal of Agricultural and Environmental Ethics* 8:1-16.

Beeson, K. E. 1923. *Common Barberry and Black Stem Rust in Indiana*. Extension Bulletin 110. Lafayette, IN: Purdue University.*

Beevers, R. 1988. *The Garden City Utopia: A Critical Biography of Ebenezer Howard*. London: Macmillan.

Bekker, A., et al. 2004. "Dating the Rise of Atmospheric Oxygen." *Nature* 427:117–20.

Bellemare, M. F., et al. 2017. "On the Measurement of Food Waste." *American Journal of Agricultural Economics* aax034.*

Bennett, H. H. 1936. "Wild Life and Erosion Control." *Bird-Lore* 38:115-21.

Berck, P., and J. Lipow. 2012 (1995). "Water and an Israel-Palestinian Peace Settlement." In *Practical Peacemaking in the Middle East*, 2 vols., ed. S. L. Spiegel, 2:139–58. NY: Routledge.

Berg, A. 1971. "Famine Contained: Notes and Lessons from the Bihar Experience." In Blix, G., et al., eds. *Famine: A Symposium Dealing with Nutrition and Relief Operations in Times of Disaster*. Uppsala: Almqvist & Wiksells.

Berg, P., and M. Singer. 1995. "The Recombinant DNA Controversy: Twenty Years Later." *PNAS* 92:9011–13.

Berg, P., et al. 1975. "Summary Statement of the Asilomar Conference on Recombinant DNA Molecules." *PNAS* 72:1981–84.

Bergandi, D., and P. Blandin. 2012. "De la Protection de la Nature au Développement Durable: Genèse d'un Oxymore Éthique et Politique." *Revue d'Histoire des Sciences* 65:103–42.

[Bernard, C.?] 1948. *International Union for the Protection of Nature*. Brussels: M. Hayez.

Berner, R. A. 1995. "A. G. Högbom and the Development of the Concept of the Geochemical Carbon Cycle." *American Journal of Science* 295:491–95.

Best, D., and E. Levina. 2012. *Facing China's Coal Future: Prospects and Challenges for Carbon Capture and Storage*. Paris: IEA.*

Bhat, S. 2015. "India's Solar Power Punt." *Forbes India*, 23 April.*

Bhattacharya, D., et al. 2004. "Photosynthetic Eukaryotes Unite: Endosymbiosis Connects the Dots." *BioEssays* 26:50–60.

Bhoothalingam, S. 1993. *Reflections on an Era: Memoirs of a Civil Servant*. Delhi: Affiliated East-West Press.

Bickel, L. 1974. *Facing Starvation: Norman Borlaug and the Fight Against Hunger*. NY: Reader's Digest Press.

Biello, D. 2016. *The Unnatural World: The Race to Remake Civilization in Earth's Newest Age*. NY: Scribner.

Black, R. D. C., ed. 1972–81. *Papers and Correspondence of William Stanley Jevons*. 7 vols. London: Macmillan.

Bliven, B. 1948. "Forty Thousand Frightened People." *The New Republic*, 4 Oct.

Blizzard, R. 2003. "Genetically Altered Foods: Hazard or Harmless?" Gallup.com, 12 Aug.*

Block, D. R. 2009. "Public Health, Cooperatives, Local Regulation, and the Development of Modern Milk Policy: The Chicago Milkshed, 1900–1940." *Journal of Historical Geography* 35:128–53.

Blum, B. 1980. "Coal and Ecology." *EPA Journal*, September.

Boidt, D. R. 1970. "Colder Winters Held Dawn of New Ice Age." *WP*, 11 Jan.

Bolen, E. G. 1975. "In Memoriam: Clarence Cottam." *The Auk* 92:118–25.

Bolinger, M., and J. Seel. 2015. *Utility-Scale Solar 2014: An Empirical Analysis of Project Cost, Performance, and Pricing Trends in the United States*. Berkeley, CA: Lawrence Berkeley Laboratory.*

Bolt, J., and J. L. van Zanden. 2013. *The First Update of the Maddison Project; Re-estimating Growth Before 1820*. Maddison Project Working Paper 4. Database at http://www.ggdc.net/maddison/maddison-project/data/mpd_2013–01.xlsx.*

Bond, T. C., et al. 2013. "Bounding the Role of Black Carbon in the Climate System: A Scientific Assessment." *Journal of Geophysical Research: Atmospheres* 118:5380–552.

Bontemps, C. 1876. "La Diffusion de la Force: La Machine Solaire de M. Mouchot." *La Nature* 4:102–107.*

Bordot, L. 1958. "La Vie et l'Oeuvre d'Augustin Mouchot." In *XXVIIIe Congrés de l'Association Bourguignonne des Sociétés Savantes*, ed. Anon. Châtillon-sur-Seine, France: Société Historique et Archéologique de Châtillon.

Borlaug, N. E. 2007. "Sixty-Two Years of Fighting Hunger: Personal Recollections." *Euphytica* 157:287–97.

———. 1997. "Feeding a World of 10 Billion People: The Miracle Ahead." *Biotechnology and Biotechnological Equipment* 11:3–13.

———. 1994. "Preface." In Rajaram, S., and G. P. Hettel, eds. *Wheat Breeding at CIMMYT: Commemorating 50 Years of Research in Mexico for Global Wheat Improvement.* México, D.F.: CIMMYT.

———. 1988. "Challenges for Global Food and Fiber Production." *Kungliga Skogs-och Lantbruksakademiens Tidskrift* (Supplement) 21:15–55.

———. 1972. "Statement on Agricultural Chemicals." *Clinical Toxicology* 5:295–97.

———. 1968. "Wheat Breeding and Its Impact on World Food Supply." *Proceedings of the Third International Wheat Genetics Symposium.* Canberra: Australian Academy of Science, 1–36.*

———. 1958. "The Impact of Agricultural Research on Mexican Wheat Production." *Transactions of the New York Academy of Sciences* 20:278–95.*

———. 1957. "The Development and Use of Composite Varieties Based on the Mechanical Mixing of Phenotypically Similar Lines Developed Through Backcrossing." In *Report of the Third International Wheat Rust Conference*, ed. Oficina de Estudios Especiales. Saltsville, MD: Plant Industry Station, 12–18.*

———. 1950a. *Métodos Empleados y Resultados Obtenidos en el Mejoramiento del Trigo en México.* Misc. Bull. 3. México, D.F.: Oficina de Estudios Especiales.

———. 1950b. "Summary of Sources of Stem Rust Resistance Found in Rockefeller Foundation Wheat Breeding Program in Mexico." In *Report of the Wheat Stem Rust Conference at University Farm, St. Paul, Minnesota*, eds. E. C. Stakman, et al. St. Paul: University of Minnesota Agricultural Experiment Station.*

———. 1945 (1942). *Variation and Variability of* Fusarium lini. Technical Bulletin 168. Minneapolis: University of Minnesota Agricultural Experiment Station.

———. 1941. "Red Stain of Box Elder Trees." M.S. thesis, University of Minnesota.

Borlaug, N. E., and R. G. Anderson. 1975. "Defence of Swaminathan." *New Scientist* 65:280–81.

Borlaug, N. E., and J. A. Rupert. 1949. "The Development of New Wheat Varieties for Mexico." In *Forty-First Annual Meeting of the American Society of Agronomy and the Soil Science Society of America.* Abstracts. Mimeograph. Milwaukee, WI: ASA/SSA.*

Borlaug, N. E., et al. 1953. "The Rapid Increase and Distribution of Stem Rust Race 49 Further Complicates the Program of Developing Stem Rust Resistant Wheats for Mexico." In *Second International Wheat Stem Rust Conference*, ed. Anon., 10–11. Beltsville, MD: Plant Industry Station.*

Borlaug, N. E., et al. 1952. "Mexican Varieties of Wheat Resistant to Race 15B of Stem Rust." *Plant Disease Reporter* 36:147–50.

Borlaug, N. E., et al. 1950. *El Trigo como Cultivo de Verano en los Valles Altos de Mexico.* Folleto de Divulgacion 10. Mexico, D.F.: Oficina de Estudios Speciales.

Botero, F. 2017 (1589). *The Reason of State*, trans. Robert Bireley. CUP.

Botkin, D. 2016. *Twenty-Five Myths That Are Destroying the Environment: What Many Environmentalists Believe and Why They Are Wrong.* NY: Taylor Trade.

———. 2012. *The Moon in the Nautilus Shell: Discordant Harmonies Reconsidered.* OUP.

———. 1992 (1990). *Discordant Harmonies: A New Ecology for the Twenty-First Century.* OUP.

Boulding, K. E. 1964. "The Economist and the Engineer: Economic Dynamics and Public Policy in Water Resource Development." In *Economics and Public Policy in Water Resource Development*, ed. S. C. Smith and E. M. Castle, 82–92. Ames: Iowa State University Press.

———. 1963. "Agricultural Organizations and Policies: A Personal Evaluation." In Iowa

State University Center for Agricultural and Economic Development, ed. *Farm Goals in Conflict: Farm Family Income, Freedom and Security*. Ames: Iowa State University Press, 156–66.

———. 1944. "Desirable Changes in the National Economy After the War." *Journal of Farm Economics* 26:95–100.

Bouzouggar, A., et al. 2007. "82,000-Year-Old Shell Beads from North Africa and Implications for the Origins of Modern Human Behavior." *PNAS* 104:9964–69.

Bowen, M. 2005. *Thin Ice: Unlocking the Secrets of Climate in the World's Highest Mountains*. NY: Henry Holt.

Bowler, P. 1976. "Malthus, Darwin, and the Concept of Struggle." *Journal of the History of Ideas* 37:631–50.

Bowles, S. 2009. "Did Warfare Among Ancestral Hunter-Gatherers Affect the Evolution of Human Social Behaviors?" *Science* 324:1293–98.

Bowman, G. A. 1950. "Tests Show Chemical Fertilizer Unharmful." *San Bernardino Sun*, 17 Sept.

Bowman, M., et al. 2010. *Lyster's International Wildlife Law*. 2nd ed. CUP.

Bowring, S. P. K., et al. 2014. "Applying the Concept of 'Energy Return on Investment' to Desert Greening of the Sahara/Sahel Using a Global Climate Model." *Earth Systems Dynamics* 5:43–53.

Boyd Orr, J. 1948. *Soil Fertility: The Wasting Basis of Human Society*. London: Pilot Press.

BP (British Petroleum). 2015. *BP Energy Outlook 2035*. London: BP.*

Brand, S. 2010 (2009). *Whole Earth Discipline: An Ecopragmatist Manifesto*. NY: Penguin Books.

Brandes, O. M., and D. B. Brooks. 2007. *The Soft Path for Water in a Nutshell*. Ottawa: Friends of the Earth.*

Brass, P. R. 1986. "The Political Uses of Crisis: The Bihar Famine of 1966-1967." *Journal of Asian Studies* 45:245-67.

Bratspies, R. 2007. "Some Thoughts on the American Approach to Regulating Genetically Modified Organisms." *Kansas Journal of Law and Public Policy* 16:101–31.

Braudel, F. 1981 (1979). *The Structures of Everyday Life: The Limits of the Possible*. Trans. S. Reynolds. Vol. 1 of *Civilization and Capitalism, 15th–18th Century*. NY: Harper & Row.

Brauer, M., et al. 2016. "Ambient Air Pollution Exposure Estimation for the Global Burden of Disease 2013." *Environmental Science and Technology* 50:79–88.

Bray, A. J. 1991. "The Ice Age Cometh." *Policy Review* 58:82–84.

Brazhnikova, M. G. 1987. "Obituary: Gyorgyi Frantsevich Gause." *Journal of Antibiotics* 60:1079–80.

Brenchley, R., et al. 2012. "Analysis of the Bread Wheat Genome Using Whole-Genome Shotgun Sequencing." *Nature* 491:705–10.

Bristow, L. A., et al. 2017. "N_2 Production Rates Limited by Nitrite Availability in the Bay of Bengal Oxygen Minimum Zone." *Nature Geoscience* 10:24–29.

Brock, W. H. 2002. *Justus von Liebig: The Chemical Gatekeeper*. CUP.

Brook, B. W., and , C. J. A. Bradshaw. 2015. "Key Role for Nuclear Energy in Global Biodiversity Conservation." *Conservation Biology* 29:707–12.

Brooke, J. H. 2001. "The Wilberforce-Huxley Debate: Why Did It Happen?" *Science and Christian Belief* 13:127–41.*

Brooks, D. B. 1993. "Adjusting the Flow: Two Comments on the Middle East Water Crisis." *Water International* 18:35–39.

Brooks, D. B., and O. M. Brandes. 2011. "Why a Soft Water Path, Why Now and What Then?" *International Journal of Water Resources Management* 27:315–44.

Brooks, D. B., and S. Holtz. 2009. "Water Soft Path Analysis: From Principles to Practice." *Water International* 34:158–69.

Brooks, D. B., et al. 2010. "A Book Conversation Between the Editors and a Reviewer: 'The Soft Path Approach.'" *Water International* 35:336–45.

Brooks, D. B., et al., eds. 2009. *Making the Most of the Water We Have: The Soft Path Approach to Water Management*. London: Earthscan.

Brown, J. M. 1999. *Nehru*. NY: Routledge.

Browne, J. 1999. *Darwin's* Origin of Species. London: Atlantic Books.

———. 2002. *Charles Darwin: The Power of Place*. NY: Alfred A. Knopf.

———. 1995. *Charles Darwin: Voyaging*. NY: Alfred A. Knopf.

Bryce, R. 2008. *Gusher of Lies: The Dangerous Delusions of "Energy Independence."* NY: Public Affairs.

Bundesverband der Energie- und Wasserwirtschaft. 2015. *VEWA Survey: Comparison of European Water and Wastewater Prices*. Bonn: WVGW.*

Burgchardt, C. 1989. "The Saga of Pithole City." In *History of the Petroleum Industry Symposium*, ed. S. T. Pees, et al. Tulsa, OK: American Association of Petroleum Geologists, 78–83.

Burger, J. C., et al. 2008. "Molecular Insights into the Evolution of Crop Plants." *American Journal of Botany* 95:113–122.

Burgherr, P., and B. Hirschberg. 2014. "Comparative Risk Assessment of Severe Accidents in the Energy Sector." *Energy Policy* 74:S45-S56.

Burkhardt, F., et al., eds. 1985–. *The Correspondence of Charles Darwin*. Multiple vols. CUP.*

Burroughs, J. 1903. "Real and Sham Natural History." *AM* 91:298–309.

Butchard, E. 1936. "Mosquito Control in Nassau County." In *New Jersey Mosquito Extermination Association 1936*:194–96.

Byers, M. R. 1934. "The Distressful Dairyman." *North American Review* 237:215–33.

Byres, T. 1972. "The Dialectics of India's Green Revolution." *South Asian Review* 5:99-116.

Caiazzo, F., et al. 2013. "Air Pollution and Early Deaths in the United States. Part I: Quantifying the Impact of Major Sectors in 2005." *Atmospheric Environment* 79:198–208.

Cain, L. P. 1972. "Raising and Watering a City: Ellis Sylvester Chesbrough and Chicago's First Sanitation System." *Technology and Culture* 13:353-72.

Caldeira, K., and L. Wood. 2008. "Global and Arctic Climate Engineering: Numerical Model Studies." *PTRSA* 366:4039-56.

Callendar, G. S. 1939. "The Composition of the Atmosphere Through the Ages." *Meteorological Magazine* 74:33–39.

———. 1938. "The Artificial Production of Carbon Dioxide and Its Influence on Temperature." *QJRMS* 64:223–40.

Callicott, J. B. 2002. "From the Balance of Nature to the Flux of Nature: The Land Ethic in a Time of Change." In *Aldo Leopold and the Ecological Conscience*, ed. R. L. Knight and S. Riedel, 90–105. OUP.

Campbell, C. L., and D. L. Long. 2001. "The Campaign to Eradicate the Common Barberry in the United States." In *Stem Rust of Wheat: From Ancient Enemy to Modern Foe*, ed. P. D. Peterson, 16–50. St. Paul, MN: APS Press.

Canfield, D. E., et al. 2010. "The Evolution and Future of Earth's Nitrogen Cycle." *Science* 230:192–96.

Canham, H. O., and K. S. King. 1999. *Just the Facts: An Overview of New York's Wood-Based Economy and Forest Resource*. Albany: Empire State Forest Products Association.

Carefoot, G. L., and E. R. Sprott. 1969 (1967). *Famine on the Wind: Plant Diseases and Human History*. London: Angus and Robertson.

Carll, J. F. 1890. *Seventh Report on the Oil and Gas Fields of Western Pennsylvania.* Harrisburg, PA: Board of Commissioners for the Geological Survey.*

Carlsson, N. O. L., et al. 2011. "Biotic Resistance on the Increase: Native Predators Structure Invasive Zebra Mussel Populations." *Freshwater Biology* 56:1630–37.

Carlson, D. 2007. *Roger Tory Peterson: A Biography.* Austin: University of Texas Press.

Carlton, J. T. 2008. "The Zebra Mussel *Dreissena polymorpha* Found in North America in 1986 and 1987." *Journal of Great Lakes Research* 34:770–73.

Carnegie, A. 1920. *Autobiography of Andrew Carnegie.* London: Constable & Co.*

Caro, R. 1975 (1974). *The Power Broker: Robert Moses and the Fall of New York.* NY: Vintage.

Carranza, L. 1892. "Contra-Corriente Maritime, Observada en Paita y Pacasmayo." *Boletín de la Sociedad Geográfica de Lima* 1:344–45.

Carrillo, C. N. 1893. "Hidrografía Oceánica: Las Corrientes Oceánicas y Estudios de la Corriente Peruana o de Humbolt." *Boletín de la Sociedad Geográfica de Lima* 2:72–110.

Carter, J. 1977a. "The Energy Problem (Address to the Nation, 18 April)." In *United States. Public Papers of the Presidents of the United States: Jimmy Carter, 1977–1981.* 4 vols. Washington, DC: Government Printing Office. 1:656–72.*

———. 1977b. "National Energy Program: Fact Sheet on the President's Program (20 April)." In *United States. Public Papers of the Presidents of the United States: Jimmy Carter, 1977–1981.* 4 vols. Washington, DC: Government Printing Office. 1:672–90.*

Carter, L. D. 2011. *Enhanced Oil Recovery and CCS.* Washington, DC: United States Carbon Sequestration Council.*

Case, J. F. 1975. *Biology.* 2nd ed. NY: Macmillan.

Case, J. F., and V. E. Stiers. 1971. *Biology: Observation and Concept.* NY: Macmillan.

Caugant, D. A., et al. 1981. "Genetic Diversity and Temporal Variation in the *E. coli* Population of a Human Host." *Genetics* 98:467–90.

Caughley, G. 1970. "Eruption of Ungulate Populations, with Emphasis on Himalayan Thar in New Zealand." *Ecology* 51:53.

Cavett, D. 2007. "When That Guy Died on My Show." *NYT,* 3 May.*

Cebrucean, D., et al. 2014. "CO_2 Capture and Storage from Fossil Fuel Power Plants." *Energy Procedia* 63:18–26.

Central Intelligence Agency (Office of Political Research). 1974. "Potential Implications of Trends in World Population, Food Production, and Climate." Typescript, Washington, DC.*

Centro Internacional de Mejoramiento de Maíz y Trigo. 1992. *Enduring Designs for Change: An Account of CIMMYT's Research, Its Impact, and Its Future Directions.* México, D.F.: CIMMYT.

Ceppi, P., et al. 2017. "Cloud Feedback Mechanisms and their Representation in Global Climate Models." *WIREs Climate Change* 8:4.*

Cerruti, M., and G. Lorenzana. 2009. "Irrigación, Expansión de la Frontera Agrícola y Empresariado en el Yaqui." *América Latina en la Historia Económica* 31:7–36.

Chandler, R. F., Jr. 1992. *An Adventure in Applied Science: A History of the International Rice Research Institute.* Manila: IRRI.

Chaplin, J. E. 2006. *Benjamin Franklin's Political Arithmetic: A Materialist View of Humanity.* Washington, DC: Smithsonian Institution.*

———. 1995. "Climate and Southern Pessimism: The Natural History of an Idea, 1500–1800." In *The South as an American Problem,* ed. L. J. Griffin and D. H. Doyle, 57–101. Athens: University of Georgia Press.

Chapman, H. H. 1935. *Professional Forestry Schools Report*. Washington, D.C.: Society of American Foresters.

Chapman, M. 1981. *A History of Wrestling in Iowa: From Gotch to Gable*. Ames: University of Iowa Press.

Chapman, R. N. 1928. "The Quantitative Analysis of Environmental Factors." *Ecology* 9:111–22.

———. 1926. *Animal Ecology, with Especial Reference to Insects*. Minneapolis: Burgess-Roseberry.*

Charlton, L. 1973. "Onion Shortage Stirs Consumers." *NYT*, 17 April.

Charney, J. G., et al. 1979. *Carbon Dioxide and Climate: A Scientific Assessment*. Woods Hole, MA: Ad Hoc Study Group on Carbon Dioxide and Climate.*

Chase, A. 1977. *The Legacy of Malthus: The Social Costs of Scientific Racism*. NY: Alfred A. Knopf.

Chen, Y., et al. 2013. "Evidence on the Impact of Sustained Exposure to Air Pollution on Life Expectancy from China's Huai River Policy." *PNAS* 110:12936–41.

Chernow, R. 2004 (1998). *Titan: The Life of John D. Rockefeller, Sr.* NY: Vintage.

Chesler, E. 1992. *Woman of Valor: Margaret Sanger and the Birth Control Movement in America*. NY: Simon and Schuster.

Chichilnisky, G. 1996. "An Axiomatic Approach to Sustainable Development." *Social Choice and Welfare* 13:231–57.

Cho, C. H., et al. 1993. "Origin, Dissemination, and Utilization of Wheat Semi-Dwarf Genes in Korea." In *Proceedings of the 8th International Wheat Genetic Symposium*, ed. T. E. Miller and R. M. D. Koebner, 223–31. Bath, UK: Bath Press.

Chopra, V. L. 2005. "Mutagenesis: Investigating the Process and Processing the Outcome for Crop Improvement." *Current Science* 89:353–59.

Choudhury, N. 2013. "India Unveils Plans for Massive Concentrated Solar Power." *Climate Home*, 18 July.*

Chowdhury, S., and S. Dey. 2016. "Cause-specific Premature Death from Ambient PM2.5 Exposure in India: Estimate Adjusted for Baseline Mortality." *Environment International* 91:283–90.

Christensen, C. M. 1992. "Elvin Charles Stakman, 1885–1979." *Biographical Memoirs of the National Academy of Sciences* 61:331–49.

Christian, D. 2005. *Maps of Time: An Introduction to Big History*. UCP.

Christianson, G. E. 1999. *Greenhouse: The 200-Year Story of Global Warming*. NY: Walker.

Church, J. A., et al. 2013. "Sea Level Change." In *Climate Change: The Physical Science Basis*, ed. T. F. Stocker et al., 1137–1216. Working Group I Contribution to the Fifth Assessment Report of the Intergovernmental Panel on Climate Change. CUP.

Church, W. C. 1911 (1890). *The Life of John Ericsson*. 2 vols. NY: Charles Scribner's Sons.

Churchill, W. S. 2005 (1931). *The World Crisis*. NY: The Free Press.

Clack, C. T. M., et al. 2017. "Evaluation of a Proposal for Reliable Low-Cost Grid Power with 100% Wind, Water, and Solar." *PNAS* 114: 6722–27.

Clark, G. 2007. *A Farewell to Alms: A Brief Economic History of the World*. Princeton, NJ: Princeton University Press.

Clark, P. 2013. "UK Solar Power Rush Sparks Local Protest." *Financial Times*, 25 Aug.*

Clark, R. W. 1968. *The Huxleys*. NY: McGraw-Hill.

Clark, W. 2003. "Ebenezer Howard and the Marriage of Town and Country." *Organization and Environment* 16:87–97.

Clark, W. C., ed. 1982. *Carbon Dioxide Review*. OUP.

Clarke, R. 1972. "Soft Technology: Blueprint for a Research Community." *Undercurrents*, May.

Clayton, B. C. 2015. *Market Madness: A Century of Oil Panics, Crises, and Crashes.* OUP.

Cleaver, H. 1972. "The Contradictions of the Green Revolution." *American Economic Review* 62:177–88.

Clements, F. E. 1916. *Plant Succession: An Analysis of the Development of Vegetation.* Washington, DC: Carnegie Institution.*

———. 1905. *Research Methods in Ecology.* Lincoln, NE: Jacob North and Co.*

Clements, F. E., and V. E. Shelford. 1939. *Bio-Ecology.* NY: John Wiley.

Cleveland, H. 2002. *Nobody in Charge: Essays on the Future of Leadership.* San Francisco: Jossey-Bass.

Clodfelter, M. 2006 (1998). *The Dakota War: The United States Army Versus the Sioux, 1862–1865.* Jefferson, NC: McFarland and Co.

Coburn, K., and Christenson, M., eds. 1958–2002. *The Notebooks of Samuel Taylor Coleridge.* 5 vols. Princeton, NJ: Princeton University Press.

Cochet, A. 1841. *Disertación Sobre el Orijen del Huano de Iquique, su Defectibilidad Influencia que Tiene en la Formación del Nitrate de Soda de Tarapac.* Lima: J. M. Monterola.

Cockburn, A. 2007. "Al Gore's Peace Price." Counterpunch.org, 13 Oct.*

Cohen, A. J., et al. 2017. "Estimates and 25-year Trends of the Global Burden of Disease Attributable to Ambient Air Pollution: An Analysis of Data from the Global Burden of Diseases Study 2015." *Lancet* 389: 1907-18.

Cohen, I. B. 1985. "Three Notes on the Reception of Darwin's Ideas on Natural Selection." In *The Darwinian Heritage,* ed. D. Kohn, 589–607. Princeton, NJ: Princeton University Press.

Cohen, N. 2008. "Israel's National Water Carrier." *Present Environment and Sustainable Development* 2:15–27.

Cohen, S. A. 1976. "The Genesis of the British Campaign in Mesopotamia, 1914." *Middle Eastern Studies* 12:119–32.

Cohen, Y., and J. Glater. 2010. "A Tribute to Sidney Loeb: The Pioneer of Reverse Osmosis Desalination Research." *Desalination and Water Treatment* 15:222-27.

Cohn, V. 1971. "U.S. Scientist Sees New Ice Age Coming." *WP*, 9 July.

Coker, R. E. 1908a. "The Fisheries and the Guano Industry of Peru." *Bulletin of the Bureau of Fisheries* 28:333–65.*

———. 1908b. "Condición en que se Encuentra la Pesca Marina desde Paita hasta Bahía de la Independencia." *Boletín del Ministerio de Fomento* (Lima) 6(2):89–117; 6(3):54–95; 6(4):62–99; and 6(5):53–114.

———. 1908c. "La Industria del Guano." *Boletín del Ministerio de Fomento* (Lima) 6(4):25–34.

Cole, L. W. 2016. "The Evolution of Per-Cell Organelle Number." *Frontiers in Cell and Developmental Biology* 4:85.*

Coletta, A., et al. 2007. *Case Studies on Climate Change and World Heritage.* Paris: UNESCO.*

Colligan, D. 1973. "Brace Yourself for Another Ice Age." *Science Digest* 57:57– 61.

Collins, P. 2002. "The Beautiful Possibility." *Cabinet,* Spring.*

Collins, R. M. 2000. *More: The Politics of Economic Growth in Postwar America.* OUP.

Commoner, B. 1976. *The Poverty of Power: Energy and the Economic Crisis.* NY: Alfred A. Knopf.

Comprehensive Assessment of Water Management in Agriculture. 2007. *Water for Food, Water for Life: A Comprehensive Assessment of Water Management in Agriculture.* Colombo: International Water Management Institute.*

Cone, A., and W. B. Johns. 1870. *Petrolia: A Brief History of the Pennsylvania Petroleum Region.* NY: D. Appleton and Company.*

Conford, P. 2011. *The Development of the Organic Network: Linking People and Themes, 1945–95.* Edinburgh: Floris Books.

———. 2001. *The Origins of the Organic Movement.* Edinburgh: Floris Books.

Connelly, M. 2008. *Fatal Misconception: The Struggle to Control World Population.* HUP.

Connor, D. J. 2008. "Organic Agriculture Cannot Feed the World." *Field Crops Research* 106:187–90.

Considine, T. J., Jr., and T. S. Frieswyk. 1982. *Forest Statistics for New York—1980.* Broomall, PA: U.S. Department of Agriculture (Resources Bulletin of the Northeast NE-71).

Cooley, H., and N. Ajami. 2014. "Key Issues for Seawater Desalination in California: Cost and Financing." In Gleick, P.H., et al. *The World's Water: Volume 8,* 93–121. Washington, DC: Island Press.

Cooley, H., et al. 2006. *Desalination, with a Grain of Salt: A California Perspective.* Oakland, CA: Pacific Institute.

Coolidge, H. J., Jr. 1948. "Conférence pour l'Établissement de l'Union Internationale pour la Protection de la Nature." Typescript, NS/UIPN/9, UNESCO Archives.*

Cormos, C.-C. 2012. "Integrated Assessment of IGCC Power Generation Technology with Carbon Capture and Storage (CCS)." *Energy* 42:434–45.

Cottam, C., et al. 1938. "What's Wrong with Mosquito Control?" *Transactions of the Third North American Wildlife Conference,* 81–107. 14–17 Feb. Washington, DC: American Wildlife Institute.

Cotter, J. 2003. *Troubled Harvest: Agronomy and Revolution in Mexico, 1880–2002.* Westport, CT: Praeger Publishers.

———. 1994. "The Origins of the Green Revolution in Mexico: Continuity or Change?" In Latin America in the 1940s: War and Postwar Transitions, ed. D. Rock, 224–47. UCP.*

Courtney, L. H. 1897. "Jevons's Coal Question: Thirty Years After." *Journal of the Royal Statistical Society* 60:789–810.

Cox, M. 2015. *The Politics and Art of John L. Stoddard: Reframing Authority, Otherness and Authenticity.* NY: Lexington Books.

Cox, T. S., et al. 2006. "Prospects for Developing Perennial Grain Crops." *BioScience* 56:649-59.

Crabb, A. R. 1947. *The Hybrid-Corn Makers: Prophets of Plenty.* New Brunswick, NJ: Rutgers University Press.

Crawford, E. 1997. "Arrhenius' 1896 Model of the Greenhouse Effect in Context." *Ambio* 26:6–11.

Crease, R., and C. C. Mann. 1996 (1986). *The Second Creation: Makers of the Revolution in Twentieth-Century Physics.* New Brunswick, NJ: Rutgers University Press.

Crews, T. E., and L. R. DeHaan. 2015. "The Strong Perennial Vision: A Response." *Agroecology and Sustainable Food Systems* 39:500-15.

Crisp, A., et al. 2015. "Expression of Multiple Horizontally Acquired Genes Is a Hallmark of Both Vertebrate and Invertebrate Genomes." *Genome Biology* 16:50.*

Crocus [C. C. Leonard]. 1867. *The History of Pithole.* Pithole City, PA: Morton, Longwell & Co.

Crookes, W., et al. 1900. *The Wheat Problem: Based on Remarks Made in the Presidential Address to the British Association at Bristol in 1898.* NY: G. P. Putnam's Sons.*

Crova, M. A. 1884. "Rapport sur les Expériences Faites a Montpellier pendant l'Année 1881 par la Commission des Apparelis Solaires." *Académie Des Sciences et Lettres de Montepellier (Sciences)* 10:289–329.

Crow, J. F. 1994. "Hitoshi Kihara, Japan's Pioneer Geneticist." *Genetics* 137:891–94.

Crutzen, P. J. 2006. "Albedo Enhancement by Stratospheric Sulfur Injections: A Contribution to Resolve a Policy Dilemma?" *Climatic Change* 77:211–19.

———. 2002. "Geology of Mankind." *Nature* 415:23.

Crutzen, P. J., and E. F. Stoermer. 2000. "The 'Anthropocene.'" *Global Change News Letter* (IGBP) 41:17–18.

Cruz, M., et al. 2015. *Ending Extreme Poverty and Sharing Prosperity: Progress and Policies*. Policy Research Note 15/03. Washington, DC: World Bank.*

Cullather, N. 2010. *The Hungry World: America's Cold War Battle Against Famine in Asia.* HUP.

Culver, J. C., and J. Hyde. 2001. *American Dreamer: A Life of Henry A. Wallace.* NY: W. W. Norton.

Curry, C. L., et al. 2014. "A Multimodel Examination of Climate Extremes in an Idealized Geoengineering Experiment." *Journal of Geophysical Research: Atmospheres* 119:3900–23.*

Curry, J. A., and P. J. Webster. 2011. "Climate Science and the Uncertainty Monster." *Bulletin of the American Meteorological Society* 92:1667-82.

Curwen-McAdams, C., and S. S. Jones. 2017. "Breeding Perennial Grain Crops Based on Wheat." *Crop Science* 57:1172-88.

Curwen-McAdams, C., et al. 2016. "Toward a Taxonomic Definition of Perennial Wheat: A New Species x*Tritipyrum aaseae* described." *Genetic Resources and Crop Evolution* 1-9.*

Cushman, G. T. 2014 (2013). *Guano and the Opening of the Pacific World: A Global Ecological History.* CUP.

———. 2006. "The Lords of Guano: Science and the Management of Peru's Marine Environment, 1800–1973." Ph.D. dissertation, University of Texas at Austin.

———. 2004. "Enclave Vision: Foreign Networks in Peru and the Internationalization of El Niño Research During the 1920s." *Proceedings of the International Commission on the History of Meteorology* 1:65–74.

———. 2003. "Who Discovered the El Niño–Southern Oscillation?" Paper given at Presidential Symposium on the History of the Atmospheric Sciences, 83rd Annual Meeting of the American Meteorological Society, Long Beach, CA.*

Czaplicki, A. 2007. "'Pure Milk Is Better Than Purified Milk': Pasteurization and Milk Purity in Chicago, 1908–1916." *Social Science History* 31:411-33.

Dabdoub, C. 1980. *Breve Historia del Valle del Yaqui.* México, D.F.: Editores Asociados Mexicanos.

Dahl, E. J. 2001. "Naval Innovation: From Coal to Oil." *Joint Force Quarterly* 27:50–56.*

Dalrymple, D. G. 1986 (1969). *Development and Spread of High-Yielding Wheat Varieties in Developing Countries.* 7th ed. Washington, DC: Bureau for Science and Technology.

Damodoran, H. 2016. "After the Revolution." *Indian Express,* 6 Dec.

Darrah, W. C. 1972. *Pithole, the Vanished City: A Story of the Early Days of the Petroleum Industry.* Gettysburg, PA: William Culp Darrah.

Darwin, C. 1872. *On the Origin of Species.* 6th ed. London: John Murray.*

———. 1859. *On the Origin of Species.* 1st ed. London: John Murray.*

Darwin, C., and A. Wallace. 1858. "On the Tendency of Species to Form Varieties." *Journal of the Proceedings of the Linnean Society of London (Zoology)* 3:45–50.*

Darwin, F., ed. 1887. *The Life and Letters of Charles Darwin, Including an Autobiographical Chapter.* 3 vols. London: John Murray.

Dasgupta, B. 1977. *Agrarian Change and the New Technology in India.* Geneva: U.N. Research Institute for Social Development.

Davis, D. B. 2006. *Inhuman Bondage: The Rise and Fall of Slavery in the New World.* OUP.

Dawe, D. 2000. "The Contribution of Rice Research to Poverty Alleviation." In Sheehy, J.E., et al., eds. *Redesigning Rice Photosynthesis to Increase Yield.* Amsterdam: Elsevier, 3-12.

Dawkins, R. 2004. *The Ancestor's Tale: A Pilgrimage to the Dawn of Life.* Boston: Houghton Mifflin.

Day, D. T. 1909. "Petroleum Resources of the United States." In United States National Conservation Commission. *Report of the National Conservation Commission, with Accompanying Papers,* 3:446–64. Washington, DC: Government Printing Office (60th Cong., 2nd Sess., Doc. 676).*

DeConto, R. M., and D. Pollard. 2016. "Contribution of Antarctica to Past and Future Sea-Level Rise." *Nature* 531:591–97.

Deese, R. S. 2015. *We Are Amphibians: Julian and Aldous Huxley on the Future of Our Species.* UCB.

Defoe, D. 1719. *The Life and Strange Surprizing Adventures of Robinson Crusoe, of York, Mariner.* London: W. Taylor.*

———. 1711. *An Essay upon the Trade to Africa.* [London].*

De Gans, H. 2002. "Law or Speculation? A Debate on the Method of Forecasting Population Size in the 1920s." *Population* 57:83–108.

Delacorte, G. T. 1929. "Dell Publications." *Writer's Digest,* Jan.

De Las Rivas, J., et al. 2002. "Comparative Analysis of Chloroplast Genomes: Functional Annotation, Genome-Based Phylogeny, and Deduced Evolutionary Patterns." *Genome Research* 12:567–83.*

DeLillo, D. 1989 (1982). *The Names.* NY: Vintage Books.

Delyannis, E. 2005. "Historic Background of Desalination and Renewable Energies." *Solar Energy* 75:357–66.

Delyannis, E., and V. Belessiotis. 2010. "Desalination: The Recent Development Path." *Desalination* 264:206–13.

Dennery, É. 1931 (1930). *Asia's Teeming Millions, and Its Problems for the West,* trans. J. Peile. London: Jonathan Cape.

DeNovo, J. A. 1955. "Petroleum and the United States Navy before World War I." *Mississippi Valley Historical Review* 41:641–56.

Depew, D. J. 2010. "Darwinian Controversies: An Historiographical Recounting." *Science and Education* 19:323–66.

de Ponti, T., et al. 2012. "The Crop Yield Gap Between Organic and Conventional Agriculture." *Agricultural Systems* 108:1–9.

Desrochers, P., and C. Hoffbauer. 2009. "The Post War Intellectual Roots of the Population Bomb: Fairfield Osborn's 'Our Plundered Planet' and William Vogt's 'Road to Survival' in Retrospect." *The Electronic Journal of Sustainable Development* 1:37-61.*

Devereux, M. W. 1941. "Reasons for the Replacement of Children in Foster Home Care Placed by the Boston Children's Aid Association in 1938, 1939, and 1940." M.A. thesis, Boston University School of Social Work.*

Devereux, S. 2000. *Famine in the Twentieth Century.* Institute of Development Studies Working Paper 105. Brighton, UK: University of Sussex.*

Devlin, J. C., and G. Naismith. 1977. *The World of Roger Tory Peterson.* NY: New York Times Books.

Dey, D. 1995. *Acorn Production in Red Oak.* Sault Ste. Marie: Ontario Forest Research Institute.*

Diamond, J. 2012. *The World Until Yesterday: What Can We Learn from Traditional Societies?* New York: Viking.

———. 2006. *Collapse: How Societies Choose to Fail or Succeed.* NY: Viking.

DiFonzo, J. H., and R. C. Stern. 2007. "Addicted to Fault: Why Divorce Reform Has Lagged in New York." *Pace Law Review* 27:559–603.*

Dil, A.. 2004. "Life and Work of M. S. Swaminathan: An Introductory Essay." In *Life and Work of M. S. Swaminathan: Toward a Hunger-Free World*, ed. A. Dil, 29–64. Madras: EastWest Books.

Dileva, F. D. 1954. "Iowa Farm Price Revolt." *Annals of Iowa* 32:171–202.

DiMichele, W. A., et al. 2007. "Ecological Gradients within a Pennsylvanian Mire Forest." *Geology* 35:415–18.

Dmitri, C., et al. 2005. *The 20th Century Transformation of U.S. Agriculture and Farm Policy.* Washington, D.C.: USDA (Economic Information Bulletin 3).

Dodson, J., et al. 2014. "Use of Coal in the Bronze Age in China." *The Holocene* 24:525–30.

Doel, R. E. 2009. "Quelle Place pour les Sciences de l'Environnement Physique dans l'Histoire Environnementale?" *Revue d'Histoire Moderne et Contemporaine* 56(4): 137–64.*

Donnelly, K. 2014. "The Red Sea–Dead Sea Project Update." In Gleick, P. H., et al. *The World's Water, Volume 8.* Washington: Island Press, 153-58.

Doughty, R. W. 1988. *Return of the Whooping Crane.* Austin: University of Texas Press.

Dowie, M. 2009. *Conservation Refugees: The Hundred-Year Conflict Between Global Conservation and Native Peoples.* MIT.

Drescher, S. 2009. *Abolition: A History of Slavery and Antislavery.* CUP.

Dréze, J. 1991. "Famine Prevention in India." In Sen, A., and J. Drèze, eds. *The Political Economy of Hunger*, vol. 2. Oxford: Clarendon Press, 13-124.

Dubin, H. J., and J. P. Brennan. 2009. *Combating Stem and Leaf Rust of Wheat: Historical Perspective, Impacts, and Lessons Learned.* IFPRI Discussion Paper 910. Washington, DC: International Food Policy Research Institute.

Duffy, D. C. 1994. "The Guano Islands of Peru: The Once and Future Management of a Renewable Resource." *BirdLife Conservation Series* 1:68–76.

———. 1989. "William Vogt: A Pilgrim on the Road to Survival." *American Birds* 43:1256–57.

Durham, W. H. 1979. *Scarcity and Survival in Central America: Ecological Origins of the Soccer War.* Stanford, CA: Stanford University Press.

Dworkin, S. 2009. *The Viking in the Wheat Field: A Scientist's Struggle to Preserve the World's Harvest.* NY: Walker Publishing Company.

Dwyer, J. J. 2002. "Diplomatic Weapons of the Weak: Mexican Policymaking during the U.S.-Mexican Agrarian Dispute, 1934–41." *Diplomatic History* 26:375–95.

Dyson, T. 2005 (1996). *Population and Food: Global Trends and Future Prospects.* NY: Routledge.

Dyson, T., and A. Maharatna. 1992. "Bihar Famine, 1966-67 and Maharashtra Drought, 1970-73: The Demographic Consequences." *Economic and Political Weekly* 27:1325-32.

E.A.L. 1948. "Is Starvation Ahead?" *Boston Globe*, 5 Aug.

East, E. M. 1923. *Mankind at the Crossroads.* NY: Charles Scribner's Sons.*

Easterbrook, G. 1997. "Forgotten Benefactor of Humanity." *AM* 279:75–82.

Ebelot, A. 1869. "La Chaleur Sociale et les Applications Industrielles." *Revue des Deux Mondes* 83:1019–21.

Ebenstein, A. 2010. "The 'Missing Girls' of China and the Unintended Consequences of the One Child Policy." *Journal of Human Resources* 45:87–115.

Edwards, P. N. 2011. "History of Climate Modeling." *WIREs Climate Change* 2:128–39.

Edwards, R. D. 2011. "Trends in World Inequality in Life Span Since 1970." *Population and Development Review* 37:499–528.

Egerton, F. E. 1973. "Changing Concepts of the Balance of Nature." *Quarterly Review of Biology* 48:322–50.

Eguiguren Escudero, V. 1894. "Las Lluvias en Piura." *Boletín de la Sociedad Geográfica de Lima* 4:241–58.

Ehrlich, P. R. 2008. "Population, Environment, War, and Racism: Adventures of a Public Scholar." *Antipode* 40:383-88.

———. 1969. "Eco-Catastrophe!" *Ramparts*, September.*

———. 1968. *The Population Bomb*. NY: Ballantine Books.

Ehrlich, P. R., and A. H. Ehrlich. 1981. *Extinction: The Causes and Consequences of the Disappearance of Species*. NY: Random House.

———. 1974. *The End of Affluence: A Blueprint for Your Future*. NY: Ballantine Books.

Ehrlich, P. R., and J. P. Holdren. 1969. "Population and Panaceas: A Technological Perspective." *BioScience* 19:1065–71.

Eiberg, H., et al. 2008. "Blue Eye Color in Humans May Be Caused by a Perfectly Associated Founder Mutation in a Regulatory Element Located Within the HERC2 Gene Inhibiting OCA2 Expression." *Human Genetics* 123:177–87.

Ellegard, A. 1958. "Public Opinion and the Press: Reactions to Darwinism." *Journal of the History of Ideas* 19:379–87.

Ellman, M. 1981. "Natural Gas, Restructuring and Re-industrialisation: The Dutch Experience of Industrial Policy." In *Oil or Industry? Energy, Industrialisation and Economic Policy in Canada, Mexico, the Netherlands, Norway and the United Kingdom*, ed. T. Barker and V. Brailovsky, 149–66. London: Academic Press.

Ellis, R. J. 1979. "Most Abundant Protein in the World." *Trends in Biochemical Sciences* 4:241–44.

Elton, C. S. 1930. *Animal Ecology and Evolution*. Oxford: Clarendon Press.

Emerson, R. W. 1860. *The Conduct of Life*. Boston: Ticknor and Fields.*

Energy Information Administration (U.S.). 2016. *Annual Energy Outlook*. Washington, DC: Department of Energy.*

———. 2013. *Updated Capital Cost Estimates for Utility Scale Electricity Generating Plants*. Washington, DC: Department of Energy.*

———. 1990–. *Electric Power Monthly*. Washington, DC: Department of Energy.*

———. 1970–. *Electric Power Annual 2014*. Washington, DC: Department of Energy.*

Epstein, P. R., et al. "Full Cost Accounting for the Life Cycle of Coal." *Annals of the New York Academy of Sciences* 1219:73–98.

Erdélyi, A. 2002. *The Man Who Harvests Sunshine: The Modern Gandhi; M. S. Swaminathan*. Budapest: Tertia Kiadó.

Ericsson, J. 1888. "The Sun Motor." *Scientific American Supplement* 26:10592.*

———. 1870. "Ericsson's Solar Engine." *Engineering* (London), 14 Oct.*

Errington, P. 1938. "No Quarter." *Bird-Lore* 40:5-6.

Esfandiary, F. M. 1973. *Up-Wingers: A Futurist Manifesto*. NY: John Day Co.

Esteva, G. 1983. *The Struggle for Rural Mexico*. South Hadley, MA: Bergin & Garvey.

European Environment Agency. 2011. *Europe's Environment: An Assessment of Assessments*. Luxembourg: Publications Office of the European Union.*

Evans, H. B. 1997. *Water Distribution in Ancient Rome: The Evidence of Frontinus*. Ann Arbor: University of Michigan Press.

Evans, S. M. 1997 (1989). *Born for Liberty: A History of Women in America*. 2nd ed. NY: Free Press.

Eve, A. S., and C. H. Creasey. 1945. *Life and Work of John Tyndall*. London: Macmillan & Co.

Evenson, R. E., and D. Gollin. 2003. "Assessing the Impact of the Green Revolution, 1960 to 2000." *Science* 300:758–62.

Ezkurdia, I., et al. 2014. "Multiple Evidence Strands Suggest That There May Be as Few as 19,000 Human Protein-Coding Genes." *Human Molecular Genetics* 23:5866–78.

Fagan, B. 2009 (1999). *Floods, Famines, and Emperors: El Niño and the Fate of Civilizations*. 2nd ed. NY: Basic Books.

Fagundes, N. J. R., et al. 2007. "Statistical Evaluation of Alternative Models of Human Evolution." *PNAS* 104:17614–19.

Fairbarn, R. H. 1919. *History of Chickasaw and Howard Counties, Iowa*. 2 vols. Chicago: S. J. Clarke.

Famiglietti, J. S. 2014. "The Global Groundwater Crisis." *Nature Climate Change* 4:945–48.

Farley, J. 2004. *To Cast Out Disease: A History of the International Health Division of the Rockefeller Foundation*. OUP.

Fatondji, D., et al. 2001. "Zai: A Traditional Technique for Land Rehabilitation in Niger." *ZEF News*: 1–2.

Feder, E. 1976. "McNamara's Little Green Revolution: World Bank Scheme for Self-Liquidation of Third World Peasantry." *Economic and Political Weekly* 11:532–41.

Ferguson, R. B. 2013a. "Pinker's List: Exaggerating Prehistoric War Mortality." In Fry, ed. 2013:112–31.

———. 2013b. "The Prehistory of War and Peace in Europe and the Near East." In Fry, ed. 2013:191–240.

———. 1995. *Yanomami Warfare: A Political History*. Santa Fe, NM: School of American Research Press.

Ferling, J. 2013. *Jefferson and Hamilton: The Rivalry That Forged a Nation*. NY: Bloomsbury.

Ferrier, R. W. 2000 (1982). *The History of the British Petroleum Company*. Vol. 1. CUP.

Feuer, L. S. 1975. "Is the 'Darwin–Marx Correspondence' Authentic?" *Annals of Science* 32: 1–12.

Fialka, J. 1974. "Solar Energy's Big Push into the Marketplace." *Washington Star*, 17 July.

Fischer, T., et al. 2014. *Crop Yields and Global Food Security: Will Yield Increase Continue to Feed the World?* Canberra: Australian Centre for International Agricultural Research.

Fitzgerald, D. 1986. "Exporting American Agriculture: The Rockefeller Foundation in Mexico, 1943–53." *Social Studies of Science* 16:457–83.

Flader, S. L. 1994. *Thinking like a Mountain: Aldo Leopold and the Evolution of an Ecological Attitude Toward Deer, Wolves, and Forests*. Madison: University of Wisconsin Press.

Flandreau, C. E. 1900. *The History of Minnesota and Tales of the Frontier*. St. Paul, MN: E. W. Porter.*

Flanner, J. 1949. "Letter from Paris." *The New Yorker*, 7 May.

Fleming, J. R. 2010. *Fixing the Sky: The Checkered History of Weather and Climate Control*. CUP.

———. 2007. *The Callendar Effect: The Life and Work of Guy Stewart Callendar (1898–1964)*. Boston: American Meteorological Society.

———. 1998. *Historical Perspectives on Climate Change*. OUP.

Floudas, D., et al. 2012. "The Paleozoic Origin of Enzymatic Lignin Decomposition Reconstructed from 31 Fungal Genomes." *Science* 336:1715–19.

Foley, J. 2014. "A Five-Step Plan to Feed the World." *National Geographic* 225:4–21.

Foote, E. 1856. "Circumstances Affecting the Heat of the Sun's Rays." *American Journal of Science and Arts* 22:382-83.

Forest Europe. 2015. *State of Europe's Forests 2015*. Madrid: Ministerial Conference on the Protection of Forests in Europe.*

Fosdick, R. B. 1988. *The Story of the Rockefeller Foundation*. NY: Transaction Publishers.

Foskett, D. J. 1953. "Wilberforce and Huxley on Evolution." *Nature* 172:920.

Fourcroy, A. F., and L. N. Vauquelin. 1806. "Mémoire sur le Guano, ou sur l'Engrais Naturel des Îlots de la Mer du Sud, près des Côtes du Pérou." *Memoires de l'Institut des Sciences, Lettres et Arts: Sciences Mathématiques et Physiques* 6:369–81.*

Fourier, J. 1824. "Remarques Générales sur les Températures du Globe Terrestre et des Espaces Planétaires." *Annales de Chemie et de Physique* 27:136–67.*

———. 1827. "Mémoire sur les Températures du Globe Terrestre et des Espaces Planétaires." *Mémoires de l'Académie Royale des Sciences* 7:569–604.*

Fowler, G. 1972. "Hugh Moore, Industrialist, Dies." *NYT*, 26 Nov.

Fox, S. 1981. *The American Conservation Movement: John Muir and His Legacy*. Madison: University of Wisconsin Press.

Frank, C. R., Jr. 2014. *The Net Benefits of Low and No-Carbon Electricity Technologies*. Washington, DC: Brookings Institution.*

Franklin, B. 1755. *Observations Concerning the Increase of Mankind, Peopling of Countries, &c.* Boston: S. Kneeland.*

Frederickson, D. S. 1991. "Asilomar and Recombinant DNA: The End of the Beginning." In *Biomedical Politics*, ed. K. E. Hanna, 258–307. Washington, DC: National Academies Press.

Freebairn, D. K. 1995. "Did the Green Revolution Concentrate Incomes? A Quantitative Study of Research Reports." *World Development* 23:265-79.

Freeman, M. C., et al. 2015. "Climate Sensitivity Uncertainty: When Is Good News Bad?" *PTRS* 373(2055).

Freese, B. 2004 (2003). *Coal: A Human History*. NY: Penguin Books.

Freire, P. 2014 (1968). *Pedagogy of the Oppressed, trans. M. B. Ramos*. NY: Bloomsbury.

Fritts, C. E. 1883. "On a New Form of Selenium Cell, and Some Electrical Discoveries Made by Its Use." *American Journal of Science* 126:465–72.*

———. 1885. "On the Fritts Selenium Cells and Batteries." *Proceedings of the American Association for the Advancement of Science* 33:97–108.*

Froeb, A. C. 1936. "Accomplishments in Mosquito Control in Suffolk County, Long Island." In *Proceedings of the New Jersey Mosquito Extermination Association 1936*:128–29.

Fromartz, S. 2006. *Organic, Inc.: Natural Foods and How They Grew*. NY: Harvest Books.

Frost, P. 2006. "European Hair and Eye Color: A Case of Frequency-Dependent Sexual Selection?" *Evolution and Human Behavior* 27:85–103.

Fry, D. P., ed. 2013. *War, Peace, and Human Nature: The Convergence of Evolutionary and Cultural Views*. OUP.

Fuchs, R. J. 2010. *Cities at Risk: Asia's Coastal Cities in an Age of Climate Change*. Honolulu: East-West Center.*

Fukuyama, F. 1998. "Women and the Evolution of World Politics." *Foreign Affairs* 77:24–40.

Fuller, D. Q., et al. 2007. "Dating the Neolithic of South India: New Radiometric Evidence for Key Economic, Social, and Ritual Transformations." *Antiquity* 81: 755–78.

Furbank, R. T., et al. 2015. "Improving Photosynthesis and Yield Potential in Cereal Crops by Targeted Genetic Manipulation: Prospects, Progress, and Challenges." *Field Crops Research* 182:19–29.

Fussell, G. E. 1972. *The Classical Tradition in West European Farming*. Cranbury, NJ: Fairleigh Dickinson Press.

F.W.V. and C.A. 1901. "Knut Angstrom on Atmospheric Absorption." *Monthly Weather Review* 29:268.

Gaillard, G. 2004 (1997). *The Routledge Dictionary of Anthropologists*, trans. P. J. Bowman. NY: Routledge.

Gall, Y. M. (Я. М. Галл). 2011. Г.Ф. Гаузе (1910–1986): Творческий Образ. Экология И Теория Эволюции [G. F. Gause [1910–1986]: Creative Image. Ecology and Evolutionary Theory]. Биосфера [Biosphere] 3:423–44.

Gall, Y. M., and M. B. Konashev. 2001. "The Discovery of Gramicidin S: The Intellectual Transformation of G. F. Gause from Biologist to Researcher of Antibiotics and on Its Meaning for the Fate of Russian Genetics." *History and Philosophy of the Life Sciences* 23:137–50.

Gallagher, W. 2006. *House Thinking: A Room-by-Room Look at How We Live*. NY: HarperCollins.

Gallman, R. E. 1966. "Gross National Product in the United States, 1834–1909." In *Output, Employment, and Productivity in the United States After 1800*, ed. D. S. Brady, 3–90. Washington, DC: National Bureau of Economic Research.*

Galloway, J. N., et al. 2003. "The Nitrogen Cascade." *Bioscience* 53:341–56.

Gandhi, I. 1975. "The Challenge of Drought." In Indira Gandhi Abhinandan Samiti, ed. *The Spirit of India*. New Delhi: Asia Publishing House, 4 vols., 1:67-69.

Gardiner, S. M. 2011. *A Perfect Moral Storm: The Ethical Tragedy of Climate Change*. OUP.

Garnet, T. 2013. "Food Sustainability: Problems, Perspectives, and Solutions." *Proceedings of the Nutrition Society* 72:29–39.

Garrett, H. E., et al. 1991. "Black Walnut (*Juglans nigra* L.) Agroforestry—Its Design and Potential as a Land-use Alternative." *The Forestry Chronicle* 67:213-18.

Gaskell, G., et al. 2006. *Europeans and Biotechnology in 2005: Patterns and Trends*. Special Eurobarometer 244b.*

Gat, A. 2013. "Is War Declining—and Why?" *Journal of Peace Research* 50:149–57.

———. 2006. *War in Human Civilization*. OUP.

Gauld, C. 1992a. "The Historical Anecdote as a 'Caricature': A Case Study." *Research in Science Education* 22:149–56.

———. 1992b. "Wilberforce, Huxley, and the Use of History in Teaching About Evolution." *American Biology Teacher* 54:406–10.

Gause, G. F. 1934. *The Struggle for Existence*. Baltimore: Williams & Wilkins.*

———. 1930. "Studies on the Ecology of the Orthoptera." *Ecology* 11:307–25.

General Education Board. 1916 (1915). *The General Education Board: An Account of Its Activities, 1902–1914*. 3rd ed. NY: General Education Board.*

George, S. 1986 (1976). *How the Other Half Dies: The Real Reasons for World Hunger*. NY: Penguin.

Gifford, T., ed. 1996. *John Muir: His Life and Letters and Other Writings*. Seattle: The Mountaineers.

Gill, E. 2010. "Lady Eve Balfour and the British Organic Food and Farming Movement." Ph.D. dissertation, Aberystwyth University.*

Gilley, S. 1981. "The Huxley-Wilberforce Debate: A Reconsideration." In *Religion and Humanism*, ed. K. Robbins, 325–40. Oxford: Basil Blackwell/Ecclesiastical History Society.

Gimpel, J. 1983 (1976). *The Medieval Machine: The Industrial Revolution of the Middle Ages*. NY: Penguin.

Glacken, C. J. 1976 (1967). *Traces on the Rhodian Shore: Nature and Culture in Western Thought from Ancient Times to the End of the Eighteenth Century*. UCP.

Gleick, J. 1988 (1987). *Chaos: Making a New Science*. NY: Penguin Books.

Gleick, P. H. 2003. "Global Freshwater Resources: Soft-Path Solutions for the 21st Century." *Science* 302:1524–28.

———. 2002. "Soft Water Paths." *Nature* 418:373.

———. 2000. "The Changing Water Paradigm: A Look at Twenty-First Century Water Resources Development." *Water International* 25:127–38.

———. 1998. *The World's Water 1998–1999: The Biennial Report on Freshwater Resources*. Washington, DC: Island Press.

———. 1996. "Water Resources." In *Encyclopedia of Climate and Weather*, ed. S. H. Schneider, 2:817–23. OUP.

Gleick, P. H., and M. Palaniappan. 2010. "Peak Water Limits to Freshwater Withdrawal and Use." *PNAS* 107:11155–62.

Glick, T. F., ed. 1988. *The Comparative Reception of Darwinism*. Chicago: University of Chicago Press.

Godefroit, P., et al. 2014. "A Jurassic Ornithischian Dinosaur from Siberia with Both Feathers and Scales." *Science* 345:451–55.

Godfray, H. C. J., et al. 2010. "Food Security: The Challenge of Feeding 9 Billion People." *Science* 327:812–18.

Goh, P. S., et al. 2017. "The Water-Energy Nexus: Solutions towards Energy-Efficient Desalination." *Energy Technology* 5:1136–55.

Gohdes, A., and M. Price. 2013. "First Things First: Assessing Data Quality Before Model Quality." *Journal of Conflict Resolution* 57:1090–1108.

Gold, E. 1965. "George Clarke Simpson, 1878–1965." *Biographical Memoirs of Fellows of the Royal Society* 11:156–75.

Goldsmith, E., et al. 1972. "A Blueprint for Survival." *The Ecologist* 2:1–43.*

Goldstein, J. S. 2011. *Winning the War on War: The Decline of Armed Conflict Worldwide*. NY: Dutton.

Gómez-Baggethun, E., et al. 2009. "The History of Ecosystem Services in Economic Theory and Practice: From Early Notions to Markets and Payment Schemes." *Ecological Economics* 69:1209–18.

González, B. P. 2006. "La Revolución Verde en México." *Agrária* (São Paulo) 4:40–68.

González-Paleo, L., et al. 2016. "Back to Perennials: Does Selection Enhance Tradeoffs Between Yield and Longevity?" *Industrial Crops and Products* 91:272–78.

Goodall, A. H. 2008. "Why Have the Leading Journals in Management (and Other Social Sciences) Failed to Respond to Climate Change?" *Journal of Management Inquiry* 20:1–14.

Goodell, A. R. S. 1975. "The Visible Scientists." Ph.D. dissertation, Stanford University.

Goodell, J. 2011 (2010). *How to Cool the Planet: Geoengineering and the Audacious Quest to Fix Earth's Climate*. NY: Mariner Books.

Goodisman, M. A. D., et al. 2007. "Genetic and Morphological Variation over Space and Time in the Invasive Fire Ant *Solenopsis invicta*." *Biological Invasions* 9:571–84.

Gopalkrishnan, G. 2002. *M. S. Swaminathan: One Man's Quest for a Hunger-Free World*. Chennai, India: Sri Venkatesa Printing House.

Gossett, R. F. 1997. *Race: The History of an Idea in America*. OUP.

Gorman, H. S. 2013. *The Story of N: A Social History of the Nitrogen Cycle and the Challenge of Sustainability*. New Brunswick, NJ: Rutgers University Press.

Government of India. Ministry of Food and Agriculture. 1959. *Report on India's Food Crisis and Steps to Meet It*. Delhi: Government of India.

Government of India. Ministry of Science and Technology. 1958. Scientific Policy Resolution official memorandum (4 March).*

Gowans, A. 1986. *The Comfortable House: North American Suburban Architecture, 1890–1930.* MIT.

Graham, F., Jr., and C. W. Buchheister. 1990. *The Audubon Ark: A History of the Audubon Society.* NY: Alfred A. Knopf.

Graham, J. A. 1904. "Sun Motor Solves Mystery of Electricity's Source." *Chicago Daily Tribune,* 6 Nov.

Grant, M. 1916. *The Passing of the Great Race, or, The Racial Basis of European History.* NY: Charles Scribner's Sons.

Grassini P., et al. 2013. "Distinguishing Between Yield Advances and Yield Plateaus in Historical Crop Production Trends." *Nature Communications* 4:2918.

Grattan-Guinness, I. 1972. *Joseph Fourier, 1768–1830: A Survey of His Life and Work.* MIT.

Great Britain House of Commons. 1913. *The Parliamentary Debates.* Vol. 6, 7–25 July. 5th ser., v.55. London: Her Majesty's Stationery Office.*

Great Britain House of Lords. 1830. *Report from the Select Committee of the House of Lords Appointed to Take into Consideration the State of the Coal Trade in the United Kingdom.* London: House of Commons.

Green, C. C. 2006. "Forestry Education in the United States." *Issues in Science and Technology Librarianship* 46 (supp.).*

Greenfield, G. 1934 "News of Activities with Rod and Gun." *NYT,* 24 Oct.

Greenhalgh, S. 2008. *Just One Child: Science and Policy in Deng's China.* UCP.

Griffin, D. 2014. "Thousands Protest Against Pylons and Wind Turbines." *Irish Times,* 15 April.*

Griffin, K. 1974. *The Political Economy of Agrarian Change: An Essay on the Green. Revolution.* HUP.

Grove, N. 1974. "Oil, the Dwindling Treasure." *National Geographic* 145:792–825.

Guillemot, H. 2014. "Les Désaccords sur le Changement Climatique en France: Au-delà d'un Climat Bipolaire." *Natures Sciences Sociétés* 22–340–50.

Guo, Z., et al. 2010. "Discovery, Evaluation and Distribution of Haplotypes of the Wheat Ppd-D1 Gene." *New Phytologist* 185:841–51.

Gwynne, P. 1975. "The Cooling World." *Newsweek,* 28 April.

———. 2014. "My 1975 'Cooling World' Story Doesn't Make Today's Climate Scientists Wrong." *Insidescience.org,* 21 May.*

Gyllenborg, G. A. (J. G. Wallerius). 1770 (1761). *The Natural and Chemical Elements of Agriculture,* trans. J. Mills. London: John Bell.

Hailman, J. 2006. *Thomas Jefferson and Wine.* Oxford: University Press of Mississippi.

Hall, S. S. 2016. "Editing the Mushroom." *Scientific American* 314:56–63.

———. 1987. "One Potato Patch That Is Making Genetic History." *Smithsonian* 18:125–36.

Hallegatte, S., et al. 2013. "Future Flood Losses in Major Coastal Cities." *Nature Climate Change* 3:802–6.

Hambler, C. 2013. "Wind Farms vs. Wildlife." *The Spectator,* 5 Jan.*

Hamilton, J. D. 2013. "Historical Oil Shocks." In *Routledge Handbook of Major Events in Economic History,* ed. R. E. Parker and R. Whaples, 239–65. NY: Routledge.

Hamilton, T. J. 1949. "Estimate of 500-Year Oil Supply Draws Criticism in U.N. Parley." *NYT,* 23 Aug.

Hanlon, J. 1974. "Top Food Scientist Published False Data." *New Scientist* 64:436–37.

Hansen, J. E., et al. 1992. "Potential Climate Impact of Mount Pinatubo Eruption." *Geophysical Research Letters* 19: 215-18

Hansen, J. E., et al. 1978. "Mount Agung Eruption Provides Test of a Global Climatic Perturbation." *Science* 199:1065–67.

Hanson, E. P. 1949. "Mankind Need Not Starve." *The Nation* 169:464–67.

Harari, Y. N. 2015. *Sapiens: A Brief History of Humankind*. NY: HarperCollins.

Hardin, G. 1976. "Carrying Capacity as an Ethical Concept." *Soundings* 59:120–37.

Hare, F. K. 1988. "World Conference on the Changing Atmosphere: Implications for Security, held at the Toronto Convention Centre, Toronto, Ontario, Canada, During 27–30 June 1988." *Environmental Conservation* 15:282–83.

Harper, J. A., and C. L. Cozart. 1990. *Oil and Gas Developments in Pennsylvania in 1990 with Ten-Year Review and Forecast*. Harrisburg: Pennsylvania Geological Survey.*

Harrington, W., et al. 2000. "On the Accuracy of Regulatory Cost Estimates." *Journal of Policy Analysis and Management* 19:297–322.

Hartmann, B. 2017. *The America Syndrome: Apocalypse, War, and Our Call to Greatness*. NY: Seven Stories Press.

———. 1995 (1987). *Reproductive Rights and Wrongs: The Global Politics of Population Control*. Rev. ed. Boston: South End Press.

———. and J. Boyce. 2013 (1983). *A Quiet Violence: View from a Bangladesh Village*. NY: CreateSpace.

Harwood, J. 2009. "Peasant Friendly Plant Breeding and the Early Years of the Green Revolution in Mexico." *Agricultural History* 83:384–410.

Hasson, D. 2010. "In Memory of Sidney Loeb." *Desalination* 261:203-04.

Haszeldine, R. S. 2009. "Carbon Capture and Storage: How Green Can Black Be?" *Science* 325:1647–52.

Haub, C. 1995. "How Many People Have Ever Lived on Earth?" *Population Today*, Feb.

Hawkes, L. 1940. "Prof. A. G. Högbom." *Nature* 145:769.

Hay, W. H. 2013. *Experimenting on a Small Planet: A Scholarly Entertainment*. NY: Springer.

Hayes, D. 1977. *Rays of Hope: The Transition to a Post-Petroleum World*. NY: W. W. Norton.

Hayes, H. K., et al. 1936. *Thatcher Wheat*. St. Paul: University of Minnesota Agricultural Experiment Station Bulletin 325.*

Hayes, H. K., and R. J. Garber. 1921. *Breeding Crop Plants*. NY: McGraw-Hill.

Hayes, R. C., et al. 2012. "Perennial Cereal Crops: An Initial Evaluation of Wheat Derivatives." *Field Crops Research* 133:68–89.

Hazell, P. B. R. 2009. *The Asian Green Revolution*. Washington, D.C.:International Food Policy Research Institute.

Heilbroner, R. L. 1995 (1953). *The Worldly Philosophers*. 7th ed. NY: Touchstone.

Helms, J. D. 1984. "Walter Lowdermilk's Journey: Forester to Land Conservationist." *Environmental Review* 8:132–45.

Hempel, L. C. 1983. "The Politics of Sunshine: An Inquiry into the Origin, Growth, and Ideological Character of the Solar Energy Movement in America." Ph.D. dissertation, Claremont Graduate School.

Henshilwood, C. S., et al. 2011. "A 100,000-Year-Old Ochre-Processing Workshop at Biombos Cave, South Africa." *Science* 334:219–22.

Henshilwood, C. S., and F. d'Errico. 2011. "Middle Stone Age Engravings and Their Significance to the Debate on the Emergence of Symbolic Material Culture." In Henshilwood, C. S., and F. d'Errico, eds. *Homo Symbolicus: The Dawn of Language, Imagination, and Spirituality*, 75–96. Amsterdam: John Benjamins.

Herivel, J. 1975. *Joseph Fourier: The Man and the Physicist*. Oxford: Clarendon Press.

Hernandez, R. R., et al. 2014. "Environmental Impacts of Utility-Scale Solar Energy." *Renewable and Sustainable Energy Reviews* 29:766–79

Hertzman, H. 2017. *Atrazine in European Groundwater: The Distribution of Atrazine and its Relation to the Geological Setting.* M.S. Thesis, Umeå University (Sweden).*

Hesketh, I. 2009. *Of Apes and Ancestors: Evolution, Christianity, and the Oxford Debate.* Toronto: University of Toronto Press.

Hesser, L. 2010. *The Man Who Fed the World.* NY: Park East Press.

Hettel, G. 2008. "Luck Is the Residue of Design." *Rice Today*, Jan.*

Heun, M., et al. 1997. "Site of Einkorn Wheat Domestication Identified by DNA Fingerprinting." *Science* 278:1312–14.

Hewitt de Alcántara, C. 1978. *La Modernización de la Agricultura Mexicana, 1940–1970.* México, D.F.: Siglo XXI.

Hibberd, J., et al. 2008. "Using C4 Photosynthesis to Increase the Yield of Rice—Rationale and Feasibility." *Current Opinion in Plant Biology* 11:228–31.

Hildahl, K. 2001. *Saude: A Brief History of a Village in Northeast Iowa.* Privately printed.*

Hinkel, J., et al. 2014. "Coastal Flood Damage and Adaptation Costs Under 21st Century Sea-Level Rise." *PNAS* 111:3292–97.

Hisard, P. 1992. "Centenaire de l'Observation du Courant Côtier El Niño, Carranza, 1892: Contributions de Krusenstern et de Humboldt à l'Observation du Phénomène 'ENSO.'" In *Paleo-ENSO Records International Symposium: Extended Abstracts*, ed. L. Ortlieb and J. Macharé, 133–41. Lima: ORSTOM/CONCYTEC.

Hitchens, C. 2005. "Equal Time." *Slate*, Aug 23.*

Hoddeson, L. 1981. "The Discovery of the Point-Contact Transistor." *Historical Studies in the Physical Sciences* 12:41–76.

Hoffman, E. 1896. *La Vie et les Travaux de Charles le Maout (1805–1887).* Le Havre: Imprimerie François le Roi.

Hoglund, A. W. 1961. "Wisconsin Dairy Farmers on Strike." *Agricultural History* 35:24–34.

Holdgate, M. 2013 (1999). *The Green Web: A Union for World Conservation.* NY: Earthscan.

———. 2001. "A History of Conservation." *In Our Fragile World: Challenges and Opportunities for Sustainable Development*, ed. M. K. Tolba, 1:341–53. Oxford: EOLSS Publishers.

Holland, J. 1835. *The History and Description of Fossil Fuel, the Colleries, and Coal Trade of Great Britain.* London: Whitaker and Co.*

Hollerson, W. 2010. "Popular Protests Put Brakes on Renewable Energy." *Der Spiegel*, 21 Jan.*

Hollett, D. 2008. *More Precious than Gold: The Story of the Peruvian Guano Trade.* Teaneck, NJ: Fairleigh Dickinson University Press.

Holthaus, E. 2015. "Stop Vilifying Almonds." *Slate*, 17 April.*

Holway, D. A., et al. 2002. "The Causes and Consequences of Ant Invasions." *Annual Review of Ecology and Systematics* 33:181–233.

Hoopman, D. 2007. "The Faithful Heretic: A Wisconsin Icon Pursues Tough Questions." *Wisconsin Energy Cooperative News*, May.*

Hopkins, D. P. 1948. *Chemicals, Humus, and the Soil.* London: Faber and Faber.

Hossard, L., et al. 2016. "A Meta-Analysis of Maize and Wheat Yields in Low-Input vs. Conventional and Organic Systems." *Agronomy Journal* 108:1155–67.

Howard, A. 1945. *Farming and Gardening for Health or Disease.* London: Faber and Faber.

———. 1940. *An Agricultural Testament.* OUP.

Howard, A., and Y. D. Wad. 1931. *The Waste Products of Agriculture: Their Utilization as Humus.* OUP.

Howard, E. 1902. *Garden Cities of To-Morrow*. London: Swan Sonnenschein.*

———. 1898. *To-Morrow: A Peaceful Path to Real Reform*. London: Swan Sonnenschein.*

Howard, L. E. 1953. *Sir Albert Howard in India*. London: Faber and Faber.*

Howeler, R. H., ed. 2011. *The Cassava Handbook*. Cali, Colombia: CIAT.

Huang, C. Y., et al. 2003. "Direct Measurement of the Transfer Rate of Chloroplast DNA into the Nucleus." *Nature* 422:72–76.

[Hubbert, M. K.] 2008 (1934). *Technocracy Study Course*. 5th ed. NY: Technocracy, Inc.*

———. 1962. *Energy Resources*. Washington, DC: National Academy of Sciences–National Research Council (Publication 1000-D).

———. 1956. *Nuclear Energy and the Fossil Fuels*. Houston: Shell Development Company Publication No. 95.

———. 1951. "Energy from Fossil Fuels." In *Smithsonian Institution. Annual Report of the Board of Regents of the Smithsonian Institution (1950)*, 255–72. Washington, DC: Government Printing Office.*

———. 1949. "Energy from Fossil Fuels." *Science* 109:103–9.

———. 1938. "Determining the Most Probable." *Technocracy* 12:4–10.

Hugo, V. 2000 (1862). *Les Misérables*, trans. C. E. Wilbour. NY: Random House.*

Hulme, M. 2011. "Reducing the Future to Climate: A Story of Climate Determinism and Reductionism." *Osiris* 26:245–66.

———. 2009a. "On the Origin of 'the Greenhouse Effect': John Tyndall's 1859 Interrogation of Nature." *Weather* 64:121–23.

———. 2009b. *Why We Disagree About Climate Change: Understanding Controversy, Inaction, and Opportunity*. CUP.

———. 2001. "Climatic Perspectives on Sahelian Desiccation: 1973–1998." *GEC* 11:21–23.

Hunt, S. J. 1973. "Growth and Guano in 19th-Century Peru." Research Program in Economic Development, Woodrow Wilson School, Princeton University, Discussion Paper No. 34.*

Hunter, M. C., et al. 2017. "Agriculture in 2050: Recalibrating Targets for Sustainable Intensification." *BioScience* bix010. doi: 10.1093/biosci/bix010.

Hunter, R. 2009. "Positionality, Perception, and Possibility in Mexico's Valle del Mezquital." *Journal of Latin American Geography* 8:49–69.

Hutchens, J. K. 1946. "People Who Read and Write." *NYT*, 17 March.

Hutchins, J. 2000. "Warren Weaver and the Launching of MT: Brief Biographical Note." In *Early Years in Machine Translation: Memoirs and Biographies of Pioneers*, ed. J. Hutchins, 17–20. Amsterdam: John Benjamins.

Hutchinson, G. E. 1950. *The Biogeochemistry of Vertebrate Excretion*. NY: American Museum of Natural History (Bulletin 96).

———. 1948. "On Living in the Biosphere." *Scientific Monthly* 67:393–97.

Huxley, A. 1993 (1939). *After Many a Summer Dies the Swan*. Chicago: Ivan R. Dee.

Huxley, J. 1946. *UNESCO: Its Purpose and Its Philosophy*. London: Preparatory Commission of the United Nations Educational, Scientific, and Cultural Organisation.*

———. 1937 (1926). *Essays in Popular Science*. London: Penguin Books.

———. 1933 (1931). *What Dare I Think? The Challenge of Modern Science to Human Action and Belief*. London: Chatto and Windus.

Huxley, J., and A. C. Haddon. 1939 (1931). *We Europeans: A Survey of "Racial" Problems*. Harmondsworth, Middlesex: Penguin Books.

Huxley, L. 1901. *Life and Letters of Thomas Henry Huxley*. 2 vols. New York: D. Appleton.*

Huxley, T. H. 1887. "On the Reception of the 'Origin of Species.'" In *The Life and Letters* *of Charles Darwin, Including an Autobiographical Chapter,* ed. F. Darwin, 2:179–204. London: John Murray.*

Inarejos Muñoz, J. A. 2010. "De la Guerra del Guano a la Guerra del Godo. Condicionantes, Objetivos y Discurso Nacionalista del Conflicto de España con Perú y Chile (1862–1867)." *Revista de Historia Social y de las Mentalidades* 14:137–70.

Ingram, F. W., and G. A. Ballard. 1935. "The Business of Migratory Divorce in Nevada." *Law and Contemporary Problems* 2:302–8.

Inman, M. 2016. *The Oracle of Oil: A Maverick Geologist's Quest for a Sustainable Future.* NY: W. W. Norton.

Institute for Economics and Peace. 2017. *Global Peace Index 2017.* IEP Report 48. Sydney: Institute for Economics and Peace.*

Instituto Nacional de Estadística y Geografía (México). 2015. *Estadísticas Históricas de México 2014.* México, D.F.: INEGI.*

International Atomic Energy Agency. 2008. *Estimation of Global Inventories of Radioactive Waste and Other Radioactive Materials.* IAEA-TECDOC-1591. Vienna: IAEA.*

International Desalination Agency. 2017. *IDA Desalination Yearbook 2016–2017.* Topsfield, MA: IDA.

International Energy Agency (IEA). 2016. *World Energy Outlook 2016.* Paris:IEA.

———. 2015a. *CO_2 Emissions from Fuel Combustion.* Paris: IEA.*

———. 2015b. *Key World Energy Statistics 2015.* Paris: IEA.*

———. 2014a. *Energy Technology Perspectives 2014: Harnessing Electricity's Potential.* Paris: IEA.*

———. 2014b. *World Energy Outlook 2014.* Paris: IEA.*

———. 2012. *Technology Roadmap: High-Efficiency, Low-Emissions Coal-Fired Power Generation.* Paris: IEA.

International Energy Agency Coal Industry Advisory Board. 2010. *Power Generation from Coal: Measuring and Reporting Efficiency Performance and CO2 Emissions.* Paris: IEA.*

International Labor Organization. 2017. *Global Estimates of Modern Slavery: Forced Labor and Forced Marriage.* Geneva: ILO.*

Isaacson, W. 2014. *The Innovators: How a Group of Hackers, Geniuses, and Geeks Created the Digital Revolution.* NY: Simon and Schuster.

Israel, G., and A. M. Gasca. 2002. *The Biology of Numbers: The Correspondence of Vito Volterra on Mathematical Biology.* Basel: Birkhäuser.

Ittersum, M. K. v., et al. 2013. "Yield Gap Analysis with Local to Global Relevance—A Review." *Field Crops Research* 143:4–17.*

Iyer, R. D. 2002. *Scientist and Humanist: M. S. Swaminathan.* Mumbai: Bharatiya Vidya Bhavan.

Jack, M. 1968. "The Purchase of the British Government's Shares in the British Petroleum Company, 1912–1914." *Past & Present* 39:139–68.

Jacobson, M. Z., et al. 2015a. "Low-Cost Solution to the Grid Reliability Problem with 100% Penetration of Intermittent Wind, Water, and Solar for All Purposes." *PNAS* 112:15060–65.

Jacobson, M. Z., et al. 2015b. "100% Clean and Renewable Wind, Water, and Sunlight (WWS) All-Sector Energy Roadmaps for the 50 United States." *Energy and Environmental Science* 8:2093–117.

Jacobson, M. Z., and M. A. Delucchi. 2011a. "Providing All Global Energy with Wind, Water, and Solar Power, Part I: Technologies, Energy Resources, Quantities and Areas of Infrastructure, and Materials." *Energy Policy* 39: 1154–69.

———. 2011b. "Providing All Global Energy with Wind, Water, and Solar Power, Part II: Reliability, System and Transmission Costs, and Policies." *Energy Policy* 39: 1170–90.

————. 2009. "A Path to Sustainable Energy by 2030." *Scientific American* 301:58–65.

Jagadish, S. V. K, et al. 2015. "Rice Responses to Rising Temperatures—Challenges, Perspectives, and Future Directions." *Plant, Cell, and Environment* 38:1686-98.

James, P. 2006 (1979). *Population Malthus: His Life and Times*. Oxford: Routledge.

Jamieson, D. 2014. *Reason in a Dark Time: Why the Struggle Against Climate Change Failed—and What It Means for Our Future*. OUP.

Jarrige, F. 2010. "'Mettre le Soleil en Bouteille': Les Appareils de Mouchot et l'Imaginaire Solaire au Début de la Troisième République." *Romantisme* 150:85–96.

Jaureguy, F., et al. 2008. "Phylogenetic and Genomic Diversity of Human Bacteremic *Escherichia coli* Strains." *BMC Genetics* 9:560.*

Jennings, P. R. 1964. "Plant Type as a Rice Breeding Objective." *Crop Science* 4:13–15.

Jensen, J. V. 1991. *Thomas Henry Huxley: Communicating for Science*. Cranbury, NJ: Associated University Presses.

————. 1988. "Return to the Wilberforce-Huxley Debate." *British Journal for the History of Science* 21:161–79.

Jeon, J.-S., et al. 2011. "Genetic and Molecular Insights into the Enhancement of Rice Yield Potential." *Journal of Plant Biology* 54:1-9.

Jesness, O. B., et al. 1936. *The Twin City Milk Market*. Bulletin 331. Minneapolis: University of Minnesota Agricultural Experiment Station.

Jevons, W. S. 1866. *The Coal Question; An Inquiry Concerning the Progress of the Nation, and the Probable Exhaustion of Our Coal-Mines*. 2nd ed. London: Macmillan and Co.*

Jez, J. M., et al. 2016. "The Next Green Movement: Plant Biology for the Environment and Sustainability." *Science* 353:1241-44.

Ji, Q., and S. Ji. 1996. "On the Discovery of the Earliest Fossil Bird in China (Sinosauropteryx gen. nov.) and the Origin of Birds." *Chinese Geology* 233:30–33.*

Jiang, Q., et al. 2016. "Rational Persuasion, Coercion or Manipulation? The Role of Abortion in China's Family Policies." *Annales Scientia Politica* 5:5–16.

Johansson, Å., et al. 2012. *Looking to 2060: Long-Term Global Growth Prospects*. OECD Economic Policy Papers 3. Paris: Organisation for Economic Co-operation and Development.*

Johnson, C. E. 2015. "'Turn on the Sunshine': A History of the Solar Future." Ph.D. dissertation, University of Washington.

Jonas, H. 1984. *The Imperative of Responsibility*. Chicago: University of Chicago Press.

Jones, G., and L. Bouamane. 2012. "'Power from Sunshine': A Business History of Solar Energy." Harvard Business School Working Paper 12-105.*

Jongman, B., et al. 2012. "Global Exposure to River and Coastal Flooding: Long Term Trends and Changes." *GEC* 22:823–35.

Jonsson, F. A. 2014. "The Origins of Cornucopianism: A Preliminary Genealogy." *Critical Historical Studies* 1:151–68.

————. 2013. *Enlightenment's Frontier: The Scottish Highlands and the Origins of Environmentalism*. YUP.

Josey, C. C. 1923. *Race and National Solidarity*. NY: Charles Scribner's Sons.

Joshi, S. R., et al. 2015. "Physical and Economic Consequences of Sea-Level Rise: A Coupled GIS and CGE Analysis Under Uncertainties." *Environmental and Resource Economics* 65(4):813–39.

Joyce, C. 1985. "Strawberry Field Will Test Man-Made Bacterium." *New Scientist*, 14 Nov.

Jundt, T. 2014a. *Greening the Red, White, and Blue: The Bomb, Big Business, and Consumer Resistance in Postwar America*. OUP.

————. 2014b. "Dueling Visions for the Postwar World: The UN and UNESCO Conferences on Resources and Nature, and the Origins of Environmentalism." *Journal of American History* 101:44–70.

Jungk, A. 2009. "Carl Sprengel—the Founder of Agricultural Chemistry: A Re-appraisal Commemorating the 150th Anniversary of His Death." *Journal of Plant Nutrition and Soil Science* 172:633–36.

Kabashima, J. N., et al. 2007. "Aggressive Interactions Between *Solenopsis invicta* and *Linepithema humile* (Hymenoptera: Formicidae) Under Laboratory Conditions." *Journal of Economic Entomology* 100:148–54.

Kabore, D., and C. Reij. 2004. *The Emergence and Spreading of an Improved Traditional Soil and Water Conservation Practice in Burkina Faso.* Washington, DC: International Food Policy Research Institute.

Kaessman, H., et al. 2001. "Great Ape DNA Sequences Reveal a Reduced Diversity and an Expansion in Humans." *Nature Genetics* 27:155-56.

Kale, S. 2014. *Electrifying India: Regional Political Economies of Development.* Palo Alto, CA: Stanford University Press.

Karatayev, A. Y., et al. 2014. "Twenty-Five Years of Changes in *Dreissena* spp. Populations in Lake Erie." *Journal of Great Lakes Research* 40:550–59.

Kasting, J. F. 2014. "Atmospheric Composition of Hadean–Early Archean Earth: The Importance of CO." In *Earth's Early Atmosphere and Surface Environment*, ed. G. H. Shaw, 19–28. Geological Society of America Special Paper 504.

Katz, Y. 1994. "The Extension of Ebenezer Howard's Ideas on Urbanization Outside the British Isles: The Example of Palestine." *GeoJournal* 34:467–73.

Kauppi, P. E., et al. 1992. "Biomass and Carbon Budget of European Forests, 1971 to 1990." *Science* 256:70-74.

Kay, L. E. 1993. *The Molecular Vision of Life: Caltech, the Rockefeller Foundation, and the Rise of the New Biology.* OUP.

Kean, S. 2012. *The Violinist's Thumb: And Other Lost Tales of Love, War, and Genius, as Written by Our Genetic Code.* Boston: Little, Brown.

Keeley, L. H. 1996. *War Before Civilization: The Myth of the Peaceful Savage.* OUP.

Keeling, C. D. 1998. "Rewards and Penalties of Monitoring the Earth." *Annual Review of Energy and the Environment* 23:25–82.

———. 1978. "The Influence of Mauna Loa Observatory on the Development of Atmospheric CO_2 Research." In *Mauna Loa Observatory: A 20th Anniversary Report*, ed. J. Miller, 36–54. Washington, DC: National Oceanic and Atmospheric Administration.

———. 1960. "The Concentration and Isotopic Abundances of Carbon Dioxide in the Atmosphere." *Tellus* 12:200–3.

Keirn, T. 1988. "Daniel Defoe and the Royal African Company." *Historical Research* 61:243–47.

Keith, D. W. 2013. *A Case for Climate Engineering.* Boston: Boston Review Press.

———. 2000. "Geoengineering the Climate: History and Prospect." *Annual Review of Energy and the Environment* 25:245–84.

———et al. 2016. "Stratospheric Solar Geoengineering without Ozone Loss." *PNAS* 113:14910-14.

Kemp, J. 2013. "Peak Oil, Not Climate Change Worries Most Britons." Reuters, 18 July.*

Keynes, R. 2002 (2001). *Darwin, His Daughter, and Human Evolution.* NY: Riverhead Books.

Kincer, J. B. 1933. "Is Our Climate Changing? A Study of Long-Time Temperature Trends." *Monthly Weather Review* 61:251–59.

King, J. R., and W. R. Tschinkel. 2008. "Experimental Evidence That Human Impacts Drive Fire Ant Invasions and Ecological Change." *PNAS* 105:20339–43.

King, R. J. 2013. *The Devil's Cormorant: A Natural History.* Lebanon, N.H.: University Press of New England.

Kingsbury, N. 2011 (2009). *Hybrid: The History and Science of Plant Breeding.* Chicago: University of Chicago Press.

Kingsland, S. E. 2009. "Frits Went's Atomic Age Greenhouse: The Changing Labscape on the Lab-Field Border." *Journal of the History of Biology* 42:289–324.

———. 1995 (1985). *Modeling Nature: Episodes in the History of Population Ecology.* 2nd ed. Chicago: University of Chicago Press.

———. 1986. "Mathematical Figments, Biological Facts: Population Ecology in the Thirties." *Journal of the History of Biology* 19:235–56.

Kinkela, D. 2011. *DDT and the American Century: Global Health, Environmental Politics, and the Pesticide That Changed the World.* Chapel Hill: University of North Carolina Press.

Kintisch, E. 2010. *Hack the Planet: Science's Best Hope—or Worst Nightmare—for Averting Climate Catastrophe.* Hoboken, NJ: John Wiley and Sons.

Kirchmann, H., et al. 2016. "Flaws and Criteria for Design and Evaluation of Comparative Organic and Conventional Cropping Systems." *Field Crops Research* 186:99–106.

Kislev, M. E. 1982. "Stem Rust of Wheat 3300 Years Old Found in Israel." *Science* 216:993–94.

Kittler, R., et al. 2003. "Molecular Evoution of *Pediculus humanus* and the Origin of Clothing." *Current Biology* 14:1414–17 (erratum, 14:2309).

Klein, T. M., et al. "High-Velocity Microprojectiles for Delivering Nucleic Acids into Living Cells." *Nature* 327:70-73.

Kluckhohn, F. L. 1947. "$28,000,000 Urged to Support M.I.T." *NYT*, 15 June.

Knickerbocker, Jerry, and J. A. Harper. 2009. "Anatomy of a Ghost Town—Pithole." In *History and Geology of the Oil Regions of Northwestern Pennsylvania*, ed. J. A. Harper, 108–19. 74th Annual Field Conference of Pennsylvania Geologists. Middletown, PA: Field Conference of Pennsylvania Geologists.*

Knipling, E. F. 1945. "The Development and Use of DDT for the Control of Mosquitoes." *Journal of the National Malaria Society* 4:77-92.

Kniss, A. R., et al. 2016. "Commercial Crop Yields Reveal Strengths and Weaknesses for Organic Agriculture in the United States." *PLoS ONE* 11:e0161673.*

Knoll, E., and J. N. McFadden., eds. 1970. *War Crimes and the American Conscience.* NY: Holt, Rinehart and Winston.

Kolmer, J. A., et al. 2011. "Expression of a Thatcher Wheat Adult Plant Stem Rust Resistance QTL on Chromosome Arm 2BL Is Enhanced by Lr34." *Crop Science* 51:526–33.

Krajewski, C., and D. G. King. 1996. "Molecular Divergence and Phylogeny: Rates and Patterns of Cytochrome b Evolution in Cranes." *Molecular Biology and Evolution* 13:21–30.*

Krajewski, C., and J. W. Fetzner. 1994. "Phylogeny of Cranes (Gruiformes: Gruidae) Based on Cytochrome-*B* DNA Sequences." *The Auk* 111:351–65.

Kravitz, B., et al. 2014. "A Multi-Model Assessment of Regional Climate Disparities Caused by Solar Geoengineering." *Environmental Research Letters* 9:074013.*

Kremen, C., and A. Miles. 2012. "Ecosystem Services in Biologically Diversified Versus Conventional Farming Systems: Benefits, Externalities, and Trade-Offs." *Ecology and Society* 17:40.*

Kromdijk, J., et al. 2016. "Improving Photosynthesis and Crop Productivity by Accelerating Recovery from Photoprotection." *Science* 354:657–61.

Kropotkin, P. 1901 (1898). *Fields, Factories and Workshops; or, Industry Combined with Agriculture and Brain Work with Manual Work.* NY: G. P. Putnam's Sons.

Kryza, F. T. 2003. *The Power of Light: The Epic Story of Man's Quest to Harness the Sun.* NY: McGraw-Hill.

Kuhlwilm, M., et al. 2016. "Ancient Gene Flow from Early Modern Humans into Eastern Neanderthals." *Nature* 530:429–433.

Kunkel Water Efficiency Consulting. 2017. "Report on the Evaluation of Water Audit Data for Pennsylvania Water Utilities." Memorandum, 15 Feb.*

Kunstler, J. H. 2005. *The Long Emergency: Surviving the Converging Catastrophes of the Twenty-First Century.* NY: Grove Press.

Labban, M. 2008. *Space, Oil, and Capital.* London: Routledge.

Lacina, B., and N. P. Gleditsch. 2013. "The Waning of War Is Real: A Response to Gohdes and Price." *Journal of Conflict Resolution* 57:1109–27.

Lacina, B., et al. 2006. "The Declining Risk of Death in Battle." *International Studies Quarterly* 50:673–80.

Lane, C. S., et al. 2013. "Ash from the Toba Supereruption in Lake Malawi Shows No Volcanic Winter in East Africa at 75 ka." *PNAS* 110:8025–29.

Lane, N. 2002. *Oxygen: The Molecule that Made the World.* OUP.

Langley, S. P. 1888. "The Invisible Solar and Lunar Spectrum." *LPMJS* 26:505–20.

Lankford, B. 2012. "Fictions, Fractions, Factorials and Fractures: On the Framing of Irrigation Efficiency." *Agricultural Water Management* 108:27-38.

Laplantine, R. 2014. "Thinking Between Shores: Georges Devereux." *Books and Ideas*, 27 Oct.*

Larcher, D., and J.-M. Tarascon. 2015. "Towards Greener and More Sustainable Batteries for Electrical Energy Storage." *Nature Chemistry* 7:19–29.

Large, E. C. 1946 (1940). *Advance of the Fungi.* London: Jonathan Cape.

Larkin, P. J., and M. T. Newell. "Perennial Wheat Breeding: Current Germplasm and a Way Forward for Breeding and Global Cooperation." In Batello, C., et al., eds. *Perennial Crops for Food Security.* Rome: FAO, 39-53.

Lavalle y Garcia, J. A. d. 1917. "Informe Preliminar Sobre la Causa de la Mortalidad Anormal de las Aves Ocurrida en el Mes de Marzo del Presente Año." *Memoria del Directorio de la Compañía Administradora del Guano* 8:61–88.

Layton, R. 2005. "Sociobiology, Cultural Anthropology, and the Causes of Warfare." In *Warfare, Violence, and Slavery in Prehistory: Proceedings of a Prehistoric Society Conference at Sheffield University*, ed. M. P. Pearson and I. J. N. Thorpe, 41–48. Oxford: Archaeopress.

Lear, L. 2009 (1997). *Rachel Carson: Witness for Nature.* NY: Mariner Books.

LeBlanc, S. A., and K. E. Register. 2003. *Constant Battles: The Myth of the Peaceful, Noble Savage.* NY: St. Martin's Press.

Lebergott, S. 1993. *Pursuing Happiness: American Consumers in the Twentieth Century.* Princeton, NJ: Princeton University Press.

———. 1976. *The American Economy: Income Wealth and Want.* Princeton, NJ: Princeton University Press.

Leffler, E. M., et al. 2012. "Revisiting an Old Riddle: What Determines Genetic Diversity Levels within Species?" *PLoS Biology* 10:e1001388.

Leflaive, X., et al. 2012. "Water." In: *Organisation for Economic Cooperation and Development. OECD Environmental Outlook to 2050.* Paris: OECD, 207-.*

Lehmann, J. 2007. "A Handful of Carbon." *Nature* 447:143–44.

Lehmann, J., et al. 2006. "Bio-char Sequestration in Terrestrial Ecosystems—an Overview." *Mitigation and Adaptation Strategies for Global Change* 11:403–27.

Lelieveld, J., et al. 2015. "The Contribution of Outdoor Air Pollution Sources to Premature Mortality on a Global Scale." *Nature* 525:367–71.

Le Maout, C. 1902. *Lettres au Ministre de l'Agriculture sur le Tir du Canon et ses Conséquences au Point de Vue Agricole.* Le Havre: Imprimerie François le Roi.

Leonard, K. J. 2001. "Stem Rust—Future Enemy?" In *Stem Rust of Wheat: From Ancient Enemy to Modern Foe*, ed. P. D. Peterson, 119–46. St. Paul, MN: APS Press.

Leonard, K. J., and L. J. Szabo. 2005. "Stem Rust of Small Grains and Grasses Caused by *Puccinia graminis*." *Molecular Plant Pathology* 6:99–111.

Leopold, A. 1999. *The Essential Aldo Leopold: Quotations and Commentaries* (eds. C. D. Meine and R. L. Knight). Madison: University of Wisconsin Press.

———. 1993 (1953). *The Round River: A Parable*. In *Round River: From the Journals of Aldo Leopold*, ed. L. Leopold, 158–65. OUP.

———. 1991 (1941). "Ecology and Politics." In *The River of the Mother of God: and Other Essays by Aldo Leopold*, ed. S. L. Flader and J. B. Callicott, 281–86. Madison: University of Wisconsin Press.

———. 1979 (1923). "Some Fundamentals of Conservation in the Southwest." *Environmental Ethics* 1:131–41.

———. 1949. *A Sand County Almanac and Sketches Here and There*. OUP.

———. 1938. "Conservation Esthetic." *Bird-Lore* 40:101–9.

———. 1933. *Game Management*. NY: Charles Scribner's Sons.

———. 1924. "Grass, Brush, Timber, and Fire in Southern Arizona." *Journal of Forestry* 22:1–10.

Leridon, H. 2006. "Demographic Effects of the Introduction of Steroid Contraception in Developing Countries." *Human Reproduction Update* 12:603-16.

Leslie, S., ed. 1901. *Letters of John Richard Green*. London: Macmillan.*

Levenson, T. 1989. *Ice Time: Climate, Science, and Life on Earth*. NY: Harper & Row.

Levorsen, A. I. 1950. "Estimates of Undiscovered Petroleum Reserves." In United Nations 1950, Vol. 1:94–110.

Lev-Yadun, S., et al. 2000. "The Cradle of Agriculture." *Science* 288:1602–3.

Lewis, C. H. 1991. "Progress and Apocalypse: Science and the End of the Modern World." Ph.D. dissertation, University of Minnesota.

Lewis, S. 2009. "A Force of Nature: Our Influential Anthropocene Period." *The Guardian*, 23 July.*

Li, J. Z., et al. 2008. "Worldwide Human Relationships Inferred from Genome-Wide Patterns of Variation." *Science* 319:1100–44.

Li, W.-H., and L. A. Sadler. 1991. "Low Nucleotide Diversity in Man." *Genetics* 129:513–23.

Liang, Z., et al. 2016. "Review on Current Advances, Future Challenges and Consideration Issues for Post-Combustion CO_2 Capture Using Amine-Based Absorbents." *Chinese Journal of Chemical Engineering* 24:278–88.

Liao, P. V., and J. Dollin. 2012. "Half a Century of the Oral Contraceptive Pill: Historical Review and View to the Future." *Canadian Family Physician* 58:e757–e760.*

Liebig, J. v. 1859. *Letters on Modern Agriculture*, ed. and trans. J. Blyth. NY: John Wiley.*

———. 1855. *Die Grundsätze der agricultur-chemie mit Rücksicht auf die in Englend angestellten Untersuchungen*. Braunschweig: Friedrich Vieweg and Sohn.*

———. 1840. *Die Organische Chemie in ihrer Anwendung auf Agricultur und Physiologie*. Braunschweig: Friedrich Vieweg und Sohn.*

Lin, Q. F. 2014. "Aldo Leopold's Unrealized Proposals to Rethink Economics." *Ecological Economics* 108:104–14.

Lindow, S. E., et al. 1982. "Bacterial Ice Nucleation: A Factor in Frost Injury to Plants." *Plant Physiology* 70:1084–89.

Linnér, B.-O. 2003. *The Return of Malthus: Environmentalism and Post-war Population-Resource Crises*. Isle of Harris, UK: White Horse Press.

Liu, W., et al. 2015. "The Earliest Unequivocally Modern Humans in Southern China." *Nature* 526:696–700.

Lixil Group. 2016. *The True Cost of Poor Sanitation.* Tokyo: Lixil Group.*

Loftus, A. C. 2011. *Tel Aviv, Israel. Treating Wastewater for Reuse Using Natural Systems.* SWITCH Training Kit Case Study. Freiburg: ICLEI European Secretariat GmbH.*

Lord, R. 1948. "The Ground from Under Your Feet." *Saturday Review,* 7 Aug.

Lorence, J. T. 1988. "Gerald T. Boileau and the Politics of Sectionalism: Dairy Interests and the New Deal, 1933–1938." *Wisconsin Magazine of History* 71:276–95.

Lorenz, E. N. 1972. "Predictability: Does the Flap of a Butterfly's Wings in Brazil Set Off a Tornado in Texas?" Paper presented to American Association for the Advancement of Science, Washington, DC, 29 Dec.*

———. 1963. "Deterministic Nonperiodic Flow." *Journal of Atmospheric Sciences* 20:130–41.

Lotka, A., 1925. *Elements of Physical Biology.* Baltimore: Williams and Wilkins.

———. 1907. "Relation Between Birth Rates and Death Rates." *Science* 26:21–22.

Lovins, A. 1976. "Energy Strategy: The Road Not Taken?" *Foreign Affairs* 55:65–96.

Lowdermilk, W. C. 1948. *Conquest of the Land Through Seven Thousand Years.* Washington, DC: Soil Conservation Service.*

———. 1944. *Palestine: Land of Promise.* London: Victor Gollancz.

———. 1942. "Conquest of the Land through Seven Thousand Years." Mimeograph. Washington, DC: Soil Conservation Service.*

———. 1940. "Tracing Land Use Across Ancient Boundaries." Mimeograph. Washington, DC: Soil Conservation Service.*

———. 1939. "Reflections in a Graveyard of Civilizations." *Christian Rural Fellowship Bulletin* 45.

Lowdermilk, W. C., and M. Chall. 1969. *Soil, Forest, and Water Conservation and Reclamation in China, Israel, Africa, and the United States.* 2 vols. Typescript. Berkeley: University of California Regional Oral History Office.*

Lowe, K. M. 2016. *Baptized with the Soil: Christian Agrarians and the Crusade for Rural America.* OUP.

Lubofsky, E. 2016. "The Promise of Perennials." *CSA News,* Nov.

Lucas, J. R. 1979. "Wilberforce and Huxley: A Legendary Encounter." *Historical Journal* 22:313–30.

Luckey, T. D. 1972. "Introduction to Intestinal Microecology." *American Journal of Clinical Nutrition* 25:1292–94.

———. 1970. "Gnotobiology Is Ecology." *American Journal of Clinical Nutrition* 23:1533–40.

Lüdeke-Freund, F. 2013. "BP's Solar Business Model: A Case Study on BP's Solar Business Case and Its Drivers." *International Journal of Business Environment* 6:300-28.

Lumpkin, T. A. 2015. "How a Gene from Japan Revolutionized the World of Wheat: CIMMYT's Quest for Combining Genes to Mitigate Threats to Global Food Security." In *Advances in Wheat Genetics: From Genome to Field; Proceedings of the 12th International Wheat Genetics Symposium,* ed. Y. Ogihara, et al., 13–20. NY: Springer Open.

Luten, D. B. 1986 (1980). Ecological Optimism in the Social Sciences. In *Progress Against Growth: Daniel B. Luten on the American Landscape,* ed. T. R. Vale, 314–35. New York: Guilford Press.

Lynas, M. 2011. *The God Species: Saving the Planet in the Age of Humans.* Washington, DC: National Geographic Society.

Lynch, M. C. 2016. *The "Peak Oil" Scare and the Coming Oil Flood.* Santa Barbara, CA: Praeger.

Lyons, T. W., et al. 2014. "The Rise of Oxygen in Earth's Early Ocean and Atmosphere." *Nature* 506:307–14.

MacDonald, G. F., et al. 1979. *The Long Term Impact of Atmospheric Carbon Dioxide on Climate*. JASON Technical Report JSR-78-07. Alexandria, VA: SRI International.*

MacDowell, N., et al. 2010. "An Overview of CO_2 Capture Technologies." *Energy and Environmental Science* 3:1645–69.

Macekura, S. J. 2015. *Of Limits and Growth: The Rise of Global Sustainable Development in the Twentieth Century*. CUP.

Madrigal, A. 2011. *Powering the Dream: The History and Promise of Green Technology*. NY: Da Capo Press.

Madureira, N. L. 2012. "The Anxiety of Abundance: William Stanley Jevons and Coal Scarcity in the Nineteenth Century." *Environment and History* 18: 395–421.

Magiels, G. 2010. *From Sunlight to Insight: Jan IngenHousz, the Discovery of Photosynthesis, and Science in the Light of Ecology*. Brussels: VUBPress.

Magistro, L. 2015. *Amministrazione Straordinaria: Obiettivi e Primi Resultati*. Venice: Consorzio Venezia Nuova.*

Mahrane, Y., et al. 2012. "De la Nature à la Biosphère: l'Invention Politique de l'Environnement Global, 1945–1972." *Vingtième Siècle* 113:127–41.*

Maji, K. J., et al. 2017. "Disability-adjusted Life Years and Economic Cost Assessment of the Health Effects Related to PM2.5 and PM10 Pollution in Mumbai and Delhi, in India from 1991 to 2015." *Environmental Science and Pollution Research* 24:4709–30.

Malcolm, A. H. 1974. "The 'Shortage' of Bathroom Tissue: A Classic Study in Rumor." *NYT*, 3 Feb.

Mallet, J. 2012. "The Struggle for Existence: How the Notion of Carrying Capacity, K, Obscures the Links Between Demography, Darwinian Evolution, and Speciation." *Evolutionary Ecology Research* 14:627–65.

Malm, A. 2016. *Fossil Capital: The Rise of Steam Power and the Roots of Global Warming*. Brooklyn: Verso.

Malthus, T. R. 1872. *An Essay on the Principle of Population; or, A View of its Past and Present Effects on Human Happiness*. 7th ed. London: Reeves and Turner.*

———. 1826. *An Essay on the Principle of Population; or, A View of Its Past and Present Effects on Human Happiness*. 6th ed. London: John Murray.

[———]. 1798. *An Essay on the Principle of Population, as It Affects the Future Improvement of Society*. London: J. Johnson.*

Manabe, S., and S. T. Wetherald. 1975. "The Effects of Doubling the CO_2 Concentration on the climate of a General Circulation Model." *Journal of the Atmospheric Sciences* 32:3-15.

———. 1967. "Thermal Equilibrium of the Atmosphere with a Given Distribution of Relative Humidity." *Journal of the Atmospheric Sciences* 24:241-59.

Mané, A. 2011. "Americans in Haifa: The Lowdermilks and the American-Israeli Relationship." *Journal of Israeli History* 30:65–82.

Manlay, R. J., et al. 2006. "Historical Evolution of Soil Organic Matter Concepts and Their Relationships with the Fertility and Sustainability of Cropping Systems." *Agriculture, Ecosystems and Environment* 119:217–33.

Mann, C. C. 2015. "Solar or Coal? The Energy India Picks May Decide Earth's Fate." *Wired*, Nov.*

———. 2014. "Coal: It's Dirty, It's Dangerous, and It's the Future of Clean Energy." *Wired*, April.*

———. 2013. "What If We Never Run Out of Oil?": *AM* 311:48–61.*

———. 2011a. *1493: Uncovering the New World Columbus Created*. NY: Alfred A. Knopf.

———. 2011b. "The Birth of Religion." *National Geographic* 219:34–59.*

———. 2008. "Our Good Earth." *National Geographic* 214:80–106.*

———. 2007. "The Rise of Big Water." *Vanity Fair*, May.*

———. 2006. "The Long, Strange Resurrection of New Orleans." *Fortune*, 21 Aug.*

———. 2005. *1491: New Revelations of the Americas Before Columbus*. NY: Alfred A. Knopf.

———. 2004. *Diversity on the Farm: How Traditional Crops Around the World Help to Feed Us All*. NY: Ford Foundation/Political Economy Research Institute.*

———. 1999. "Genetic Engineers Aim to Soup Up Crop Photosynthesis." *Science* 283:314-16.

Mann, C. C., and M. L. Plummer. 1998. *Noah's Choice: The Future of Endangered Species*. NY: Alfred A. Knopf.

Mann, J. 2014. "Why Narendra Modi Was Banned from the U.S." *Wall Street Journal*, 2 May.

Mao, J.-D., et al. 2012. "Abundant and Stable Char Residues in Soils: Implications for Soil Fertility and Carbon Sequestration." *Environmental Science and Technology* 46:9571–76.

Marchant, J. 1917. *Birth-Rate and Empire*. London: Williams and Norgate.

Marchetti, C. 1977. "On Geoengineering and the CO_2 Problem." *Climatic Change* 1:59-68.

Margulis, L., and E. Dobb. 1990. "Untimely Requiem." *The Sciences* 30:44–49.

Margulis, L., and D. Sagan. 2003 (2002). *Acquiring Genomes: A Theory of the Origins of Species*. NY: Basic Books.

———.1997 (1986). *Microcosmos: Four Billion Years of Evolution from Our Microbial Ancestors*. UCP.

Marsh, G. P. 1864. *Man and Nature; or, Physical Geography as Modified by Human Action*. NY: Charles Scribner.*

Martin, R. 2013. *Earth's Evolving Systems: The History of Planet Earth*. Burlington, MA: Jones and Bartlett Learning.

Martin, W., et al. 2002. "Evolutionary Analysis of Arabidopsis, Cyanobacterial, and Chloroplast Genomes Reveals Plastid Phylogeny and Thousands of Cyanobacterial Genes in the Nucleus." *PNAS* 99:12246–51.*

Martin, W., et al. 1998. "Gene Transfer to the Nucleus and the Evolution of Chloroplasts." *Nature* 393:162–65.

Marx, K. 1906–1909 (1867–94). *Capital: A Critique of Political Economy*, trans. S. Moore and E. Aveling. 3 vols. Chicago: Charles H. Kerr & Co.

Marzano, A. 1996. "La Politica Inglese in Mesopotamia e il Ruolo del Petrolio (1900–1920)." *Il Politico* 61:629–650.

Matchett, K. 2006. "At Odds over Inbreeding: An Abandoned Attempt at Mexico/United States Collaboration to 'Improve' Mexican Corn, 1940–1950." *Journal of the History of Biology* 39:345–72.

Matelski, M. J. 1991. *Variety Sourcebook II: Film-Theater-Music*. Stoneham, MA: Focal Press.

Mathews, S. W. 1976. "What's Happening to Our Climate?" *National Geographic* 150:576–615.

Maugeri, L. 2006. *The Age of Oil: The Mythology, History, and Future of the World's Most Controversial Resource*. NY: Praeger Publishers.

Maugh, T. H., II. 1987a. "Frost Failed to Damage Sprayed Test Crop, Company Says." *Los Angeles Times*, 9 June.

———. 1987b. "Plants Used in UC's Genetic Test Uprooted." *Los Angeles Times*, 27 May.

Maurel, C. 2010. *Histoire de l'UNESCO: Les Trente Premières Années, 1945–1974*. Paris: l'Harmattan.

Mayer, K., et al. 2015. *Elite Capture: Subsidizing Electricity Use by Indian Households*.

Washington, DC: International Bank for Reconstruction and Development/The World Bank.

Mayhew, R. J. 2014. *Malthus: The Life and Legacies of an Untimely Prophet*. HUP.

Mazur, S., ed. 2010 (2009). *The Altenberg 16: An Exposé of the Evolution Industry*. Berkeley, CA: North Atlantic Books.

McCarrison, R. 1944 (1936). *Nutrition and National Health*. London: Faber and Faber.

———. 1921. *Studies in Deficiency Disease*. London: Frowde, Hodder & Stoughton.*

McCarrison, R., and B. Viswanath. 1926. "The Effect of Manurial Conditions on the Nutritive and Vitamin Values of Millet and Wheat." *Indian Journal of Medical Research* 14:351–78.

McClellan, J., et al. 2012. "Cost Analysis of Stratospheric Albedo Modification Delivery Systems." *Environmental Research Letters* 7:034019.*

McClellan, J., et al. 2011. *Geoengineering Cost Analysis*. Cambridge, MA: Aurora Flight Sciences.*

McCormick, M. A. 2005. "Of Birds, Guano, and Man: William Vogt's Road to Survival." Ph.D. dissertation, University of Oklahoma.

McCulloch, J. R. 1854 (1837). *A Descriptive and Statistical Account of the British Empire: Exhibiting Its Extent, Physical Capacities, Population, Industry, and Civil and Religious Institutions*. 4th ed. 2 vols. London: Longman, Brown, Green, and Longmans.*

McCusker, K. E., et al. 2014. "Rapid and Extensive Warming Following Cessation of Solar Radiation Management." *Environmental Research Letters* 9: 024005.*

McDonald, R. I., et al. 2009. "Energy Sprawl or Energy Efficiency: Climate Policy Impacts on Natural Habitat for the United States of America." *PLoS ONE* 4:e6802.*

McFadden, G. I. 2014. "Origin and Evolution of Plastids and Photosynthesis in Eukaryotes." *Cold Spring Harbor Perspectives in Biology* 6:a016105.*

McGee, W. J., ed. 1909. *Proceedings of a Conference of Governors in the White House, Washington, DC, May 13–15, 1908*. Washington, DC: Government Printing Office.*

McGrory, M. 1948. "Apostle of Conserving World's Resource Says Education Is 1 of 3 Main Factors." *Washington Star*, 22 Aug.

McKelvey, J. J. 1987. "J. George Harrar, 1906–1982." *Biographical Memoirs of the National Academy of Sciences* 57:27–56.*

McKibben, B. 2007. "Green from the Ground Up." *Sierra* 92:42–46, 73–75.

———. 1989. *The End of Nature*. New York: Anchor Books.

McLaurin, J. J. 1902 (1896). *Sketches in Crude-Oil: Some Accidents and Incidents of the Petroleum Development in All Parts of the Globe*. 3rd ed. Franklin, PA: J. J. McLaurin.*

McMahon, S., and J. Parnell. 2014. "Weighing the Deep Continental Biosphere." *FEMS Microbiology* 87:113–20.

McNeill, J. R. 2001. *Something New Under the Sun: An Environmental History of the Twentieth Century World*. NY: W. W. Norton.

Meacham, S. 1970. *Lord Bishop: The Life of Samuel Wilberforce, 1805–1873*. HUP.

Meadows, D. H., et al. 1972. *The Limits to Growth*. NY: Universe Books.

Meine, C. D. 2013. "Notes on the Texts and Illustrations." In *Aldo Leopold: A Sand County Almanac and Other Writings on Ecology and Conservation*, by A. Leopold, 859–72. NY: Library of America.

———. 2010 (1988). *Aldo Leopold: His Life and Work*. 2nd ed. Wisconsin: University of Wisconsin Press.

Meine, C. D., and G. W. Archibald., eds. 1996. *The Cranes: Status Survey and Conservation Action Plan*. Gland, Switzerland: IUCN.

Mejcher, H. 1972. "Oil and British Policy Towards Mesopotamia, 1914–1918." *Middle Eastern Studies* 8:377–91.

Mekonnen, M. M. and A. Y. Hoekstra. 2016. "Four Billion People Facing Severe Water Scarcity." *Science Advances* 2(2): e1500323.

Melillo, E. M. 2012. "The First Green Revolution: Debt Peonage and the Making of the Nitrogen Fertilizer Trade, 1840–1930." *American Historical Review* 117:1028–60.

Mellor, J. W., and S. Gavian. 1987. "Famine: Causes, Prevention, and Relief." *Science* 235:539–45.

Mence, T., ed. 1981. *IUCN: How It Began, How It Is Growing Up*. Mimeograph. International Union for Conservation of Nature and Natural Resources.*

Mendoza García, M. E., and G. Tapia Colocia. 2010. "La Situación Demográfica de México, 1910–2010." In *Consejo Nacional de Población. La Situación Demográfica de México 2010*. México, D.F.: CONAPO.*

Menon, S., et al. 2010. "Black Carbon Aerosols and the Third Polar Ice Cap." *Atmospheric Chemistry and Physics* 10:4559–71.

Mercader, J. 2009. "Mozambican Grass Seed Consumption During the Middle Stone Age." *Science* 326:1680–3.

Metzger, R. L., and B. A. Silbaugh. 1970. "Location of Genes for Seed Coat Color in Hexaploid Wheat, Triticumaestivum L." *Crop Science* 10:495–96.

Miller, C. 2003. "Rough Terrain: Forest Management and Its Discontents, 1891–2001." *Food, Agriculture and Environment* 1:135–38.

———. 2001. *Gifford Pinchot and the Making of Modern Environmentalism*. Washington, DC: Island Press.

Miller, C., and J. G. Lewis. 1999. "A Contested Past: Forestry Education in the United States, 1898–1998." *Journal of Forestry* 97:38–43.

Miller, R. 2003. "Bible and Soil: Walter Clay Lowdermilk, the Jordan Valley Project and the Palestine Debate." *Middle Eastern Studies* 39:55–81.

Miller, R., and R. Greenhill. 2006. "The Fertilizer Commodity Chains: Guano and Nitrate, 1840–1930." In *From Silver to Cocaine: Latin American Commodity Chains and the Building of the World Economy, 1500–2000*, ed. S. Topik, et al. Durham, NC: Duke University Press.

Miller, S. W. 2007. *An Environmental History of Latin America*. CUP.

Milman, O., and J. Glenza. 2016. "At Least 33 US Cities Used Water Testing 'Cheats' over Lead Concerns." *The Guardian*, 2 Jun.*

Mitchell, T. 2011. *Carbon Democracy: Political Power in the Age of Oil*. NY: Verso.

Mlot, N. J., et al. 2011. "Fire Ants Self-Assemble into Waterproof Rafts to Survive Floods." *PNAS* 108:7669–73.*

Moffett, A. 1937. "Audubon's 'Birds of America': His Monumental Work Becomes Available to the General Public." *NYT*, Dec. 5.

Moffett, M. W. 2012. "Supercolonies of Billions in an Invasive Ant: What Is a Society?" *Behavioral Ecology* 23:925–33.

Mohr, J. C. 1970. "Academic Turmoil and Public Opinion: The Ross Case at Stanford." *Pacific Historical Review* 39:39–61.

Mokyr, J., ed. 2003. *The Oxford Encyclopedia of Economic History*. 4 vols. OUP.

Möller, F. 1963. "On the Influence of Changes in the CO_2 Concentration in Air on the Radiation Balance of the Earth's Surface and on the Climate." *Journal of Geophysical Research* 68: 3877–86.

Montgomery, J. 2013. "K Road Gives Up on Calico Solar Project." *Renewable Energy World*, 1 July.*

Moore-Colyear, R. J. 2002. "Rolf Gardiner, English Patriot and the Council for the Church and Countryside." *The Agricultural History Review* 49:187-209.

More, A. 1917 (1916). *Uncontrolled Breeding, or Fecundity Versus Civilization*. NY: Critic and Guide Co.

Morris, I. 2014. *War! What Is It Good For? Conflict and the Progress of Civilization from Primates to Robots.* NY: Farrar, Straus, and Giroux.

Morris, W. 1914 (1881). "Art and the Beauty of the Earth." In *The Collected Works of William Morris*, ed. M. Morris, 22:155–74. CUP.

Morrisette, P. M. 1988. "The Evolution of Policy Responses to Stratospheric Ozone Depletion." *Natural Resources Journal* 29:793-820.

Morton, O. 2015. *The Planet Remade: How Geoengineering Could Change the World.* Princeton, NJ: Princeton University Press.

———. 2009 (2007). *Eating the Sun: How Plants Power the Planet.* NY: Harper Perennial.

Mosher, S. 2008. *Population Control: Real Costs, Illusory Benefits.* NY: Transaction Publishers.

Mouchot, A. 1875. "Résultats Obtenus dans les Essais d'Applications Industrielles de la Chaleur Solaire." *Comptes Rendus Hebdomadaires des Séances de l'Académie des Sciences* 81:571–74.*

———. 1869a. *La Chaleur Sociale et les Applications Industrielles.* Paris: Gauthier-Villars.*

———. 1869b. "La Chaudière Solaire." *Annales de la Société d'Agriculture, Sciences, Arts et Belles-lettres d'Indre-et-Loire* 48:114–115.

———. 1864. "Sur les Effets Mécaniques de l'Air Confine Échauffé par les Rayons du Soleil." *Comptes Rendus Hebdomadaires des Séances de l'Académie des Sciences* 39:527.

Modi, N. 2011. *Convenient Action: Gujarat's Response to Challenges of Climate Change.* Delhi: Macmillan.

Mudge, F. B. 1997. "The Development of the 'Greenhouse' Theory of Global Climate Change from Victorian Times." *Weather* 52:13–17.

Mumford, L. 1964. "Authoritarian and Democratic Technics." *Technology and Culture* 5:1–8.

Muir, L. M., ed. 1938. *John of the Mountains: The Unpublished Journals of John Muir.* Boston: Houghton Mifflin.

Muir, J. 1901. *Our Natural Parks.* Boston: Houghton Mifflin.*

Muirhead, J. 2014. "Concentrating Solar Power in India: An Outlook to 2024." *Business Standard*, 15 Sept.*

Mukherji, A., et al. 2009. *Revitalizing Asia's Irrigation to Sustainably Meet Tomorrow's Food Needs.* Colombo: IWMI.

Murchie, E. H., et al. 2009. "Agriculture and the New Challenges for Photosynthesis Research." *New Phytologist* 181:532–52.

Murphy, P. G., et al. 1935. *The Drought of 1934: A Report of the Federal Government's Assistance to Agriculture.* Typescript. Washington, DC: Drought Coordinating Committee.

Murphy, R. C. 1954. "Informe Sobre el Viaje de Estudios Realizado por el Dr. R. Cushman Murphy en el Año 1920." *Boletín de la Compañia Administradora del Guano* 30:16–20.*

———. 1936. *Oceanic Birds of South America: A Study of Species of the Related Coasts and Seas.* 2 vols. NY: Macmillan.*

———. 1926. "Oceanic and Climatic Phenomena along the West Coast of South America During 1925." *Geographical Review* 16:26–54.

———. 1925a. "Equatorial Vignettes." *Natural History* 25:431–49.

———. 1925b. *Bird Islands of Peru: The Record of a Sojourn on the West Coast.* NY: G. P. Putnam's Sons.

Murphy, R. C., and W. Vogt. 1933. "The Dovekie Influx of 1932." *The Auk* 50:325–49.

Murray, S. O. 2009. "The Pre-Freudian Georges Devereux, the Post-Freudian Alfred Kroeber, and Mohave Sexuality." *Histories of Anthropology Annual* 5:12–27.

Myren, D. T. 1969. "The Rockefeller Foundation Program in Corn and Wheat in Mexico." In *Subsistence Agriculture and Economic Development*, ed. C. R. Wharton, 438–52. Chicago: Aldine Publishing Co.

Naam, R. 2013. *The Infinite Resource: The Power of Ideas on a Finite Planet*. NY: UPNE.

Nabokov, P., and L. Loendorf. 2002. *American Indians and Yellowstone National Park: A Documentary Overview*. Yellowstone National Park, WY: U.S. National Park Service.*

Nabokov, V. 1991 (1952). *The Gift*, trans. M. Scammell. NY: Vintage Books.

Nahm, J. S. 2014. "Varieties of Innovation: The Creation of Wind and Solar Industries in China, Germany, and the United States." Ph.D. dissertation, MIT.*

Nasaw, D. 2006. *Andrew Carnegie*. NY: Penguin Press.

Nash, R. 1973 (1967). *Wilderness and the American Mind*. YUP.

Nehru, J. 1994 (1946). *The Discovery of India*. OUP.

Nelsen, M. P., et al. 2016. "Delayed Fungal Evolution Did Not Cause the Paleozoic Peak in Coal Production." *PNAS* 113:2442–47.

Newhall, C. G., and S. Punongbayan, eds. 1997. *Fire and Mud. Eruptions and Lahars of Mount Pinatubo, Philippines*. Seattle: University of Washington Press.

New York Herald Tribune Forum. 1948. *Our Imperiled Resources: Report of the 17th Annual New York Herald Tribune Forum*. NY: Herald Tribune.

Nichols, H. B. 1948. "Greed held Check to Stretching Natural Resources." *Christian Science Monitor*, 15 Sep.

Nicholson, S. E., et al. 1998. "Desertification, Drought, and Surface Vegetation: An Example from the West African Sahel." *Bulletin of the American Meteorological Society* 79:815–29.

Nienaber, M. 2015. "Power Line Standoff Holds Back Germany's Green Energy Drive." Reuters, 3 June.*

Nierenberg, N., et al. 2010. "Early Climate Change Consensus at the National Academy: The Origins and Making of Changing Climate." *Historical Studies in the Natural Sciences* 40:318-49.

Nixon, E. B., ed. 1957. *Franklin D. Roosevelt and Conservation, 1911–1945*. 2 vols. Washington, DC: General Services Administration, National Archives and Records Service.*

Nixon, R. 2011. *Slow Violence and the Environmentalism of the Poor*. HUP.

Nordhaus, W. D. 2013. *The Climate Casino: Risk, Uncertainty, and Economics for a Warming World*. YUP.

North, S. 1948. "Three Billion Coolies in A.D. 2000." *WP*, 8 Aug.

Northbourne, Lord. 1940. *Look to the Land*. London: Dent.

Norwine, J. 1977. "A Question of Climate." *Environment* 19:6–27.

Núñez, E., and G. Petersen. 2002. *Alexander von Humboldt en el Perú: Diario de Viaje y Otros Escritos*. Lima: Banco Central de Reserva del Perú, Fondo Editorial.

Oberg, B. B., ed. 2002. *The Papers of Thomas Jefferson*. Vol. 29, *1 March 1796–31 December 1797*. Princeton, NJ: Princeton University Press.*

Odum, E. P. 1953. *Fundamentals of Ecology*. Philadelphia: W. B. Saunders.

Ó Gráda, C. 2015. *Eating People Is Wrong, and Other Essays on Famine, Its Past and Its Future*. Princeton, NJ: Princeton University Press.

Okimori, Y., et al. 2003. "Potential of CO_2 Emission Reductions by Carbonizing Biomass Waste from Industrial Tree Plantation in South Sumatra, Indonesia." *Mitigation and Adaptation Strategies for Global Climate Change* 8:261–80.

Oldfield, S. 2004. "Howard, Louise Ernestine, Lady Howard (1880–1969)." In *Oxford Dictionary of National Biography*. OUP.

Olien, D. D., and R. M. Olien. 1993. "Running Out of Oil: Discourse and Public Policy, 1909–29." *Business and Economic History* 22:36–66.

Olsson, T. C. 2013. "Agrarian Crossings: The American South, Mexico, and the Twentieth-Century Remaking of the Rural World." Ph.D. dissertation, University of Georgia.

O'Neill, W. L. 1969. *Everyone Was Brave: The Rise and Fall of Feminism in America*. Chicago: Quadrangle Books.

Oreskes, N. 2000. "Why Predict? Historical Perspectives on Prediction in Earth Science." In *Prediction: Science, Decision Making, and the Future of Nature*, ed. D. Sarewitz, et al. Washington, DC: Island Press, 23–40.

——and E. Conway. 2010. *Merchants of Doubt: How a Handful of Scientists Obscured the Truth on Issues from Tobacco Smoke to Global Warming*. NY: Bloomsbury.

Ornstein, L., et al. 2009. "Irrigated Afforestation of the Sahara and Australian Outback to End Global Warming." *Climatic Change* 97:409–37.

Ortiz, R., et al. 2007. "High Yield Potential, Shuttle Breeding, Genetic Diversity, and a New International Wheat Improvement Strategy." *Euphytica* 157:365–84.

Osborn, F. 1948. *Our Plundered Planet*. Boston: Little, Brown.

Osprovat, D. 1995 (1981). *The Development of Darwin's Theory: Natural History, Natural Theology, and Natural Selection, 1838–1859*. CUP.

O'Sullivan, E. 2015. "Warren Weaver's *Alice in Many Tongues*: A Critical Appraisal." In *Alice in a World of Wonderlands: The Translations of Lewis Carroll's Masterpiece*, ed. J. A. Lindseth and A. Tannenbaum, 3 vols, 1:29–41. New Castle, DE: Oak Knoll.

O'Sullivan, R. 2015. *American Organic: A Cultural History of Farming, Gardening, Shopping, and Eating*. Lawrence: University Press of Kansas.

Otterbein, K. F. 2004. *How War Began*. College Station: Texas A&M University Press.

Ouellet, V. 2016. "$80M Temple Project in Rural Ontario Threatened by Wind Turbines." *CBC News*, 27 June.*

Paarlberg, D. 1970. *Norman Borlaug—Hunger Fighter*. Washington, DC: Government Printing Office.

Pais, A. 1982. *"Subtle Is the Lord . . .": The Science and the Life of Albert Einstein*. OUP.

Pal, B. P., et al. 1958. "Frequency and Types of Mutations Induced in Bread Wheat by Some Physical and Chemical Mutagens." *Wheat Information Service* 7:14–15.

Palmer, I. 1972. *Food and the New Agricultural Technology*. Geneva: UN Research Institute for Social Development.

Palmer, M. A., et al. 2010. "Mountaintop Mining Consequences." *Science* 327:148-49.

Parisi, A. J. 1977. "'Soft' Energy, Hard Choices." *NYT*, 16 Oct.

Parker, H. A. 1906. "The Reverend Francis Doughty." *Transactions of the Colonial Society of Massachusetts* 10:261–75.

Parthasarathi, A. 2007. *Technology at the Core: Science and Technology with Indira Gandhi*. Delhi: Pearson Longman.

Paterson, O. 2014. "I'm Proud of Standing Up to the Green Lobby." *The Telegraph*, 20 July 2014.*

Patterson, G. 2009. *The Mosquito Crusades: A History of the American Anti-Mosquito Movement from the Reed Commission to the First Earth Day*. New Brunswick, NJ: Rutgers University Press.

Paull, J. 2014. "Lord Northbourne, the Man Who Invented Organic Farming, a Biography." *Journal of Organic Systems* 9:31–53.

Pearl, R. 1927. "The Growth of Populations." *Quarterly Review of Biology* 2:532–48.

——. 1925. *The Biology of Population Growth*. NY: Alfred A. Knopf.

Pearl, R., and L. J. Reed. 1920. "On the Rate of Growth of the Population of the United States Since 1790 and its Mathematical Representation." *PNAS* 6:275-88.

Pearse, A. 1980. *Seeds of Plenty, Seeds of Want: Social and Economic Implications of the Green Revolution.* Oxford: Clarendon Press.

Pedersen, J. S., et al. 2006. "Native Supercolonies of Unrelated Individuals in the Invasive Argentine Ant." *Evolution* 60:782–91.

Peixoto, J. P., and A. H. Oort. 1992. *The Physics of Climate.* NY: Springer-Verlag.

Pereira, M. C. 2005. "A Highly Innovative, High Temperature, High Concentration, Solar Optical System at the Turn of the Nineteenth Century: The Pyrheliophoro." In *Proceedings of EuroSun2004, the 14th International Sunforum*, ed. A. Goetzburger et al., 3 vols., 3:661–72. Freiburg, Germany: PSE GmbH.*

Pérez-Fodich, A., et al. 2014. "Climate Change and Tectonic Uplift Triggered the Formation of the Atacama Desert's Giant Nitrate Deposits." *Geology* 42:251–54.

Perkins, J. H. 1997. *Geopolitics and the Green Revolution: Wheat, Genes, and the Cold War.* OUP.

Perkins, V. L. 1965. "The AAA and the Politics of Agriculture: Agricultural Policy Formulation in the Fall of 1933." *Agricultural History* 39:220–29.

Perlin, J. 2013. *Let It Shine: The 6,000-Year Story of Solar Energy.* Novato, CA: New World Library.

———. 2002 (1999). *From Space to Earth: The Story of Solar Electricity.* HUP.

Petersen, R. H. 1974. "The Rust Fungus Life Cycle." *Botanical Review* 40:453–513.

Peterson, C. 1989. "Experts, OMB Spar on Global Warming; "Greenhouse Effect" May Be Accelerating, Scientists Tell Hearing." *WP*, 9 May.

Peterson, J. P. 1936. "The CCC in Mosquito Work in North Jersey." In *New Jersey Mosquito Extermination Association 1936*:134–37.

Peterson, T. C., et al. 2008. "The Myth of the 1970s Global Cooling Scientific Consensus." *Bulletin of the American Meteorological Society* 89:1325-37.

Peterson, R. T. 1989. "William Vogt: A Man Ahead of His Time." *American Birds* 43:1254–55.

———. 1973. "The Evolution of a Magazine." *Audubon*, January.

Peterson, S. M., et al. 2016. *Groundwater-Flow Model of the Northern High Plains Aquifer in Colorado, Kansas, Nebraska, South Dakota, and Wyoming.* U.S. Geological Survey Scientific Investigations Report 2016-5153. Washington, DC: USGS.*

Pettit, P. 2013. *The Paleolithic Origins of Human Burial.* NY: Routledge.

Pew Research Center. 2015. *Public and Scientists' Views on Science and Society*, 29 Jan.*

Philips, M. 1886. "The Petroleum Pet." *Rocky Mountain News*, 14 March.

Phillips, R., and R. Milo. 2009. "A Feeling for the Numbers in Biology." *PNAS* 106:21465–71.*

Picton, L. 1949. *Nutrition and the Soil: Thoughts on Feeding.* NY: Devin-Adair.

Pielke, R., Jr. 2011 (2010). *The Climate Fix: What Scientists and Politicians Won't Tell You About Global Warming.* NY: Basic Books.

Pierrehumbert, R. T. 2016. "How to Decarbonize? Look to Sweden." *Bulletin of the Atomic Scientists* 72:105–11.

———. 2011. "Infrared Radiation and Planetary Temperature." *Physics Today* 64:33–38.

———. 2004. "Warming the World." *Nature* 432:677.

Pifre, A. 1880. "Nouveaux Résultats d'Utilisation de la Chaleur Solaire Obtenue à Paris." *Comptes Rendus des Travaux de l'Académie des Sciences* 91:388–89.*

Pignon, C. P., et al. 2017. "Loss of Photosynthetic Efficiency in the Shade: An Achilles Heel for the Dense Modern Stands of Our Most Productive C4 Crops?" *Journal of Experimental Biology* 68:335–45.

Pilat, O. R. 1935. "Where Millions Play." *BDE*, 31 May.*

Pinchot, G. 1909. "Conservation." In *Addresses and Proceedings of the First National Con-*

servation Congress, Held at Seattle, Washington, August 26–28, ed. B. N. Baker, et al., 70–78. Washington, DC: Executive Committee of the National Conservation Congress.*

———. 1905. *A Primer of Forestry, Part 2*. Practical Forestry 24. Washington, DC: U.S. Department of Agriculture.

Pinker, S. 2011. *The Better Angels of Our Nature: Why Violence Has Declined*. NY: Allan Lane.

Pittock, A. B. 2009 (2005). *Climate Change: The Science, Impacts and Solutions*. 2nd ed. NY: Routledge.

Placzek, S. et al. 2016. "BRENDA in 2017: New Perspectives and New Tools in BRENDA." *Nucleic Acids Research* 45:D380-88.

Platt, J. 1969. "What We Must Do." *Science* 166:116.

Plumer, B. 2017. "U.S. Nuclear Comeback Stalls as Two Reactors are Abandoned." *NYT*, 31 Jul.

Pohlman, J. W., et al. 2017. "Enhanced CO_2 Uptake at a Shallow Arctic Ocean Seep Field Overwhelms the Positive Warming Potential of Emitted Methane." *PNAS* 114:5355-60.

Pollan, M. 2007 (2006). *The Omnivore's Dilemma: A Natural History of Four Meals*. NY: Penguin.

Pollock, A. 1929. "Alexander Woollcott's Play, 'The Channel Road,' Opens at the Plymouth Theater Without Disturbance." *BDE*, 18 Oct.

Pomeranz, K. 2000. *The Great Divergence: China, Europe, and the Making of the Modern World Economy*. Princeton, NJ: Princeton University Press.

Ponisio, L. C., et al. 2015. "Diversification Practices Reduce Organic to Conventional Yield Gap." *Proceedings of the Royal Society B* 282:20141396.*

Ponte, L. 1976. *The Cooling*. Englewood Cliffs, NJ: Prentice-Hall.

Pope, C. H. 1903. *Solar Heat: Its Practical Applications*. Boston: C. H. Pope.*

Portis, A. R., and M. E. Salvucci. 2002. "The Discovery of Rubisco Activase–Yet Another Story of Serendipity." *Photosynthesis Research* 73:257–64.

Postel, S. L., et al. 1996. "Human Appropriation of Renewable Fresh Water." *Science* 271:785–88.

Pottier, G. F. 2014. "Augustin Mouchot, Pionnier de l'Énergie Solaire à Tours en 1864." Unpublished ms. Tours: Archives Départementales d'Indre-et-Loire. *

Pouillet, C. 1838. *Mémoire sur la Chaleur Solaire: Sur les Pouvoirs Rayonnants et Absorbants de l'Air Atmosphérique et sur la Température de l'Espace*. Paris: Bachelier.*

Powell, M. A. 2016. *Vanishing America: Species Extinction, Racial Peril, and the Origins of Conservation*. HUP.

Power Grid Corporation of India. 2013. *Desert Power India—2050*. Gurgaon, India: Power Grid Corporation of India.*

Prado-Martinez, J., et al. 2013. "Great Ape Genetic Diversity and Population History." *Nature* 499:471–75.

Pratt, W. E. 1952. "Toward a Philosophy of Oil-Finding." *AAPG Bulletin* 36:2231–36.

Priest, T. 2014. "Hubbert's Peak: The Great Debate over the End of Oil." *Historical Studies of the Physical Sciences* 44:37–79.

Prieto, P. A., and C. A. Hall. S. 2013. *Spain's Photovoltaic Revolution: The Energy Return on Investment*. NY: Springer.

Priore, E. R. 1979. "Warren Weaver." *Physics Today* 72:72.

Prud'homme, A. 2012. *The Ripple Effect: The Fate of Freshwater in the Twenty-First Century*. New York: Simon and Schuster.

Purdy, J. 2015. "Environmentalism's Racist History." *New Yorker*, Aug. 13.

Quinnez, B. 2007–2008. "Augustin-Bernard Mouchot (1825–1912): Un Missionnaire de l'Énergie Solaire." *Mémoires de l'Académie des Sciences, Arts et Belles-lettres de Dijon* 142:297–321.

Radkau, J. 2008 (2002). *Nature and Power: A Global History of the Environment*, trans. T. Dunlap. CUP.

Rajaram, S. 1999. "Wheat Germplasm Improvement: Historical Perspectives, Philosophy, Objectives, and Missions." In *Wheat Breeding at CIMMYT: Commemorating 50 Years of Research in Mexico for Global Wheat Improvement* (Wheat Special Report 29), ed. S. Rajaram and G. P. Hettel, 1–10. México, D.F.: CIMMYT.

Ramalingaswami, V., et al. "Studies of the Bihar Famine of 1966-67." In Blix, G., et al., eds. *Famine: A Symposium Dealing with Nutrition and Relief Pperations in Times of Disaster*. Uppsala: Almqvist & Wiksells.

Ransom, R., and S. Sutch. 1990. "Who Pays for Slavery?," in America, R. F., ed. *The Wealth of Races: The Present Value of Benefits from Past Injustices*. Westport, CT: Greenwood Press, 31–54.

Rao, N. 2015. *M. S. Swaminathan in Conversation with Nitya Rao*. New Delhi: Academic Foundation.

Rasky, F. 1949. "Vogt and Osborn: Our Fighting Conservationists." *Tomorrow* 9:5–10.

Rasool, S. I., and S. H. Schneider. 1971. "Atmospheric Carbon Dioxide and Aerosols: Effects of Large Increases on Global Climate." *Science* 173:138–41.

Raven, J. A. 2013. "Rubisco: Still the Most Abundant Protein of Earth?" *New Phytologist* 198:1–3.

Raven, J. A., and J. F. Allen. 2003. "Genomics and Chloroplast Evolution: What Did Cyanobacteria Do for Plants?: *Genome Biology* 4:209.

Ray, D. K., et al. 2013. "Yield Trends Are Insufficient to Double Global Crop Production by 2050." *PLoS One* 8: e66428.

———. 2012. "Recent Patterns of Crop Yield Growth and Stagnation." *Nature Communications* 3:1293.*

Reed, E. W. 1992. *American Women in Science before the Civil War*. Minneapolis: University of Minnesota Press.

Reed, J. 1983 (1978). *The Birth Control Movement and American Society: From Private Vice to Public Virtue*. Princeton, NJ: Princeton University Press.

Rees, M. 1987. "Warren Weaver." *Biographical Memoirs of the National Academy of Sciences* 57:493–530.*

Regal, B. 2002. *Henry Fairfield Osborn: Race and the Search for the Origins of Man*. Burlington, Vt.: Ashgate Publishing.

Reidel, B. 2015. *JFK's Forgotten Crisis: Tibet, the CIA, and Sino-Indian War*. Washington, DC: Brookings Institution Press.

Reij, C. 2014. "Re-Greening the Sahel: Linking Adaptation to Climate Change, Poverty Reduction, and Sustainable Development in Drylands." In *The Social Lives of Forests: Past, Present, and Future of Woodland Resurgence*, ed. S. Hecht et al., 303–11. Chicago: University of Chicago Press.

Reij, C., et al. 2005. "Changing Land Management Practices and Vegetation on the Central Plateau of Burkina Faso (1968–2002)." *Journal of Arid Environments* 63:648–55.

Reiley, F. A. 1936. "The CCC in Mosquito Work in Southern New Jersey." *Proceedings of the New Jersey Mosquito Extermination Association 1936*, 129–34.

Reinton, O. 1973. "The Green Revolution Experience." *Instant Research on Peace and Violence* 3:58–73.

Reisner, M. 1993 (1986). *Cadillac Desert: The American West and its Disappearing Water*. Rev. ed. N.Y.: Penguin.

Reitz, L. P., and S. C. Salmon. 1968. "Origin, History, and Use of Norin 10 Wheat." *Crop Science* 8:686–89.

Revelle, R., and H. E. Suess. 1957. "Carbon Dioxide Exchange Between Atmosphere and Ocean and the Question of an Increase of Atmospheric CO_2 During the Past Decades." *Tellus* 9:18–27.

Richards, J. F. 2005 (2003). *The Unending Frontier: An Environmental History of the Modern World.* Berkeley: UCP.

Richardson, L. F. 1960. *Statistics of Deadly Quarrels.* Pacific Grove, CA: Boxwood Press.

Richetti, J. J. 2005. *The Life of Daniel Defoe: A Critical Biography.* Malden, MA: Blackwell Publishing.

Richter, D., et al. 2017. "The Age of the Hominin Fossils from Jebel Irhoud, Morocco, and the Origins of the Middle Stone Age." *Nature* 546:293-96.

Richter, B. 2014 (2010). *Beyond Smoke and Mirrors: Climate Change and Energy in the 21st Century.* 2nd ed. CUP.

Riley, J. C. 2005. "Estimates of Regional and Global Life Expectancy, 1800–2001." *Population and Development Review* 31:537–43.

Riordan, M., and L. Hoddeson. 1998. *Crystal Fire: The Invention of the Transistor and the Birth of the Information Age.* NY: W. W. Norton.

Rist, G. 2009 (1997). *The History of Development: From Western Origins to Global Faith.* 3rd ed. New Delhi: Academic Foundation.

Rivers Cole, J., and S. McCoskey. 2013. "Does Global Meat Consumption Follow an Environmental Kuznets Curve?" *Sustainability* 9:26-36.

Roadifer, R. E. 1986. "Size Distributions of the World's Largest Known Oil, Tar Accumulations." *OGJ*, 24 Feb.

Roberts, D. G., and D. Barrett. 1984. "Nightsoil Disposal Practices of the 19th Century and the Origin of Artifacts in Plowzone Proveniences." *Historical Archaeology* 18:108–15.

Robertson, T. R. 2012a. *The Malthusian Moment: Global Population Growth and the Birth of American Environmentalism.* NY: Routledge.

———. 2012b. "Total War and the Total Environment: Fairfield Osborn, William Vogt, and the Birth of Global Ecology." *Environmental History* 17:336–64.

———. 2009. "Conservation After World War II: The Truman Administration, Foreign Aid, and the 'Greatest Good.'" In *The Environmental Legacy of Harry S. Truman*, ed. K. B. Brooks, 32–47. Kirksville, MO: Truman State University Press.

———. 2005. "The Population Bomb: Population Growth, Globalization, and American Environmentalism, 1945–1980." Ph.D. dissertation, University of Wisconsin.

Robin, L., et al. 2013. *The Future of Nature: Documents of Global Change.* New Haven, CT: Yale University Press.

Robinson, T. L., and Lakso, A. N. 1991. "Bases of yield and production efficiency in apple orchard systems." *Journal of the American Society for Horticultural Science* 116:188-94.

Robock, A. 2008. "Twenty Reasons Why Geoengineering Might Be a Bad Idea." *Bulletin of the Atomic Scientists* 64:14–18, 59.

Robock, A., et al. 2009. "Benefits, Risks, and Costs of Stratospheric Geoengineering." *Geophysical Research Letters* 36:L19703.*

Robock, A., and O. B. Toon. 2010. "Local Nuclear War, Global Suffering." *Scientific American* 302:74–81.

Rockefeller Foundation. 1916–. Annual Reports. NY: Rockefeller Foundation.

Rockström, J. 2009. "A Safe Operating Space for Humanity." *Nature* 461:472–75.

Rockström, J., et al. 2009. "Planetary Boundaries: Exploring the Safe Operating Space for Humanity." *Ecology and Society* 14:32.

Rodale, J. I. 1952. "Looking Back, Part IV." *The Beginning of Our Experimental Farm* 20: 11–12, 37–38.

———. 1948. *The Healthy Hunzas*. Emmaus, PA: Rodale Press.

———. 1947. "With the Editor: The Principle of Eminent Domain." *Organic Gardening* 1:16–18.

Rodrigues, J. 1999. *A Conspiração Solar do Padre Himalaya: Esboço Biográfico dum Português da Ecologia*. Porto: Árvore.

Rodríguez, J., et al. 1957. "The Rust Problems of the Important Wheat Producing Areas of Mexico." In *Report of the Third International Wheat Rust Conference*, ed. Oficina de Estudios Especiales, 126–28. Saltsville, MD: Plant Industry Station.*

Roe, G. H., and M. B. Baker. 2007. "Why Is Climate Sensitivity So Unpredictable?" *Science* 318:629–32.

Roelfs, A. P., et al. 1992. *Rust Diseases of Wheat: Concepts and Methods of Disease Management*. Mexico, D.F.: CIMMYT.*

———. 1982. "Effects of Barberry Eradication on Stem Rust in the United States." *Plant Diseases* 66:177–82.*

Rook, R. E. 1996. "Blueprints and Prophets: Americans and Water Resource Planning for the Jordan River Valley, 1860–1970." Ph.D. dissertation, Kansas State University.

Rosenfeld, S. S. 1968. "The Food Squeeze." *WP*, 19 Sept.

Ross, E. A. 1927. *Standing Room Only?* NY: The Century Co.

Ross, W. 2015. "The Death of Venice." *The Independent*, 14 May.*

Royal Society (U.K.). 2009. *Reaping the Benefits: Science and the Sustainable Intensification of Global Agriculture*. London: Royal Society.*

Rubin, O. 2009. "The Merits of Democracy in Famine Protection—Fact or Fallacy?" *European Journal of Development Research* 21:699–717.

Rudwick, M. J. S. 2008. *Worlds Before Adam: The Reconstruction of Geohistory in the Age of Reform*. Chicago: University of Chicago Press.

Rupert, J. A. 1951. *Resistencia al Chahuixtle como Factor en el Mejoramiento del Trigo en México*. Folleto Technico 7. México, D.F.: Oficina de Estudios Speciales.*

Russell, E. 2001. *War and Nature: Fighting Humans and Insects with Chemicals from World War I to "Silent Spring."* CUP.

Rybczynski, W. 1986. *Home: A Short History of an Idea*. NY: Viking.

Sabin, P. 2013. *The Bet: Paul Ehrlich, Julian Simon, and Our Gamble over Earth's Future*. YUP.

Sagan, L. 1967. "On the Origin of Mitosing Cells." *Journal of Theoretical Biology* 14:225–74.

Sagan, D., ed. 2012. *Lynn Margulis: The Life and Legacy of a Scientific Rebel*. White River Junction, VT: Chelsea Green.

Sage, R. F., et al. 1987. "The Nitrogen Use Efficiency of C3 and C4 Plants. III: Leaf Nitrogen Effects on the Activity of Carboxylating Enzymes in Chenopodium album (L.) and Amaranthus retroflexus (L.)." *Plant Physiology* 85:355–59.

Saha, M. 2012. "State Policy, Agricultural Research And Transformation Of Indian Agriculture With Reference To Basic Food-Crops, 1947-75." Ph.D. dissertation, Iowa State University.*

Salinas-Zavala, C. A., et al. 2006. "Historic Development of Winter-Wheat Yields in Five Irrigation Districts in the Sonora Desert, Mexico." *Interciencia* 31:254–61.*

Salisbury, B. A., et al. 2003. "SNP and Haplotype Variation in the Human Genome." *Mutation Research* 526:53–61.

Sanderson, E. W. 2009. *Mannahatta: A Natural History of New York City*. NY: Abrams.

Sanger, M. 1950. "Lasker Award Address." *The Malthusian*, 25 Oct.

———. 1949. "A Question of Privilege." *Women United*, Oct.

Saussure, H.-B. 1786. *Voyages dans les Alpes*. 4 vols. Geneva: Barde, Manget and Comp.*

———. 1784. "Lettre de M. Saussure aux Auteurs de Journal." *Journal de Paris* 108:475–79 (Supp., 17 April.).

Sauvy, A. 1972. "La Population du Monde et les Ressources de la Planète: Un Projet de Recherches." *Population* 27:967–77.*

Sayre, N. F. 2008. "The Genesis, History, and Limits of Carrying Capacity." *Annals of the Association of American Geographers* 98:120–34.

Scheffler, S. 2013. *Death and the Afterlife*. OUP.

Scheinost, P. L., et al. "Perennial Wheat: The Development of a Sustainable Cropping System for the U.S., Pacific Northwest." *American Journal of Alternative Agriculture* 16:147-51.

Schirrmeister, B. E., et al. 2016. "Cyanobacterial Evolution During the Precambrian." *International Journal of Astrobiology* 15(3):187–204.

Schmidt, G., and S. Rahmsdorf. 2005. "11°C Warming, Climate Crisis in 10 Years?" *Realclimate.org*, 29 Jan.*

Schneider, C., and J. Banks. 2010. *The Toll From Coal: An Updated Assessment of Death and Disease from America's Dirtiest Energy Source*. Boston: Clean Air Task Force.*

Schneider, S. H. 2009. *Science as a Contact Sport: Inside the Battle to Save Earth's Climate*. Washington, DC: National Geographic Society.

———. 1975. "On the Carbon Dioxide–Climate Confusion." *Journal of the Atmospheric Sciences* 32:2060–66.

Schoijet, M. 1999. "Limits to Growth and the Rise of Catastrophism." *Environmental History* 4:515–30.

Schuler, F. E. 1998. *Mexico Between Hitler and Roosevelt: Mexican Foreign Relations in the Age of Lázaro Cárdenas, 1934–1940*. Albuquerque: University of New Mexico Press.

Scott, J. C. 2017. *Against the Grain: Plants, Animals, Microbes, Captives, Barbarians, and a New Story of Civilization*. YUP.

Sears, P. B. 1948. "We Survive or Perish as Part of the Earth." *New York Herald Tribune*, 28 March.

Seaton, C. W. 1877. *Census of the State of New York for 1875*. Albany: Weed, Parsons and Company.*

Sebben, M. L., et al. 2015. "Seawater Intrusion in Fractured Coastal Aquifers: A Preliminary Numerical Investigation Using a Fractured Henry Problem." *Advances in Water Resources* 85:93–108.

Secretaria de Agricultura (México). 1946. *Informe de Labores de la Secretaria de Agricultura del 10 de Septiembre de 1945 al 31 de Agosto de 1946*. México, D.F.: Editorial Cultura.

Self, S., et al. 1981. "The Possible Effects of Large 19th and 20th Century Volcanic Eruptions on Zonal and Hemispheric Surface Temperatures." *Journal of Vulcanology and Geothermal Research* 11:41–60.

Self, S., and M. R. Rampino. 2012. "The 1963–1964 Eruption of Agung Volcano (Bali, Indonesia)." *Bulletin of Vulcanology* 74:1521–36.

Semple, E. C. 1911. *Influences of Geographic Environment, on the Basis of Ratzel's System of Anthropo-Geography*. NY: Henry Holt and Company.*

Sender, R., et al. 2016. "Are We Really Vastly Outnumbered? Revisiting the Ratio of Bacteria to Host Cells in Humans." *Cell* 164:337–40.

Serna-Chavez, H. M., et al. 2013. "Global Drivers and Patterns of Microbial Abundance in Soil." *Global Ecology and Biography* 22:1162–72.

Sesno, F. 2006. "Poll: Most Americans Fear Vulnerability of Oil Supply." CNN, 5 July.*

Seton, E. T. 1898. *Wild Animals I Have Known*. NY: Charles Scribner's Sons.

Seufert, V., et al. 2012. "Comparing the Yields of Organic and Conventional Agriculture." *Nature* 485:229–32.

Shabecoff, P. 1993. *A Fierce Green Fire: The American Environmental Movement*. NY: Hill and Wang.

———. 1988. "Global Warming Has Begun, Expert Tells Senate." *NYT*, 24 June.

Shah, S. 2010. *The Fever: How Malaria Has Ruled Humankind for 500,000 Years*. NY: Farrar, Straus and Giroux.

Shakerian, F. et al. 2015. "A Comparative Review Between Amines and Ammonia as Sorptive Media for Post-Combustion CO2 Capture." *Applied Energy* 148:10–22.

Shang, Y., et al. 2013. "Systematic Review of Chinese Studies of Short-Term Exposure to Air Pollution and Daily Mortality." *Environment International* 54:100–11.

Shannon, C. E., and W. Weaver. 1949. *The Mathematical Theory of Communication*. Chicago: University of Illinois Press.

Shelley, P. B. 1920 (1820). *A Philosophical View of Reform*. OUP.

Shiklomanov, I. A. 2000. "Appraisal and Assessment of World Water Resources." *Water International* 25:11–32.

———. 1993. "World Freshwater Resources." In *Water in Crisis: A Guide to the World's Freshwater Resources*, ed. P. H. Gleick, 13–24. OUP.

Shiklomanov, I. A., and J. A. Balonishnikova. 2003. "World Water Use and G. Water Availability: Trends, Scenarios, Consequences." In *Water Resources Systems: Hydrological Risk, Management, and Development (Proceedings of Symposium at IUGG2003, Sapporo)*, ed. G. Blöschl et al., 358–64. IAHS Publication 281. Wallingford, UK: International Association of Hydrological Sciences.

Shiva, V. 1991. *The Violence of the Green Revolution: Third World Agriculture, Ecology,. And Politics*. London: Zed Books.

Shuman, F. 1914. "Feasibility of Utilizing Power from the Sun." *Scientific American* 110:179.

Shwadran, B. 1977. *Middle East Oil: Issues and Problems*. Cambridge, MA: Schenkman Publishing Co.

Siebert, S., et al. 2010. "Groundwater Use for Irrigation—A Global Inventory." *Hydrological and Earth Systems Science* 14:1863–80.

Siegel, S. M. 2015. *Let There be Water: Israel's Solution for a Water-Starved World*. NY: St. Martin's Press.

Sills, D. L. 1975. "The Environmental Movement and Its Critics." *Human Ecology* 3:1-41.

Silvi, C. 2010. "Storia del Vapore e dell'Elettricitá dal Calore del Sole con Specchi Piani o quasi Piani." *Energia, Ambiente e Innovazione* 56:34–47.

Simberloff, D. 2014. "The 'Balance of Nature'—Evolution of a Panchreston." *PLoS Biology* 12: e1001963.*

Simmons, M. 2006 (2005). *Twilight in the Desert: The Coming Saudi Oil Shock and the World Economy*. NY: John Wiley.

Simonin, L. 1876. "L'Emploi Industriel de la Chaleur Solaire." *Revue des Deux Mondes* 15:200–13.*

Simpson, R. D. 2014. "Do Regulators Overestimate the Costs of Regulation?" *Journal of Benefit-Cost Analysis* 5:315–32.

Singer, N. 1999. "Sandia Geothermal Technology Plays Key Role in Killing Out-of-Control Natural Gas Well." *Sandia Lab News*, 19 Nov.*

Singh, B. P. 2014. "Science Communication in India: Policy Framework." *Journal of Scientific Temper* 2:141-51.

Singh, R. P., et al. 1994. "Rust Diseases of Wheat." In *Guide to the CIMMYT Wheat Crop*

Protection Subprogram (Wheat Special Report 24), ed. E. E. Saari and G. P. Hettel, 19–33. Mexico, D.F.: CIMMYT.

Skocpol, T., and K. Finegold. 1982. "State Capacity and Economic Intervention in the Early New Deal." *Political Science Quarterly* 97:255–78.

Skousen, M., ed. 2007. *The Completed Autobiography of Benjamin Franklin*. 2 vols. Washington, DC: Regnery Publishing.

Smaje, C. 2015. "The Strong Perennial Vision: A Critical Review." *Agroecology and Sustainable Food Systems* 39: 471-99.

Smil, V. 2016. "Harvesting the Biosphere." *The World Financial Review*, Jan.-Feb.

———. 2013a. *Harvesting the Biosphere: What We Have Taken from Nature*. MIT.

———. 2013b. *Should We Eat Meat? Evolution and Consequences of Modern Carnivory*. Ames, IA: Wiley-Blackwell.

———. 2011a. "Global Energy: The Latest Infatuations." *American Scientist* 99:212–19.

———. 2011b. "Nitrogen Cycle and World Food Production." *World Agriculture* 2:9-13.

———. 2008. *Energy in Nature and Society: General Energetics of Complex Systems*. MIT.

———. 2001. *Enriching the Earth: Fritz Haber, Carl Bosch, and the Transformation of World Food Production*. MIT.

Smith, A. 1776. *An Inquiry into the Nature and Causes of the Wealth of Nations*. 2 vols. London: W. Strahan and T. Cadell.*

Smith, B. G., ed. 2008. *The Oxford Encyclopedia of Women in World History*. 4 vols. OUP.

Smith, J. 2013. "The Huxley-Wilberforce 'Debate' on Evolution, 30 June 1860." In *BRANCH: Britain, Representation, and Nineteenth-Century History*, ed. D. F. Falluga. Extension of *Romanticism and Victorianism on the Net*. Web.*

Smith, K. L., and L. M. Polvani. 2017. "Spatial Patterns of Recent Antarctic Surface Temperature Trends and the Importance of Natural Variability: Lessons from Multiple Reconstructions and the CMIP5 Models." *Climate Dynamics* 48:2653-70.

Smith, M. B. 1998. "The Value of a Tree: Public Debates of John Muir and Gifford Pinchot." *The Historian* 60:757–78.

Smith, W. 1891. *The Rise and Extension of Submarine Telegraphy*. London: J. S. Virtue & Co.*

———. 1873a. "Effect of Light on Selenium During the Passage of an Electric Current." *Nature* 7:303.

———. 1873b. "The Action of Light on Selenium." *Journal of the Society of Telegraph Engineers* 2:31–33.

Snyder, T. 2015. *Black Earth: The Holocaust as History and Warning*. NY: Tim Duggan Books.

Soares, P., et al. 2012. "The Expansion of mtDNA Haplogroup L3 Within and Out of Africa." *Molecular Biology and Evolution* 29:915–27.

Soden, D. L., ed. 1999. *The Environmental Presidency*. New York: SUNY Press.

Song, J. 1985. "Systems Science and China's Economic Reforms." In *Control Science and Technology for Development*, ed. J. Yang, 1–8. NY: Pergamon Press.

Sonnenfeld, D. A. 1992. "Mexico's 'Green Revolution,' 1940–1980: Towards an Environmental History." *Environmental History Review* 16:28–52.

Sørensen, B. 2011 (1979). *Renewable Energy: Physics, Engineering, Environmental Impacts, Economics and Planning*. 4th ed. Burlington, MA: Academic Press.

Sorenson, R. P. 2011. "Eunice Foote's Pioneering Research on CO_2 and Climate Warming." *Search and Discovery* 70092.*

Sörlin, S. 2016 (2013). "Ice Diplomacy and Climate Change: Hans Ahlmann and the Quest for a Nordic Region Beyond Borders." In *Science, Geopolitics and Culture in the Polar Region: Norden Beyond Borders*, ed. S. Sörlin. NY: Routledge.

Sorrell, S., and J. Speirs. 2010. "Hubbert's Legacy: A Review of Curve-Fitting Methods to Estimate Ultimately Recoverable Resources." *Natural Resources Research* 19:209–30.

Sparks, D. L. 2006. "Historical Aspects of Soil Chemistry." In *Footprints in the Soil: People and Ideas in Soil History*, ed. B. P. Warkentin, 307–38. NY: Elsevier Science.

Specter, M. 2015. "The Gene Hackers." *The New Yorker*, 16 Nov.

Spiro, J. P. 2009. *Defending the Master Race: Conservation, Eugenics, and the Legacy of Madison Grant*. Burlington: University of Vermont Press.

Sprengel, C. 1828. "Von den Substanzen der Ackerkrume und des Untergrundes, insbesondere, wie solche durch die chemische Analyse." *Journal für Technische und Ökonomische Chemie* 2:423–74, 3:42–99, 313–52, and 397–421.

Stakman, E. C. 1957. "Problems in Preventing Plant Disease Epidemics." *American Journal of Botany* 44:259–67.

———. 1937. "The Promise of Modern Botany for Man's Welfare Through Plant Protection." *The Scientific Monthly* 44:117–30.

———. 1923 (1919). *Destroy the Common Barberry*. Farmer's Bulletin 1058. Washington, DC: U.S. Department of Agriculture.

Stakman, E. C., et al. 1940. "Observations on Stem Rust Epidemiology in Mexico." *American Journal of Botany* 27:90–99.

Stakman, E. C., and D. G. Fletcher. 1930 (1927). *The Common Barberry and Black Stem Rust*. Farmer's Bulletin 1544. Washington, DC: U.S. Department of Agriculture.

Standage, T. 2013 (1998). *The Victorian Internet: The Remarkable Story of the Telegraph and the Nineteenth Century's On-line Pioneers*. NY: Bloomsbury USA.

Steffen, W., et al. 2015. "Planetary Boundaries: Guiding Human Development on a Changing Planet." *Science* 347:1259855.

Steffen, W., et al. 2011. "The Anthropocene: Conceptual and Historical Perspectives." *PTRSA* 369:842–67.

Stenni, B., et al. Forthcoming. "Antarctic Climate Variability at Regional and Continental Scales over the Last 2,000 Years." *Climate of the Past*.*

Sterba, J. 2012. *Nature Wars: The Incredible Story of How Wildlife Comebacks Turned Backyards into Battlegrounds*. NY: Crown.

Sterling, C. 1972. "Club of Rome Tackles the Planet's 'Problematique.'" *WP*, 2 March.

———. 1971. "Doomsday Prophecies About Climate Are Eased Some." *WP*, 21 July.

Stern, R. 2013. *Oil Scarcity Ideology in U.S. National Security Policy, 1909–1980*. Stanford, CA: Freeman Spogli Institute for International Studies.*

Stevenson, F. B. 1925. "The Top of the News." *BDE*, April 9.

Stoddard, L. 1920. *The Rising Tide of Color Against White World-Supremacy*. NY: Charles Scribner's Sons.*

Stouffer, R. J., and S. Manabe. 2017. "Assessing Temperature Pattern Projections Made in 1989." *Nature Climate Change* 7:163–65.

Stouffer, R. J., et al. 1989. "Interhemispheric Asymmetry in Climate Response to a Gradual Increase of Atmospheric CO_2." *Nature* 342:660–62.

Stracher, G., et al. 2011–15. *Coal and Peat Fires: A Global Perspective*. 4 vols. Waltham, MA: Elsevier.

Strayer, D.L., et al. 2014. "Decadal-Scale Change in a Large-River Ecosystem." *BioScience* 64:496-510.

———. 2011. "Long-term Changes in a Population of an Invasive Bivalve and its Effects." *Oecologia* 165:1063–72.

Streets, D. G., et al. 2013. "Radiative Forcing Due to Major Aerosol Emitting Sectors in China and India." *Geophysical Research Letters* 40:4409–14.

Study of Critical Environmental Problems. 1970. *Man's Impact on the Global Environment: Assessment and Recommendations for Action*. MIT.

Suarez, A. V., et al. 1999. "Behavioral and Genetic Differentiation Between Native and Introduced Populations of the Argentine Ant." *Biological Invasions* 1:43–53.

Sugiura, M. 1995. "The Chloroplast Genome." *Essays in Biochemistry* 30:49–57.

Sunamura, E., et al. 2009. "Intercontinental Union of Argentine Ants: Behavioral Relationships Among Introduced Populations in Europe, North America, and Asia." *Insectes Sociaux* 56:143–47.

Surridge, C. 2002. "The Rice Squad." *Nature* 416:676-78.

Swaminathan, M. S. 2015. *Combating Hunger and Achieving Food Security.* CUP

———. 2010a. *From Green to Evergreen Revolution: Indian Agriculture: Performance and Challenges.* New Delhi: Academic Foundation.

———. 2010b. Introduction. In *Science and Sustainable Food Security: Selected Papers of M. S. Swaminathan,* ed. M. S. Swaminathan, 1–26. Hackensack, NJ: World Scientific.

———. 2006. "An Evergreen Revolution." *Crop Science* 46:2293–2303.

———. 2000. "An Evergreen Revolution." *The Biologist* 47:85–89.

———. 1996. *Sustainable Agriculture: Towards an Evergreen Revolution.* Delhi: Konark.

———. 1969. "Scientific Implications of HYV Programme." *Economic and Political Weekly* 4:69–75.

———. 1965. "The Impact of Dwarfing Genes on Wheat Production." *Journal of the IARI Post-Graduate School* 2:57–62.

Swaminathan, M. S., and A. T. Natarajan. 1956. "Effects of Fast Neutron Radiation on Einkorn, Emmer and Bread Wheats." *Wheat Information Service* 4:5–6.

Swartz, M. 2008. "The Gospel According to Matthew." *Texas Monthly,* Feb.*

Tabita, F. R., et al. 2008. "Phylogenetic and Evolutionary Relationships of RubisCO and the RubisCO-like Proteins and the Functional Lessons Provided by Diverse Molecular Forms." *PTRSB* 363:2629-40.

Takahashi, K. I., and D. L. Gautier. 2007. "A Brief History of Oil and Gas Exploration in the Southern San Joaquin Valley of California." In *Petroleum Systems and Geological Assessment of Oil and Gas in the San Joaquin Basin Province: California,* ed. A. H. Scheirer. Professional Paper 1713. Washington, DC: U.S. Geological Survey.*

Tanno, K., and G. Willcox. 2006. "How Fast Was Wild Wheat Domesticated?" *Science* 311:1886.

Taubes, G. 2007. *Good Calories, Bad Calories: Challenging the Conventional Wisdom on Diet, Weight Control, and Disease.* NY: Knopf.

Taylor, F. H. 1884. *The Derrick's Hand-Book of Petroleum.* Oil City, PA: Derrick Publishing Co.

Taylor, S. H., and S. P. Long. 2017. "Slow Induction of Photosynthesis on Shade to Sun Transitions May Cost at Least 21% in Productivity." *PTRSB* 372:20160543.

Teller, E. 1960. "We're Going to Work Miracles." *Popular Mechanics,* March.

Teltsch, K. 1949. "Population Gains Held to Be a Danger." *NYT,* 18 Aug.

Tennyson, M. E., et al. 2012. *Assessment of Remaining Recoverable Oil in Selected Major Oil Fields of the San Joaquin Basin, California.* Washington, DC: U.S. Geological Survey.*

Themnér, L., and P. Wallensteen. 2013. "Armed Conflicts, 1946-2012." *Journal of Peace Research* 50:509-21.

Theophrastus. 1916 (3rd cent. B.C.). *Enquiry into Plants and Minor Works on Odours and Weather Signs,* trans. A. Hort. 2 vols. NY: G. P. Putnam's Sons.*

Thomas, L. 2002. *Coal Geology.* NY: John Wiley & Sons.

Thompson, D. 1989. "The Most Hated Man in Science." *Time,* 4 Dec.

Thompson, S. P. 1910. *The Life of William Thomson, Baron Kelvin of Largs.* 2 vols. London: Macmillan and Company.*

Thompson, W. S. 1929. *Danger Spots in World Population.* NY: Alfred A. Knopf.

Thomson, K. 2000. "Huxley, Wilberforce and the Oxford Museum." *American Scientist* 88:210–13.

Thomson, W. (Lord Kelvin). 1881. "On the Sources of Energy in Nature Available to Man for the Production of Mechanical Effects." *Nature* 14:433–36

Thorpe, I. J. N. 2005. "The Ancient Origins of Warfare and Violence." In Pearson and Thorpe eds. 2005:1–18.

Throckmorton, R. I. 1951. "The Organic Farming Myth." *Country Gentleman* 121:21, 103–5.

Thurston, R. H. 1901. "Utilising the Sun's Energy." *Cassier's Magazine* 20:283–88.*

Tierney, J. 1990. "Betting the Planet." *NYT*, 2 Dec.

Tilman, D., et al. 2011. "Global Food Demand and the Sustainable Intensification of Agriculture." *PNAS* 108: 20260–64.

Tinoco, A. 2012. "Portugal na Exposição Universal de 1904—O Padre Himalaia e o Pirelióforo." *Cadernos de Sociomuseologia* 42:113–27.

Tomory, L. 2012. *Progressive Enlightenment: The Origins of the Gaslight Industry, 1780–1820*. Cambridge, MA: MIT Press.

Tooby, J., and L. Cosmides. 2010. "Groups in Mind: The Coalitional Roots of War and Morality." In *Human Morality and Sociality: Evolutionary & Comparative Perspectives*, ed. H. Høgh-Olesen, 91–234. NY: Palgrave Macmillan.

Toups, M.A., et al. 2011. "Origin of Clothing Lice Indicates Early Clothing Use by Anatomically Modern Humans in Africa." *Molecular Biology and Evolution* 28:29–32.

Travis, J. 2003. "Lice Hint at a Recent Origin of Clothing." *Science News*, 23 Aug.*

Tröstl, J., et al. 2016. "The Role of Low-Volatility Organic Compounds in Initial Particle Growth in the Atmosphere." *Nature* 533:527–31.

Tunzelman, N. v. 2003 (1986). "Coal and Steam Power." In *Atlas of Industrializing Britain, 1780–1914*, ed. J. Langton and R. J. Morris. NY: Methuen & Co.

Tsutsui, N. D., and A. V. Suarez. 2003. "The Colony Structure and Population Biology of Invasive Ants." *Conservation Biology* 17:48–58.

Tsitsin, N. V., and V. F. Lubimova. 1959. "New Species and Forms of Cereals Derived from Hybridization Between Wheat and Couch Grass." *American Naturalist* 93:181–91.

Tuckwell, W. 1900. *Reminiscences of Oxford*. NY: Cassell and Company.*

Tyndall, J. 1861. "On the Absorption and Radiation of Heat by Gas and Vapours, and on the Physical Connexion of Radiation, Absorption, and Conduction." *LPMJS* 22:169–94, 273–85.

Union Internationale pour la Protection de la Nature. 1950. *International Technical Conference on the Protection of Nature*. Paris: UNESCO.*

United Kingdom. British Command Papers. 1939. *British White Paper: Statement of Policy, 17 May*. London: Her Majesty's Stationer's Office.*

UNESCO [United Nations Educational, Scientific and Cultural Organization]. 1949. *Documents Préparatoires à la Conférence Technique Internationale pour la Protection de la Nature*. Paris: UNESCO. (partly *)

United Nations. Economic and Social Commission for Western Asia and Bundesanstalt für Geowissenschaften und Rohstoffe. 2013. *Inventory of Shared Water Resources in Western Asia*. Beirut: UN-ESCWA and BGR.

United Nations. 1950. *Proceedings of the United Nations Scientific Conference on the Conservation and Utilization of Natural Resources, 17 Aug–6 Sep 1949*. 8 vols. Lake Success, NY: U.N. Dept. of Economic Affairs.

———. 1947. *Yearbook of the United Nations, 1946–47*. Lake Success, NY: U.N. Dept. of Public Information.

United Nations Department of Economic and Social Affairs. 2006. *World Urbanization Prospects: The 2005 Revision*. ESA/P/WP/200. New York: United Nations.*

United Nations. Food and Agriculture Organization. 2017. *The State of Food Insecurity and Nutrition in the World 2017*. Rome: FAO.*

———. 2013. *FAO Statistical Yearbook 2013*. Rome: FAO.*

———. 2011a. *The State of the World's Land and Water Resources for Food and Agriculture (SOLAW)*. London: FAO/Earthscan.*

———. 2011b. *Global Food Losses and Food Waste—Extent, Causes, and Prevention*. Rome: FAO.*

———. 2009. *The State of Food Insecurity in the World*. Rome: FAO.*

———. 2004. *The State of Food and Agriculture, 2003–2004*. Rome: FAO.

United Nations. Scientific Committee on the Effects of Atomic Radiation. 2016. *Developments Since the 2013 UNSCEAR Report on the Levels and Effects of Radiation Exposure due to the Nuclear Accident following the Great East-Japan Earthquake and Tsunami*. NY: United Nations.*

United Nations. World Health Organization. 2016. *Ambient Air Pollution: A Global Assessment of Exposure and the Burden of Disease*. Geneva: WHO.*

United States Bureau of the Census. 1909. *A Century of Population Growth from the First Census of the United States to the Twelfth, 1790-1900*. Washington, DC: Government Printing Office.*

United States Committee for Global Atmospheric Research Program. 1975. *Understanding Climatic Change*. Washington, DC: National Academy of Sciences.

United States Department of Agriculture, Bureau of Agricultural Economics. 1933. "Farmers' Strikes and Riots in the United States, 1932–33." Unpub. ms.

United States Department of State. 1976. *Foreign Relations of the United States, 1949*. 9 vols. Washington, DC: Government Printing Office.*

———. [1949.] *Proceedings of the Inter-American Conference on Conservation of Renewable Natural Resources*. [Washington, D.C.]: Department of State.

United States President's Science Advisory Committee. 1965. *Restoring the Quality of Our Environment*. Washington, DC: Government Printing Office.

United States Senate. 1987–88. *The Greenhouse Effect and Climate Change. Hearings Before the Committee on Energy and Natural Resources*. 100th Cong., 1st Sess. 2 vols. Washington, DC: Government Printing Office.

———. 1966. *Population Crisis. Hearings Before the Subcommittee on Foreign Aid Expenditures of the Committee on Government Operations*. 89th Cong., 2nd Sess. Washington, DC: Government Printing Office.

———. 1922. *Agricultural Appropriations Bill, 1923. Hearings Before the Subcommittee of the Committee on Appropriations*. 67th Cong., 2nd Sess. Washington, DC: Government Printing Office.

Usher, P. 1989. "World Conference on the Changing Atmosphere: Implications for Security (Review)." *Environment* 31:25–28.

Vain, P., et al. 1995. "Foreign Gene Delivery into Monocotyledonous Species." *Biotechnology Advances* 13:653-71.

Van der Heijden, M. G. A., et al. 2008. "The Unseen Majority: Soil Microbes as Drivers of Plant Diversity and Productivity in Terrestrial Ecosystems." *Ecology Letters* 11:296–310.

Van der Ploeg, R. R., et al. 1999. "On the Origin of the Theory of Mineral Nutrition of Plants and the Law of the Minimum." *Soil Science Society of America Journal* 63:1055–62.

Van der Veen, C. J. 2000. "Fourier and the 'Greenhouse Effect.'" *Polar Geography* 24:132-52.

Van Wees, H. 2004. *Greek Warfare: Myths and Realities*. London: Gerald Duckworth & Co.

Van Wilgenburg, E., et al. 2010. "The Global Expansion of a Single Ant Colony." *Evolutionary Applications* 3:136–43.

Varshney, A. 1998. *Democracy, Development, and the Countryside: Urban-Rural Struggles in India.* CUP.

Varughese, G., and M. S. Swaminathan. 1967. "Sharbati Sonora: A Symbol of the Age of Algeny." *Indian Farmer* 17:8–9.

Vassilou, M. S. 2009. *Historical Dictionary of the Petroleum Industry.* Lanham, MD: Scarecrow Press.

Venezian, E. L., and W. K. Gamble. 1969. *The Agricultural Development of Mexico: Its Structure and Growth Since 1950.* NY: Praeger Publishers.

Verhulst, P.-F. 1838. "Notice sur la Loi que la Population suit dans son Accroissement." *Correspondance Mathématique et Physique* 10:113–21.

Vettel, E. J. 2006. *Biotech: The Countercultural Origins of an Industry.* Philadelphia: University of Pennsylvania Press.

Viator [pseud.]. 1865. "Our Special Correspondence." *Wisconsin State Register*, 2 Dec.

Victor, D. G. 2008. "On the Regulation of Geoengineering." *Oxford Review of Economic Policy* 24:322–36.

Vietmeyer, N. 2009–10. *Borlaug.* 3 vols. Lorton, VA: Bracing Books.

Viswanath, B. 1953. "Organic Versus Inorganic Manures in Land Improvement and Crop Production." *Proceedings of the National Institute of Sciences of India* 19:23–25.

Viswanath, B., and M. Suryanarayana. 1927. "The Effect of Manuring a Crop on the Vegetative and Reproductive Capacity of the Seed." *Memoirs of the Department of Agriculture in India (Chemical Series)* 9:85–124.

Vitousek, P. M., et al. 1997. "Human Domination of Earth's Ecosystems." *Science* 277:494–99.

Vitousek, P. M., et al. 1986. "Human Appropriation of the Products of Photosynthesis." *BioScience* 36:368–73.

Vizcarra, C. 2009. "Guano, Credible Commitments, and Sovereign Debt Repayment in 19th-Century Peru." *Journal of Economic History* 69:358–87.

Vogt, J. A. 1941. "To Sea Lions!" *AM* 167:18–25.

———. 1940. "The White Island." *AM* 166:265–73.

Vogt, G. 2007. "The Origins of Organic Farming." In Organic Farming: An International History, ed. W. Lockeretz. Cambridge, MA: CABI Publishing.

Vogt, W. 1966. "Statement of William Vogt," in United States Senate. *Hearings Before the Subcommittee on Foreign Aid Expenditures of the Committee on Government Operations On S. 1676* (Eighty-Ninth Congress, Second Session). Washington, D.C.: Government Publishing Office, 3:718-50.

———. 1965. "We Help Build the Population Bomb." *NYT*, 4 April.

———. 1963. *Comments on a Brief Reconnaissance of Resource Use, Progress, and Conservation Needs in Some Latin American Countries.* NY: Conservation Foundation.

———. 1961. "From Bird-Watching, a Feeling for Nature." *NYT*, 11 Jun.

———. 1960. *People! Challenge to Survival.* NY: William Sloan Associates.

———. 1950. "Getting Sex Appeal into Editorials on Conservation." *The Masthead* 2:42.

———. 1949. "Let's Examine Our Santa Claus Complex." *Saturday Evening Post* 222:17–19, 76–78.

———. 1948a. "A Continent Slides to Ruin." *Harper's Magazine* 196:481–88.

———. 1948b. *The Road to Survival.* NY: William Sloan Associates.

[———]. 1946a. *Report on Activities of Conservation Section, Division of Agricultural Cooperation, Pan American Union (1943–1946).* Washington, DC: Pan American Union.

———. 1946b. *The Population of Venezuela and Its Natural Resources.* Washington, DC: Pan American Union.

———. 1946c. *The Population of El Salvador and Its Natural Resources.* Washington, DC: Pan American Union.

———. 1945a. "Unsolved Problems Concerning Wildlife in Mexican National Parks." In *Transactions of the 10th North American Wildlife Conference,* ed. E. M. Quee, 355–58. Washington, DC: American Wildlife Institute.

———. 1945b. "Hunger at the Peace Table." *Saturday Evening Post* 217:17, 109–10.

———. 1944. *El Hombre y la Tierra.* Biblioteca Enciclopedia Popular No. 32. México, D.F.: Secretaria de Educación Pública.

———. 1943. "Road to Beauty." *Bulletin of the Pan-American Union* 77:661–71.

———. 1942a. "An Ecological Depression on the Peruvian Coast." In *Proceedings of the Eighth American Scientific Congress, May 10–18, 1940,* ed. Anon. 3:507–27. Washington, DC: Department of State.

———. 1942b. "Informe Sobre las Aves Guaneras." *BCAG* 18:3–132.

———. 1942c. "Influencia de la Corriente de Humboldt en la Formación de Depósitos Guaníferos." *Simiente* 12:3–14.

———. 1940. "Una Depresión Ecológica en la Costa Peruana." *BCAG* 16:307–29.

———. 1939. "Enumeración Preliminar de Algunos Problemas Relacionados con la Producción del Guano en el Perú." *BCAG* 15:285–301.

———. 1938a. "Birding Down Long Island." *Bird-Lore* 40:331–40.

———. 1938b. "Preliminary Notes on the Behavior and the Ecology of the Eastern Willet." *Proceedings of the Linnaean Society of New York* 49:8–42.

———. 1937a. *Thirst on the Land: A Plea for Water Conservation for the Benefit of Man and Wildlife.* Circular 32. New York: National Association of Audubon Societies.

[———]. 1937b. Editorial. *Bird-Lore* 39:296.

[———]. 1935. Editorial. *Bird-Lore* 37:127.

———. 1931. "The Birds of a Cat-Tail Swamp." *Bulletin to the Schools: University of the State of New York* 17:162.

[Vogt, W., et al.] 1965. *Human Conservation in Central America.* Washington, DC: Conservation Foundation.

Vogt, W., et al. 1939. "Report of the Committee on Bird Protection, 1938." *The Auk* 56:212–19.

Von Czerny, F. 1881. *Die Veränderlichkeit des Klimas und ihre Ursachen.* Vienna: A. Hartleben's Verlag.*

Von Humboldt, A., and H. Berghaus. 1863. *Briefwechsel Alexander von Humboldt's mit Heinrich Berghaus aus den Jahren 1825 bis 1858.* Leipzig: Hermann Costenoble.*

Von Laue, M. 1934. "Fritz Haber Gestorben." *Naturwissenschaften* 22:97.

Von Neumann, J. 1955. "Can We Survive Technology?" *Fortune,* June.

Voosen, P. 2016. "Climate Scientists Open Up their Black Boxes to Scrutiny." *Science* 354:401-02.

———. 2012. "Ocean Clouds Obscure Warming's Fate, Create 'Fundamental Problem' for Models." *E&E News,* 26 Nov.*

Vorontsov, N. N., and J. M. Gall. 1986. "Georgyi Frantsevich Gause 1910–1986." *Nature* 323:113.

Vranken, L., et al. 2014. "Curbing Global Meat Consumption: Emerging Evidence of a Second Nutrition Transition." *Environmental Science & Policy* 39:95-106.

Wadland, J. H. 1978. *Ernest Thompson Seton: Man in Nature and the Progressive Era, 1880–1915.* NY: Arno Press.

Wagner, G., and M. L. Weitzman. 2015. *Climate Shock: The Economic Consequences of a Hotter Planet.* Princeton, NJ: Princeton University Press.

———. 2012. "Playing God." *Foreign Policy*, 24 Oct.*

Wagner, G., and R. J. Zeckhauser. 2017. "Confronting Deep and Persistent Climate Uncertainty." Harvard Kennedy School Faculty Research Working Paper Series RWP16-025.*

Wagoner, P., and J. R. Schaeffer. 1990. "Perennial Grain Development: Past Efforts and Potential for the Future." *Critical Reviews in Plant Sciences* 9:381–408.

Wald, M. 2013. "Former Energy Secretary Chu Joins Board of Canadian Start-Up." *NYT*, 17 Dec.

Waldner, E. 2006. "Exploration Boom After 1998 Blowout Sputtering Along." *Bakersfield Californian*, 10 Feb.

Walker, B. J., et al. 2016. "The Costs of Photorespiration to Food Production Now and in the Future." *Annual Review of Plant Biology* 67:107–29.

Wallace, H. A. 1941. "Wallace in Mexico." *Wallace's Farmer and Iowa Homestead*, 22 Feb.

Wang, P., et al. 2016. "Finding the Genes to Build C4 Rice." *Current Opinion in Plant Biology* 31:44–50.

Wang, T., and J. Watson. 2010. "Scenario Analysis of China's Emission Pathways in the 21st Century for Low-Carbon Transition." *Energy Policy* 38:3537–46.

Warde, P. 2007. *Energy Consumption in England and Wales, 1560–2000*. Rome: Consiglio Nazionale delle Ricerche.*

Warde, P., and S. Sörlin. 2015. "Expertise for the Future: The Emergence of Environmental Prediction, c. 1920–1970." In *The Struggle for the Long-Term in Transnational Science and Politics: Forging the Future*, ed. J. Andersson and E. Rindzevičiūtė, 38–62. NY: Routledge.

Water Integrity Network. 2016. *Water Integrity Global Outlook*. Berlin: Water Integrity Network.*

Waterhouse, A. C. 2013. *Food and Prosperity: Balancing Technology and Community in Agriculture*. NY: Rockefeller Foundation.*

Weart, S. R. 2008 (2003). *The Discovery of Global Warming*. Rev. and expanded ed. HUP.*
———. 1997. "Global Warming, Cold War, and the Evolution of Research Plans." *Historical Studies in the Physical and Biological Sciences* 27:319–56.

Weaver, W. 1970. *Scene of Change: A Lifetime in American Science*. NY: Charles Scribner's Sons.
———. 1951. "Alice's Adventures in Wonderland, Its Origin, Its Author." *Princeton University Library Chronicle* 13:1–17.

Webb, J. L. A., Jr. 2009. *Humanity's Burden: A Global History of Malaria*. CUP.

Weber, B. 2011. "Lynn Margulis, Evolution Theorist, Dies at 73." *NYT*, 24 Nov.

Webster, B. 1969. "End Papers." *NYT*, 8 Feb.

Weidensault, S. 2007. *Of a Feather: A Brief History of American Birding*. NY: Harcourt.

Weimerskirch, H., et al. 2012. "Foraging in Guanay Cormorant and Peruvian Booby, the Major Guano-Producing Seabirds in the Humboldt Current System." *Marine Ecology Progress Series* 458:231–45.*

Weinberg, G. L., ed. 2006 (1928). *Hitler's Second Book: The Unpublished Sequel to Mein Kampf*, trans. K. Smith. NY: Enigma Books.

Weinzettel, J. et al. 2013. "Affluence Drives the Global Displacement of Land Use." *GEC* 23:433–38.

Weisskopf, M. 1988. "Scientist Says Greenhouse Effect Is Setting In." *WP*, 24 June.

Weitzman, M. L. 2007. "A Review of the Stern Review on the Economics of Climate Change." *Journal of Economic Literature* 45:703–24.

Wertime, T. A. 1983. "The Furnace versus the Goat: The Pyrotechnologic Industries and Mediterranean Deforestation in Antiquity." *Journal of Field Archaeology* 10:445–52.

Whitaker, A. P. 1960. "Alexander von Humboldt and Spanish America." *Proceedings of the American Philosophical Society* 104:317–22.

White, A. D. 1897. *A History of the Warfare of Science with Theology in Christendom*. 2 vols. New York: D. Appleton.

White, A. F. 2015. *Plowed Under: Food Policy Protests and Performance in New Deal America*. Bloomington: Indiana University Press.

White, D. 1919. "The Unmined Supply of Petroleum in the United States." *Automotive Industries* 40:361, 376, 385.

White, R. 1995. *The Organic Machine: The Remaking of the Columbia River*. NY: Hill and Wang.

White, D., and M. Rabago-Smith. 2011. "Genotype-phenotype Associations and Human Eye Color." *Journal of Human Genetics* 56:5-7.

Whitman, W. B., et al. 1998. "Prokaryotes: The Unseen Majority." *PNAS* 95:6578–83.

Wigley, T. M. L. 2006. "A Combined Mitigation/Geoengineering Approach to Climate Stabilization." *Science* 314:452–54.

Wik, R. M. 1973 (1972). *Henry Ford and Grass-Roots America*. Ann Arbor: University of Michigan Press.

Wilberforce, R. 1888. *Life of Samuel Wilberforce*. London: Kegan Paul, Trench, & Co.*

[Wilberforce, S.] 1860. "Review of On the Origin of Species." *Quarterly Review* 108:225–64.

Wildman, S. G. 2002. "Along the Trail from Fraction I Protein to Rubisco (Ribulose Bisphosphate Carboxylase-oxygenase)." *Photosynthesis Research* 73:243–50.

Wilkinson, T. J., and L. Raynes. 2010. "Hydraulic landscapes and imperial power in the Near East." *Water History* 2:115–44.

Will, G. F. 1975. "A Change in the Weather." *WP*, 24 Jan.

Willberg, D. 2017. "CCS Facility at Boundary Dam Returning to Normal Operations." *Estevan Mercury*, 21 Jul.*

Willcox, G. 2007. "The Adoption of Farming and the Beginnings of the Neolithic in the Euphrates Valley: Cereal Exploitation Between the 12th and 8th Millennia cal BC." In *The Origins and Spreads of Domestic Plants in Southwest Asia and Europe*, ed. S. Colledge and J. Connolly. Walnut Creek, CA: Left Coast Press.

Williams, J. 1789. *The Natural History of the Mineral Kingdom*. 2 vols. Edinburgh: Thomas Ruddiman.*

Williams, M. 2006. *Deforesting the Earth: From Prehistory to Global Crisis, an Abridgment*. Chicago: University of Chicago Press.

Williamson, S. H., and L. P. 2015. "Measuring Slavery in 2011 Dollars." MeasuringWorth.com.*

Wilson, E. O. 1995 (1994). *Naturalist*. New York: Warner Books.

Wineke, W. R. 2008. "Global Cooling Advocate Reid Bryson, 88." *Wisconsin State Journal*, 13 June.

Wines, R. A. 1986. *Fertilizer in America: From Waste Recycling to Resource Exploitation*. Philadelphia: Temple University Press.

Winslow, M., et al. 2004. *Desertification, Drought, Poverty, and Agriculture: Research Lessons and Opportunities*. Rome: ICARDA, ICRISAT, and the UNCCD Global Mechanism.

Wöbse, 2011. "'The World After All Was One': The International Environmental Network of UNESCO and IUPN, 1945–1950." *Contemporary European History* 20:331–48.

Wolf, A. 2015. *The Invention of Nature: Alexander von Humboldt's New World*. NY: Alfred A. Knopf.

Wolf, A. T. 1972. *Hydropolitics Along the Jordan River: Scarce Water and Its Impact on the Arab-Israeli Conflict.* NY: United Nations University Press.

Wolf, E. R. 1982. *Europe and the People Without History.* UCP.

Woody, T. 2012. "Sierra Club, NRDC Sue Feds to Stop Big California Solar Power Project." *Forbes,* 27 March.

Woolf, D., et al. 2010. "Sustainable Biochar to Mitigate Global Climate Change." *Nature Communications* 1:1–9.*

World Bank. 2008. *World Development Report 2008: Agriculture for Development.* Washington, DC: World Bank.

World Energy Council. 2013. *World Energy Scenarios: Composing Energy Futures to 2050.* London: World Energy Council.*

Worster, D. 1997 (1994). *Nature's Economy: A History of Ecological Ideas.* 2nd ed. CUP.

Wrench, G. 2009 (1938). *The Wheel of Health: A Study of the Hunza People and the Keys to Health.* Australia: Review Press.

Wrigley, E. A. 2010. *Energy and the English Industrial Revolution.* CUP.

Wylie, K. 1941. "Agricultural Relations with Mexico." *Foreign Agriculture* 6:365–73.*

Xu, Y.-S., et al. 2009. "Geo-hazards with Characteristics and Prevention Measures Along the Coastal Regions of China." *Natural Hazards* 49:479–500.

Yang, C.-C., et al. 2010. "Loss of Microbial (Pathogen) Infections Associated with Recent Invasions of the Red Imported Fire Ant *Solenopsis invicta.*" *Biological Invasions* 12:3307–18.

Yates, P. L. 1981. Mexico's Agricultural Dilemma. Tucson: University of Arizona Press.

Yergin, D. 2012 (2011). *The Quest: Energy, Security, and the Remaking of the Modern World.* Rev. ed. New York: Penguin Books.

———. 2008 (1991). *The Prize: The Epic Quest for Oil, Money, and Power.* New York: Free Press.

Yong, E. 2016. *I Contain Multitudes: The Microbes Within Us and a Grander View of Life.* NY: Ecco.

Yulish, C. B., ed. 1977. *Soft vs. Hard Energy Paths: 10 Critical Essays on Amory Lovins' "Energy Strategy: The Road Not Taken?"* NY: Charles Yulish Associates.

Zeng, Z., et al. 2017. "Climate Mitigation from Vegetation Biophysical Feedbacks during the Past Three Decades." *Nature Climate Change* 10.1038/NCLIMATE3299.

Zhang, X., et al. 2016. "Establishment and Optimization of Genomic Selection to Accelerate the Domestication and Improvement of Intermediate Wheatgrass." *The Plant Genome* 9.*

Zhu, X.-G., et al. 2010 "Improving Photosynthetic Efficiency for Greater Yield." *Annual Review of Plant Biology* 61:235–61.

Zhu, Z., et al. 2016. "Greening of the Earth and its Drivers." *Nature Climate Change* 6:791-95.

Zimmer, C. 2011. *A Planet of Viruses.* Chicago: University of Chicago Press.

Zirinsky, M. P. 1992. "Imperial Power and Dictatorship: Britain and the Rise of Reza Shah, 1921–1926." *International Journal of Middle East Studies* 24:639–63.

Zirkle, C. 1957. "Benjamin Franklin, Thomas Malthus and the United States Census." *Isis* 48:58–62.

Zon, R., and W. N. Sparhawk. 1923. *Forest Resources of the World.* 2 vols. NY: McGraw-Hill Book Company.

INDEX

Page numbers in *italics* refer to illustrations.

Map and Illustration Credits